Nanotechnology

Ram Prasad • Manoj Kumar • Vivek Kumar
Editors

Nanotechnology

An Agricultural Paradigm

Springer

Editors
Ram Prasad
Amity Institute of Microbial Technology
Amity University
Noida, Uttar Pradesh, India

Manoj Kumar
Amity Institute of Microbial Technology
Amity University
Noida, Uttar Pradesh, India

Vivek Kumar
Amity Institute of Microbial Technology
Amity University
Noida, Uttar Pradesh, India

ISBN 978-981-10-4572-1 ISBN 978-981-10-4573-8 (eBook)
DOI 10.1007/978-981-10-4573-8

Library of Congress Control Number: 2017943328

Printed on acid-free paper

This Springer imprint is published by Springer Nature
The registered company is Springer Nature Singapore Pte Ltd.
The registered company address is: 152 Beach Road, #21-01/04 Gateway East, Singapore 189721, Singapore

Contents

About the Editors

Dr. Ram Prasad, Ph.D. is assistant professor at the Amity Institute of Microbial Technology, Amity University, Uttar Pradesh, India. His research interest includes plant-microbe interactions, sustainable agriculture, and microbial nanobiotechnology. Dr. Prasad has more than a hundred publications to his credit, including research papers and book chapters and five patents issued or pending, and edited or authored several books. Dr. Prasad has 11 years of teaching experience, and he has been awarded the Young Scientist Award (2007) and Prof. J.S. Datta Munshi Gold Medal (2009) by the International Society for Ecological Communications, a fellowship (2010) by the Society for Applied Biotechnology, the Outstanding Scientist Award (2015) in the field of microbiology by Venus International Foundation, and the American Cancer Society UICC International Fellowship for Beginning Investigators (USA, 2014). In 2014–2015, Dr. Prasad served as visiting assistant professor in the Department of Mechanical Engineering at Johns Hopkins University, USA.

Dr. Manoj Kumar, Ph.D. is a scientist with sanguine behavior who is adoring about research and development, with a commitment to lifelong learning. He is determined on high-quality science that contributes broadly to both increasing intellectual knowledge of plant development and increasing the ecological niche. He has a high level of professional desire and intellectual hunt and the potential to fulfill the dream of his high-impact publications and the future recognition of these by academic peers. Dr. Kumar has pursued his Ph.D. in plant biotechnology from prestigious Jawaharlal Nehru University and then was awarded two postdoctoral fellowships consecutively: (i) DBT-PDF from IISc Bangalore in 2005 and then NRF-PDF from the University of Pretoria. Dr. Manoj Kumar is a researcher of plant biotechnology in the Amity Institute of Microbial Technology at Amity

University, Uttar Pradesh, India. Dr. Kumar's research is the integration of microbial genetics with a breadth of plant physiological approaches to enable novel gene discovery and conferring metabolites.

Dr. Vivek Kumar, Ph.D. is a scientist involved in teaching, research, and guidance, with a pledge to enduring knowledge. Dr. Kumar is working in the Amity Institute of Microbial Technology at Amity University, Noida, Uttar Pradesh, India. He obtained his master's and doctoral degree from CCS Haryana Agricultural University, Hisar, Haryana, India. He is serving in the editorial board of reputed international journals, viz., *EnvironmentAsia*, the *International Journal of Biological and Chemical Sciences*, the *Journal of Advanced Botany and Zoology*, and the *Journal of Ecobiotechnology*. He has published 61 research papers, 19 book chapters, 6 review articles, and 2 books. Dr. Kumar has also served as microbiologist for 8 years in the Department of Soil and Water Research, Public Authority of Agricultural Affairs and Fish Resources, Kuwait. Dr. Kumar has organized a number of conferences/workshops as a convener/organizing secretary.

Agricultural Nanotechnology: Concepts, Benefits, and Risks

1

Jeyabalan Sangeetha, Devarajan Thangadurai, Ravichandra Hospet, Prathima Purushotham, Gururaja Karekalammanavar, Abhishek Channayya Mundaragi, Muniswamy David, Megha Ramachandra Shinge, Shivasharana Chandrabanda Thimmappa, Ram Prasad, and Etigemane Ramappa Harish

Abstract

Nanotechnology is one of the utmost significant tools in modern agriculture is predicted to become a driving cost-effective force in the near future. Nanotechnology in agriculture has gained drive in the last decade with an abundance of public funding, but the step of development is uncertain, even though many disciplines come under the agriculture system. This could be attributed to a unique nature of farm production whereby energy and matter are exchanged freely, the scale of demand of input materials constantly being enormous in contrast with industrial nanoproducts. The nanotechnologic intervention in farming has prospects for improving the efficiency of nutrient use through nanoformulations of fertilizers, surveillance and control of pests and diseases, improvement of new-generation pesticides, biosensors (which are exclusively used in remote

J. Sangeetha
Department of Environmental Science, Central University of Kerala, Kasaragod, Kerala 671316, India

D. Thangadurai (✉) • R. Hospet • P. Purushotham • A.C. Mundaragi
Department of Botany, Karnatak University, Dharwad, Karnataka 580003, India
e-mail: drthanga.kud@gmail.com

G. Karekalammanavar • M. David • E.R. Harish
Department of Zoology, Karnatak University, Dharwad, Karnataka 580003, India

M.R. Shinge • S.C. Thimmappa
Department of Microbiology and Biotechnology, Karnatak University, Dharwad, Karnataka 580003, India

R. Prasad
Amity Institute of Microbial Technology, Amity University, Noida, Uttar Pradesh 201313, India

R. Prasad et al. (eds.), *Nanotechnology*, DOI 10.1007/978-981-10-4573-8_1

sensing devices for precision farming), clay-based nanoresources for precision water management, and reclamation of salt-affected lands.

Keywords
Conventional farming • Nanotools • Nanoprocesses • Nanomaterials • Nanofertilizers • Nanopesticides

1.1 Introduction

Nanotechnology refers to the engineering and restructuring of functional systems on the scale of molecules and atoms. It is an interdisciplinary field which has the potentiality for the drastic changes in the fields of medicine, food, pharmacology, and agriculture since the eighteenth and nineteenth centuries. Inadequate supply of food is the current major problem in developing countries, due to increasing population as a result of an impact on agricultural practices and productivity. The agricultural field broadly gets benefited by the nanotech-based devices to inspect diseases in a rapid manner, to enhance the capacity of the crop plants for the possible intake of minerals, and to lead molecular treatment of diseases (Huang et al. 2007). Nanotechnology for agricultural applications predominantly shows growth from theoretical possibilities to applicable realms. An appreciation of the experimental tool is designed to operate at nanometric levels to boost research in molecular and cellular biology. The current global population is nearly about 7 billion, with 50% living in Asia. Due to increase in proportion, those living in developing countries face extreme food scarcity as a result of environmental impacts on agriculture, including storms, droughts, and flood. In a similar way, the agricultural production is constrained by a certain number of biotic and abiotic factors, and, the insect pests, diseases, and weeds cause the high loss for the potential production of agricultural crops (Dhaliwal et al. 2010). Nanomaterial-based pesticides and insecticides were manufactured exclusively for the management of insect pests using bioconjugated nanoparticles (encapsulation), agricultural productivity was enhanced for slow release of nutrients and water molecules, nanoparticle-mediated gene or DNA was incorporated in the plants for developing insect pest-resistant varieties, and also nanomaterials were used for preparing different kind of biosensors, which are exclusively used in remote sensing devices for precision farming (Prasad 2014, 2016; Prasad et al. 2014). The use of traditional methods of integrated pest management in agriculture has some major limits, and also applying chemical pesticides like DDT imposed major effects on soil fertility and major drawbacks – its adverse effects on animals and human beings. Thus, it has been concluded that nanotechnology provides an efficient and eco-friendly approach for insect pest's management of plant crops without harming the environment (Rai and Ingle 2012). Insects were found in all the types of environment, and it occupies more than two-thirds of known species of animals in the world. Insects infest to injure the plants and stored products directly or indirectly; the damage that causes less than 5% is not considered (Dhaliwal et al. 2010). Pimentel (2009) estimated that throughout the world approximately, 14% of the crops were damaged by pests; plant pathogens cause 13% loss; it has been estimated that overall, US $2000 billion per year causes loss by pests, to overcome this major issue; the nanomaterials and nanotechnology play a major role in the management of insect pests in agriculture (Rai and Ingle 2012). The agricultural innovation incorporates a few difficulties to overcome, such as increasing negative impacts in the agricultural outputs and potential risks on plant-related diseases; the current method makes proper utilization of the nanotech-empowered smart devices that can perform a dual role of being a preventive device and cautioning the system. These sensors recognize the health-related issues before they become visible to farmers and at the same time help by giving remedial measures. The nanobased tools could likewise be used to screen the delivery of agrochemicals.

The phenomenal difference between the bulk material and nanoparticles is that a high proportion of the atoms are present on the surface of nanoparticles (Maurice and Hochella 2008). When compared to particle sizes, nanoparticles have different compositions on the surface, different densities and types of sites, and also different reactivity with respect to processes such as adsorption and redox reaction which may significantly be used in synthesizing nanomaterials for use in agriculture (Waychunas et al. 2005; Hochella et al. 2008). Hence, there exist some gray areas while making farm production system respond to desired productivity levels, environmental quality assessment, and obligations of societal ethics. The nanotechnology also poses serious risks and problem toward the health and environment. An existing study performed by nanomaterials caused serious health hazards and showed toxic effects and also when entered into a human body leads to tissue damage by reaching all the vital organs. Utilization of another emerging technique, silver nanoparticles for delivery of fertilizers to plants because of their antimicrobial properties, but the studies poses a serious threat to the ecosystem by causing membrane damage; it reduces the annual growth of grass and depletes the photosynthesis in the algal body (*Chlamydomonas reinhardtii*). The presence of silver nanoparticles requires much time for recovery and tends to accumulate in its tissue exceeding the limit. In most of the countries, the important cash crop soybean was produced using manufactured nanomaterials with fossil fuel equipment which allow manufactured nanomaterials for local deposition on the crop. The continuous use of wastewater treatment to plants impacts on the plant-microbe interaction which directly affects N_2-fixing symbiosis for which some metals are sensitive (Rameshaiah et al. 2015). Thus, in the field of agriculture, there are a few conceivable outcomes to investigate and to manage the potential upcoming products and procedures. Hence, the broad study is required to comprehend the mechanism of nanoparticle compounds toxicity and their effects on the environment.

1.2 Conventional Farming: Issues and Limitations

Agriculture is one of the essential key commodities for the development of countries economic as well social wellness. At the same time, farming system, which is incorporated in crop production, plays a crucial role in achieving eco-friendly sustainable agricultural needs (Rolf 1996). Presently, organic and conventional farming methods are vigorously used for quality and high-output production of crops (Verena et al. 2012). Even though these methods show somewhat promising economic yield, still there are few barriers to rethink about using such farming systems. The limitations of the intensive conventional farming system depends on applications of chemical-based species invasion control, synthetic fertilizers, and genetically modified organisms (Health Research Funding 2015). Conventional farming is a modern industrial-oriented farming system. This farming system involves the production of essential vegetation and products of meat that typically we eat. Every year, tons of food crops and economic crops were produced in a mass quantity. Conventional farming shares many features like large farming area, huge capital investment for production of crops, large-scale hybrid crop yield, high labor dependent, single crop grown over many seasons, extensive utilization of fertilizers, pesticides, and the use of other chemical agents for protection from weeds.

1.2.1 Conventional Farming Leads to Health Issues

For any method of farming, it should first give prior importance to health aspects. Nowadays, the health concern of the conventional farming method is extensively debated. The improper methods used for farming practices may affect many health disorders on people and living beings. The conventional farming method can affect the surrounding environment by introducing many hazardous water pollutants such as salts, sediments, herbicides, fertilizers (nitrates and phosphorus), and pesticides (Bouwmeester et al. 2009). We can observe many pesticides from almost all chemical groups in groundwater especially water bodies beneath and nearby areas of agricultural land. Such low-quality, contaminated water body can highly impact on agricultural production, fishery production, and drinking water system. The subtherapeutic use of antibiotics in poultry and in the other animal breeding process has many health problems. Farmers, laborers working in a field, which are fumigated by toxic inorganic chemicals, facing various neurological, cardiac, and physical disorders and certain animal and human disease-causing viruses and bacteria, have developed resistance to presently used antibiotics.

1.2.2 Intensive Conventional Farming Affects Environment

Agricultural ecosystems contribute habitats for important wild species of flora and fauna. Such environment can mainly observe in the traditional farming region that cultivates many diverse species. However, increasing demand for food and agricultural products with respect to the drastic rise in a population may lead to the clearing of large-scale natural habitats to make room for intensive conventional farming. In the conventional farming method, fertilizers from an inorganic source may not protect soil structure, and they may cause variations in pH and ion concentrations of the soil sample, further substantially minimizing some floral and faunal populations, which are an integral part of the agricultural ecosystem. The use of inorganic fertilizer fails to balance the soil organic matter and inhibits crop rooting, resulting in low water retention capacity of soil. The highly soluble inorganic fertilizers directly effect on the loss of many elite micro and macronutrients and may cause a nutrient deficiency in crop plants. Nutrient cycling efficiency is another major concern in the conventional farming system. The potential for improving nutrient cycling between human population and agriculture as a whole has to be seriously considered, and it can be overcome successfully by incorporating modern innovations in agricultural research like using nano-agrotechnology, to promote a more resource efficient and sustainable approach toward maintaining the fertility of soils and surround environment. All around, 40% of yield generation originates from the 16% of irrigated farm fields (Gleick 1993; Postel et al. 1996). However, long haul watering system and seepage hones have quickened the rate of weathering of soil minerals, turned soils acidic, or created salt developments and possible surrender of a portion of the best cultivating lands (Österholm and Åström 2004; Presley et al. 2004). Intensive culturing, watering system, and compost dressing have additionally brought about more broad harm to the carbon profile in soils than early agrarian practices did (Knorr et al. 2005).

Recent trends in agricultural practices connected with the green revolution have significantly expanded the worldwide nourishment supply. They have additionally had an incidental, unfavorable effect on the earth and on biological community administrations, highlighting the requirement for more practical sustainable agricultural methods (Tillman et al. 2002). It is very much reported that over-the-top and unseemly utilization of manures and pesticides has expanded supplements and poisons in groundwater and surface waters, causing well-being and water refinement costs and diminishing fishery and other aquatic populations. Farming practices that corrupt soil quality adds to eutrophication of seagoing natural surroundings and may require the cost of expanded treatment, watering system, and vitality to keep up profitability on debased soils. Groundwater levels are withdrawing in ranges where more water is being pumped out for the watering system than can be renewed by the rains (IFPRI 2002; Rodell et al. 2009).

The impediments of conventional advancements could be judged by the way that advocates of an alternative farming method such as conservation agriculture (Hobbs et al. 2008). The proposed conservation techniques that are neither new nor reasonable on the grounds that cultivating works in an open framework and subsequently conservation of agribusiness are thermodynamically not exceptionally justifiable in such a framework (Knowler and Bradshaw 2007). All laws relating to conservation just work with detached frameworks. Likewise, organic farming depends on the affirmation of the destructive impacts of green revolution innovations, yet it can neither fulfill high efficiency nor guarantee a superior situation and better sustenance products.

1.3 Current Agricultural Production Systems

Agriculture is considered to be the backbone of several developing countries and still it continues to be. Traditional agricultural practices involved techniques such as crop rotation, intercropping, and irrigation (Horwith 1985). Over a time frame, agricultural practices have changed remarkably due to expanding worldwide population, atmospheric variability, and food demand (Aggarwal et al. 2004; Ericksen 2009). Traditional agricultural practices were replaced by conventional farming systems subsequently leading to multiple glitches affecting both mankind and the environment. Though agricultural intensification was successful at a certain point of time; however, recent studies confirm its serious consequences. Extensive uses of synthetic chemical fertilizers and pesticides have affected soil quality and microflora leading to poor crop yield and nutrient-deficient crops. Nevertheless, recent innovations in biosciences have made possible to overcome these issues, and more efforts are being laid globally to achieve sustainable agriculture. Presently, more emphasis is being laid on environment safe farming techniques and conservation and recycling of bioresources that involves the use of eco-friendly chemicals and bioresources that can easily be get degraded. Novel approaches such as organic farming, vertical farming, and precision farming are currently being practiced and promoted for future use. Various agriculture production systems are practiced such as conventional, integrated, and organic (Pretty 1997; Gosling and Shepherd 2005; Gomiero et al. 2008).

Conventional farming also known as intensive farming is the high-input technique wherein machinery and synthetic chemicals are applied to crops to gain better crop productivity (Pretty and Bharucha 2014). The use of genetically engineered products and pesticides has indeed affected the environment (Pimentel and Raven 2000; Fontes et al. 2002). Furthermore, it involves the use of mono-cropping and high-level nonrenewable natural resources such as potassium and phosphorous (Childers et al. 2011). Thus, intensive farming has caused tremendous socioeconomic and environmental problems in a momentum to maximize better crop yield. On the other hand, organic farming is an environmentally safe internationally acclaimed agricultural production system that ensures quality food, enriched with nutrients without affecting soil quality. It involves the holistic approach of farming that principally excludes the use of chemicals and aims toward achieving sustainable agriculture (Ortiz Escobar and Hue 2007). It is largely based on utilization of natural resources and traditional agriculture practices. Previous studies have shown that crop yield of organic farming is better than conventional farming under various environmental stresses. Moreover, organic farming increases water holding capacity, organic carbon, and mycorrhizal associations, including biodiversity in soil (Liebig and Doran 1999; Wells et al. 2000; Lotter et al. 2003); organic farming is effective in the conservation of biodiversity (Garg and Balodi 2014). Several approaches such as biodynamic, biological, and permaculture agricultural are the few among other agricultural production systems that are in analogous to organic farming with slight variations. In the last decade, several comparative analyses, among organic, integrated, and conventional farming over different geographical areas, have conducted reports suggesting that organic farming and integrated farming yield are comparatively lower than the conventional farming (Lotter et al. 2003; Swezey et al. 2007). Nevertheless, earlier studies confirmed that when gross profit is taken into consideration, organic and integrated farming are much higher than conventional farming (Pacini et al. 2003; Parra-Lopez et al. 2006). More recently, comparative analysis between organic and conventional farming revealed that organic farming yields of individual crops are on an average of 80% of conventional yields (Ponti et al. 2012).

Precision farming also known as satellite farming is a next-generation technology that makes use of information science. It makes use of global positioning system (GPS) and global navigation satellite system (GNSS) that permits an agriculturist to screen and maintain farm on a real-time basis. Wherein, the advanced technological sensors are fixed in the field to gain information on soil, pests, weeds, and weather conditions through which crop protection and productivity are achieved. Studies confirm that it allows the efficient use of fertilizers and pesticides (Van Alphen and Stoorvogel 2000; Jensen 2015).

Nanotechnology is a prospective tool that can be used to transform agricultural sectors; it aids in understanding the biochemical mechanisms of crops by improving the conventional practices for assessing environmental factors and in application of fertilizers and pesticides. In a current scenario, it has wide potential application in agriculture and food processing industries, especially nanobiosensors that are efficient and ultrasensitive (Rai et al. 2012; Jampílek and Kráľová 2015; Prasad et al. 2017). Nanomaterials and nanostructures such as carbon nanotubes, quantum dots, and nanofibers are currently being utilized in agriculture as biosensors for monitoring soil quality and fertilizer delivery. Furthermore, the nanoparticles synthesized from biopolymers such as cellulose and starch are gaining importance since they are nontoxic to humans and environmentally considered safe. Across the globe, several emerging countries are embracing nanotechnology advancements and its potential application in agriculture and food sectors. In future, this novel technology anticipates to becoming an economic driving force for a long term.

1.4 Nanotools, Nanoprocesses, and Nanomaterials

Nanoscience and nanotechnology are the fastest growing in the fields of energy, medicine, environmental science, biotechnology, and many other fields. The hallmark feature of these fields is an incorporation of traditional subjects like chemistry, physics, and biology in order to study and explore phenomenon at the nanosize. The characteristic feature of nanosized materials has different properties and smaller than their counterparts. Indeed, materials with nanosize dimensions fascinate scientist, researchers, and engineers; there is a growing and most impactful field in global scientific research (Jackman et al. 2016; Prasad et al. 2016). Nowadays, nanomaterials have become an inflection point in material science, emphasizing the advances and innovative trends in research and in applying to solve scientific problems in a relevant way. Engineering nanomaterial is a key term that opens many doors to the development of nanotools, which passes information in the form of nanoworld (López-Lorente and Valcárcel 2016; Valcárcel and López-Lorente 2016). The use of nanomaterials as nanotools has increased substantially between the twentieth and twenty-first centuries due to their simplicity of fabrication and functionalization. The development of new and more precise nanotools has greatly changed the field of nanotechnology by performing rapid and exact and in delivering the results in less than 1 day. Indeed, the manipulation of their properties by manipulating the material, size, and condition of the manicure could pave the way to the use of nanomaterials as analytical tools (Dinarelli et al. 2016; López-Lorente and Valcárcel 2016; Penon et al. 2016). Nanocomponents, tools, and methods are under continuous evaluation, and each generation is providing a sounder base for the following generation. (https://en.wikibooks.org/wiki/Nanotechnology/Perspective#Nanocomponents.2C_Tools.2C_and_Methods). In special, metallic nanoparticles serves as an important branch in nanotechnology. Noble metal nanoparticles have significant impact across a diverse range of areas, including catalysis, sensing, optoelectronics, energy conversion, and medicine (Maier et al. 2001; Jin et al. 2003; Heiz and Landman 2007; Anker 2008; Noginov et al. 2009; Atwater and Polman 2010; Arvizo et al. 2012; Aziz et al. 2015, 2016; Prasad et al. 2016).

The transformations in nanoscience and nanotechnology began with the developments of nanotools with an ability to see at the atomic level using scanning electron microscope (SEM) and transmission electron microscope (TEM) (Binnig and Rohrer 1982; Binnig et al. 1986; Jackson et al. 2006; Weiss 2007). Presently, the most advanced is an atomic force microscope (AFM); accentuation was mainly on the enhanced imaging resolution compared to that of optical microscopy. Rapid AFM will permit us to observe the elements of bionatural processes at the nanometer scale (Alessandrini and Facci 2005) and furthermore helps in comprehending biophysicochemical interactions at the nano-bio interface (Nel et al. 2009). These tools progress the ability to control individual atoms and particles, as well as their assemblies into precise structures (Eigler and Schweizer 1990; Piner et al. 1999; Revenko 2000; Takami et al. 2010; Claridge et al. 2011, 2013; Alivisatos et al. 2013). In the brightness of the advancements in the resolving power of microscopy, notable progress has been acknowledged in developing new tools, materials, and innovative strategies that have made up new avenues in science, technology, and medicine. Some modifications have been already addressed, problems in physical, chemical, and biological sciences via nanotools (Ducker et al. 1991; Manne et al. 1994; Strausser et al. 1997; Van Landingham et al. 1997; Magonov 2001; Cayuela et al. 2016). The reliability of more improvement and precise instruments for validating and controlling synthetic responses enables the synthesis of various nanomaterials and nanostructures with conspiring composition, size, condition, and functionalization. For example, gold, silver, fullerenes, grapheme, and carbon nanotubes (CNTs) as sorbents in solid-phase extraction are considered as potential nanotools (Rafati and Gill 2016). In biomedical applications, the tube-shaped nanostructure of DNAs acts as nanotemplates that demonstrate the exciting aspects of DNA nanotechnology and considered as novel nanotool in characterizing the nanomaterials wide range of applications owing to their excellent (Roco et al. 2011). In the years since, generous speculation has been working through a few initiative programs around the world, hoping that the sensational advances seen in nanotools and nanomaterials will proceed and will be used in different fields of skill as well as manufacturing and commercialization (Ariga et al. 2008).

Nanoprocessing involves the process and methods of developing engineered nanostructures and devices having minimum dimensions less than 100 nm. This principle is same for every aspect of nanomaterial research and development aiming at their utilization of a variety of applications from energy harvesting and storage, electronics, sensing, medicine, and human health care. The nanoprocessing technique is fascinating and facilitated by unprecedented growth of knowledge and deep understanding of the characteristics, properties, and their integration with engineered nanomaterials into multifunctional devices (Schäffer et al. 2000; Lee et al. 2003; Smith et al. 2003; Ginger et al. 2004; Gates et al. 2005; Biswas et al. 2006; Acharya et al. 2008; Ariga et al. 2007, 2008, 2010, 2011; Rogers and Lee 2008; Sakakibara et al. 2011; Ando et al. 2010; Kraemer et al. 2009; Li et al. 2009; Liddle and Gallatin 2009; Mailly 2009; Marrian and Tennant 2009; Schmid et al. 2009; Yaman et al. 2011). Advances in nanoprocessing techniques led to technological breakthroughs like storing a data chip in a single atom that represent a sequence of 8 bits processed as an exclusive unit of data known as a byte. Evolution of a novel-based nanoprocessing techniques will lead to applications in the areas such as nanomedicine with a focus on public health care, biosensors, and optics (Hoenlein et al. 2006; Prasad et al. 2016). Indeed, industries of today's world are in need of continuous refinement of manufacturing procedures and in the generation of applications that are efficient, authentic, environmentally friendly, and more frugal. In light of advances in the processing of nanomaterials, they extend the possibility for highly reproducible mass fabrication of systems with complex geometries and functionalities, similar drug delivery system, and commercial enterprise. The nanoprocessing methods are split into two major categories: "top-down" and "bottom-up" methods. In the top-down method, the nanoprocessing tools are controlled by an external force to produce the desired conditions and characteristics starting from larger dimensions and cutting them to the needed size. In the bottom-up method, atomic elements incorporated with more complex nanoscale assemblies or directed self-assemblies based on complex mechanisms and technologies (Biswas et al. 2012). The top-down method includes optical lithography, E-beam lithography, soft lithography/photolithography, nanoimprint lithography and blocks copolymer lithography and scanning probe lithography. On the other hand, the bottom-up method includes atomic layer deposition, solgel processing, molecular self-assembly, vapor phase deposition, and DNA scaffolding (Mamalis et al. 2005). Amplification of this existing nanoprocessing techniques will facilitate rapid growth of nanomaterials. Nanoprocessing technique makes the nanomaterials as the cornerstones of nanoscience and nanotechnology. Nanomaterials vary in distinct sizes and shapes such as nanoscale in one dimension (e.g., surface films), two dimensions (e.g., strands or fibers), or three dimensions (e.g., particles). They can synthesize in single, intertwined, aggregated structures with tubular, globular, and irregular shapes. Common sorts of nanomaterials incorporate nanotubes, nanocable dendrimers, and nanorods. Nanomaterials have wide applications in the field of nanoscience and other interdisciplinary and showcase distinctive physical-chemical attributes of normal chemicals (i.e., silver nano, gold, titanium dioxide, carbon nanotube, nanowires, carbon nano, silica) (Lue 2007; Ghosh 2009; NIOSH 2009; Alagarasi 2011).

1.5 Potentials and Risks of Nanotechnology in Agriculture

Effective implementation of nanotechnology applications in agriculture can be fruitful if normal procedures are re-enacted in the more sophisticated scientific way. Given that nanotechnology may allow for the precise control of assembling at the nanometer scale, various novel conceivable outcomes in elevating the precision practices are conceivable. Controlled delivery of agrochemicals – fertilizers, pesticides, herbicides, and plant growth regulators – is possible with the use of nanotechnology as it provides powerful tools to develop new methodologies for modern agriculture. One of the upsides of nanoscale delivery vehicles in agriculture applications is its enhanced strength of the payloads against degradation in the environment, consequently increasing its effectiveness while reducing the amount applied. This reduction helps address agricultural chemical runoff and alleviate the environmental consequence.

Nanotools involve numerous nanoscale carriers including encapsulation, entrapment polymers and dendrimers, surface ionic and frail bond connections such, different systems might be utilized to store, ensure, convey, and discharge by planned payloads in yield generation forms (Chaudhary et al. 2010). Nanotubes, nanowires, nanoparticles, or nanocrystals are frequently used to enhance the signal transduction determined by sensing components in response to exposure to biological and chemical compounds having comparative size. Nanomaterials show novel surface chemistry and distinct thermal, electrical, and optical properties useful to enhance sensitivity, reduce response times, and improve detection limits and can be used in multiplexed systems (Johnston 2010). Nanosensors represent powerful tool with an improved and advanced feature, compared to the existing biosensors and analytical sensors. Nanosensors are characterized as analytical devices having not less than one sensing dimension and not greater than 100 nm developed for observing physiochemical property set-up which are generally difficult to reach (Pandey et al. 2010).

Field sensing framework to screen environmental stresses and crop condition through the system of remote nanosensors positioned over cultivated fields gives crucial information for real-time monitoring moisture level, temperature, soil fertility, crop nutrient status, plant diseases, weeds, and insects. This detecting technology gives ideal time for planting and harvesting crops, the time and level of water, fertilizers, herbicides, pesticides, and other treatment need to be managed at given specific plant physiology, pathology, and other environmental conditions.

Recent studies have demonstrated that nanoscale cellulosic nanocrystals obtained from agricultural crops and trees open up a radically new market for cellulosic nanomaterials as they can be utilized as a part of the fortification polymeric matrix as nanocomposite (Leistritz et al. 2007). Carbon nanotubes and nanoparticles of SiO_2, ZnO, Au, and TiO_2 can add to an improving development of plants by enhancing component uptake and utilization of supplements (Baruah and Dutta 2009). The headway in nanofabrication and characterization tools has empowered physicochemical and natural associations of plant cell organelles and various disease-causing pathogens. A comprehensive understanding of plant pathogenic systems, for example, flagella motility and biofilm formation, will prompt enhancing treatment procedures to control the infection and secure the plant (Cursino et al. 2009). Crop genome is being widely concentrated on to focus biotechnology researchers on developing potentially enhanced plant varieties tolerant to different stress conditions such as environmental stress – droughts, salinity, diseases, and others. The advancement in nanotechnology-empowered gene sequencing is relied upon to present quick and cost-effective capabilities within a decade consequently promoting the best effective identification and utilization of plant genetic trait resources.

Nanofertilizers often contain nutrients or growth promoters encapsulated in nanoscale polymers, chelates, or emulsions which benefit in slow, targeted, efficient release into the fields. Nanopesticides often follow a similar model to nanofertilizers and contain active pesticide ingredients – insecticides, fungicides, and bactericides associated within a nanoscale product or carrier; such formulation of nanopesticide possesses increased stability and solubility and enables slow release and effective in targeted delivery in pest management (De Rosa et al. 2010; Prasad et al. 2014; Bhattacharyya et al. 2016). Target-specific nanoformulations of herbicide can be effectively used combined with the smart delivery system which blends with soil particles and inhibits the growth of weeds that have become resistant to conventional herbicides. Targeted use of nano-herbicides will be useful in effective weed control program (Ditta 2012).

Upcoming nanotechnologies in agriculture field seem quite interesting and promising. However the lack of adequate data about key risk evaluation aspects, for example, nanoparticle toxicity, bioaccumulation, exposure information or infusion risk (Yogesh et al. 2015), lethality of the biological community (Illuminato 2009), and potential residual deposit carry over with food stuff (Chaudhari and Castle 2011), and nanomaterial phytotoxicity affects the most concern toward application of nanomaterials in various agriculture sector (Lav et al. 2012). It was highlighted that health and environmental risk posed by nanomaterial cannot be assessed easily as they have diverse properties and behavior. Manufactured use and potential release of nanomaterials have preceded evaluation of risk to ecosystem including humans.

1.6 Conclusion and Future Trends

The application of nanoscience still owing to a lot of grounds in the agriculture field, for example, the unappropriate response from people toward genetically modified (GM) crops, lot of support is needed by the government agricultural research and technological units to explore nanobased instruments and new-fangled technologies (Prasad et al. 2014). As the conventional farming system works based on the use of synthetic chemical fertilizer and herbicidal compounds, it represented a typical cash grain, mono-crop farming unit. Unethical use of inorganic fertilizers, fungicide, and herbicide lead to dangerous negative impact on health concerns of human being and surrounding biotic environment. The biological constraints of conventional farming can be solved by incorporating new sustainable farming system with advanced technologies. Hence, it is very much important to rethink of the use of such controversial conventional farming systems for the protection of natural biodiversity. Though nanotechnology finds a potential role in agriculture, some issues have to be sorted out. Importantly, risk assessment and cost are involved in manufacturing of nanoparticles (Sharon et al. 2010; Sekhon 2014).

In agriculture sector, nanotechnology has phenomenal potential to facilitate and frame the next stage of precision farming techniques, but, at the same time, one has to be cautious about public acceptance of this novel technology. Like different advances low-cost nanomaterial and field application advancements are required for their application in farming systems. Specific utilization of nanomaterials would somehow minimize risk factors connected with it. The advancement of nanomaterials with intense dispersion and interaction between the fluid and solid phases, with well-understood toxicodynamics and toxic kinetics, biodegradable in soil and environment, less toxic and more photo-generative, smart and stable application in agriculture sector could be ideal for their efficient effective use in agricultural products.

References

Acharya S, Hill JP, Ariga K (2008) Soft Langmuir–Blodgett technique for hard nanomaterials. Adv Mater 21(29):2959–2981

Aggarwal PK, Joshi PK, Ingram JSI, Gupta RK (2004) Adapting food systems of the Indo-Gangetic plains to global environmental change: key information needs to improve policy formulation. Environ Sci Pol 7(6):487–498

Alagarasi A (2011) Introduction to nanomaterials. https://nccr.iitm.ac.in/2011.pdf. Accessed 3 Sep 2016

Alessandrini A, Facci P (2005) AFM: a versatile tool in biophysics. Meas Sci Technol 16:R65–R92

Alivisatos AP, Andrews AM, Boyden ES, Chun M, Church GM, Deisseroth K, Donoghue JP, Fraser SE, Lippincott-Schwartz OJ, Looger LL, Masmanidis S, McEuen PL, Nurmikko AV, Park H, Peterka DS, Reid C, Roukes ML, Scherer A, Schnitzer M, Sejnowski TJ, Shepard KL, Tsao D, Turrigiano G, Weiss PS, Xu C, Yuste R, Zhuang X (2013) Nanotools for neuroscience and brain activity mapping. ACS Nano Focus 7(3):1850–1866

Ando Y, Miyake K, Mizuno A, Korenaga A, Nakano M, Mano H (2010) Fabrication of nano stripe surface structure by multilayer film deposition combined with micropatterning. Nanotechnology 21(9):095304. doi:10.1088/0957-4484/21/9/095304

Anker JN (2008) Biosensing with plasmonic nanosensors. Nat Mater 7:442–453

Ariga K, Hill JP, Ji Q (2007) Layer-by-layer assembly as a versatile bottom-up nanofabrication technique for exploratory research and realistic application. Phys Chem Chem Phys 9(19):2319–2340

Ariga K, Hill JP, Lee MV, Vinu A, Charvet R, Acharya S (2008) Challenges and breakthroughs in recent research on self-assembly. Sci Technol Adv Mater 9:014109

Ariga K, Lee MV, Mori T, Yu X-Y, Hill JP (2010) Two-dimensional nanoarchitectonics based on self-assembly. Adv Colloid Interf Sci 154:20–29

Ariga K, Li M, Richards GJ, Hill JP (2011) Nanoarchitectonics: a conceptual paradigm for design and synthesis of dimension-controlled functional nanomaterials. J Nanosci Nanotechnol 11(1):1–13

Arvizo RR, Bhattacharyya S, Kudgus RA, Giri K, Bhattacharya R, Mukherjee P (2012) Intrinsic therapeutic applications of noble metal nanoparticles: past, present and future. Chem Soc Rev 41:2943–2970

Atwater HA, Polman A (2010) Plasmonics for improved photovoltaic devices. Nat Mater 9:205–213
Aziz N, Faraz M, Pandey R, Sakir M, Fatma T, Varma A, Barman I, Prasad R (2015) Facile algae-derived route to biogenic silver nanoparticles: synthesis, antibacterial and photocatalytic properties. Langmuir 31:11605–11612
Aziz N, Pandey R, Barman I, Prasad R (2016) Leveraging the attributes of *Mucor hiemalis*-derived silver nanoparticles for a synergistic broad-spectrum antimicrobial platform. Front Microbiol 7:1984. doi:10.3389/fmicb.2016.01984
Baruah S, Dutta J (2009) Nanotechnology applications in population sensing and degradation in agriculture: a review. Envioron Chem Lett 7:161–204
Bhattacharyya A, Duraisamy P, Govindarajan M, Buhroo AA, Prasad R (2016) Nano-biofungicides: emerging trend in insect pest control. In: Prasad R (ed) Advances and applications through fungal nanobiotechnology. Springer International Publishing, Cham, pp 307–319
Binnig G, Rohrer H (1982) Scanning tunneling microscopy. Helv Phys Acta 55:726–735
Binnig G, Quate CF, Gerber CH (1986) Atomic force microscope. Phys Rev Lett 56:930–933
Biswas A, Eilers H, Hidden F, Aktas OC, Kiran CVS (2006) Large broadband visible to infrared plasmonic absorption from Ag nanoparticles with a fractal structure embedded in a Teflon AF® matrix. Appl Phys Lett 88:013103. http://dx.doi.org/10.1063/1.2161401
Biswas A, Bayer IS, Biris AS, Wang T, Dervishi E, Faupel F (2012) Advances in top–down and bottom-up surface nanofabrication: techniques, applications and future prospects. Adv Colloid Interf Sci 170:2–27
Bouwmeester H, Dekkers S, Noordam M, Hagens W, Bulder A, de Heer C, ten Voorde S, Wijnhoven S, Sips A (2009) Health impact of nanotechnologies in food production. RIKILT Institute of Food Safety/RIVM National Institute for Public Health and the Environment, Wageningen. http://www.rivm.nl/bibliotheek/rapporten/000200704.pdf. Accessed 2 Sep 2016
Cayuela A, Benítez-Martínez S, Soriano ML (2016) Carbon nanotools as sorbents and sensors of nanosized objects: the third way of analytical nanoscience and nanotechnology. Trends Anal Chem 84:172–180. http://dx.doi.org/10.1016/j.trac.2016.02.016
Chaudhari Q, Castle L (2011) Food applications of nanotechnologies: an overview of opportunities and challenges for developing countries. Trends Food Sci Technol 22:595–603
Chaudhary Q, Castle L, Watkins R (2010) Nanotechnologies in food. RSC Publishing, London
Childers DL, Corman J, Edwards M, Elser JJ (2011) Sustainability challenges of phosphorus and food: solutions from closing the human phosphorus cycle. Bioscience 61(2):117–124
Claridge SA, Schwartz JJ, Weiss PS (2011) Electrons, photons, and force: quantitative single-molecule measurements from physics to biology. ACS Nano 5:693–729
Claridge SA, Liao WS, Thomas JC, Zhao Y, Cao H, Cheunkar S, Serino AC, Andrews AM, Weiss PS (2013) From the bottom up: dimensional control and characterization in molecular monolayers. Chem Soc Rev 42:2725–2745
Cursino L, Li Y, Zaini PA, De La Fuente L, Hoch HC, Burr TJ (2009) Twitching motility and biofilm formation are associated with tonB1 in *Xylella fastidiosa*. FEMS Microb Lett 299(2):193–199
De Rosa MC, Monreal C, Schnitzer M, Walsh R, Sultan Y (2010) Nanotechnology in fertilizers. Nat Nanotechnol 5:91. doi:10.1038/nnano.2010.2
Dhaliwal GS, Jindal V, Dhawan AK (2010) Insect pest problems and crop losses: changing trends. Indian J Ecol 37(1):1–7
Dinarelli S, Girasole M, Kasas S, Longo G (2016) Nanotools and molecular techniques to rapidly identify and fight bacterial infections. Journal of Microbiological Methods. doi:10.1016/j.mimet.2016.01.005
Ditta A (2012) How helpful is nanotechnology in agriculture? Adv Nat Sci Nanosci Nanotechnol 3(3):033002. doi:10.1088/2043-6262/3/3/033002
Ducker WA, Senden TJ, Pashley RM (1991) Direct measurement of colloidal forces using an atomic force microscope. Nature 353(6341):239–241

Eigler D, Schweizer EK (1990) Positioning single atoms with a scanning tunneling microscope. Nature 344:524–526

Ericksen PJ (2009) Conceptualizing food systems for global environmental change research. Glob Environ Chang 18(1):234–245

Fontes EMG, Pires CSS, Sujii ER, Panizzi AR (2002) The environmental effects of genetically modified crops resistant to insects neotrop. Entomologiste 31(4):497–513

Garg A, Balodi R (2014) Recent trends in agriculture: vertical farming and organic farming. Adv Plants Agric Res 1(4):00023. doi:10.15406/apar.2014.01.00023

Gates BD, Xu Q, Stewart M, Ryan D, Willson CG, Whitesides GM (2005) New approaches to nanofabrication: molding, printing, and other techniques. Chem Rev 105:1171–1196

Ghosh P (2009) Colloid and interface science. PHI Learning, New Delhi

Ginger DS, Zhang H, Mirkin CA (2004) The evolution of dip-pen nanolithography. Angew Chem Int Ed 43(1):30–45

Gleick PH (1993) Water and conflict: fresh water resources and international security. Int Secur 18(1):79–112

Gomiero T, Paoletti MG, Pimentel D (2008) Energy and environmental issues in organic and conventional agriculture. Crit Rev Plant Sci 27(4):239–254

Gosling P, Shepherd M (2005) Long-term changes in soil fertility in organic arable farming systems in England, with particular reference to phosphorus and potassium. Agric Ecosyst Environ 105(1):425–432

Health Research Funding (2015) Pros and cons of conventional farming. http://healthresearchfunding.org/6-pros-and-cons-of-conventional-farming/. Accessed 2 Sep 2016

Heiz U, Landman U (2007) Nanocatalysis. Springer-Verlag, Berlin

Hobbs PR, Sayre K, Gupta R (2008) The role of conservation agriculture in sustainable agriculture. Philos Trans R Soc Lond Ser B Biol Sci 363:543–555

Hochella MF Jr, Lower SK, Maurice PA, Penn RL, Sahai N, Sparks DL, Twining BS (2008) Nanominerals, mineral nanoparticles, and earth systems. Science 319:1631–1635

Hoenlein W, Duesberg GS, Graham AP, Kreupl F, Liebau M, Pamler W, Seidel R, Unger E (2006) Nanoelectronics beyond silicon. Microelectron Eng 83(4):619–623

Horwith B (1985) A role for intercropping in modern agriculture. Bioscience 35(5):286–291

Huang J, Li Q, Sun D, Lu Y, Su Y, Yang X, Wang H, Wang Y, Shao W, He N, Hong J, Chen C (2007) Biosynthesis of silver and gold nanoparticles by novel sundried *Cinnamomum camphora* leaf. Nanotechnology 18:105–104

IFPRI (2002) Green revolution: curse or blessing? International Food Policy Research Institute, Washington, DC. http://oregonstate.edu/instruct/css/330/three/Green.pdf. Accessed 2 Sep 2016

Illuminato IS (2009) Binding particle to patience-nanotechnology in a true context of sustainability. https://www.oecd.org/science/nanosafety/44029701.pdf. Accessed 2 Sep 2016

Jackman JA, Cho D-J, Lee J, Chen JM, Besenbacher F, Bonnell DA, Hersam MC, Weiss PS, Joon N (2016) Nanotechnology education for the global world: training the leaders of tomorrow. ACS Nano 10:5595–5599

Jackson NB, Chaurand PR, Fulghum JE, Hernandez R, Higgins DA, Hwang R, Kneipp K, Koretsky AP, Larabell CA, Stranick SJ, Webb WW, Weiss PS, Woodbury N, Xie XS, Yeung ES (2006) Visualizing chemistry: the progress and promise of advanced chemical imaging. National Academies Press, Washington, DC

Jampílek J, Kráľová K (2015) Application of nanotechnology in agriculture and food industry, its prospects and risks. Ecol Chem Eng S22(3):321–361

Jensen PK (2015) Target precision and biological efficacy of two nozzles used for precision weed control. Precis Agric 16(6):705–717

Jin R, Cao C, Hao E, Métraux GS, Schatz GC1, Mirkin CA (2003) Controlling anisotropic nanoparticle growth through plasmon excitation. Nature 425:487–490

Johnston CT (2010) Probing the nanoscale architecture of clay minerals. Clay Miner 45:245–279

Knorr W, Prentice IC, House JI, Holland EA (2005) Long-term sensitivity of soil carbon turnover to warming. Nature 433:298–302

Knowler D, Bradshaw B (2007) Farmers' adoption of conservation agriculture: a review and synthesis of recent research. Food Policy 32:25–48
Kraemer S, Fuierer RR, Gorman CB (2009) Scanning probe lithography using self-assembled monolayers. Chem Rev 103:4367–4418
Lav RK, Sindhuja S, Joe MM, Reza E, Edmund WS (2012) Applications of nanomaterials in agricultural production and crop protection: a review. Crop Prot 35:64–70
Lee KB, Lim JH, Mirkin CA (2003) Protein nanostructures formed via direct-write dip-pen nanolithography. J Am Chem Soc 125:5588–5589
Leistritz FL, Hodur NM, Senechal DM, Stowers MD, Calla D, Saffron CM (2007) Biorefineries using agricultural residue feedstock in the great plains. http://ageconsearch.umn.edu/bitstream/7323/2/ae070001.pdf. Accessed 2 Sep 2016
Li L, Rafael RG, Gershgoren E, Hwang H, Fourkas JT (2009) Achieving lambda/20 resolution by one-color initiation and deactivation of polymerization. Science 324:910–913
Liddle JA, Gallatin GM (2009) Lithography, metrology and nanomanufacturing. Nano Scale 3(7):2679–2688
Liebig MA, Doran JW (1999) Impact of organic production practices on soil quality indicators. J Environ Qual 28(5):1601–1609
López-Lorente ÁI, Valcárcel M (2016) The third way in analytical nanoscience and nanotechnology: involvement of nanotools and nanoanalytes in the same analytical process. Trends Anal Chem 75:1–9
Lotter DW, Seidel R, Liebhardt W (2003) The performance of organic and conventional cropping systems in an extreme climate year. Am J Altern Agric 18(2):1–9
Lue JT (2007) Physical properties of nanomaterials. In: Nalwa HS (ed) Encyclopedia of nanoscience and nanotechnology. American Scientific Publishers, Stevenson Ranch, pp 1–46
Magonov SN (2001) Visualization of polymers at surfaces and interfaces with atomic force microscopy. In: Nalwa HR (ed) Handbook of surfaces and interfaces of materials, 2nd edn. Academic, Cambridge, pp 393–429
Maier SA, Brongersma ML, Kik PG, Meltzer S, Requicha AAG, Atwater HA (2001) Plasmonics – a route to nanoscale optical devices. Adv Mater 13:1501–1505
Mailly D (2009) Nanofabrication techniques. Eur Phys J Special Topics 172:333–342
Mamalis AG, Markopoulos A, Manolakos DE (2005) Micro and nanoprocessing techniques and applications. Nanotechnology Perceptions 1:31–52
Manne S, Cleveland JP, Gaub HE, Stucky GD, Hansma PK (1994) Direct visualization of surfactant hemimicelles by force microscopy of the electric double layer. Langmuir 10(12):4409–4413
Marrian CRK, Tennant DM (2009) Nanofabrication. J Vac Sci Technol A 21:S207–S215
Maurice PA, Hochella MF (2008) Nanoscale particles and processes: a new dimension in soil science. Adv Agron 100:123–153
Nel AE, Mädler L, Velegol D, Xia T, Hoek EMV, Somasundaran P, Klaessig F, Castranova V, Thompson M (2009) Understanding biophysicochemical interactions at the nano–bio interface nature materials. Nat Mater 8(7):543–557
NIOSH (2009) Approaches to safe nanotechnology: managing the health and safety concerns associated with engineered nanomaterials. https://www.cdc.gov/niosh/docs/2009-125/pdfs/2009-125.pdf. Accessed 2 Sep 2016
Noginov MA, Zhu G, Belgrave AM, Bakker R, Shalaev VM, Narimanov EE, Stout S, Herz E, Suteewong T, Wiesner U (2009) Demonstration of a spaser-based nanolaser. Nature 460(7259):1110–1112
Ortiz Escobar ME, Hue NV (2007) Current developments in organic farming. Recent Res Devel Soil Sci 2:29–62
Österholm P, Åström M (2004) Quantification of current and future leaching of sulfur and metals from boreal acid sulfate soils, western Finland. Aust J Soil Res 42:547–551

Pacini C, Wossink A, Giesen G, Vazzana C, Huirne R (2003) Evaluation of sustainability of organic, integrated and conventional farming systems: a farm and field-scale analysis. Agric Ecosyst Environ 95:273–288

Pandey RR, Saini KK, Dhayal M (2010) Using nano-arrayed structures in Sol-Gel delivered Mn^{2+} doped TiO_2 for high sensitivity urea biosensor. J Biosen Bioelectr 1:1–4

Parra-Lopez C, Calatrava-Requena J, de Haro-Gimenez T (2006) A multi-criteria evaluation of environmental performances of conventional, organic and integrated olive-growing systems in the south spain based on experts knowledge. Ren Agr Food Sys 22(3):189–203

Penon O, Marín MJ, Amabilino DB, Russell DA, Pérez-García L (2016) Iron oxide nanoparticles functionalized with novel hydrophobic and hydrophilic porphyrins as potential agents for photodynamic therapy. J Colloid Interface Sci 462. (2016):154–165

Pimentel D (2009) Pesticide and pest control. In: Pshin P, Dhawan AK (eds) Integrated pest management: innovation-development process. Springer, Dordrecht, pp 83–87

Pimentel DS, Raven PH (2000) Bt corn pollen impacts on non-target Lepidoptera: assessment of effects in nature. Proc Natl Acad Sci 97:8198–8199

Piner RD, Zhu J, Xu F, Hong SH, Mirkin CA (1999) "Dip-Pen" nanolithography. Science 283:661–663

Ponti T, Rijk B, Van Ittersum MK (2012) The crop yield gap between organic and conventional agriculture. Agric Syst 108:1–9

Postel SL, Daily GC, Ehrlich PR (1996) Human appropriation of renewable fresh water. Science 271:785–788

Prasad R (2014) Synthesis of silver nanoparticles in photosynthetic plants. J Nanopart Article ID 963961. http://dx.doi.org/10.1155/2014/963961

Prasad R (2016) Advances and applications through fungal nanobiotechnology. Springer, International Publishing, Cham

Prasad R, Kumar V, Prasad KS (2014) Nanotechnology in sustainable agriculture: present concerns and future aspects. Afr J Biotechnol 13(6):705–713

Prasad R, Pandey R, Barman I (2016) Engineering tailored nanoparticles with microbes: quo vadis. WIREs Nanomed Nanobiotechnol 8:316–330. doi:10.1002/wnan.1363

Prasad R, Bhattacharyya A, Nguyen QD (2017) Nanotechnology in sustainable agriculture: recent developments, challenges, and perspectives. Front Microbiol 8:1014. doi:10.3389/fmicb.2017.01014

Presley DR, Ransom MD, Kluitenberg GJ, Finnell PR (2004) Effect of thirty years of irrigation on the genesis and morphology of two semi-arid soils in Kansas. Soil Sci Soc Am J 68:1916–1926

Pretty J (1997) The sustainable intensification of agriculture. Nat Resour Forum 21:247–256

Pretty J, Bharucha ZP (2014) Sustainable intensification of agriculture systems. Ann Bot 114(8):1571–1596

Rafati A, Gill P (2016) Ultrastructural characterizations of DNA nanotubes using scanning tunneling and atomic force microscopes. J Microscop Ultrastruct 4:1–5

Rai M, Ingle A (2012) Role of nanotechnology in agriculture with special reference to management of insect pests. Appl Microbiol Biotechnol 94(2):287–293

Rai V, Acharya S, Dey N (2012) Implications of nanobiosensors in agriculture. J Biomater Nanobiotechnol 3:315–324

Rameshaiah GN, Pallavi J, Shabnam S (2015) Nano fertilizers and nano sensors – an attempt for developing smart agriculture. Int J Eng Res Gen Sci 3(1):313–320

Revenko I (2000) Probing the life sciences with atomic force microscopy. In: Proceedings of 1st symposium on micro and nanostructures of biological systems. Martin Luther University Halle-Wittenberg at Halle, pp 1–51

Roco MC, Mirkin CA, Hersam MC (2011) Nanotechnology research directions for societal needs in 2020: retrospective and outlook. Springer, Berlin

Rodell M, Velicogna I, Famiglietti JS (2009) Satellite-based estimates of groundwater depletion in India. Nature 460:999–1002

Rogers JA, Lee HH (2008) Unconventional nanopatterning techniques and applications. Wiley, Weinheim

Rolf M (1996) The role of agriculture in the economy and society. http://www.fao.org/docrep/W7440E/w7440e03.htm. Accessed 2 Sep 2016

Sakakibara K, Hill JP, Ariga K (2011) Thin-film-based nanoarchitectures for soft matter: controlled assemblies into two-dimensional worlds. Small 7(10):1288–1308

Schäffer E, Thurn-Albrecht T, Russell TP, Sakakibara K, Hill JP, Ariga K (2000) Electrically induced structure formation and pattern transfer. Let Nat 403:874–877

Schmid GM, Miller M, Brooks C, Khusnatdinov N, La Brake D, Resnick DJ, Sreenivasan SV, Gauzner G, Lee K, Kuo D, Weller D, Yang XM (2009) Step and flash imprint lithography for manufacturing patterned media. J Vac Sci Technol B 27:573

Sekhon BS (2014) Nanotechnology in agri-food production: an overview. Nanotechnol Sci Appl 7:31–53

Sharon M, Choudhary KA, Kumar R (2010) Nanotechnology in agricultural diseases and food safety. J Phytology 2(4):83–92

Smith JC, Lee KB, Wang Q, Finn MG, Johnson JE, Mrksich M, Mirkin CA (2003) Nanopatterning the chemospecific immobilization of cowpea mosaic virus capsid. Nano Lett 3(7):883–886

Strausser YE, Schroth M, Sweeney JJ (1997) Characterization of the low-pressure chemical vapor deposition grown rugged polysilicon surface using atomic force microscopy. J Vac Sci Technol A 15:1007. http://dx.doi.org/10.1116/1.580507

Swezey SL, Goldman P, Bryer J, Nieto D (2007) Six-year comparison between organic, IPM and conventional cotton production systems in the Northern San Joaquin Valley, California. Ren Agr Food Sys 22(1):30–40

Takami T, Ye T, Pathem BK, Arnold DP, Sugiura KI, Bian Y, Jiang J, Weiss PS (2010) Manipulating double-decker molecules at the liquid solid interface. J Am Chem Soc 132:16460–16466

Tillman D, Cassman KG, Matson PA, Naylor R, Polasky S (2002) Agricultural sustainability and intensive production practices. Nature 418:671–677

Valcárcel M, López-Lorente ÁI (2016) Recent advances and trends in analytical nanoscience and nanotechnology. Trends Anal Chem. doi:10.1016/j.trac.2016.05.010

Van Alphen BJ, Stoorvogel JJ (2000) A methodology for precision nitrogen fertilization in high-input farming systems. Precis Agric 2(4):319–322

Van Landingham MR, Mc Knight SH, Palmese GR, Elings JR, Huang X, Bogetti TA, Eduljee RF, Gillespie JW (1997) Nanoscale indentation of polymer systems using the atomic force microscopy. J Adhes 64(1–4):31–59

Verena S, Navin R, Jonathan AF (2012) Comparing the yields of organic and conventional agriculture. Nature 485:229–230

Waychunas GA, Kim CS, Banfield JA (2005) Nanoparticulate iron oxide minerals in soils and sediments: unique properties and contaminant scavenging mechanisms. J Nanopart Res 7:409–433

Weiss PS (2007) A conversation with Dr. Heinrich Rohrer: STM co-inventor and one of the founding fathers of nanoscience. ACS Nano 1(1):3–5

Wells AT, Chan KY, Cornish PS (2000) Comparison of conventional and alternative vegetable farming systems on the properties of a yellow earth in New South Wales. Agric Ecosyst Environ 80(1–2):47–60

Yaman M, Khudiyev T, Ozgur E, Kanik M, Aktas O, Ozgur EO, Deniz H, Korkut E, Bayindir M (2011) Arrays of indefinitely long uniform nanowires and nanotubes. Nat Mater 10:494–591

Yogesh B, Gangadhara K, Chidanand R, Gaurav C, Padmabhushan U (2015) Nanotechnology in agriculture: a review. J Pure App Microbiol 9(1):1–11

Nanotechnology in Life Science: Its Application and Risk

2

Gero Benckiser

Abstract

From the increasing number of patent registrations, the European Commission (EC) concludes that nanotechnology ranks within the six most promising key technologies. Most successful at inventing nanomaterials—measuring one billionth of a meter—is nature, which provides blueprints of self-organizing physical-chemical nanoparticles (NPs) properties and opens new dimensions for researchers in exploiting nature's NPs and developing new products for increasing the agricultural and industrial productivity.

After summarizing product processing effects, advantages and risks for agriculture, food, nutrition, and medicine, this book chapter discusses reasons why NPs use should occur balanced and carefully controlled by health and landscape policy. Only then a successful and profitable use for the overall benefit at lowest environmental pollution is achieved.

Keywords

Absorber • Biocidal nanoparticles • Encapsulated systems • Novel food • Nanopharmaceuticals • Sensor elements • Lab on a chip analytic • Toxicity • Risk assessment

G. Benckiser (✉)
Institute of Applied Microbiology, Justus Liebig University of Giessen, Giessen, Germany

Research Center for BioSystems, Land Use, and Nutrition (IFZ), Justus Liebig University of Giessen, Heinrich-Buff-Ring 26-32, 35392 Giessen, Germany
e-mail: gero.benckiser@umwelt.uni-giessen.de; gerobenckiser@yahoo.de

R. Prasad et al. (eds.), *Nanotechnology*, DOI 10.1007/978-981-10-4573-8_2

2.1 Introduction

Nanotechnology is one of the six "key technologies" that the European Commission (EC) is considering as adequate to initiate sustainable competitiveness and growth among industries and countries. In the advancement of nanotechnology supporting obligation, an increasing number of nanotechnology-related publications and patents in the field of agriculture, foods, nutrition, and medicine encourage the EC, although industry experts stress that agricultural nanotechnology does not demonstrate a sufficient economic return to counterbalance the high initial production investments (Benckiser 2012; Parisi et al. 2015; Prasad et al. 2017b). However, the research engagement beyond the increasing number of nanotechnology-related publications and patents and the rapid progress of nanotechnology key industries are convincing arguments for the EC to support nanoresearch engagement and to explore possibilities of how the bulk of publications and patents could be transferred into marketable and profitable agricultural products over time.

The most inventive at producing nanomaterials, measuring one billionth of a meter, is nature, and its blueprints of self-organizing, physical-chemical nanoparticles (NPs) properties open new dimensions to researchers as the bulk of publications and patents substantiate. Researchers have started wrapping nanoactive substances into additives, nano inside, and develop marketable products, fabricated by medicine, food, nutrition, and agriculture-related industries. Nitrification inhibitor-containing N-fertilizers, oil-applied pesticides, formulated drugs, and bags with which these products are transported to the consumers, nano outside, are examples (Ravichandran 2010; Chen et al. 2012; Horstkotte and Odoerfer 2012; Shen et al. 2012; Benckiser et al. 2013, 2016; Prasad 2014, 2016; Aristov et al. 2016; Leung et al. 2016; Scandorieiro et al. 2016).

Through 2012, more than 68 million patents were registered within the International Patent Documentation Centre (INPADOC), the World Intellectual Property Organization (WIPO), the Patent Cooperation Treaty (PCT), and the patent offices, mostly of America, Europe, Germany, and Japan and a keyword search within this wealth of patents under the headings: nanotechnology, super absorber, agriculture, nutrition, and food technology generated 28,149 positive matches (Benckiser 2012). A closer look on the first 500 patents of the 28,149 positive matches unveiled that approximately 320 of the 500 patents or 64% belong to machine-integrated NPs parts or control devices (Connor and Jacobs, 1999; Schmitt et al. 2010; Kuhlmeier et al. 2012; Banitz et al. 2013; De Lorenzo 2014). Approximately 36% of the NPs represent metal oxides, fertilizers, pesticides, and drugs, which find use in agriculture, food technology, nutrition, and medicine (Allen et al. 2011; Benckiser 2012; Benckiser et al. 2013; Prasad et al. 2014, 2016; Heo and Hwang 2016; Palomo and Filice 2016; Shamaila et al. 2016).

The slowly useable made NPs motivated the international science and technology community and administrators to invest more engagement into converting NPs into marketable products with a broad range of applications (Fig. 2.1; Buentello et al. 2005; Tzuzuki 2009; Kong and Ziegler 2012). Yet, in this context NPs side effects of already marketed NPs products in agriculture, food technology, nutrition,

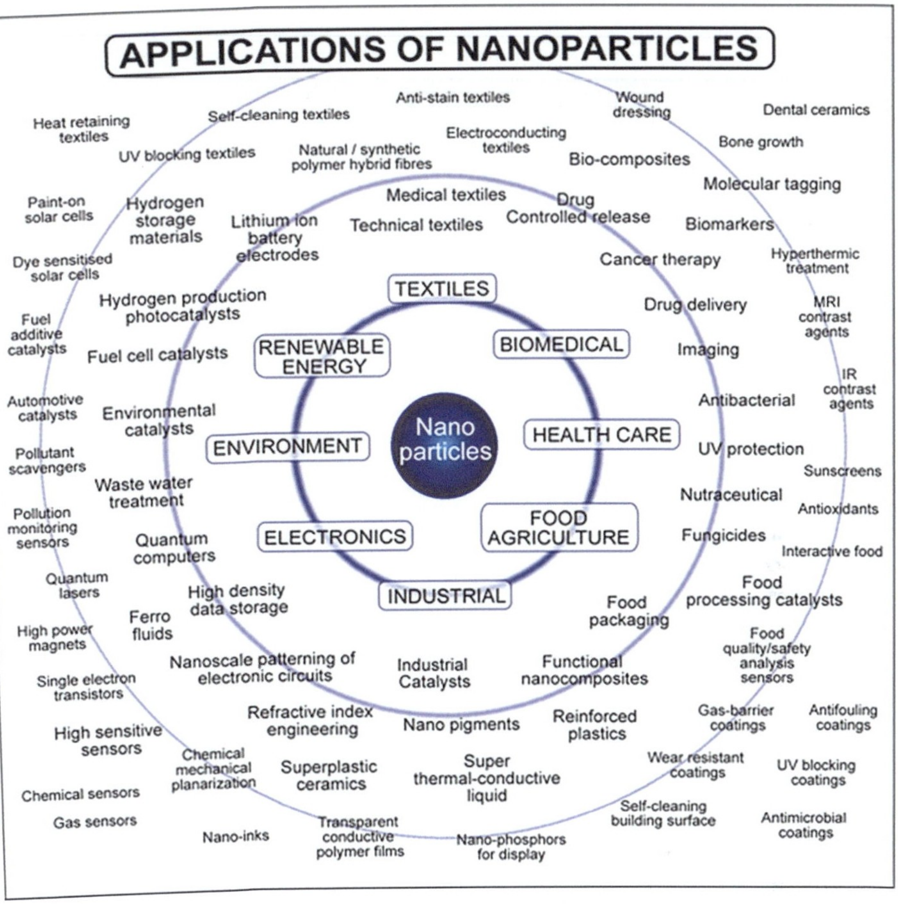

Fig. 2.1 Nanoparticle application areas (Reprint with permission of Prof Tzuzuki)

medicine are observed, because target-orientated applied NPs diffuse in nontarget environments, exhibit there unwished reactions, cause uncertainties, open debates concerning consumer perception of nanotech products, and force EC and the EU parliament on basis of little understood NPs behaviors, fates, and side effects in nontarget environments to legislate nanofood regulations (Cobb and Macoubrie 2004; Parisi et al. 2015). Continuing national and international regulations specifying debates are regularly conducted to achieve a better consumer perception of nanotech products (e.g., European Environment Agency Technical report No 4/2011; International Cooperation on Cosmetic Regulation, ICCR, Canada; the EU Biocidal Product Regulation (BPR); the provision of food information to consumers (labeling) by the European Parliament; Cobb and Macoubrie, 2004; Grobe and Rissanen 2012; Reidy et al. 2013).

NP-related inventions in agriculture, food, nutrition, and medicine marketed and applied mostly in combination with inorganic or organic super-absorbing, polymeric structured compounds, challenge not only competition and collaboration between companies and nations but also landscape policy, which has to handle NPs side effects (Grobe and Rissanen 2012; Benckiser et al. 2013; Afzal et al. 2016; Heo and Hwang 2016; Yu et al. 2016).

This book chapter discusses after summarizing NPs exploitation, product processing effects, and advantages and risks for agriculture, food, nutrition, and medicine the balancing reasons for beneficial and profitable NPs use and a successful health and landscape policy.

2.2 Nanoparticle Exploitation and Developing Products

NPs in agriculture, food, textile, packing industries, health care, medicine, and the modern electronic and renewable energy sector increase in importance, and nanostructured alumina products, controlling effectively *Sarocladium oryzae* (L.) and *Rhyzopertha dominica* (F.)—a major insect pest in stored food supplies—and the oriental fruit fly *Bactrocera dorsalis* (Hendel)—a very destructive pest of fruit effectively controlled by nanogel immobilized semiochemical pheromones are examples of NPs products with relevance in agriculture (Bhagat et al. 2013; Prasad 2014; Prasad et al. 2014; Sadeghi and Ebadollahi 2015). Encapsulated NPs products offer easy handling, transportation without refrigeration, sustained release, and stability at open ambient conditions, reducing recharging frequency and indicate ecogeniality. Pharmaceuticals and nanodrug delivery systems are in use, and for the raw material suppliers, agriculture, forestry, and the food and packing industry nano application perspectives are predicted toward the 21st century (Figs. 2.1 and 2.2; Hilder and Boulter 1999; Sticklen 2009; Ravichandran 2010; Duncan 2011; Neethirajan and Jayas 2011; Dickinson 2012; Grobe and Rissanen 2012; Horstkotte and Odoerfer 2012; Kong and Ziegler 2012; Kuhlmeier et al. 2012; Benckiser et al. 2013; Voytas and Gao 2014; Sehkon 2014; Cherukula et al. 2016; Jivani et al. 2016). The newly developed methods, DNA microarrays, nanosensing, microelectromechanical systems, and microfluidics are prerequisites for achieving NPs and encapsulated NPs products, with which protein bioseparation, biological and chemical contaminant sampling, nanosolubilization, nutrient delivery, food coloring, nutra- and pharmaceutical efficacy, food quality, the conversion of CO_2 into fuels and other chemicals, water splitting into H_2 and oxidized organic compounds, the designing of self-cleaning surfaces, water disinfection and purification, or the TiO_2 or Ni@g--PC3N4 nanophotocatalyst conversion of benzyl alcohol into benzaldehyde are improved (Fig. 2.2; Kuhlmeier et al. 2012; Deng et al. 2015; Horst et al. 2015; Liu and Lal 2015; Smith et al. 2015; Sunkara et al. 2015; Yang et al. 2015; Kumar et al. 2016; Ouyang et al. 2016; Palomo and Filice 2016). Benzaldehyde, a widely applied organic compound of economic importance in food, pharmaceutical, perfumery, and other chemical industries and anti-inflammatory, antioxidative, immunomodulating, antimicrobial properties exhibiting metal-based NPs are income promotional and broaden with products consisting of chlorohexidine and *Scutellaria baicalensis* bacteria, encapsulated mesoporous silica nanoparticles, or encapsulated baicalin—a flavonoid-based NP isolated from *Scutellaria baicalensis*—or baicalein, a traditional Chinese medicine component, offer agrifood and health safety perspectives (Fig. 2.2; Ikemoto et al. 2000; Ravichandran 2010; Jang et al. 2014; Sekhon 2014; Seneviratne et al. 2014; Cherukula et al. 2016; Leung et al. 2016; Raza et al. 2016; Shamaila et al. 2016).

2.3 Nanotechnology Advantages in Agriculture, Food, and Medicine

Self-organizing nanoprocesses, leading to novel NPs that differ in their properties from those of bulk materials, can have miniaturized devices (Kuhlmeier et al. 2012), micro-electro-mechanical systems (Jivani et al. 2016), or devices inserted in gene manipulated bacteria and archaea (de Lorenzo 2014). Such miniaturized devices broaden the diagnostic boundaries and enable researchers to develop advanced analytics and methods and to design and realize marketable NPs products, bringing innovation and benefits in societal areas, such as agriculture, food, nutrition, medicine, and related industries (Figs. 2.1 and 2.2; Horstkoote and Odoerfer 2012; Kong and Ziegler 2012). The numerous nanotechnology exploiting inventions, inter alia exemplified by the self-organized growth arrays of graphene nanoribbons on structured silicon carbide substrates, documents the continually progressing NPs success (Rogers 2010; Benckiser 2012). The wealth of ideas not only lead to nanolevel fabricated carbon nanomaterial electronics but also application possibilities in agriculture, food, nutrition, medicine, and related industries for years to come.

Miniaturized systems, integrated in technical and laboratory equipment, are such advanced that a sensitive and accurate detection cannot only be helpful to develop a broad range of technological applications but also to detect more precisely NPs, diffusing from the target site into nontarget environments and trigger unwished impacts (Afzal et al. 2016; Cherukula et al. 2016; Heo and Hwang 2016; Kumar et al. 2016; Leung et al. 2016; Palomo and Filice 2016; Ouyang et al. 2016; Raza et al. 2016; Shamaila et al. 2016; Yu et al. 2016). Nanopharmaceuticals, nanotechnology-based drugs bring a significant benefit for the patient. They have a better solubility, dissolution rate, more enhanced oral bioavailability as conventionally formulated drugs, and a pinpoint application but may concomitantly show unwished NPs side effects, nudging the need for public safety regulations (Fig. 2.2; Grobe and Rissanen 2012; Schmitt et al. 2010; Horstkotte and Odoerfer 2012; Prasad et al. 2017a; Scandorieiro et al. 2016). Nitrification inhibitors, encapsulated in nitrogen fertilizers (NIENF), control not only solubility, nanodissolution, and plant N bioavailability but concomitantly reduce N losses, NO_3^- groundwater pollution, and N_2O climate impacts (Zerulla et al. 2001; Benckiser et al. 2013, 2016; Yang et al. 2016). Decreased particle size ranges of NPs, nutra-, and nanopharmaceuticals offer the possibility of sterile filtration before application, improve dose proportionality, and reduce food effects. Continually advancing NPs detectability provides suitability in following diffusion routes that reduce administrative risk assessments.

Agriculture	Food, Nutrition, Food packing	Cosmetic, Medicine
Encapsulated nanosensors on chip basis to detect substrate-microbe-animal-plant-pathogen interactions, to monitor soil and crop growth conditions (De Lorenzo, 2014; Heo and Hwang, 2016; Aristov et al, 2016) ***Encapsulated flavor enhancers*** (nutraceuticals; Sehkon, 2014). ***Nanocapsules*** a more efficient delivering of growth hormones, vaccines, genetic engineered DNA and agrichemicals (pesticides, fertilizers targeting (Hilder, Boulter, 1999; Sticklen, 2009; Benckiser et, al., 2013; Voytas, Gao, 2014) ***Fiber crops*** for advanced textile production (Kong, Ziegler, 2012.)	***Nanostructured organics*** vitamins, antioxidants; colorants, flavouring agents ***inorganic nano-sized* nutrients, supplementing *additives*** alkaline earth metals, iron, silver, titanium dioxide, silica, non-metals calcium, magnesium, selenium for enhancing food and animal feed production, for improving absorption, taste, beverage texture bioavailability, food, or emulsions as mayonnaise by adding water nanodroplets. ***Nanosensors, biopolymer-based carriers (liposomes, nanomicelles), nanocoatings, nanofiltration*** to mask taste ingredients, protect from degradation, improve food safety by increasing contact surfaces, self cleaning properties, antimicrobial effects and achieve an otical better appearance (Duncan, 2011; Nethirajan, Javas, 2011; Dickinson, 2012).	***Analytical tools and nanoimaging*** to facilitate microfabrication of polymeric substrates, novel drug delivery, micro-electro-mechanical system devices for a controlled application, reached for a better electric light control e.g. in graphen, quantum devices by plasmon polaritons-coupled photon excitations. ***Nanomaterials based sensors, nanodevices, clinical, regulatory and toxicological issues*** for detecting and controlling pathogens and toxicants easier. ***Nanopharmaceuticals and delivery systems*** improve phiarmacological efficacy by modified physicochemical properties and targeted body distribution (Schmitt et al., 2010; Chen et al. 2012; Horstkotte, Odoerfer, 2012; Shen et al., 2013; De Lorenzo, 2014; Jivani et al, 2016)

Fig. 2.2 Nanotechnology applications in agriculture, food, nutrition, food packing, cosmetic, and medicine

The nonprofit Frost and Sullivan Institute (FSI), dedicated to leveraging innovation to address global challenges, estimated in 2012 an annual BioMEMS growth rate of more than 15% for China. Such predicted growth rates of controlled applications by MEMS demonstrate that novel products on micron, nanometer scale basis and related industrial productivities are going to be revolutionized. The study of biological systems enters into an improved treatment of diseases with new polymer-based anticancer agent delivery systems, a development of specialized tools for minimally invasive surgery, novel cell-sorting systems allowing high-throughput data collection, and precision measurement techniques for microfabricating devices (Darshana D. BioMEMS Impact on Healthcare, Report Frost & Sullivan, Mountain View, March 30, 2012, cited by Kuhlmeier et al. 2012; Horstkotte and Odoerfer 2012; Kong and Ziegler 2012; Senior et al. 2012; Prasad et al. 2016). In the nearest future, an increase of related patents is expected (Benckiser 2012: Frost and Sullivan 2016).

Micro devices, applicable for medical and food analysis and under current development, belong to three principal branches: nanoparticles, nano fluidics, and nano sensors, and consumer-oriented developed miniaturized system products find use as automotive coatings, smart textiles, sunscreens, or easy-to-clean surfaces for bathrooms (Horstkotte and Odoerfer 2012; Kuhlmeier et al. 2012; Kong and Ziegler 2012). Such products demonstrate in their breadth of possible and expected applications private and public advantage and are perceived positively by a majority of consumers. However, the progress of nanotechnology analytics visualizes NPs side effects and spread uncertainties (Cobb and Macoubrie, 2004; Grobe and Rissanen 2012; Zimmer et al. 2012; Coppola and Verneau 2014; Gupta et al. 2015). Organizations as the European Union (EU) started communication patterns concerning definition and recommendation of nanomaterials, which resemble those of the genetically modified organism (GMO) debate. The discussion will increase the more nanotechnology-based products are marketed and with increasing detected NPs side effect industries of concern silence communication. Yet, stakeholders, risk management, mistrusting consumer organizations, and policy makers should not careen in debating NPs' benefits and risk regulations for achieving improved public safety by enhancing NPs risk assessments.

2.4 Risk Assessment of Nanomaterials in Agriculture, Food, and Medicine

During processing, separation, consolidation, and deformation of materials, natural events or human activities occur atom or single molecule changes, which may lead to risky incident events, ending in an averaged descending, undesirable adverse effects (Taniguchi 1974; Renn 2008). Risk events more broadly defined are probabilities, the likelihood of event estimates, hazard multiplied by exposure, viewed as physical harm to humans, cultural artifacts, or ecosystems (Senjen 2012). Risky can be NPs side effects, caused by uncontrolled NPs diffusion from the place of requirement into surrounding landscapes, which are increasingly observed, for example, the U.S. Food and Drug Administration (FDA) drafted and published in 2012 a risk guidance document. Defined in the risk guidance document are nanotechnology factors, which particularly NP-containing foods processing manufacturers should consider, because NPs observably cause significant changes, such as:

- Affect the identity of the food substance
- Affect the safety of the use of the food substance
- Affect the regulatory status of the use of the food substance
- Warrant a regulatory submission to FDA (Grobe and Rissanen 2012).

In this context, the scientific community—the Scientific FDA Committee—identified deficiencies in characterizing, detecting, and measuring (engineered) nanomaterials in food, feed, and biological matrices and asserted after evaluation that the availability of data for oral exposure and for any consequent toxicity is extremely limited. The majority of information concerning toxicokinetics and toxicology comes from in vitro or in vivo studies having only a confined transferability to human risks. The Scientific FDA Committee reminded the lack of methods for a satisfying testing of (engineered) nanomaterials and that due to our limited current knowledge concerning environmental NPs, impacts particularly in the food area research endeavors are required to answer open questions.

Our limited knowledge concerning side effects through nanotechnology application and the not easy to be answered questions concerning NPs in health care must be debated increasingly because of detected possible connections between nanomaterials in food or feed and gut diseases (Stracke et al. 2006; Grobe and Rissanen 2012; Böhmert et al. 2015; Xiao et al. 2016). Novel (nano) foods and feeds, consumed by humans and animals, and agrochemicals, used in plant protection or cosmetics and medicines and applied in health care, are sources from which NPs are diffusing into nontarget surrounding environments. The relative newness of NPs application and health interaction and the limited transferability of NPs animal tests to human scientific key issues complicate risk assessment dialogues between academia, industries, stakeholders, regulators, and nongovernmental organizations (NGOs). Such discussions are urgently needed to guarantee an almost safe nanotechnology application in agriculture, nutrition, and medicine and that unwanted NPs side effects, requiring practical risk assessment recommendations, can be widely prevented.

Little commented information has spread by the increasing wealth of NPS-related literature and innovations in the agrifood sector since only recently and has fueled uncertainties and made consumers wary (Brown 2009; Benckiser 2012). For public safety and risk assessment, debate-engaged NGOs claimed to expand the narrow EU Cosmetics Directive risk definition that limits the scope by predominantly focusing on insoluble or bio-persistent materials (Raj et al. 2012; Senjen 2012). NGOs call for a debate, which not only pertains to insoluble or bio-persistent materials but also includes human health and environment aspects, ownership of technology and accountability, privacy, security, surveillance and human enhancement, equity of access, and impact on national economies and industries (Senjen and Hansen 2011). In such a debate of concern should be public involvement in technology development and convergence of nanotechnology, nanobiology, and nanocomputer technology, and such a debate should not end in the emergence of a global nanodivide into a nanotechnology reinforcing toward global inequalities (Schroeder et al. 2016). The gap between rich and poor countries should not increase, and also this aspect should play a weighty role in the nanotechnology risk assessment discussion. In assessing risk-related regulations, governance regimes will act reality adaptive and avoid being archaic in the discussion of concern. They mostly will start simply by focusing on already marketed, relatively low-tech nanoproducts. Learning from the process of governing relatively simple existing products, governance regimes will reflexively speed up as more complex, later-stage novel nanoproducts emerge in structuring the adaptation process (Brown 2009).

In Europe, the European Parliament and the EU Scientific Committee on Emerging and Newly Identified Health Risks (SCENIHR) have begun to take an interest in nanotechnology and have started with increased safety testing, a specific risk-assessment procedure in cosmetic and novel (nano) food industries that nanomaterials containing products can be labeled (Raj et al. 2012). NGOs suspect that the European Parliament and SCENIHR NPS concerning initiative is not only limited in scope but that the necessity of a constructive and in risk-assessment results ending dialogue will not come into effect for a long time.

2.5 Conclusions

Agriculture, food, textile, packing industries, and medicine penetrating NPs play an increasing role in health care, the sector of modern electronic and renewable energy. Public debate on NPs' benefits and risks is occurring between industries, stakeholders, process regulators, risk managing administrators, mistrusting consumer organizations, and policy makers. The described knowledge about NPs use in agriculture, food technology, nutrition, and medicine and the behavioral similarities of encapsulated NPs in marketed products in the field of applications under consideration have these in common:

(a) Enhanced NPs solubility, dissolution rate oral or plant bioavailability, and the possibility of a target-orientated application.
(b) Decreased particle size range offers sterile filtration before application.
(c) Improved dose proportionality.
(d) Concerning nanoagrifood products, a reduced food damage and N loss, NO_3^- groundwater pollution, N_2O climate impact reduction.
(e) Suitability of NP-based very sensible sensor devices enhances the detectability, and NPs diffusion routes in nontarget environments administrations can follow more easily.
(f) More satisfying reach of public safety concerning NPs side effects, risk assessments after more profound nanotechnology key-risk discussions enabled through advanced analysis, and better informed policy makers.

References

Afzal A, Rafique M, Iqbal N, Qaiser AA, Anwar AW, Iqbal SS (2016) Synergistic effect of functionalized nanokaolin decorated MWCNTs on the performance of cellulose acetate (CA) membranes spectacular. Nanomaterials 6:79. doi:10.3390/nano6040079

Allen C, Mittal G, Sung CJ, Toulson EL (2011) An aerosol rapid compression machine for studying energetic-nanoparticle-enhanced combustion of liquid fuels Proc Combustion. Proc Combust Inst 33:3367–3374. doi:10.1016/j.proci.2010.06.007

Aristov AI, Manousidaki M, Danilov A, Terzaki K, Fotakis C, Farsari M, Kabashin AV (2016) 3D plasmonic crystal metamaterials for ultra-sensitive bio-sensing. Sci Rep 6(25380):1–8. doi:10.1038/srep25380

Banitz T, Johst K, Wick LY, Schamfuß S, Harms H, Frank K (2013) Highways versus pipelines: contributions of two fungal transport mechanisms to efficient bioremediation. Environ Microbiol Rep 5:211–218. doi:10.1111/1758-2229.12002

Benckiser G (2012) Nanotechnology and patents in agriculture, food technology, nutrition and medicine – advantages and risks. Recent Pat Food Nutr Agric 4:171–175

Benckiser G, Christ E, Herbert T, Weiske A, Blome J, Hardt M (2013) The nitrification inhibitor 3,4-dimethylpyrazole-phosphat (DMPP) – quantification and effects on soil metabolism. Plant Soil 371:257–266. doi:10.1007/s11104-013-1664-6

Benckiser G, Ladha JK, Wiesler F (2016) Climate change and nitrogen turnover in soils and aquatic environments. In: Marxsen J (ed) Climate change and microbial ecology: current research and future trends. Caister Academic Press, Norfolk, ch 8, 22 p

Bhagat D, Samanta SK, Bhattacharya S (2013) Efficient management of fruit pests by pheromone nanogels. Sci Rep 3(1294):1–8. doi:10.1038/srep01294

Böhmert L, Niemann LD, Juling S, Lampen A (2015) Molecular mechanism of silver nanoparticles in human intestinal cells. Nanotoxicology 9:852–860. doi:10.3109/17435390.2014.980760

Brown S (2009) The new deficit model. Nat Nanotechnol 4:609–612. doi:10.1038/nnano.2009.278

Chen J, Badioli M, Alonso-González P, Thongrattanasiri S, Huth F, Osmond J, Spasenović M, Centeno A, Pesquera A, Godignon P, Elorza AZ, Camara N, García de Abajo FJ, Hillenbrand R, Koppens FHL (2012) Optical nano-imaging of gate-tunable graphene plasmons. Nature 487:77–81. doi:10.1038/ nature11254

Cherukula K, Lekshmi KM, Uthaman S, Cho K, Cho C-S, Park I-K (2016) Multifunctional inorganic nanoparticles: recent progress in thermal therapy and imaging. Nanomaterials 6:76. doi:10.3390/nano6040076

Cobb MD, Macoubrie J (2004) Public perceptions about nanotechnology: risks, benefits and trust. J Nanopart Res 6:395–405. doi:10.1007/s11051-004-3394-4

Conner AJ, Jacobs JM (1999) Genetic engineering of crops as potential source of genetic hazard in the human diet. Mutat Res:223–234. doi:10.1016/S1383-5742(99)00020-4

Coppola AD, Verneau F (2014) An empirical analysis on technophobia/technophilia in consumer market segmentation. Agric Food Econ 142. doi:10.1186/2193-7532-2-2

De Lorenzo V (2014) From the shelfish gene to the shelfish metabolism: revisiting the central dogma. BioEssays 36:226–235. doi:10.1002/bies.201300153

Deng S, Bei Y, Lu X, Du Z, Wang B, Wang Y, Huang J, Yu G (2015) Effect of co-existing organic compounds on adsorption of perfluorinated compounds onto carbon nanotubes. Front Environ Sci Eng 9:784–792. doi:10.1007/s11783-015-0790-1

Dickinson E (2012) Use of nanoparticles and microparticles in the formation and stabilization of food emulsions. Trends Food Sci Techn 24:4–12. doi:10.1016/ j.tifs.2011.09.006

Duncan TV (2011) Applications of nanotechnology in food packaging and food safety: barrier materials, antimicrobials and sensors. J Colloid Interface Sci 363:1–24. doi:10.1016/j.jcis.2011.07.017

Frost and Sullivan, company report (2016) Recent innovations in nano-instrumentation – Nanotech TOE. http://www.frost.com/sublib/display-report.do?id=D787-00-1F-00-00

Grobe A, Rissanen ME (2012) Nanotechnologies in agriculture and food: an overview of different fields of application, risk assessment and public perception. Recent Pat Food Nutr Agric 4:176–186

Gupta N, Fischer ARH, Frewer L (2015) Ethics, risk and benefits associated with different applications of nanotechnology: a comparison of expert and consumer perceptions of drivers of societal acceptance. NanoEthics 9:93–108. doi:10.1007/s11569-015-0222-5

Heo J, Hwang C-S (2016) Application of L-aspartic acid-capped ZnS:Mn as a colloidal nanocrystals photosensor for the detection of copper (II) ions in aqueous solution. Nanomaterials 6:82. doi:10.3390/nano6050082

Hilder VA, Boulter D (1999) Genetic engineering of crop plants for insect resistance – a critical review. Crop Prot 18:177–191. doi:10.1016/S0261-2194(99)00028-9

Horst MF, Lassalle V, Ferreira ML (2015) Nanosized magnetite in low cost materials for remediation of water polluted with toxic metals, azo- and antraquinonic dyes. Front Environ Sci Eng 9:746–769. doi:10.1007/s11783-015-0814-x

Horstkotte E, Odoerfer KI (2012) Towards improved therapies using nanopharmaceuticals: recent patents on pharmaceutical nanoformulations. Recent Pat Food Nutr Agric 4(3):220–244

Ikemoto S, Sugimura K, Yoshida N, Yasumoto R, Wada S, Yamamoto K, Kishimoto T (2000) Antitumor effects of scutellariae radix and its components baicalein, baicalin, and wogonin on bladder cancer cell lines. Urology 55:951–955

Jang EJ, Cha SM, Choi SM, Cha JD (2014) Combination effects of baicalein with antibiotics against oral pathogens. Arch Oral Biol 59:1233–1241

Jivani RR, Lakhtaria GJ, Patadiya DD, Patel LD, Jivani NP, Jhala BP (2016) Biomedical micro-electromechanical systems (BioMEMS): revolution in drug delivery and analytical techniques. Saudi Pharm J 24:1–20. doi:10.1016/j.jsps.2013.12.003

Kong L, Ziegler GR (2012) Patents on fiber spinning from starches. Recent Pat Food Nutr Agric 4(3):210–219

Kuhlmeier D, Sandetskaya N, Allelein S (2012) Application of nanotechnology in miniaturized systems and its use in medical and food analysis. Recent Pat Food Nutr Agric 4:187–199

Kumar A, Kumar P, Joshi C, Manchanda M, Boukherroub R, Jain Suman L (2016) Nickel decorated on phosphorous-doped carbon nitride as an efficient photocatalyst for reduction of nitrobenzenes. Nanomaterials 6:59. doi:10.3390/nano6040059

Leung KCF, Seneviratne CJ, Li X, Leung PC, Lau CBS, Wong CH, Pang KY, Wong CW, Wat E, Jin L (2016) Synergistic antibacterial effects of nanoparticles encapsulated with *Scutellaria baicalensis* and pure chlorhexidine on oral bacterial biofilms. Nanomaterials 6:61. doi:10.3390/nano6040061

Liu R, Lal R (2015) Effects of molecular weight and concentration of carboxymethyl cellulose on morphology of hydroxyapatite nanoparticles as prepared with one-step wet chemical method. Front Environ Sci Eng 9:804–812. doi:10.1007/s11783-015-0785-y

Neethirajan S, Jayas DS (2011) Nanotechnology for the food and bioprocessing industries. Food Bioprocess Technol 4:39–47. doi:10.1007/s11947-010-0328-2

Ouyang W, Kuna E, Yepez A, Balu AM, Romero AA, Colmenares JC, Luque R (2016) Mechanochemical synthesis of TiO_2 nanocomposites as photocatalysts for benzyl alcohol photo-oxidation. Nanomaterials 6:93. doi:10.3390/nano6050093

Palomo JM, Filice M (2016) Biosynthesis of metal nanoparticles: novel efficient heterogeneous nanocatalysts. Nanomaterials 6:84. doi:10.3390/nano6050 084

Parisi C, Vigani M, Rodríguez-Cerezo E (2015) Agricultural nanotechnologies: what are the current possibilities? NanoToday 10:124–127. doi:10.1016/j.nantod.2014.09.009

Prasad R (2014) Synthesis of silver nanoparticles in photosynthetic plants. J Nanoparticles 963961. http://dx.doi.org/10.1155/2014/963961

Prasad R (2016) Advances and applications through fungal nanobiotechnology. Springer International Publishing, Cham. ISBN:978-3-319-42989-2

Prasad R, Kumar V, Prasad KS (2014) Nanotechnology in sustainable agriculture: present concerns and future aspects. Afr J Biotechnol 13(6):705–713

Prasad R, Pandey R, Barman I (2016) Engineering tailored nanoparticles with microbes: quo vadis. WIREs Nanomed Nanobiotechnol 8:316–330. doi:10.1002/wnan.1363

Prasad R, Pandey R, Varma A, Barman I (2017a) Polymer based nanoparticles for drug delivery systems and cancer therapeutics. In: Kharkwal H, Janaswamy S (eds) Natural polymers for drug delivery. CABI, Wallingford, pp 53–70

Prasad R, Bhattacharyya A, Nguyen QD (2017b) Nanotechnology in sustainable agriculture: recent developments, challenges, and perspectives. Front Microbiol 8:1014. doi:10.3389/fmicb.2017.01014

Raj S, Jose S, Sumod US, Sabitha M (2012) Nanotechnology in cosmetics: opportunities and challenges. J Pharm Bioallied Sci 4:186–193. doi:10.4103/0975-7406.99016

Ravichandran R (2010) Nanotechnology applications in food and food processing: innovative green approaches, opportunities and uncertainties for global market. Int J Green Nanotech: Phys Chem 1:72–96. doi:10.1080/19430871003684440

Raza MA, Kanwal Z, Rauf A, Sabri AN, Riaz S, Naseem S (2016) Size- and shape-dependent antibacterial studies of silver nanoparticles synthesized by wet chemical routes. Nanomaterials 6:74. doi:10.3390/nano6040074

Reidy B, Haase A, Luch A, Dawson KA, Lynch I (2013) Mechanisms of silver nanoparticle release, transformation and toxicity: a critical review of current knowledge and recommendations for future studies and applications. Materials 6:2295–2350. doi:10.3390/ma6062295

Renn O (2008) Concepts of risk: an interdisciplinary review- part 1: disciplinary concepts. GAIA 17:50–66

Rogers JA (2010) Nanoribbons on the edge. Nat Nanotechnol 5:698–699

Sadeghi R, Ebadollahi A (2015) Susceptibility of *Rhyzopertha dominica* (F.) (Coleoptera: Bostrichidae) and *Sitophilus oryzae* (L.) (Coleoptera: Curculionidae) to Spinosad (Tracer®) as an eco-friendly biopesticide. Ecologgia Balkanica 7:39–44. http://eb.bio.uni-plovdiv.bg

Salamanca-Buentello F, Persad DL, Court EB, Martin DK, Daar AS, Singer PA (2005) Nanotechnology and the developing world. PLoS Med 2(5):e97

Scandorieiro S, de Camargo LC, Lancheros CAC, Yamada-Ogatta SF, Nakamura CV, de Oliveira AG, Andrade CGTJ, Duran N, Nakazato G, Kobayashi RKT (2016) Synergistic and additive effect of oregano essential oil and biological silver nanoparticles against multidrug-resistant bacterial strains. Front Microbiol 7:760. http://dx.doi.org/10.3389/fmicb.2016.00760

Schmitt F, Lagopoulos L, Käuper P, Rossi N, Busso N, Barge J, Wagnières G, Laue C, Wandrey C, Juillerat-Jeanneret L (2010) Chitosan-based nanogels for selective delivery of photosensitizers to macrophages and improved retention in and therapy of articular joints. J Control Release 144:242–250. doi:10.1016/j.jconrel.2010.02.008

Schroeder D, Dalton-Brown S, Schrempf B, Kaplan D (2016) Responsible, inclusive innovation and the nano-divide. NanoEthics 10:177–188. doi:10.1007/s11569-016-0265-2

Sekhon BS (2014) Nanotechnology in agri-food production: an overview. Nanotechnol Sci Appl 7:31–53. http://dx.doi.org/10.2147/NSA.S39406

Seneviratne CJ, Leung KC, Wong CH, Lee SF, Li X, Leung PC, Lau CB, Wat E, Jin LJ (2014) Nanoparticle-encapsulated chlorhexidine against oral bacterial biofilms. PLoS One 9:e103234. http://dx.doi.org/10.1371/journal.pone.0103234

Senior K, Müller S, Schacht VJ, Bunge M (2012) Antimicrobial precious-metal nanoparticles and their use in novel materials. Recent Pat Food Nutr Agric 4:200–209

Senjen R (2012) Nanotechnology and patents – how can potential risks be assessed? Recent Pat Food Nutr Agric 4:245–249

Senjen R, Hansen SF (2011) Towards a nanorisk appraisal framework. Comptes Rendus Physique12:637–6648 10.1016/j.crhy.2011.06.005

Shamaila S, Zafar N, Riaz S, Sharif R, Nazir J, Naseem S (2016) Gold nanoparticles: an efficient antimicrobial agent against enteric bacterial human pathogen. Nanomaterials 6:71. doi:10.3390/nano6040071

Shen J-M, Xu L, Lu Y, Cao H-M, Xu Z-G, Chen T, Zhang H-X (2012) Chitosan-based luminescent/magnetic hybrid nanogels for insulin delivery, cell imaging, and antidiabetic research of dietary supplements. Int J Pharm 427:400–409. doi:10.1016/j.ijpharm.2012.01.059

Smith RC, Li J, Padungthon S, Sengupta AK (2015) Nexus between polymer support and metal oxide nanoparticles in hybrid nanosorbent materials (HNMs) for sorption/desorption of target ligands. Front Environ Sci Eng 9:929–938. doi:10.1007/s11783-015-0795-9

Sticklen MB (2009) Plant genetic engineering for biofuel production: towards affordable cellulosic ethanol. Nat Rev Genet 9:433–443. doi:10.1038/nrg2336

Stracke F, Weiss B, Lehr CM, König K, Schaefer UF, Schneider M (2006) Multiphoton microscopy of dermal penetration of nanoparticle-borne drugs. J Invest Dermatol 126:2224–2233. doi:10.1038/sj.jid.5700374

Sunkara B, Su Y, Zhan J, He J, McPherson GL, John VT (2015) Iron-carbon composite microspheres prepared through a facile aerosol-based process for the simultaneous adsorption and reduction of chlorinated hydrocarbons. Front Environ Sci Eng 9:939–947. doi:10.1007/s11783-015-0807-9

Taniguchi N (1974) On the basic concept of "Nano-Technology." In: Proceedings of the international conference on production engineering, Tokyo, Part II, Japan Society of Precision Engineering, pp 18–23

Tzuzuki T (2009) Commercial scale production of inorganic nanoparticles. Int J Nanotechnol 6:567

Voytas DF, Gao C (2014) Precision genome engineering and agriculture: opportunities and regulatory challenges. PLoS Biol 12:e1001877. http://dx.doi.org/10.1371/journal.pbio.1001877

Xiao B, Ma P, Viennois E, Merlin D (2016) Urocanic acid-modified chitosan nanoparticles can confer anti-inflammatory effect by delivering CD98 siRNA to macrophages. Colloids Surf B: Biointerfaces 143:186. doi:10.1016/j. colsurfb.2016.03.035

Yang Y, Yu Z, Nosaka T, Doudrick K, Hristovski K, Herckes P, Westerhoff P (2015) Interaction of carbonaceous nanomaterials with wastewater biomass. Front Environ Sci Eng 9:823–831. doi:10.1007/s11783-015-0787-9

Yang M, Fang Y, Sun D, Shi Y (2016) Efficiency of two nitrification inhibitors (dicyandiamide and 3, 4-dimethypyrazole phosphate) on soil nitrogen transformations and plant productivity: a meta-analysis. Sci Rep 6, Article number: 22075 doi:10.1038/srep22075

Yu K, Wang M, Wu J, Qian K, Sun J, Lu X (2016) Modification of the interfacial interaction between carbon fiber and epoxy with carbon hybrid materials. Nanomaterials 6:89. doi:10.3390/nano6050089

Zerulla W, Barth T, Dressel J, Erhardt K, Horchler von Locquenqhien K, Pasda G, Raedle M, Wissemeier AH (2001) DMPP-a new nitrification inhibitor for agriculture and horticulture: an introduction. Biol Fertil Soils 34:79–84. doi:10.1007/s003740100380

Zimmer R, Hertel R, Böl G-F (eds) (2012) Risk perception of nanotechnology-analysis of media coverage. Federal Institute for Risk Assessment, Berlin. Available at: http://www.bfr.bund.de/cm/350/risk_perception_of_nanotechnology_analysis_of_media_coverage.pdf

Production of Bionanomaterials from Agricultural Wastes

3

Jeyabalan Sangeetha, Devarajan Thangadurai, Ravichandra Hospet, Prathima Purushotham, Kartheek Rajendra Manowade, Mohammed Abdul Mujeeb, Abhishek Channayya Mundaragi, Sudisha Jogaiah, Muniswamy David, Shivasharana Chandrabanda Thimmappa, Ram Prasad, and Etigemane Ramappa Harish

Abstract

Nature is gifted with numerous nanomaterials which could be simply prepared from plant materials. Agricultural waste (waste produced on a farm through various farming activities) includes both natural and nonnatural wastes. In the agricultural residues, refuse and wastes create a significant amount of worldwide agricultural productivity. It has variously been estimated that wastes can account for over 30% of worldwide agricultural productivity. The goal of this chapter is to assess the most recent trends to produce bionano nanomaterials from agricultural waste. Nanocellulose extraction from agricultural wastes is a promising substitute for waste treatment, and a few more wide applications of nanocellulose in biological science are much expected in the near future. The most salient nanocellulose applications in this chapter deal with the production and support

J. Sangeetha (✉)
Department of Environmental Science, Central University of Kerala,
Kasaragod, Kerala 671316, India
e-mail: drjsangeetha.cuk@gmail.com

D. Thangadurai • R. Hospet • P. Purushotham • A.C. Mundaragi
Department of Botany, Karnatak University, Dharwad, Karnataka 580003, India

K.R. Manowade • M. David • E.R. Harish
Department of Zoology, Karnatak University, Dharwad, Karnataka 580003, India

M.A. Mujeeb • S. Jogaiah • S.C. Thimmappa
Department of Microbiology and Biotechnology, Karnatak University,
Dharwad, Karnataka 580003, India

R. Prasad
Amity Institute of Microbial Technology, Amity University,
Noida, Uttar Pradesh 201313, India

R. Prasad et al. (eds.), *Nanotechnology*, DOI 10.1007/978-981-10-4573-8_3

matrices for enzyme immobilization, biosensors, and antimicrobial agents. Silicon nanoparticles concluded to be one of the elite compounds for the enhancement of agricultural yields.

Keywords
Nanocellulose • Agrowaste • Activate carbon • Black carbon • Graphene

3.1 Introduction

The biomaterial is a substance or fusion of substances other than drug compounds which may be synthetic or natural in origin which can be used for any timeframe, which increases or replaces mostly or absolutely any tissue, organ, or function, in order to maintain or improve the quality of life. The word biomaterial confines any substance deliberated to interact with the biotic system in order to replace living matter which has lost its function. It can serve as a vehicle, matrix, and support for stimulating new tissue growth (Williams 1999). The first generation of biomaterials began as early as 1950, under the concept that the biomaterials should be inert or elicit a minimal reaction of host tissue when implanted (Zavaglia and Prado da Silva 2016). Since then, there has been considerable research and development in the field of biomaterials leading to the production of second- and third-generation biomaterials. However, the intersection of numerous research disciplines such as chemistry, immunology, materials science, physical science, and engineering has opened a completely new dimension in science and technology called nanotechnology (Prasad 2016).

Rapid advancement in the field of nanotechnology has revolutionized every branch of sciences, including the biomaterials leading to the synthesis of bionanomaterials. Improved physical techniques like helium ion and electron microscopic techniques and nanofabrication conventions have allowed nanosized device production and analysis. They can be characterized as molecular materials composed of biological compounds (e.g., antibodies, proteins, lipids, DNA, RNA, viruses, and cellular components). The subsequent bionanomaterials may influence as novel fibers, sensors. These sorts of frameworks may allow for fabrication of complex devices by self-assembly under gentle experimental conditions (Honek 2013).

Cellulose is an important characteristic polymer derived from the residues of agricultural activities and major source for industrial purposes (Wang et al. 2014a, b). It has been almost 15 years, where the study of cellulose nanofibers as a fortifying phase in nanocomposites started. Since then, cellulose nanofibers unlock the entryway toward promising research on cellulose-based nanomaterials with expanding zone of potential applications including packaging material (Rodionova et al. 2011; Spence et al. 2011). This opens up several avenues in a country like India, where agriculture accounts for 13.7% of the GDP (as reported in 2013) and produces agricultural wastes rich in carbohydrates amounting to several million tons. For millennia, cellulose has been used in the form of wood and plant fibers as an energy source, for building materials, and for clothing. Plant cell wall is composed of microfibrils surrounded by fragile hemicellulose and lignin. Cellulose is the fibrillar segment of plant cells. Chemically, cellulose is a linear polymer of (1→4)-linked β-D-glucopyranosyl residues. Cellulose chains are usually enclosed to form compact microfibrils in which both inter- and intramolecular hydrogen bonding are stabilized (Bacic et al. 1988; Alemdar and Sain 2008). Up to 100 glucan chains are assembled together to form long thin crystallites. These crystallites are about 5 nm wide; however, this varies according to the cellulose source. They are in a well-organized manner to form microfibrils that are 8–50 nm in diameter and of lengths of a few microns (Clowes and Juniper 1968). On account of their crystal structure, these nanofibers offer better strength to the plant body. The mechanical and chemical treatments of the fibers result in chemical and structural changes on the fiber surface and the cells, which impact the properties of the fibers in composites. Recently, several researches have been performed for isolation of nanofibers from plants to use them as fillers in biocomposites (Dufresne et al. 1997a, b; Dufresne and Vignon 1998; Bhatnagar and Sain 2003, 2005).

Through chemical and mechanical methods, cellulose nanofibers are extracted from cell walls, for example, cryo-crushing (Chakraborty et al. 2006), high-pressure homogenization (Pelissari et al. 2014), granulation (Abe et al. 2007, 2009; Abe and Yano 2010), and biological treatment, such as enzyme-assisted hydrolysis (Paakko et al. 2007; Tibolla et al. 2014) and corrosive hydrolysis (Elazzouzi-Hafraoui et al. 2008; Liu et al. 2010; He et al. 2013). Use of an ultrasonic system for separation of cellulose nanofibers is a rising technique and has been broadly utilized by various specialists (Cheng et al. 2009; Wang and Cheng 2009; Cheng et al. 2010). In the process of ultrasonic treatment, ultrasound energy is exchanged to cellulose chains through a process called cavitation, which alludes to the arrangement, development, and violent collapse of cavities in the water (Chen et al. 2011). The energy provided by cavitation process in this so-called sonochemistry is roughly 10–100 kJ/mol, which is within the hydrogen bond energy scale (Tischer et al. 2010). Therefore, the ultrasonic effect can step by step break down the micron-sized cellulose filaments into nanofibers (Prerna and Sankar 2016). Several studies have been done for isolation and description of cellulose nanofibers from diverse sources, for example, potato tuber (Dufresne et al. 2000; Rezanezhad et al. 2013), cotton (de Morais et al. 2010), soybean (Wang and Sain 2007), banana rachis (Zuluaga et al. 2009), coir (Rosa et al. 2010; Rezanezhad et al. 2013), pea structure fiber (Chen et al. 2009), wood strands (Abe et al. 2007), thorny pear natural products (Habibi et al. 2008), wheat straw and soybean hulls (Alemdar and Sain 2008), citrus and corn (Rondeau-Mouro et al. 2003; Rezanezhad et al. 2013), beetroot (Dinand et al. 1999), bark of mulberry tree (Li et al. 2009), and pineapple leaf fibers (Cherian et al. 2010). Khawas et al. (2014) reported banana peel as a source of cellulosic fiber and subsequently as a biomaterial, wherein peel represents 40% of aggregate fruit weight of a banana. Unfortunately, peels are being often discarded and proceeding a serious pollution problem. The utilization of this cellulose-rich biomass would not only increase the value of this agrowaste but also help to overcome environmental pollution issues.

A few of nanobiomaterials as of now being used are the nanocellulose-based nanocomposites in combination with a range of various nanomaterials, for example, metal NP (Ag, Au, CuO, Pd, Ni, TiO_2), carbonaceous nanomaterials (graphene, carbon nanotube), and mineral nanomaterials (SiO_2, $CaCO_3$, montmorillonite) consolidated into or onto nanocellulose substrates. These nanocompounds have novel environmental applications (e.g., antibacterial, catalytic, sensing), medical devices, and structural reinforcement (Prasad 2016). Electrospun nanofibrous channels have demonstrated better execution in air and water purification contrasted with regular polymer filters (Yoon et al. 2008). In some recent researches, nanocellulose filters demonstrated the ability to remove the virus from water (Metreveli et al. 2014). Surface biofouling is one of the major difficulties for the use of membrane filter in water management. Chitosan-based antimicrobial compounds have been suggested for both storage and drinking water purification (Ghosh et al. 2011; Zhang et al. 2012; Wei et al. 2014). Water pollutant treatment by decomposition of organic contaminants is a well-featured application of nanocellulose-based nanocomposites (Kettunen et al. 2011; Zhou et al. 2013a, b; Wu et al. 2014). Nanocellulose-based nanocomposites can also be incorporated in Li-ion battery, fuel cell, and solar cell manufacturing. AuNP/nanocellulose nanocomposites have been developed successfully by many researchers which are used as biosensors to check the pollution levels (Wang and Cheng 2009).

3.2 Cellulose and Nanocellulose from Citrus and Orange Wastes

Pollution is a major problem across the world which is caused by various agricultural, industrial, and urban activities such as toxic heavy metals, pesticide residue, wastewater, food contaminants, and other environmental hazards. To overcome these problems, eco-friendly approach using agro-industrial and agricultural waste is the need of the hour. In recent years, nontoxic adsorption materials such as microcellulose and nanocellulose obtained from agricultural waste are considered as the vital source for removal of the contaminants without causing any hazardous effect on human health and environment.

Nanocellulose can be produced from various lignocellulosic sources obtained from any forest wood and agricultural wastes by three different main methods, namely, physical, chemical, and bacterial (Habibi et al. 2010; Moon et al. 2011; Aspler et al. 2013; Kardam et al. 2014). Lignocelluloses generally consist of common carbon compounds mainly found in the dead and decaying plant cell walls which are rich in organic compounds. This nonedible residual biomass found in the agricultural waste is gaining economic importance for the production of sustainable energy as well as biomaterial sectors. Nanocellulose comprises important cellulose properties such as broad chemical modification capacity, hydrophilicity, and the formation of broad semicrystalline fiber morphologies. Cellulose is a key polymeric material used in the immobilization of metal particles and metal nanoparticles as it contains six hydroxyl groups per cellobiose (repeating) unit of cellulose polymer with various advantageous characteristics, namely, low density with high mechanical features, economic value, nontoxicity, renewability, and biodegradable in nature (Klemm et al. 2011; Dufresne 2013). Cellulose is the major component for the production of nanocellulose through acid hydrolysis method. This technique introduces negative ions through acidic substances to the corresponding cellulose for hydrolyzing the amorphous to nanofibers (Zain et al. 2014).

The word nanocellulose is commonly known in various terminologies such as crystallites, nanocrystals, whiskers, nanofibrils, and nanofibers (Charreau et al. 2013). The nanocellulose is mainly classified based on their dimensions, functions, and isolation methods, which in turn depend mainly on the cellulosic source and on the processing conditions. Generally, in order to obtain qualitative nanocellulose, several investigators had concluded citrus and orange peels as the best source material. The citrus peel is categorized into two parts, such as inner part (albedo) and outer part (flavedo) composed of cellulosic tissues. Among them, a white and spongy albedo tissue is very rich in fiber content and is the principal component of citrus peel (Zain et al. 2014). Till date, the nanocelluloses have not been used in a completely uniform manner.

3.2.1 Cellulose and Nanocellulose Fibrils from Lignocellulose Fibers

Citrus peel is a significant agricultural waste material used to attain functional nanofibers (Thygesen et al. 2011; Mariño et al. 2015). Thus, the productions of nanoscale cellulose fibrils from raw materials are attracting a large economic impact. Depending upon the biomass and fiber content, different methods like biochemical, chemical, and mechanical treatments were adapted for production of nanocellulose (Mariño et al. 2015). These characters prop up the isolation of cellulose fibrils which include citrus and orange peels from various agricultural by-products. It is also documented that cellulose and cellulose nanocrystal which were successfully obtained from citrus or orange peels might be potentially used in producing functional fibers and can be explored for food application.

3.2.2 Nanocellulose for Bioethanol Production

Orange juice is one of the important health-oriented juices of the world and most people in India consume the juice as a nutrient drink. After the extraction of the juice, approximately half of the fruit weight is converted into citrus processing waste from oranges (CPWO), which is a low-cost material that is best used for the production of first-generation bioethanol (Awan et al. 2013). The authors also demonstrated that orange peel waste fermented with *Saccharomyces cerevisiae* strain yields high glucose and ethanol up to 50% m m^{-1} by a novel lab-scale direct steam injection apparatus. It is also documented that dry orange peel waste is an efficient material for bioethanol production with overall process yield (mass balance) at the bench reactor scale of 140 L (Santi et al. 2012). In addition to the bioethanol production, CPWO is also used as a precious bioresource for obtaining D-limonene essential oil (1.5% g g^{-1} of dry CPWO), nanocellulose (3% g g^{-1} of dry CPWO), and bioethanol (20% g g^{-1} of dry CPWO) determined on dry matter basis (Tsukamoto et al. 2013).

3.2.3 Nanocellulose-Based Materials for Water Treatment

The method of surface chemistry modification of nanocellulose is through the use of direct chemical modification and/or covalent attachment of molecules. Due to its high specific surface areas and numerous reactive groups, excellent adsorption performance may be achieved with different types of customized nanocelluloses which might afford an interesting alternative to conventional adsorbents such as activated carbon, ion exchange resins, or zeolite in the water. This new area of prospective use of nanocellulose in the field of water treatment was successfully achieved and documented, which showed the possibilities for preparation of adsorbents based on modified nanocellulose for removing not only heavy metal but also organic pollutants (Kalia et al. 2014). The surface modification of the nanocellulose was obtained by adding specific groups such as carboxyl (Donia et al. 2012; Yu et al. 2013), amine (Singh et al. 2014), ammonium, and xanthate (Saumya et al. 2013) to the cellulose surface.

Oxolane-2,5-dione-modified cellulose nanofibers are used for cadmium adsorption and lead ions from standard wastewater samples. It was noticed that maximum adsorption capacity was observed in cellulose-*g*-oxolane-2,5-dione nanofibers for Pb(II) and Cd(II) when compared with oxolane-2,5-dione-modified cellulose fibers. For cellulose nanofibers, adsorption capacities were 12.0 and 2.9 mmol/g for Pb(II) and Cd(II), respectively, in comparison to 0.002 mmol/g for raw cellulose (Musyoka et al. 2011; Boufi 2014).

Wastewater dyeing is an imperative division of pollutants, which can easily be identified through the naked eye. Nanocellulose hybrids consist of reactive polyhedral oligomeric silsesquioxane along with multi-N-methylol (Xie et al. 2011a, b) and fabricate porous nanocomposite gels based on partially hydrolyzed polyacrylamide and cellulose nanocrystals (Zhou et al. 2013a, b), which could be used as biosorbents at a low concentration for removal of reactive dyes like yellow B-4RFN and blue B-RN and methylene blue in wastewater treatment. In last decades, several investigations concerning the modification of cellulose with succinic anhydride for the removal of heavy metal ions have been reported (Beck-Candanedo et al. 2005). Hokkanen et al. (2013, 2016) reported the removal of heavy metals from polluted aqueous solutions by mercerized nanocellulose, a product of succinic anhydride. Application of high-performance complex nanocellulose has engrossed much attention from the industry. To meet the present demand, nanocellulose extraction from agricultural waste having significant value is the best option for wastewater treatment.

3.2.4 Orange Peel Cellulose and Nanocellulose in Textile and Food Industry

The production of orange juice on a large-scale industrial level leads to a considerable quantity of solid and liquid residues (around 8–20 million tons year^{-1} globally), which is still considered as adequate agricultural waste. Normally, orange residues have no commercial importance though the residues comprise rich soluble sugars, soluble/insoluble carbohydrates, pectin, essential oils, cellulose, and hemicellulose that can form the basis of several industrial processes (Awan et al. 2013; Tsukamoto et al. 2013). Hence, the production of orange nanofibrils from these residues not only contributes to the diminution of the cost but also plays a key role in the development of qualitative and sustainable textile. Citrus by-products are versatile in recovering eco-friendly fibers possessing better water- and oil-holding capacity, excellent fermentability, and less calorie content that are used as functional food ingredients for better human health (Lario et al. 2004; Fernando et al. 2005).

3.3 Synthesis of Graphene Oxide from Agrowastes

Graphene is a novel one-atom-thick two-dimensional substance made out of immaculate carbon, with molecules organized in a standard hexagonal shape like graphite. It is light, with a 1-m^2 sheet weighing just 0.77 mg. Graphene was discovered in 2004, and it has become an entryway for a large number of applications. Potential electronic properties (Schwierz 2010), thermal conductivity (Balandin et al. 2008), and high surface region, joined with well-characterized mechanical properties and great scattering execution, make graphene as a novel compound for basic alteration of composite materials and a few different applications (Lee et al. 2008; Si and Samulski 2008; Lu et al. 2011; Song et al. 2011; Somanathan et al. 2015). It has the combination of incredible properties which were never discovered in any of the material till date (Yadav et al. 2014). Properties like high compound steadiness, high surface region (theoretical, 2600 m^2/g), the high versatility of charge bearers (2000 m^{2}v^{-1} s^{-1}), high modulus (1Tpa), and optical transparency are essential integral part in various scientific fields (Xiao et al. 2014).

Though graphene is expensive and generally difficult to produce, notable attempts were made to discover powerful yet inexpensive approaches to make and utilize graphene or its related materials. Graphene oxide (GO) is one of those materials, which has a single atomic layered compound, made by the effective oxidation of graphite. GO has likewise influenced major interdisciplinary on account of the wide scope of visualized applications over a few logical and engineering fields, including material science, biological science, and pharmaceuticals (Georgakilas et al. 2012; Novoselov et al. 2012; Chung et al. 2013; Mao et al. 2013). The GO is generally synthesized by Staudenmaier's (ST), Hofmann's (HO), and Hummers' (HU) methods. The indigenous techniques to synthesize GO depend on the inclusion of potassium chlorate to slurry of graphite in raging nitric acid (Brodie 1859). This engineered convention can be enhanced by utilizing concentrated sulfuric acid and as well fuming nitric acid and including the chlorate in different aliquots through the span of the reaction (Staudenmaier 1898; Somanathan et al. 2015). However, comparing to other synthesis methods, graphene is more preferably produced by oxidizing graphite by Hummers' method and then chemically reduced to obtain reduced graphene oxide, which is known to be chemically prepared Graphene (Hummers and Offeman 1958).

In a previous couple of years, different methods, such as chemical vapor deposition (CVD) (May 1969), physical or chemical exfoliation procedures (Novoselov et al. 2012), the epitaxial growth on silicon carbide, and other chemical origin methods, have been used for graphene synthesis and its other derivatives like GO materials (Du et al. 2012; Singh and Vidyasagar 2014; Somanathan et al. 2015). With the use of these processes, it is possible to synthesize high-quality graphene in a large amount. However, these synthesis methods have major drawbacks also; high temperature and expensive substrates make complication in the CVD process; epitaxial growth gives a wafer-scale graphene, but it also requires high temperature more than 1500 °C and cost of silicon carbide is needed to be considered; use of toxic chemical agents and complex processing method tends to rethink for incorporating such processes. The majority of these techniques needs highly sophisticated modern approaches and they are very expensive too (Guo and Dong 2011). Almost all commercially available GO compounds were synthesized by Hummers' method or improved version of it. The expense of the commercially accessible GO per gram is around $200 (Somanathan et al. 2015). Although Hummers' strategy has been given careful consideration as a result of its high effectiveness and fulfilling reaction safety, it still has few flaws like emission of toxic compounds such as NO_2 and N_2O_4 and other environmental-related issues. In comparison with this technique, a team of researchers found a novel strategy for the synthesis of GO from agrowaste, which is said to be a more eco-friendly and cost-effective technique. Under the muffled atmosphere, sugarcane bagasse was directly oxidized for the synthesis of GO. The GO compound synthesized by this method was known as SOMA-GO (sugarcane oxidized under muffled atmosphere for graphene oxide). And also the structural characteristics of GO compound were determined by using X-ray diffraction (a preferred method for determination of crystallinity of a compound), Fourier transform infrared spectroscopy (preferred method of infrared spectroscopy), field emission scanning electron microscopy (gives topographical and elemental information), high-resolution transmission electron microscopy, and Raman spectroscopy (Du et al. 2012; Somanathan et al. 2015).

Recently, scientists developed a unique method for the synthesis of high-quality GO and reduced graphene oxide (rGO) sheets from various natural available green wastes and carbonaceous wastes, including animal wastes, vegetation wastes (leaf, bagasse, wood, and fruit wastes), and a semi-industrial waste such as a newspaper (Omid et al. 2014; Rajesh et al. 2016).

A team of scientists successfully improved the existing Hummers' method using $NaNO_3$ for the synthesis of GO. This enhanced strategy prevents the emission of toxic gases and contributes toward purifying waste liquid. The final synthesized GO products were almost similar to the products derived from the conventional

Hummers' method. Further, this environmentally friendly, improved Hummers' method can be used in the production of GO in large scale (Chen et al. 2013). Due to its abundant existence as an agrowaste, rice husk has already received much attention as a starting material in producing high-value-added compounds such as porous carbon and silica. A new synthetic strategy was demonstrated for converting rice husks into mass measures of graphene over its calcination and chemical activation. By chemically activating, agricultural waste such as rice husk ash can be used in the synthesizing bulk amounts of crystalline graphene with the nanosized realm in a reliable, rapid, scalable, and cost-effective manner (Muramatsu et al. 2014).

3.4 Production of Amorphous Silica Nanoparticles from Agrowastes

Silicon is the second most common element found in the soil and silica is a polymer of silicic acid and its general formula is SiO_2. Naturally, it exists in various forms; crystalline and amorphous are the two common types of silica, wherein the former is found in natural state and is known to be toxic and the latter is chemically synthesized and is nontoxic. Amorphous silica has been commercially exploited due to its application in electronics and industries since it acts as a good semiconductor. Several methods have been proposed in synthesizing amorphous silica that includes sol-gel method, Stober method, and thermal decomposition method (Gu et al. 2015). Amorphous silica is chemically synthesized using silicate solution or silane reagents (Van et al. 2013). From the last two decades, it is well established that application of silicon enhances the plant growth and improves crop productivity under various environmental stresses including drought (Zhang and Fang 2010); studies have also shown that silica nanoparticles (SiO_2NP) have potential application in the field of agriculture. It enhances the yield by the way of reducing the use of fertilizer and pesticides (Liu et al. 2006; Nair et al. 2010; Karimi and Mohsenzadeh 2016; Sun et al. 2016). Amorphous SiO_2NP are recognized due to their distinct properties such as nontoxic, large surface area, and physicochemical properties. Hence they act as good carrier molecules or delivery vehicles (Hussain et al. 2013). The chemical method of preparation involves the use of highly toxic chemicals such as tetraethoxysilane and tetramethoxysilane that are expensive and harmful (Suriyaprabha et al. 2014). In this context, various methodologies for synthesizing nanoparticles are proposed using agrowastes. Production of SiO_2NP has been done using different agrowastes such as corn hub (Mohanraj et al. 2012), rice husk (Kamath and Proctor 1998), and other plant sources (Affandi et al. 2009; Indhumathi et al. 2011; Kumar et al. 2012; Lu and Hsieh 2012; Hariharan and Sivakumar 2013).

Wheat straw ash contains 74.23% of silica that acts as good substrate for the synthesis and extraction of silica (Dodson et al. 2011; Javed et al. 2011; Terzioglu and Yucel 2012). Chen et al. (2010) reported the synthesis of nanosilica from a wheat straw through combustion and acid leaching. They found phytoliths which were round in shape with a diameter ranging from 14 to 22 μm in epidermal cells, whereas phytoliths in trachea were found to be oblong with a length of 18–40 μm and a width of 12–18 µm. They also hypothesized that under different temperatures, phytoliths vary in physiochemical properties, and with increased temperature, they start to aggregate. Thus, in near future, wheat husk ash can act as a potential source for silica and could also aid in reducing pollution caused.

Globally, rice stands as a second major crop. Rice husk is the predominant agrowaste after processing rice. Earlier it was used as a fuel and fertilizer additives (Mohammadinejad et al. 2016). Majority of rice husk ash is composed of SiO_2 (Yalcin and Sevinc 2001). It varies up to 90–97 % (Sidheswaran and Bhat 1996; Mansaray and Ghaly 1997; Kumar et al. 2012) and acts as good source for the production of SiO_2NP. Hassan et al. (2014) studied the synthesis of SiO_2NP by sol-gel method by way of excluding the use of surfactant so as to increase surface area (Birla et al. 2013). They could get high-purity SiO_2NP characterized by different analyses such as Fourier transform infrared spectroscopy (FTIR) and X-ray diffraction (XRD). Rafiee et al. (2012) optimized synthesis of silica, and under controlled conditions, they got highly purified silica (98.8 %) that had high surface area and high reactivity. More recently, Ghorbani et al. (2015) reported the production of SiO_2NP from rice husk using an environment safe method, wherein they could get 97% pure silicon upon hydrochloric acid pretreatment.

Corn is the third major crop produced globally; the agrowaste left after processing is tremendous; hence, it can act as a better substrate for synthesizing silica nanoparticles; corn cob ash is reported to have 47 % silica (Okoronkwo et al. 2016). Qadri et al. (2015) reported the synthesis of different polytypes of silicon carbide nanoparticles and nanorods using different residues of corn. More recently Okoronkwo et al. (2016) synthesized amorphous silica nanoparticles from the corn cob ash that was 97% pure.

Bagasse ash is a major agro-industrial waste found abundantly; it is generally used for the generation of electricity; after burning the leftover ash is known to comprise large proportions of silica. Studies have shown that it could act as a rich source for synthesis of biogenic silica (Worathanakul et al. 2009; Affandi et al. 2011). Sana et al. (2014) demonstrated in the synthesis of SiO_2NP using sugar beet bagasse ash that laser ablation-mediated synthesis was better than the chemically synthesized SiO_2NP, and it yields significantly lesser sized SiO_2NP up to 38–190 nm (Birla et al. 2013). Furthermore, they did a comparative analysis between the chemically synthesized and laser ablation methods, and their study revealed that chemically synthesized SiO_2NP had a negative effect on the growth of algae, whereas laser ablation had increased algal growth causing no harm to aquatic environment. Manjula-Rani et al. (2014) reported the synthesis of SiO_2NP utilizing economically beneficial source cow dung ash by the process of acid hydrolysis and alkali precipitation and synthesized cent percent pure SiO_2NP. Espindola-Gonzalez et al. (2010) successfully demonstrated that vermicomposting for various agrowastes such as rice husk, sugarcane bagasse, and coffee husk and further processing would yield crystalline SiO_2NP.

3.5 Carbon Nanomaterials from Agrowastes

A marked increase in contributions to the field of nanoparticle fabrication has been observed in the past decade. Technology-based nanoparticle synthesis is usually followed under maintained dimension and particle size in addition to its definition in shape under the chemical research. With the potential applications of nanoparticles (NP) which range from their utilization as a biosensor, low-cost electrodes and biomedical backgrounds (Frances et al. 2009; Staniland 2011; Antonyraj et al. 2013) are often needed in order to suffice the modern world, as it has become inevitable. A variety of nanoparticle-based products have been known, viz., nanotubes, nanowires, and nanosheets, which are comprised of at least one dimension with a size of 100 nm. These are of high importance due to their remarkable applications (Kwon and Bard 2012; Fu et al. 2013; Roy et al. 2013).

The unique properties of nanoparticles which range from physical and chemical are well known to the scientific community. In addition, NP are known to perform the function of bridging between bulk, atomic, or molecular structures which make them an appropriate tool in the field of medical, electrochemical, and biotechnological sectors (Luo et al. 2006; Murthy 2007; Wen et al. 2011; Xia et al. 2013).

Several methods that are mostly synthetic have been endorsed by many for the fabrication of NP which are of diverse structure and size. The methods despite yielding high-quality NP have certain limitations that are to be addressed. New approaches for obtaining an increased yield of NP are being sought for their utilization under industrial and commercial levels, which include electrical appliances, medicine, and agriculture products that are long lasting in addition to being safer. The techniques are thought to address the challenges that are pertaining to environmental scenario.

These are known to address many challenging issues related to environment. Advanced nanomaterial synthesis is one of the best gifts from nature which could be utilized in the right channel for betterment of mankind and the upcoming generations. Biometric approach is one of the promising fields which grant the possibility of synthesizing materials like metals and metal oxide particles through their production within biological systems which are otherwise recognized as bio-laboratory units. A variety of microbes ranging from bacteria, fungi, yeast (Stephen and Macnaughtont 1999; Kumar et al. 2011; Shivaji et al. 2011; Chan and Don 2013; Syed et al. 2013), as well as plant extracts (Akhtar et al. 2013) in addition to waste materials (Kanchi et al. 2014) have been through feasible technologies resulting in the synthesis of NP that are of potential importance and use. The effective procedures available under biological channels include diverse range of microbes that have been used in the synthesis of wide range of NP of metallic origin and are highly advantageous over chemical methodologies that are practiced for the synthesis of the same. In addition, the biological approach is known for additional features like creating greener and safer environment in addition to energy-efficient and highly cost-effective process. The biocompatibility of NP upon coating with biological molecules is believed to be much more efficient than that of results obtained from chemical methods (Mukherjee et al. 2001; Tripp et al. 2002; Hakim et al. 2005; Prasad et al. 2016).

Much of the achievements or rather applications in practice under biomedicine and allied aspects have been due to the biocompatibility of biologically coated NP (Huang et al. 2015). The methods which are much inclined to biological traits have resulted in the synthesis of NP with mesmerizing structures of different sizes and dimensions (Riddin et al. 2010; Schröfel et al. 2014); for instance, AgNP of fungal origin have been synthesized by subjecting fungal biomass to the solutions containing Ag+ ions. This fungal biomass which is often referred to as *Verticillium* has been known to produce NP that are with least toxicity, for which the reason is thought to be the continuous growth of fungi under controlled environment (Prasad 2016).

This fungal biomass which supports the growth of NP on the surface of their mycelia is thought to imperil under the influence of Ag+ ions and carboxylate group of enzymes located in the cell wall of fungi which constantly interact due to the electrostatic feature (Mukherjee et al. 2001). Cell soluble protein extracts of sulfate-reducing bacteria have been reportedly used to synthesize Pt NP that are of definite dimension (Riddin et al. 2010). The superiority of NP was noticed in those technologies where biological and enzymatic processes were employed as compared to other modes like chemical synthesis as the latter was carried out only with the help of chemicals that would be of larger expense in addition to the prior one which showed increased catalytic activity. The synthesis of Au NP has been possible through the utilization of a fungus of industrial importance, *Penicillium rugulosum*, which is reportedly feasible for handling as compared with the other species under bacteria and yeast (Mishra et al. 2012). Various extracts from variety of plant parts like leaf, bark, and root latex are known for obtaining NP which are used for stabilization or reduction purpose (Prasad 2014). One such extracts from leaves of *Jasminum sambac* have been used in synthesis of Au, Ag, and Au-Ag alloy NP (Yallappa et al. 2015). Au NP with different dimensions including those which are triangular, hexagonal, and spheroid are reportedly prepared by usage of hot water against leaf extracts of olive plant at an elevated temperature and are found to be advantageous over chemically synthesized NP (Khalil et al. 2012). In a study, poly-catalytic activity of iron-polyphenol has been reported against acid black dye in which NP were synthesized by using leaves of *Eucalyptus tereticornis*, *Melaleuca nesophila*, and *Rosmarinus officinalis* (Wang et al. 2014a, b). The mechanism of this conjugation was briefed as the reaction between polyphenols (within three plants) with $FeCl_3$ resulting in the formation of chelated ferric polyphenol NP which in addition were found to be of different structural dimensions.

In addition to the above suggested fungi and plant extracts used in NP synthesis, various algae have been noted for their ability to be used as bioremediating agents, which strongly act on metals that are of high toxicity. The ability of these algae is thought to be high in terms of bioremediation of metals and their complexes which in turn make them flexible forms to be used for the above purpose. Furthermore, high competence of these organisms is noted for their ability toward the synthesis of NP that are in conjugation with metals and their oxide (Patel et al. 2015). Thus, the biosynthesis of NP using biological substrates like fungi, yeast, plant extracts, algae, and other waste materials that are of biological origin is noted to be an emerging trend in upcoming research subjects. Carbon being a nonmetallic element is also the sixth most abundant element in the universe that belongs to P block, period 2 of periodic table. The element is commonly retrieved from coal deposits. Being the second most ample element by mass immediately after oxygen, in the human body, is known for its presence which determines the nature of a molecule to be called as organic or inorganic.

Graphite, diamond, and C-60 (better known as buckminsterfullerene or bucky-balls) are the three elemental forms in accordance with chemists. Even though fullerenes are noted for endorsing the progression of light-absorbing carbon (LAC) particles (Johnson et al. 2002), their rare availability makes the present discussion of restricting their explanation. Sp^2 bonding is observed in graphite and is of planar layers in diamond; however, sp^3 bonding is noticed which is of crystalline form and is found to be missing in atmosphere, a reason for which the aerosol scientist could not be wealthy enough.

3.5.1 Activated Carbon

The cluster of absorbing substances that are crystalline and usually have a large internal pore structure and endorse the carbon to absorb more are termed as activated carbon (AC). The terminology is derived from two important words "carbon" and "active" that refer to as the material which goes through the process of carbonization which usually is at immense temperature. The moment while a substance is active in a carbon condition, it undergoes activation protocol in order to reveal a surface area of pore to as much maximum as possible to alleviate the adsorption rate of activated carbon. AC forms an important class of porous solids with immense technological applications due to which certain qualities, viz., adsorption potential of gases and liquids has been evaluated thoroughly. Carbon could be obtained from all the organic molecules; much of the contributors belong to the kingdom of Plantae. Exploitation of agricultural products like rice husk, sugarcane bagasse, straw, and coconut shells could give rise to enormous and important amounts of carbon particles. One gram of activated carbon surface area is estimated to be about 500 m^2 which is thought to be due to extraordinary levels of microporosity which in turn is determined by nitrogen gas adsorption.

Several methods for production of AC have been suggested, one of which is known as carbonization, wherein AC has been derived from raw materials like wood, nutshells, and coal by elevating the temperatures up to 600–900 °C under the influence of inert gases like argon and nitrogen. In one more technique, "oxidation," the carbonized content is heated at a range of 600–1200 °C under oxidizing atmospheres like oxygen, carbon dioxide, or steam. Exceptional structure with remarkable properties has been the reason behind special interest regarding carbon materials with nanostructure. Carbon nanoparticles (CNP) are well known for their applications in several critical studies in addition to their practical applications and have been acknowledged in recent decades. Their utilization as components of high-performance electrode substances in batteries (Marschilok et al. 2011) and supercapacitors (Wen et al. 2014) and as excellent photoluminescent materials is well understood (Li et al. 2011). Studies have reported that carbon atom could form allotropes that are of several kinds.

The ability of carbon to form one-, two-, or zero-dimensional allotropes makes them one of the unique components to be used for practical and industrial applications. The property of carbon to form low-dimensional (2D, 1D, or 0D) allotropes collectively known as carbon nanomaterials. CNP are being explored widely for use in cancer treatment. Studies reveal that cancer treatment using radio waves can heat and destroy a tumor, lymphoma, or metastasized cancer. CNP are engineered nanomaterials and have immense applications in optics, electrical, thermal, and mechanical sectors (Hurt et al. 2006; Bennett et al. 2013; Srivastava et al. 2015). The domination of graphene family which was discovered in 2004 is well acknowledged in addition to few others like Buckminster (C60) fullerene in 1985 and carbon nanotube in 1991 (Bergmann and Machado 2015; Hong et al. 2015). The favorable configuration of sp^3 to sp^2 has been known to be presumed by carbon which further depends on the thermal generation and pressure which enthralls to give structures like graphene sheets and nanodiamonds (Mauter and Elimelech 2008).

Worldwide production of carbon nanomaterial (CNM) was estimated to be around 3500 tons and was thought to alleviate at an annual growth of 30.6%. The applications of carbon nanotubes (CNT) and carbon nanofibers (CNF) (De Jong and Geus 2000; Ajayan and Zhou 2001; Hammel et al. 2004; Hanus and Harris 2010) have been known already in addition to applications of carbon nanoropes (CNR) which could be identified (Lu 1997). The generation of CNM through available technology is known to require extensive utilization of high-class feedstocks besides high-cost catalysts and needs energy to drive CNM synthesis (Healy et al. 2008). Chemical vapor deposition (CVD) is a typical method of obtaining large-scale production of CNM which are under industrial scale (See and Harris 2007; Tessonnier et al. 2009) and is known to consume diverse and expensive supplies of ethylene (Cassell et al. 1999; Hata et al. 2004), carbon monoxide (Chiang et al. 2001; Zheng et al. 2002; Bachilo et al. 2003), and hydrogen (H_2) (Zheng et al. 2002).

Agricultural fiber is contributed by most of the developing countries for which a huge amount of agricultural waste is being utilized as fuel. Nearly 400 tons of agricultural waste are contributed by India alone, with the utilization of products like rice husk, bagasse, groundnut shell, and coconut fiber for the above said purpose.

The carbon materials with nanostructures are known to be generated by the abovementioned agricultural waste products. Uses of CNP range from being used as capacitors to high-performance electrode materials in batteries (Dominko et al. 2003); in addition to this, they are known for being used as remarkable photoluminescent materials (Li et al. 2011). Efficient binding of the molecules can be possible with the CNP that is of larger surface area.

Due to the chemical inertness of photoluminescent CNP, they are known to be highly advantageous over traditional quantum dots and organic dyes (Ray et al. 2009). With remarkable struggles being put toward the CNP production, many other bottom-up processes which are highly predominant include particle graphitization through microemulsion polymerization (Jang et al. 2002), pyrolysis (Gherghel et al. 2002; Larciprete et al. 2002; Ding and Olesik 2004, 2005; Neabo et al. 2013), treatment in supercritical water (Yang et al. 2004; Li et al. 2011), electrolysis in molten salt (Hsu et al. 1996), microwave plasma-enhanced chemical vapor deposition

(Wang et al. 2001; Yu et al. 2002), and laser vaporization of a carbon pellet (Asano et al. 2010). Nonetheless, it is obvious and is to be recorded that the above processes could only generate the crude products which are to be subjected to purification processes by employing techniques like laser ablation and arc discharge in addition to others like chemical vapor deposition which help in separating the catalyst with high efficiency.

Certain crucial parameters for considering carbon types include the considerations like pore structure, particle size, and void space between particles. Soon after the selection of the raw materials, they are processed for dehydration, carbonization, and activation processes. The processes often involve slow and gradual heating under anaerobic conditions, for which certain chemicals like zinc chloride and phosphoric acids are used additionally to alleviate these chemical reactions.

3.5.2 Black Carbon

Carbon measured by light absorption could be operationally defined as the black carbon (BC). Different sources are known for obtaining different forms of carbon; for instance, BC is obtained mainly through combustion engines (diesel engine) and burning of wood and coal under domestic and industrial scale through the utilization of heavy oils. It is obtained from incineration of agricultural waste products derived from forest fires. Synthesis of carbon nanoparticles through ion beam radiations depends on formation of point defect in diamond particle (Gruber et al. 1997; Yu et al. 2005; Neugart et al. 2007; Batalov et al. 2009), laser ablation of graphite continued with oxidation and functionalization (Sun et al. 2006; Cao et al. 2007), and thermal decomposition of organic compounds (Selvi et al. 2008; Bourlinos et al. 2008; Liu et al. 2009).

A variety of carbon nanoparticles methodologies that make use of synthetic approach are known; however, the techniques are known to yield lesser amount of product. For instance, carbon nanoparticles of fluorescent nature are known to be synthesized with different colors and can be possible through approaches like octadecylamine functionalized diamond nanoparticle that shows blue color (Gruber et al. 1997; Glinka et al. 1999; Zyubin et al. 2009), nitrogen doped with diamond showed red, and so on. Nevertheless, the magnitude of the total yield from these processes is less (<1%) (Liu et al. 2007; Zhao et al. 2008). Other methods like thermal decomposition-based methods yield less amount of product that is soluble and significantly high yield of insoluble fraction which is of lesser importance to the field. Thus synthetic methods that are involved are tedious and clumsy. Some of the other techniques despite showing significant increase in quantum yield (4–15%) and however fail to address the mechanism that is very poorly understood (Liu et al. 2009). It would, therefore, be expedient to quote that the production of CNP in a simple and efficient way under large-scale range would be quite challenging, in addition to the wide constraints in its isolation, purification, and functionalization under the chemical approach. However, utilization of biological raw materials for the same would be highly feasible in addition to other factors like cost-effective strategy and large-scale manufacturing.

3.6 Conclusion and Future Perspectives

The increasing urge for sustainability of environment is gaining research focus into biodegradable and biocompatible materials with the concept of "green" materials; in this context, nanobiomaterials from agricultural waste may be considered as an attractive alternative having biodegradable, renewable, or biocompatible properties besides being mechanically strong, stiff, and highly crystalline with outstanding thermal stability. Biodegradability, simple accessibility, and outstanding mechanical property of nanocellulose have pulled in a lot of curiosity as a novel source of nanometer-sized materials. Prospective uses of nanocellulose in fortification for biodegradable materials and wide applications of nanocellulose in particularly environmental field make nanocellulose as a key player in pollution management. Nanocellulose extraction from agricultural wastes, like citrus and orange, seems to be promising substitutes for waste treatment. Furthermore, an effective technique for extraction and few more wide applications of nanocellulose in biological science is much expected in the near future. The high surface area, the rich abundance of functional groups, and other impressive properties make graphene oxide as one of the elite carbon compounds. The potential application of grapheme oxide (GO) in various fields of science has been observed. Synthesis of GO from sugarcane bagasse and rice husk ash has made notable development in mass production of GO. However, available protocols for the synthesis of GO from agrowastes are in the very beginning stage; further, the modern methods for synthesis of eco-friendly, cost-effective, agrowaste origin GO compounds need to be developed effectively. Production of amorphous silicon nanoparticles has been found using different agrowastes such as corn hub, rice husk, and other plant sources. The enhancement in the crop yield by reducing the use of fertilizer and pesticides (SiO_2NP) concluded to be one of the elite compounds for the enhancement of agricultural yields.

Carbon being a vital component has been explored for its potential use in the field of medical and industrial purposes. The exploitation of synthetic raw materials for the production of nanostructured materials like carbon nanotubes and carbon nanofibers is of crucial importance, and the utilization of natural substances that are of secondary or lesser importance is being highly encouraged. Currently, one such approach is by the utilization of agricultural waste for the production of carbon nanomaterials which has been elucidated above. The discussion also emphasizes the importance of types of carbon (activated carbon and black carbon) and their uses in the field of medical and nanoscience. The discussion also deals with the sections of agricultural waste that could be utilized as the promising tools for construction of carbon nanoparticles thus obtaining the best of possible yield out of waste from agricultural sector. In spite of these potential points of interest, synthesis of bionanomaterials from agricultural wastes is still comparably ignored and has not yet made it to the market sector to any extent in the examination with other modern industrial divisions. The influx of research revelations seems to be mainly asserted by the small enterprises or the academic sector; hence, there should be some positive considerations to follow in the future for the commercialization of these elite bionanomaterials in large scale.

References

Abe K, Yano H (2010) Comparison of the characteristics of cellulose microfibril aggregates isolated from fiber and parenchyma cells of Moso bamboo (*Phyllostachys pubescens*). Cellulose 17(2):271–277

Abe K, Iwamoto S, Yano H (2007) Obtaining cellulose nanofibers with a uniform width of 15 nm from wood. Biomacromolecules 8(10):3276–3278

Abe K, Nakatsubo F, Yano H (2009) High-strength nanocomposite based on fibrillated chemi-thermomechanical pulp. Compos Sci Technol 69(14):2434–2437

Affandi S, Setyawan H, Winardi S, Purwanto A, Balgis R (2009) A facile method for the production of high purity silica xerogels from bagasse ash. J Adv Powder Technol 20:468–472

Affandi MM, Julianto T, Majeed A (2011) Development and stability evaluation of astaxanthin nanoemulsion. Asian J Pharma Clin Res 4:143–148

Ajayan P, Zhou O (2001) Applications of carbon nanotubes. In: Dresselhaus M, Dresselhaus G, Avouris P (eds) Carbon nanotubes. Springer, Berlin, pp 391–425

Akhtar MS, Panwar J, Yun YS (2013) Biogenic synthesis of metallic nanoparticles by plant extracts. ACS Sustain Chem Eng 1(6):591–602

Alemdar A, Sain M (2008) Isolation and characterization of nanofibers from agricultural residues – wheat straw and soy hulls. Bioresour Technol 99(6):1664–1671

Antonyraj CA, Jeong J, Kim B, Shin S, Kim S, Lee KY, Cho JK (2013) Selective oxidation of HMF to DFF using Ru/γ-alumina catalyst in moderate boiling solvents toward industrial production. J Ind Eng Chem 19(3):1056–1059

Asano H, Muraki S, Endo H, Bandow S, Iijima S (2010) Strong magnetism observed in carbon nanoparticles produced by the laser vaporization of a carbon pellet in hydrogen-containing Ar balance gas. J Phys Condens Matter 22:1–6

Aspler J, Bouchard J, Hamad W, Berry R, Beck S, Drolet F, Zou X (2013) Review of nanocellulosic products and their applications. In: Dufresne A, Thomas S, Pothan LA (eds) Biopolymer nanocomposites. Wiley, Hoboken, pp 461–508

Awan AT, Tsukamoto J, Tasic L (2013) Orange waste as a biomass for 2G-ethanol production using low cost enzymes and co-culture fermentation. RSC Adv 3:25071–25078

Bachilo S, Balzano L, Herrera J, Pompeo F, Resasco D, Weisman R (2003) Narrow (n,m)- distribution of single-walled carbon nanotubes grown using a solid supported catalyst. J Am Chem Soc 125:11186–11187

Bacic A, Harris PJ, Stone BA (1988) Structure and function of plant cell walls. In: Preiss J, Stumpf PK, Conn EE (eds) The biochemistry of plants. Academic Press, New York, pp 297–371

Balandin AA, Ghosh S, Bao W, Calizo I, Teweldebrhan D, Miao F, Lau CN (2008) Superior thermal conductivity of single-layer graphene. Nano Lett 8:902–907

Batalov A, Jacques V, Kaiser F, Siyushev P, Neumann P, Rogers LJ, McMurtrie RL, Manson NB, Jelezko F, Wrachtrup J (2009) Low temperature studies of the excited-state structure of negatively charge nitrogen-vacancy color centers in diamond. Phys Rev Lett 102:195506. http://dx.doi.org/10.1103/PhysRevLett.102.195506

Beck-Candanedo S, Roman M, Gray DG (2005) Effect of reaction conditions on the properties and behavior of wood cellulose nanocrystal suspensions. Biomol Ther 6:1048–1054

Bennett SW, Adeleye A, Ji Z, Keller AA (2013) Stability, metal leaching, photoactivity and toxicity in freshwater systems of commercial single wall carbon nanotubes. Water Res 47:4074–4085

Bergmann CP, Machado F (2015) Carbon nanomaterials as adsorbents for environmental and biological applications. Springer, Berlin, pp 1–122

Bhatnagar, A. Sain, M (2003) US Patent Pending, Application No. 60/512, 912

Bhatnagar A, Sain M (2005) Processing of cellulose nanofiber-reinforced composites. J Reinf Plast Compos 24:1259–1268

Birla SS, Gaikwad SC, Gade AK, Rai MK (2013) Rapid synthesis of silver nanoparticles from *Fusarium oxysporum* by optimizing physicocultural conditions. Sci World J 2013:796018. http://dx.doi.org/10.1155/2013/796018

Boufi S (2014) Nanofibrillated cellulose: sustainable nanofiller with broad potentials use. In: Hakeem KR, Jawaid M, Rashid U (eds) Biomass and bioenergy. Springer, Berlin, pp 267–305

Bourlinos AB, Stassinopoulos A, Anglos D, Zboril R, Georgakilas V, Giannelis EP (2008) Photoluminescent carbogenic dots. Chem Mater 20:4539–4541

Brodie BC (1859) On the atomic weight of graphite. Philos Trans R Soc Lond 149:249–259

Cao L, Wang X, Meziani MJ, Lu F, Wang H, Luo PG, Lin Y, Harruff BA, Veca LM, Murray D, Xie SY, Sun YP (2007) Carbon dots for multiphoton bioimaging. J Am Chem Soc 129(37):11318–11319

Cassell A, Raymakers J, Kong J, Dai H (1999) Large scale CVD synthesis of single-walled carbon nanotubes. J Phys Chem B 103:6484–6492

Chakraborty A, Sain M, Kortschot M (2006) Reinforcing potential of wood pulp derived microfibres in a PVA matrix. Holzforschung 60(1):53–58

Chan YS, Don MM (2013) Biosynthesis and structural characterization of Ag nanoparticles from white rot fungi. Mater Sci Eng C 33(1):282–288

Charreau H, Foresti ML, Vazquez A (2013) Nanocellulose patents trends: a comprehensive review on patents on cellulose nanocrystals, microfibrillated cellulose and bacterial cellulose. Recent Pat Nanotechnol 7(1):56–80

Chen Y, Liu C, Chang PR, Cao X, Anderson DP (2009) Bionanocomposites based on pea starch and cellulose nano whiskers hydrolyzed from pea hull fibre: effect of hydrolysis time. Carbohydr Polym 76(4):607–615

Chen H, Wang F, Zhang C, Shi Y, Jin G, Yuan S (2010) Preparation of nano-silica materials: the concept from wheat straw. J Non-Cryst Solids 356(50–51):2781–2785

Chen W, Yu H, Liu Y, Chen P, Zhang M, Hai Y (2011) Individualization of cellulose nanofibers from wood using high-intensity ultrasonication combined with chemical pretreatments. Carbohydr Polym 83(4):1804–1811

Chen J, Yao B, Li C, Shi G (2013) An improved hummers method for eco-friendly synthesis of graphene oxide. Carbon 64:225–229

Cheng Q, Wang S, Rials TG (2009) Poly(vinyl alcohol) nanocomposites reinforced with cellulose fibrils isolated by high intensity ultrasonication. Compos Part A: Appl Sci Manuf 40:218–224

Cheng Q, Wang S, Han Q (2010) Novel process for isolating fibrils from cellulose fibers by high-intensity ultrasonication II. Fibril characterization. J Appl Polym Sci 115(5):2756–2762

Cherian BM, Leao AL, de Souza SF, Thomas S, Pothan LA, Kottaisamy M (2010) Isolation of nanocellulose from pineapple leaf fibres by steam explosion. Carbohydr Polym 81(3):720–725

Chiang I, Brinson B, Huang A, Willis P, Bronikowski M, Margrave J, Smalley R, Hauge R (2001) Purification and characterization of single-wall carbon nanotubes (SWNTs) obtained from the gas-phase decomposition of CO (HiPCO process). J Phys Chem B 105:8297–8301

Chung C, Kim YK, Shin D, Ryoo SR, Hong BH, Min DH (2013) Biomedical applications of graphene and graphene oxide. Acc Chem Res 46:2211–2224

Clowes FAL, Juniper BE (1968) Plant cells. Blackwell Scientific Publications Ltd, Oxford, pp 203–297

De Jong KP, Geus JW (2000) Carbon nanofibers: catalytic synthesis and applications. Catal Rev 42:481–510

De Morais TE, Correa AC, Manzoli A, de Lima LF, de Oliveira CR, Mattoso LHL (2010) Cellulose nanofibers from white and naturally colored cotton fibers. Cellulose 17(3):595–606

Dinand E, Chanzy H, Vignon RM (1999) Suspensions of cellulose micro fibrils from sugar beet pulp. Food Hydrocoll 13(3):275–283

Ding LH, Olesik SV (2004) Synthesis of polymer nanospheres and carbon nanospheres using the monomer 1,8-dihydroxymethyl-1,3,5,7-octatetrayne. Nano Lett 4:2271–2276

Ding LH, Olesik SV (2005) Carbon micro beads produced through synthesis and pyrolysis of poly(1,8-dibutyl-1,3,5,7-octatetrayne). Chem Mater 17:2353–2360

Dodson JR, Hunt AJ, Matharu AS, Budarin VL, Clark JH (2011) The chemical value of wheat straw combustion residues. RSC Adv 1:523–530

Dominko R, Gaberscek M, Drofenik J, Bele M, Jamnik J (2003) Influence of carbon black distribution on performance of oxide cathodes for Li ion batteries. Electrochim Acta 48:3709–3716

Donia AM, Atia AA, Abouzayed FI (2012) Preparation and characterization of nano-magnetic cellulose with fast kinetic properties towards the adsorption of some metal ions. Chem Eng J 191:22–30

Du Y, Yin Z, Zhu J, Huang X, Wu X, Zeng Z, Yan Q, Zhang H (2012) A general method for the large-scale synthesis of uniform ultrathin metal sulphide nanocrystals. Nat Commun 3:1177. doi:10.1038/ncomms2181

Dufresne A (2013) Nanocellulose: a new ageless bionanomaterial. Mater Today 16(6):220–227

Dufresne A, Vignon M (1998) Improvement of starch film performances using cellulose microfibrils. Macromolecules 31:2693–2696

Dufrense A, Cavaille JY, Helbert W (1997a) Thermoplastic nanocomposites filled with wheat straw cellulose whiskers. Part II: effect of processing and modeling. Polym Compos 18:199

Dufresne A, Caville J, Vignon M (1997b) Mechanical behavior of sheets prepared from sugar beet cellulose microfibrils. J Appl Polym Sci 64:1185–1194

Dufresne A, Dupeyre D, Vignon MR (2000) Cellulose microfibrils from potato tuber cells: processing and characterization of starch-cellulose microfibril composites. J App Polym Sci 76(14):2080–2092

Elazzouzi-Hafraoui S, Nishiyama Y, Putaux JL, Heux L, Dubreuil F, Rochas C (2008) The shape and size distribution of crystalline nanoparticles prepared by acid hydrolysis of native cellulose. Biomacromolecules 9(1):57–65

Espindola-Gonzalez A, Martinez-Hernadez AL, Angeles-Chavez C, Castano VM, Velasco-Santos C (2010) Novel crystalline SiO_2 nanoparticles via annelids bioprocessing of agro-industrial wastes. Nanoscale Res Lett 5(9):1408–1417

Fernando F, María LH, Ana ME, Italo C, Fernando A (2005) Fibre concentrates from apple pomace and citrus peel as potential fibre sources for food enrichment. Food Chem 91:395–401

Frances N, Nikolay AP, Michael JFB, Tim G, Paul AM (2009) Novel one-pot synthesis and characterization of bioactive thiol-silicate nanoparticles for biocatalytic and biosensor applications. Nanotechnology 20(5):055612. doi:10.1088/0957-4484/20/5/055612

Fu H, Yang X, Jiang X, Yu A (2013) Bimetallic Ag–Au nanowires: synthesis, growth mechanism, and catalytic properties. Langmuir 29(23):7134–7142

Georgakilas V, Otyepka M, Bourlinos AB, Chandra V, Kim N, Kemp KC, Hobza P, Zboril R, Kim KS (2012) Functionalization of graphene: covalent and non-covalent approaches, derivatives and applications. Chem Rev 112:6156–6214

Gherghel L, Kubel C, Lieser G, Rader HJ, Mullen K (2002) Pyrolysis in the mesophase: a chemist's approach toward preparing carbon nano- and microparticles. J Am Chem Soc 124:13130–13138

Ghorbani F, Sanati AM, Malek M (2015) Production of silica nanoparticles from rice husk as agricultural waste by environmental friendly technique. Environ Stud Persian Gulf 2(1):56–65

Ghosh S, Ranebennur TK, Vasan H (2011) Study of antibacterial efficacy of hybrid chitosan-silver nanoparticles for prevention of specific biofilm and water purification. Int J Carbohydr Chem. 693759. doi:10.1155/2011/693759

Glinka YD, Lin KW, Chang HC, Lin SHJ (1999) Multiphoton-excited luminescence from diamond nanoparticles. J Phys Chem B 103(21):4251–4263

Gruber A, Dräbenstedt A, Tietz C, Fleury L, Wrachtrup J, von Borczyskowski C (1997) Scanning confocal optical microscopy and magnetic resonance on single defect centers. Science 276(5321):2012–2014

Gu S, Zhou J, Yu C, Luo Z, Wang Q, Shi Z (2015) A novel two-staged thermal synthesis method of generating nanosilica from rice husk via pre-pyrolysis combined with calcination. Ind Crop Prod 65:1–6

Guo S, Dong S (2011) Graphene nanosheet: synthesis, molecular engineering, thin film, hybrids, and energy and analytical applications. Chem Soc Rev 40:2644–2672

Habibi Y, Heux L, Mahrouz M, Vignon MR (2008) Morphological and structural study of seed pericarp of *Opuntia ficus-indica* prickly pear fruits. Carbohydr Polym 72(1):102–112

Habibi Y, Lucia LA, Rojas OJ (2010) Cellulose nanocrystals: chemistry, self assembling, and applications. Chem Rev 110:3479–3500

Hakim LF, Portman JL, Casper MD, Weimer AW (2005) Aggregation behavior of nanoparticles in fluidized beds. Powder Technol 160(3):149–160

Hammel E, Tang X, Trampert M, Schmitt T, Mauthner K, Eder A, Potschke P (2004) Carbon nanofibers for composite applications. Carbon 42:1153–1158

Hanus MJ, Harris AI (2010) Synthesis, characterisation and applications of coiled carbon nanotubes. J Nanosci Nanotechnol 10:2261–2283

Hariharan V, Sivakumar G (2013) Studies on synthesized nanosilica obtained from bagasse ash. Int J Chem Tech Res 5(3):1263–1266

Hassan AF, Abdelghny AM, Elhadidy H, Youssef AM (2014) Synthesis and characterization of high surface area nanosilica from rice husk ash by surfactant-free sol–gel method. J Sol-Gel Sci Technol 69:465–472

Hata K, Futaba D, Mizuno K, Namai T, Yumura M, Iijima S (2004) Water-assisted highly efficient synthesis of impurity-free single-walled carbon nanotubes. Science 306:1362

He W, Jiang S, Zhang Q, Pan M (2013) Nanofiber from *Bambusa*. Bioresources 8(4):5678–5689

Healy ML, Dahlben LJ, Isaacs JA (2008) Environmental assessment of single-walled carbon nanotube processes. J Ind Ecol 12:376–393

Hokkanen S, Repo E, Suopajärvi T, Liimatainen H, Niinimäki J, Sillanpää M (2013) Removal of heavy metals from aqueous solutions by succinic anhydride modified mercerized nanocellulose. Chem Eng J 223:40–47

Hokkanen S, Bhatnagar A, Sillanpää M (2016) A review on modification methods to cellulose-based adsorbents to improve adsorption capacity. Water Res 91:156–173

Honek JF (2013) Bionanotechnology and bionanomaterials: John Honek explains the good things that can come in very small packages. BMC Biochem 14:29. doi:10.1186/1471-2091-14-29

Hong G, Diao S, Antaris AL, Dai H (2015) Carbon nanomaterials for biological imaging and nanomedicinal therapy. Chem Rev 115:10816–10906

Hsu WK, Terrones M, Hare JP, Terrones H, Kroto HW, Walton DRM (1996) Electrolytic formation of carbon nanostructures. Chem Phys Lett 262:161–166

Huang J, Lin L, Sun D, Chen H, Yang D, Li Q (2015) Bioinspired synthesis of metal nanomaterials and applications. Chem Soc Rev 44(17):6330–6374

Hummers WS, Offeman RE (1958) Preparation of graphitic oxide. J Am Chem Soc 80:1339

Hurt RH, Monthioux M, Kane A (2006) Toxicology of carbon nanomaterials: status, trends, and perspectives on the special issue. Carbon 44:1028–1033

Hussain HI, Yi Z, Rookes JE, Kong LX, Cahill DM (2013) Mesoporous silica nanoparticles as a biomolecule delivery vehicle in plants. J Nanopart Res 15:1676. doi:10.1007/s11051-013-1676-4

Indhumathi P, Shabhudeen SSP, Saraswathy CP (2011) Synthesis and characterization of nano silica from the pods of *Delonix regia* ash. Int J Adv Eng Technol 2(4):421–426

Jang J, Oh JH, Stucky GD (2002) Fabrication of ultrafine conducting polymer and graphite nanoparticles. Angew Chem Int Ed 41:4016–4019

Javed SH, Shah FH, Manasha M (2011) Extraction of amorphous silica from wheat husk using $KMNO_4$. J Fac Eng Technol 18(1):39–46

Johnson MP, Donnet JB, Wang TK, Wang CC, Locke RW, Brinson BE, Marriott T (2002) A dynamic continuum of nano structured carbons in the combustion furnace. Carbon 40:189–194

Kalia S, Boufi S, Celli A, Kango S (2014) Nanofibrillated cellulose: surface modification and potential applications. Colloid Polym Sci 292(1):5–31
Kamath SR, Proctor A (1998) Silica gel from rice husk ash: preparation and characterization. Cereal Chem 75:484–487
Kanchi S, Kumar G, Lo AY, Tseng CM, Chen SK, Lin CY, Chin TS (2014) Exploitation of de-oiled Jatropha waste for gold nanoparticles synthesis: a green approach. Arab J Chem. http://dx.doi.org/10.1016/j.arabjc.2014.08.006
Kardam A, Raj KR, Srivastava S, Srivastava MM (2014) Nanocellulose fibers for biosorption of cadmium, nickel, and lead ions from aqueous solution. Clean Technol Environ Policy 16:385–393
Karimi J, Mohsenzadeh S (2016) Effects of silicon oxide nanoparticles on growth and physiology of wheat seedlings. Russ J Plant Physiol 63:119–123
Kettunen M, Silvennoinen RJ, Houbenov N, Nykänen A, Ruokolainen J, Saini J, Pore V, Kemell M, Ankerfors M, Lindström T (2011) Photoswitchable super absorbency based on nanocellulose aerogels. Adv Funct Mater 21(3):510–517
Khalil MMH, Ismail EH, El-Magdoub F (2012) Biosynthesis of Au nanoparticles using olive leaf extract: 1st nano updates. Arab J Chem 5(4):431–437
Khawas P, Das AJ, Dash KK, Deka SC (2014) Thin-layer drying characteristics of Kachkal banana peel (Musa ABB) of Assam, India. Int Food Res J 21(3):1011–1018
Klemm D, Kramer F, Moritz S, Lindström T, Ankerfors M, Gray D, Dorris A (2011) Nanocelluloses: a new family of nature-based materials. Angew Chem Int 50:5438–5466
Kumar KS, Amutha R, Arumugam P, Berchmans S (2011) Synthesis of gold nanoparticles: an ecofriendly approach using *Hansenula anomala*. ACS Appl Mater Interfaces 3(5):1418–1425
Kumar A, Mohanta K, Kumar D, Parkash O (2012) Properties and industrial applications of rice husk: a review. Int J Emerg Technol Adv Eng 2(10):86–90
Kwon SJ, Bard AJ (2012) DNA analysis by application of Pt nanoparticle electrochemical amplification with single label response. J Am Chem Soc 134(26):10777–10779
Larciprete R, Lizzit S, Botti S, Cepek C, Goldoni A (2002) Structural reorganization of carbon nanoparticles into single-wall nanotubes. Phys Rev B 66:12140–12142
Lario YE, Sendra E, Fuentes C, Pérez-Álvarez JA (2004) Preparation of high dietary fiber powder from lemon juice by-products. Innovative Food Sci Emerg Technol 5(1):113–117
Lee C, Wei X, Kysar JW, Hone J (2008) Measurement of the elastic properties and intrinsic strength of monolayer grapheme. Science 321:385–388
Li R, Fei J, Cai Y, Li Y, Feng J, Yao J (2009) Cellulose whiskers extracted from mulberry: a novel biomass production. Carbohydr Polym 76(1):94–99
Li HT, He XD, Liu Y, Huang H, Lian SY, Lee ST, Kang ZH (2011) One-step ultrasonic synthesis of water-soluble carbon nanoparticles with excellent photoluminescent properties. Carbon 49:605–609
Liu F, Wen LX, Li ZZ, Yu W, Sun HY, Chen JF (2006) Porous hollow silica nanoparticles as controlled delivery system for water-soluble pesticide. Mater Res Bull 41:2268–2275
Liu H, Ye T, Mao C (2007) Fluorescent carbon nanoparticles derived from candle soot. Angew Chem Int 46(34):6473–6485
Liu R, Wu D, Liu S, Koynov K, Knoll W, Li Q (2009) An aqueous route to multicolor photoluminescent carbon dots using silica spheres as carriers. Angew Chem Int 48(25):4598–4601
Liu H, Liu D, Yao F, Wu Q (2010) Fabrication and properties of transparent polymethylmethacrylate/cellulose nanocrystals composites. Bioresour Technol 101(14):5685–5692
Lu JP (1997) Elastic properties of carbon nanotubes and nanoropes. Phys Rev Lett 79:1297–1300
Lu P, Hsieh YL (2012) Highly pure amorphous silica nano-disks from rice straw. Powder Technol 225:149–155
Lu G, Park S, Yu K, Ruoff RS, Ocola LE, Rosenmann D, Chen J (2011) Toward practical gas sensing with highly reduced graphene oxide: a new signal processing method to circumvent run-to-run and device-to-device variations. ACS Nano 5:1154–1164

Luo X, Morrin A, Killard AJ, Smyth MR (2006) Application of nanoparticles in electrochemical sensors and biosensors. Electroanalysis 18(4):319–326

Manjula-Rani K, Palanisamy PN, Sivakumar P (2014) Synthesis and characterization of amorphous nano-silica from biomass ash. Int J Adv Technol Eng Sci 2(10):71–76

Mansaray KG, Ghaly AE (1997) Physical and thermochemical properties of rice husk. Energy Sources 19(9):989–1004

Mao HY, Laurent S, Chen W, Akhavan O, Imani M, Ashkarran AA, Mahmoudi M (2013) Graphene: promises, facts, opportunities, and challenges in nanomedicine. Chem Rev 113:3407–3424

Mariño M, da Silva LL, Durán N, Tasic L (2015) Enhanced materials from nature: nanocellulose from citrus waste. Molecules 20:5908–5923

Marschilok A, Lee CY, Subramanian A, Takeuchi KJ, Takeuchi ES (2011) Carbon nanotube substrate electrodes for lightweight, long-life rechargeable batteries. Energy Environ Sci 4(8):2943–2951

Mauter MS, Elimelech M (2008) Environmental applications of carbon-based nanomaterials. Environ Sci Technol 42:5843–5859

May JW (1969) Platinum surface LEED rings. Surf Sci 17:267–270

Metreveli G, Wågberg L, Emmoth E, Belák S, Strømme M, Mihranyan A (2014) A size-exclusion nanocellulose filter paper for virus removal. Adv Healthc Mater 10(3):1546–1550

Mishra A, Tripathy SK, Yun SI (2012) Fungus mediated synthesis of gold nanoparticles and their conjugation with genomic DNA isolated from *Escherichia coli* and *Staphylococcus aureus*. Process Biochem 47(5):701–711

Mohammadinejad R, Karimi S, Iravani S, Varma RS (2016) Plant-derived nanostructures: types and applications. Green Chem 18(1):20–52

Mohanraj K, Kannan S, Barathan S, Sivakumar G (2012) Preparation and characterization of nano SiO_2 from corn cob ash by precipitation method. Optoelectron Adv Mater 6:394–397

Moon RJ, Martini A, Nairn J, Simonsen J, Youngblood J (2011) Cellulose nanomaterials review: structure, properties and nanocomposites. Chem Soc Rev 40:3941–3994

Mukherjee P, Ahmad A, Mandal D, Senapati S, Sainkar SR, Khan MI, Parishcha R, Ajaykumar PV, Alam M, Kumar R, Sastry M (2001) Fungus-mediated synthesis of silver nanoparticles and their immobilization in the mycelial matrix: a novel biological approach to nanoparticle synthesis. Nano Lett 1(10):515–519

Muramatsu H, Kim YA, Yang KS, Cruz-Silva R, Toda I, Yamada T, Terrones M, Endo M, Hayashi T, Saitoh H (2014) Rice husk-derived graphene with nano-sized domains and clean edges. Small 10(14):2766–2770

Murthy SK (2007) Nanoparticles in modern medicine: state of the art and future challenges. Int J Nanomedicine 2(2):129–141

Musyoka S, Ngila C, Moodley B, Kindness A, Petrik L, Greyling C (2011) Oxolane-2,5-dione modified electrospun cellulose nanofibers for heavy metals adsorption. J Hazard Mater 192:922–927

Nair R, Varghese SH, Nair BG, Maekawa T, Yoshida Y, Kumar DS (2010) Nanoparticulate material delivery to plants. Plant Sci 179:154–163

Néabo JR, Gagné SR, Carrière CV, Morin JF (2013) Soluble conjugated one-dimensional nanowires prepared by topochemical polymerization of a butadiynes-containing star-shaped molecule in the xerogel state. Langmuir 29:3446–3452

Neugart F, Zappe A, Jelezko F, Tietz C, Boudou JP, Krueger A, Wrachtrup J (2007) Dynamics of diamond nanoparticles in solution and cells. Nano Lett 7(12):3588–3591

Novoselov KS, Fal'ko VI, Colombo L, Gellert PR, Schwab MG, Kim K (2012) A roadmap for grapheme. Nature 490:192–200

Okoronkwo EA, Imoisili PE, Olubayode SA, Olusunle SOO (2016) Development of silica nanoparticle from corn cob ash. Adv Nanoparticles 5:135–139

Omid A, Keyvan B, Ali M (2014) Synthesis of graphene from natural and industrial carbonaceous wastes. RSC Adv 4:20441–20448

Paakko M, Ankerfors M, Kosonen H, Nykanen A, Ahol S, Osterberg M, Ruokolainen J, Laine J, Larsson PT, Ikkala O, Lindström T (2007) Enzymatic hydrolysis combined with mechanical shearing and high-pressure homogenization for nanoscale cellulose fibrils and strong gels. Biomacromolecules 8(6):1934–1941

Patel V, Berthold D, Puranik P, Gantar M (2015) Screening of cyanobacteria and microalgae for their ability to synthesize silver nanoparticles with antibacterial activity. Biotechnol Rep 5:112–119

Pelissari MF, Sobral PJA, Menegalli FC (2014) Isolation and characterization of cellulose nanofibers from banana peels. Cellulose 21(1):417–432

Prasad R (2014) Synthesis of silver nanoparticles in photosynthetic plants. J Nanoparticles:963961. http://dx.doi.org/10.1155/2014/963961

Prasad R (2016) Advances and applications through fungal nanobiotechnology. Springer, International Publishing, Cham. ISBN:978-3-319-42989-2

Prasad R, Pandey R, Barman I (2016) Engineering tailored nanoparticles with microbes: quo vadis. WIREs Nanomed Nanobiotechnol 8:316–330. doi:10.1002/wnan.1363

Prerna K, Sankar CD (2016) Isolation and characterization of cellulose nanofibers from culinary banana peel using high-intensity ultrasonication combined with chemical treatment. Carbohydr Polym 137:608–616

Qadri SB, Gorzkowski E, Rath BB, Feng J, Qadri SN, Kim H, Caldwell JD, Imam MA (2015) Nanoparticles and nanorods of silicon carbide from the residues of corn. J Appl Phys 117:044306. http://dx.doi.org/10.1063/1.4906974

Rafiee E, Shahebrahimi S, Feyzi M, Shaterzadeh M (2012) Optimization of synthesis and characterization of nanosilica produced from rice husk (a common waste material). Int Nano Lett 2(29):1–9

Rajesh K, Rajesh KS, Dinesh PS (2016) Natural and waste hydrocarbon precursors for the synthesis of carbon based nanomaterials: graphene and CNTs. Renew Sust Energ Rev 58:976–1006

Ray SC, Saha A, Jana NR, Sarkar NR (2009) Fluorescent carbon nanoparticles: synthesis, characterization, and bioimaging application. J Phys Chem C 113:18546–18551

Rezanezhad S, Nazanezhad N, Asadpur G (2013) Nanocellulose from straw. Lignocellulose 2(1):282–291

Riddin T, Gericke M, Whiteley CG (2010) Biological synthesis of platinum nanoparticles: effect of initial metal concentration. Enzym Microb Technol 46(6):501–505

Rodionova G, Saito T, Lenes M, Eriksen O, Gregersen O, Fukuzumi H, Isogai A (2011) Mechanical and oxygen barrier properties of films prepared from fibrillated dispersions of TEMPO-oxidized Norway spruce and eucalyptus pulps. Cellulose 19(3):705–711

Rondeau-Mouro C, Bouchet B, Pontoire B, Robert P, Mazoyer J, Buleon A (2003) Structural features and potential texturising properties of lemon and maize cellulose microfibrils. Carbohydr Polym 53(3):241–252

Rosa MF, Medeiros ES, Malmonge JA, Gregorski KS, Wood DF, Mattoso LHC, Glennb G, Ortsb WJ, Imam SH (2010) Cellulose nano whiskers from coconut husk fibers: effect of preparation conditions on their thermal and morphological behavior. Carbohydr Polym 81(1):83–92

Roy N, Gaur A, Jain A, Bhattacharya S, Rani V (2013) Green synthesis of silver nanoparticles: an approach to overcome toxicity. Environ Toxicol Pharmacol 36(3):807–812

Sana NO, Kurs C, Tümtas UY, Yas Ö, Ortac B, Tekinay T (2014) Novel one-step synthesis of silica nanoparticles from sugarbeet bagasse by laser ablation and their effects on the growth of freshwater algae culture. Particuology 17:29–35

Santi G, Crognale S, Moresi M, Petruccioli M, D'Annibale A (2012) Improved orange peel waste pretreatments for bioethanol production. Environ Eng Manag J 11(3):S55

Saumya S, Pillai B, Abraham DE, Girija N, Geetha P, Jacob L, Koshy M (2013) Biosorption of Cd(II) from aqueous solution using xanthated nano banana cellulose: equilibrium and kinetic studies. Ecotoxicol Environ Saf 98:352–360

Schröfel A, Kratošová G, Šafařík I, Šafaříková M, Raška I, Shor LM (2014) Applications of biosynthesized metallic nanoparticles- a review. Acta Biomater 10(10):4023–4042

Schwierz F (2010) Graphene transistors. Nat Nanotechnol 5:487–496
See CH, Harris ATA (2007) Review of carbon nanotube synthesis via fluidized-bed chemical vapor deposition. Int Eng Chem Res 46:997–1012
Selvi BR, Jagadeesan D, Suma BS, Nagashankar G, Arif M, Balasubramanyam K, Eswaramoorthy M, Kundu TK (2008) Intrinsically fluorescent carbon nanospheres as a nuclear targeting vector: delivery of membrane-impermeable molecule to modulate gene expression in vivo. Nano Lett 8(10):3182–3188
Shivaji S, Madhu S, Singh S (2011) Extracellular synthesis of antibacterial silver nanoparticles using psychrophilic bacteria. Process Biochem 46(9):1800–1807
Si YC, Samulski ET (2008) Exfoliated graphene separated by platinum nanoparticles. Chem Mater 20:6792–6797
Sidheswaran P, Bhat AN (1996) Recovery of amorphous silica in pure form from rice husk. Trans Indian Ceram Soc 55:93–96
Singh PS, Vidyasagar GM (2014) Biosynthesis, characterization, and antidermatophytic activity of silver nanoparticles using Raamphal plant (*Annona reticulata*) aqueous leaves extract. Indian J Mater Sci. 412452. http://dx.doi.org/10.1155/2014/412452
Singh K, Arora JK, Sinha JMT, Srivastava S (2014) Functionalization of nanocrystalline cellulose for decontamination of Cr(III) and Cr(VI) from aqueous system: computational modelling approach. Clean Techn Environ Policy 16:1179–1191
Somanathan T, Prasad K, Ostrikov K, Saravanan A, Krishna VM (2015) Graphene oxide synthesis from agro waste. Nanomaterials 5(2):826–834
Song H, Zhang L, He C, Qu Y, Tian YF, Lv Y (2011) Graphene sheets decorated with SnO_2 nanoparticles: in situ synthesis and highly efficient materials for cataluminescence gas sensors. J Mater Chem 21:5972–5977
Spence KL, Venditti RA, Rojas OJ, Pawlak JJ, Hubbe MA (2011) Water vapor barrier properties of coated and filled microfibrillated cellulose composite films. Bioresources 6(4):4370–4388
Srivastava V, Gusain D, Sharma YC (2015) Critical review on the toxicity of some widely used engineered nanoparticles. Ind Eng Chem Res 54:6209–6233
Staniland SS (2011) Magnetosomes: bacterial biosynthesis of magnetic nanoparticles and potential biomedical applications. In: Nanotechnologies for the life sciences. Wiley, Hoboken, doi: 10.1002/9783527610419.ntls0173
Staudenmaier L (1898) Verfahren zur darstellung der graphitsäure. Ber Dtsch Chem Ges 31:1481–1487
Stephen JR, Macnaughtont SJ (1999) Developments in terrestrial bacterial remediation of metals. Curr Opin Biotechnol 10(3):230–233
Sun YP, Zhou B, Lin Y, Wang W, Fernando KAS, Pathak P, Meziani MJ, Harruff BA, Wang X, Wang H, Luo PG, Yang H, Kose ME, Chen B, Veca LM, Xie SY (2006) Quantum-sized carbon dots for bright and colorful photoluminescence. J Am Chem Soc 128(24):7756–7757
Sun D, Hussain HI, Yi Z, Rookes JE, Kong L, Cahill DM (2016) Mesoporous silica nanoparticles enhance seedling growth and photosynthesis in wheat and lupin. Chemosphere 152:81–91
Suriyaprabha R, Karunakaran G, Kavitha K, Yuvakkumar R, Rajendran V, Kannan N (2014) Application of silica nanoparticles in maize to enhance fungal resistance. IET Nanobiotechnol 8(3):133–137
Syed A, Saraswati S, Kundu GC, Ahmad A (2013) Biological synthesis of silver nanoparticles using the fungus *Humicola* sp. and evaluation of their cytotoxicity using normal and cancer cell lines Spectrochim. Acta A Mol Biomol Spectrosc 114:144–147
Terzioglu P, Yucel S (2012) Synthesis of magnesium silicate from wheat husk ash: effects of parameters on structural and surface properties. Bioresources 7(4):5435–5447
Tessonnier JP, Rosenthal D, Hansen TW, Hess C, Schuster ME, Blume R, Girgsdies F, Pfänder N, Timpe O, Su DS, Schlögl R (2009) Analysis of the structure and 187 chemical properties of some commercial carbon nanostructures. Carbon 47:1779–1798
Thygesen LG, Hidayat BJ, Johansen KS, Felby C (2011) Role of supramolecular structures in enzymatic hydrolysis of plant cell walls. J Ind Microbiol Biotechnol 38:975–983

Tibolla H, Pelissari FM, Menegalli FC (2014) Cellulose nanofibers produced from banana peel by chemical and enzymatic treatment. LWT-Food Sci Technol 59(2):1311–1318
Tischer PCSF, Sierakowski MR, Westfahl H, Tischer CA (2010) Nanostructural reorganization of bacterial cellulose by ultrasonic treatment. Biomacromolecules 11(5):1217–1224
Tripp SL, Pusztay SV, Ribbe AE, Wei A (2002) Self-assembly of cobalt nanoparticle rings. J Am Chem Soc 124(27):7914–7915
Tsukamoto J, Durán N, Tasic L (2013) Nanocellulose and bioethanol production from orange waste using isolated microorganisms. J Braz Chem Soc 24:1537–1543
Van HL, Thuc CNH, Thuc HH (2013) Synthesis of silica nano particles from Vietnamese rice husk by sol–gel method. Nanoscale Res Lett 8:58. doi:10.1186/1556-276X-8-58
Wang S, Cheng Q (2009) A novel process to isolate fibrils from cellulose fibers by high-intensity ultrasonication part 1: process optimization. J Appl Polym Sci 113(2):1270–1275
Wang B, Sain M (2007) Isolation of nanofibers from soybean source and their reinforcing capability on synthetic polymers. Compos Sci Technol 67(11–12):2521–2527
Wang XZ, Hu Z, Chen X, Chen Y (2001) Preparation of carbon nanotubes and nanoparticles by microwave plasma-enhanced chemical vapor deposition. Sci Mater 44:1567–1570
Wang Y, Zhang X, He X, Zhang W, Zhang X, Lu C (2014a) *In situ* synthesis of MnO_2 coated cellulose nanofibers hybrid for effective removal of methylene blue. Carbohydr Polym 110:302–308
Wang Z, Fang C, Megharaj M (2014b) Characterization of iron polyphenol nanoparticles synthesized by three plant extracts and their fenton oxidation of azo dye. ACS Sustain Chem Eng 2(4):1022–1025
Wei H, Rodriguez K, Renneckar S, Vikesland PJ (2014) Environmental science and engineering applications of nanocellulose-based nanocomposites. Environ Sci Nano 1:302–316
Wen ZQ, Li G, Ren D (2011) Detection of trace melamine in raw materials used for protein pharmaceutical manufacturing using surface-enhanced Raman spectroscopy (SERS) with gold nanoparticles. Appl Spectrosc 65(5):514–521
Wen X, Chen X, Tian N, Gong J, Liu J, Rümmeli MH, Chu PK, Mijiwska E, Tang T (2014) Nanosized carbon black combined with Ni_2O_3 as "Universal" catalysts for synergistically catalyzing carbonization of polyolefin wastes to synthesize carbon nanotubes and application for supercapacitors. Environ Sci Technol 48:4048–4055
Williams DF (1999) The Williams dictionary of biomaterials. Liverpool University Press, Liverpool
Worathanakul P, Payubnop W, Muangpet A (2009) Characterization for post treatment effect of bagasse ash for silica extraction. Int J Chem Mol Nucl Mater Metall Eng 3(8):398–400
Wu X, Lu C, Zhou Z, Yuan G, Xiong R, Zhang X (2014) Green synthesis and formation mechanism of cellulose nanocrystal-supported gold nanoparticles with enhanced catalytic performance. Environ Sci Nano 1(1):71–79
Xia Y, Yang H, Campbell CT (2013) Nanoparticles for catalysis. Acc Chem Res 46(8):1671–1672
Xiao Y, Long C, Zheng M, Dong H, Lei B, Zhang H, Liu Y (2014) High-capacity porous carbons prepared by KOH activation of activated carbon for super capacitors. Chin Chem Lett 25(6):865–868
Xie K, Zhao W, He X (2011a) Adsorption properties of nano-cellulose hybrid containing polyhedral oligomeric silsesquioxane and removal of reactive dyes from aqueous solution. Carbohydr Polym 83(4):1516–1520
Xie K, Jing L, Zhao W, Zhang Y (2011b) Adsorption removal of Cu^{2+} and Ni^{2+} from waste water using nano-cellulose hybrids containing reactive polyhedral oligomeric silsesquioxanes. J Appl Polym Sci 122(5):2864–2868
Yadav S, Kumar D, Sinha S (2014) Chemical carbonization of papaya seed originated charcoals for sorption of Pb(II) from aqueous solution. J Environ Chem Eng 2:9–14
Yalcin N, Sevinc V (2001) Studies on silica obtained from rice husk. Ceram Int 27:219–224

Yallappa S, Manjanna J, Dhananjaya BL (2015) Phytosynthesis of stable Au, Ag and Au–Ag alloy nanoparticles using *J. sambac* leaves extract, and their enhanced antimicrobial activity in presence of organic antimicrobials. Spectrochim Acta A Mol Biomol Spectrosc 137:236–243

Yang XG, Li C, Wang W, Yang BJ, Zhang SY, Qian YTA (2004) Chemical route from PTFE to amorphous carbon nanospheres in supercritical water. Chem Commun 3:342–343

Yoon K, Hsiao BS, Chu B (2008) Functional nanofibers for environmental applications. J Mater Chem 18(44):5326–5334

Yu J, Ahn J, Zhang Q, Yoon SF, Rusli LYJ, Gan B, Chew K, Tan KH (2002) Catalyzed growth of carbon nanoparticles by microwave plasma chemical vapor deposition and their field emission properties. J Appl Phys 91:433–436

Yu SJ, Kang MW, Chang HC, Chen KM, Yu YC (2005) Bright fluorescent nanodiamonds: no photobleaching and low cytotoxicity. J Am Chem Soc 127(50):17604–17605

Yu X, Tong S, Ge M, Wu L, Zuo J, Cao C, Song (2013) Adsorption of heavy metal ions from aqueous solution by carboxylated cellulose nanocrystals. J Environ Sci 25:933–943

Zain NFM, Yusop SM, Ahmad I (2014) Preparation and characterization of cellulose and nanocellulose from pomelo (*Citrus grandis*) albedo. J Nutr Food Sci 5(1):334. doi:10.4172/2155-9600.1000334

Zavaglia CA, Prado da Silva MH (2016) Biomaterials. Ref Module Mater Sci Mater Eng. doi:10.1016/B978-0-12-803581-8.04109-6

Zhang L, Fang M (2010) Nanomaterials in pollution trace detection and environmental improvement. Nano Today 5:128–142

Zhang L, Bai X, Tian H, Zhong L, Ma C, Zhou Y, Chen S, Li D (2012) Synthesis of antibacterial film CTS/PVP/TiO_2/Ag for drinking water system. Carbohydr Polym 89(4):1060–1066

Zhao QL, Zhang ZL, Huang BH, Peng J, Zhang M, Pang DW (2008) Facile preparation of low cytotoxicity fluorescent carbon nanocrystals by electrooxidation of graphite. Chem Commun 7(41):5116–5118

Zheng B, Lu C, Gu G, Makarovski A, Finkelstein G, Liu J (2002) Efficient CVD growth of single-walled carbon nanotubes on surfaces using carbon monoxide precursor. Nano Lett 2:895–898

Zhou C, Lee S, Dooley K, Wu Q (2013a) A facile approach to fabricate porous nanocomposite gels based on partially hydrolyzed polyacrylamide and cellulose nanocrystals for adsorbing methylene blue at low concentrations. J Hazard Mater 263:334–341

Zhou Z, Lu C, Wu X, Zhang X (2013b) Cellulose nanocrystals as a novel support for CuO nanoparticles catalysts: facile synthesis and their application to 4-nitrophenol reduction. RSC Adv 3(48):26066–26073

Zuluaga R, Putaux JL, Cruz J, Velez J, Mondragon I, Ganan P (2009) Cellulose microfibrils from banana rachis: effect of alkaline treatments on structural and morphological features. Carbohydr Polym 76(1):51–59

Zyubin AS, Mebel AM, Hayashi M, Chang HC, Lin SHJ (2009) Quantum chemical modeling of photo adsorption properties of the nitrogen-vacancy point defect in diamond. J Comput Chem 30(1):119–131

Nanomaterials: Implications on Agroecosystem

4

M.R. Davari, S. Bayat Kazazi, and O. Akbarzadeh Pivehzhani

Abstract

Nanotechnology is one of the novel discoveries which are being explored in all fields. Nowadays the use of nanotechnology in various industries including agriculture and pharmaceuticals has attracted the attention of many researchers. The role of this technology in agriculture sector (crop management, crop improvement, nanofertilizer, nanoherbicides, and nanopesticides) is examined in this study; nanofertilizer has played a more effective role than others. They can increase nutrition and reduce soil toxicity. The growth of conventional herbicide-resistant weed species can be prevented using nanoherbicides.

Keywords

Nanotechnology • Nano fertilizer • Agriculture • Crop management • Nano herbicides

4.1 Introduction

Nano comes from the Greek word "nanos," meaning "dwarf." Particles between 1 and 100 nanometers in size are considered as nanoparticles (Thakkar et al. 2010). It has been reported that nanoparticles have unique properties including high surface area to volume ratio and nanometer regime. These characteristics have developed

M.R. Davari (✉)
PNU, Tehran, I.R. of Iran
e-mail: mrdavari@gmail.com

S. Bayat Kazazi
Young Professionals for Agricultural Development (YPARD Iran), Arak, Iran

O. Akbarzadeh Pivehzhani
PETRONAS University, Ipoh, Malaysia

R. Prasad et al. (eds.), *Nanotechnology*, DOI 10.1007/978-981-10-4573-8_4

considerably the use of these materials. New interdisciplinary venture into agriculture and food sciences is provided by nanotechnology where science and engineering have converged together. A significant contribution to agricultural research has been promised by it, which can lead to propose useful solutions for solving numerous agricultural problems. Nanoparticles are widely used in various fields like agriculture, viz., detection of pollutants, plant diseases, pests, and pathogens and controlled delivery of pesticides, fertilizers, nutrients, and genetic material, and can act as nano-architects in formation and binding of soil structure (Ghormade et al. 2011; Prasad et al. 2014, 2017).

One of the materials considered at the leading edge of rapidly developing field of nanotechnology is nanomaterials. National Nanotechnology Initiative (NNI), as a research and development (R&D) program has reported that nanotechnology research and development is directed toward understanding and creating improved materials, devices, and systems that exploit nanoscale properties. Nanotechnology is considered as a novel technology which can be widely used in various fields like electronics energy, remediation, automobile, and space technology. It was reported that nanotechnology has a great potential for using in biological and medical fields; gene studies and drug making industries, tissue engineering, biosciences or biotechnology and diagnostic studies (Borm et al. 2006; Oberdörster et al. 2005; Prasad 2014; Aziz et al. 2015, 2016; Prasad et al. 2016). New interdisciplinary venture into agriculture and food sciences is provided by nanotechnology where science and engineering have converged together. A significant contribution to agricultural research has been promised by it, which can lead to propose useful solutions for solving numerous agricultural problems (Rai et al. 2016).

According to the European Commission, there are six key enabling technologies (KETs) that can be commonly used in various fields like medicine, biotechnology, electronics, materials science, and energy technologies, and nanotechnology is considered as one of these technologies. At atomic, molecular, and macromolecular scales, nanotechnology makes possible the use of phenomena and fine-tuning of materials. The unique properties of nanomaterials, attracting large-scale research investments, have caused manufacturers and producers trying to be innovative in this sector. High importance of technical innovation in agriculture is discussed in this regard. These innovations can be directed toward addressing global challenges like population growth, changes in climate, and the limited accessibility of key plant nutrients like phosphorus. For this purpose, nanotechnology used to agricultural production could play a key role, and in recent decades, research on agricultural applications increased. Using technology-oriented small- and medium-sized enterprises, like soil-enhancer products, it can be seen that some particular nano-products for the agricultural field have been put on the market. However, since high cost is needed to development it, the commercial market application of these products is so far only achieved at small scale. Although, it has been reported that higher efficiency could compensate these costs in some sectors like medical but is not true for the agricultural sector. In order to evaluate useful properties in future, the number of research in the commercial agrichemical sector is increasing (Vigani and Rodríguez-Cerezo 2014).

Today, nanotechnology has been widely used in many agricultural fields. It can be wieldy used in different stages such as production, processing, storing, packaging, and transport of agricultural products (Mousavi and Rezai 2011; Ditta 2012). Today, nanotechnology can be used to produce fertilizer which has attracted many attentions in agriculture sector. By improving fertilizer products, Nanotechnology can have a significant impact on energy, the economy and environment (DeRosa et al. 2010).

4.2 Nanotechnology and Agroecosystem

In the twenty-first century, agriculture has faced many challenges to produce food and fiber needed to meet growing population needs. Given that the number of labor force in villages has become smaller and climate is changing, agricultural production is associated with some problems. Since the population will be increased over nine billion by 2050, human beings will face with more problems in satisfying their needs. In different fields, including agriculture, nanotechnology has extraordinary potential to make easier the next stage of precision farming methods. Agricultural potential will be increased using nanotechnology in such a way that we might expect higher yields to harvest in the future in eco-friendly way even in challenging environment (Prasad 2014). At the global level, the potential of nanotechnology is recognized by many countries in the agri-food sector and is investing a significant amount on it. Nanotechnology plays a key role to meet the ever increasing population needs with declining natural resources (Ali et al. 2014).

4.2.1 Crop Improvement

There is a need for novel tools for molecular and cellular biology that are specifically designed for separation, identification, and quantification of individual genes and molecules (Warad and Dutta 1995). The genes can be delivered to specific sites at cellular levels using nanotechnology. Also it has been reported that the atoms can attain a new arrangement in the DNA of the same organism to achieve the desired feature using this technology. The genetic constitution of the crop plants can be modified using nanotechnology; therefore it can help further improvement. Mutations, both natural and induced mutations, have played an important role in crop improvement since now. In the past, certain chemical compounds and physical mutagens played a key role for conventionally induced mutation studies, but now using nanotechnology, we can examine a novel aspect in this field (Chinnamuthu and Boopathi 2009a, b).

According to Khodakovskaya et al. (2009), penetrated carbon nanotubes have a positive effect on tomato seeds as an increase in their germination efficiencies is reported. The seed germination was increased due to water uptake ability of CNTs.

It was reported that the speed of spinach growth can be increased by Nanoparticles of TiO_2 which could increase Rubisco activase activity and improving light absorbance (Yang et al. 2006). The growth of spinach can be improved using TiO_2 nanoparticles by improving nitrogen metabolism. According to DeRosa et al. (2010), inhibition of seed germination in corn and rye grass was reported using nanoparticles of ZnO. But these left porous domains in plant roots, hence letting a potential nutrient delivery system to be explored.

4.2.2 Potential Applications in Plant Systems

Nanoparticles are widely used in various fields including medical sciences. Nobel Prize winner P. Ehrlich introduced nanoparticles in 1906 as "magic bullets." Using nanoparticles, the damage to nontarget plant tissues can be reduced and release of nonspecific chemicals into the environment can be minimized (González-Melend et al. 2008). It has been reported that delivery of proteins or codelivery of proteins and DNA to plant cells can play a key role in the biological field to improve genetic transformation and gene targeting in plants (Wu et al. 2011). Similar to medical sciences, identical rules could be considered for different applications, in particular to make determined efforts to deal with phytopathological infections (Park et al. 2006) and sustained release of nutrients and growth promoters (Cui et al. 2011). Advancements in techniques in nanotechnology make it possible to deliver agrochemicals into targeted plants, tissues, and organs by using nanoparticles as carriers.

4.2.3 Crop Management

There are many important fields to increase the productivity of crops like site-specific crop management. It benefits from using inputs in necessary quantity and in essential time. Nanotechnology can enable fine sensors and monitoring systems, and these systems have significant effect on future precision agriculture methodologies. Precision agriculture is used, aimed to maximize output using at least input (i.e., fertilizers). It monitors environmental variables and can help reduce the agricultural waste and minimize environmental pollution (Doug et al. 2009). The use of autonomous sensors linked into a GPS system for real-time monitoring is considered as one of the key roles for nanotechnology-enabled devices. Distribution of these nanosensors could be done all over the field where soil conditions and growth of crop can be monitored (Chinnamuthu and Boopathi 2009a, b).

4.3 Nanomaterials and Nanofertilizers

Application of nanotechnology in plant science and agriculture is significantly increasing. As a result of progress in nanotechnology, novel methods are proposed for production nanoparticles of physiologically important metals in large scale. These methods can be used to modify fertilizer formulations to increase uptake in plant cells in such a way that nutrient loss be minimized. High surface area, sorption capacity, and controlled-release kinetics to targeted sites are some unique features of nanoparticles. These are making these materials to be considered as "smart delivery system." The nutrient use efficiency can be increased using nano-structured fertilizers through mechanisms such as targeted delivery and slow or controlled release. It has been reported that their active ingredients are precisely released in response to environmental triggers and biological demands. According to the lab scale investigations, improving crop productivity by increasing the seed germination rate, seedling growth, photosynthetic activity, nitrogen metabolism, and carbohydrate and protein synthesis is considered as some result of using nanofertilizers (Rai et al. 2016).

Nanofertilizer is defined as a product in nanometer structure that delivers nutrients to crops. For example, a thin protective polymer film or in the form of particles or emulsions of nanoscale dimensions can be used to cover encapsulation inside nanomaterials (DeRosa et al. 2010). Adding nanomaterials in order to create appropriate coating on the surface of fertilizer particles has caused the material to be stronger because it has higher surface tension than the conventional surfaces and release can be done in controlled manner (Brady and Weil 1999). As a key aspect of the use of nanotechnology in agriculture, we can note the delivery of agrochemical substance to the plants. According to Table 4.1, release of agrochemicals in a controlled manner, site-targeted delivery, reduced toxicity, and improving nutrient utilization of delivered fertilizers are some of advantages of using nanofertilizers (Cui et al. 2010; Prasad et al. 2017).

4.3.1 Importance and Role of Nanofertilizers in Improvement of Nutrient Use Efficiency

In recent years, application of nanotechnology in the practical areas has attracted more attention compared with the experimental fields (Baruah and Dutta 2009). For example, the development of release fertilizers slowly and in a controlled manner and conditional release of pesticides and herbicides, based on nanotechnology, has been considered as a key factor to improve the development of environment-friendly and sustainable agriculture. Indeed, nanotechnology has provided the feasibility of exploiting nanoscale or nanostructured materials as fertilizer carriers or controlled-release vectors for building of so-called smart fertilizer as novel facilities to improve nutrient use efficiency and minimize costs of environmental protection (Cui et al. 2011; Chinnamuthu and Boopathi 2009a, b). Fertilizers can be encapsulated within a nanoparticle which is considered as one of these new facilities and done in three ways: (a) encapsulation of the nutrient can be done within nanoporous materials, (b) a thin polymer film can be used for coating, or (c) it can be delivered as particle or emulsions of nanoscale dimensions (Rai et al. 2012). It should be noted that combined nanodevices will be used by nanofertilizers in such a way that the release of fertilizer N and P and their uptake by crops be done at the same time and it can prevent undesirable nutrient losses to soil, water, and air via direct internalization by crops and avoid the interaction of nutrients with soil, microorganisms, water, and air (DeRosa et al. 2010).

Table 4.1 Comparison of nanotechnology-based formulations and conventional fertilizers applications (Cui et al. 2010)

S. No	Properties	Nanofertilizers-enabled technologies	Conventional technology
1.	Solubility and dispersion of mineral micronutrients	Nano-sized formulation of mineral micronutrients may improve solubility and dispersion of insoluble nutrients in soil, reduce soil absorption and fixation, and increase the bioavailability	Less bioavailability to plants due to large particle size and less solubility
2.	Nutrient uptake efficiency	Nanostructured formulation might increase fertilizer efficiency and uptake ratio of the soil nutrients in crop production and save fertilizer resource	Bulk composite is not available for roots and decrease efficiency
3.	Controlled-release modes	Both release rate and release pattern of nutrients for water soluble fertilizers might be precisely controlled through encapsulation in envelope forms of semipermeable membranes coated by resin-polymer, waxes, and sulfur	Excess release of fertilizers may produce toxicity and destroy ecological balance of soil
4.	Effective duration of nutrient release	Nanostructured formulation can extend effective duration of nutrient supply of fertilizers into soil	Used by the plants at the time of delivery, the rest is converted into insoluble salts in the soil
5.	Loss rate of fertilizer nutrients	Nanostructured formulation can reduce loss rate of fertilizer nutrients into soil by leaching and/or leaking	High loss rate by leaching, rain off, and drift

4.3.2 Advantage of Nanofertilizer

Many applications in various areas of agriculture and biotechnology have been reported due to the appearances of nanotechnology. Nanostructured formulization could release their active ingredients in responding to environmental triggers and biological demands in exact terms by mechanisms such as targeted delivery or release mechanisms in controlled manner. These mechanisms can be used for nanofertilizer design. Using fertilizers in nano-size, their efficiency can be increased, soil toxicity can be reduced, and the potential negative effects associated with over-dosage will be minimized. The use of fertilizers in nano-size mainly delays the release of the nutrients and extends the fertilizer effect period (Naderi and Danesh-Shahkari 2013).

4.3.3 Different Types of Nanofertilizer

4.3.3.1 Nitrogen Fertilizers

The necessary elements for plant growth other than carbon, oxygen, and hydrogen, from the soil, must be provided. Nutrients refer to these necessary elements, and macronutrients refer to those needed in the greatest amount, whereas those needed in lesser amounts are called micronutrients. Nitrogen, phosphorus, and potassium are considered as macronutrients. These three elements can be removed from the soil by plants quickly. There are many commercial plant fertilizers which can supply these three necessary elements. N-P-K numbers indicate the amount of each element. There are many sources of nitrogen used in fertilizers, including ammonia (NH_3), diammonium phosphate ($(NH_4)_2HPO_4$), ammonium nitrate (NH_4NO_3), ammonium sulfate ($(NH_4)_2SO_4$), calcium cyanamide ($CaCN_2$), calcium nitrate ($Ca(NO_3)_2$), sodium nitrate ($NaNO_3$), and urea (N_2H_4CO). Phosphorus is generally supplied as a phosphate, such as diammonium phosphate ($(NH_4)_2HPO_4$) or calcium dihydrogen phosphate ($Ca(H_2PO_4)_2$). Potassium comes from potassium sulfate (K_2SO_4) or potassium chloride (KCl), which is also called muriate of potash (Shakhashiri 2010).

4.3.3.2 Potash Fertilizers

Potash fertilizer as natural substance can be taken by the plants as K+, and it can help in photosynthesis process, controlling water storage, and stomata opening in leaves. Polyacrylamide-based coating of pellets was used for release of potash fertilizer slowly. Mixture of potash and clay was dried for an hour and a toothpaste was used for coating it, in order to have suitable binding to the polymer. This polymer was dipped in polyacrylamide polymer (Rameshaiah et al. 2015).

4.3.3.3 Zinc Nanofertilizer

Micronutrient zinc deficiency has been reported as a serious issue in the world. Daily food can supply very less amount of zinc; therefore there are least chances of indirect supply to human using fertilizer containing zinc, for the same nanoparticles can be used to coat zinc in order to get diffused and soluble zinc (Milani et al. 2010). According to studies, solubility of zinc decreases by increasing the pH (Bickel and Killorn 2001). For design of nanoparticles, identical ratios between their surface area should be considered designed, if not, it can have a negative effect on total solubility of the zinc.

4.3.3.4 Nanoporous Zeolite

There are many strategies to increase the efficiency of the fertilizer usage, using Nano clays and zeolites are some strategies, which they are a group of naturally minerals with a honeycomb-like layered crystal structure for increasing efficiency (Chinnamuthu and Boopathi 2009a, b). Nitrogen, potassium, phosphorous, calcium, and a complete set of minor and trace nutrients can be used to fill its network. They can be as nutrients which are released slowly "according to the demand of nutrients". However, as leggo (2000) explains, the main application of zeolites in agriculture is related to nitrogen capture, storage, and release. There are many reasons for groundwater contamination, including application of soluble N fertilizers. Nitrogen-releasing dynamics of the ionic form is much faster than absorbed form (in zeolites).

4.4 Nanoherbicides

Fist, nanotechnology only has been just used in medicine and pharmacology fields, and after that, its application has just developed in crop protection. Technologies related to release in a controlled and encapsulated manner have changed fundamentally the use of herbicides and pesticides. "Smart Seed" refers to seeds imbibed with nanoencapsulations with specific bacterial strain. Nanoparticles as smart delivery systems can play a key role to target and upload substances at specific areas within whole plants (González-Melendi et al. 2008; Corredor et al. 2009). The control of parasitic weeds with nanocapsulated herbicides reducing the phytotoxicity of herbicides was reported by Pérez-de-Luque and Rubiales (2009). Various types of herbicide formulations, with emphasis on controlled-release formulations, microencapsulation, and systemic application, are discerned to increase the possibilities of their various modes of action including in conjunction with nanoparticle carrier against parasitic weed (Dhillon and Mukhopadhyay 2015).

4.5 Nanopesticide

One of the major factors to limit crop yields is plant pests. There are some methods to control conventional pest including the use of over-the-counter pesticides in large quantity where an additional cost in crop production is added. The use of pesticides higher than necessary also can increase environmental and water pollution. It should be noted that minimum amount of pesticides should be used in order to save the environment and reduce the cost in crop production (Sharon et al. 2010) For this purpose, an increase in the retention time of pesticides to control insect pests of cotton, rice soybeans, and peanuts can be raised. Synthetic insecticide lambda-cyhalothrin is considered as active ingredient of this product which is released on contact with leaves.

Pesticides in nano-size can include either very small particles of pesticidal active ingredients or other small engineered structures with useful pesticidal properties (Bergeson 2010b). The dispersion of agricultural formulations and unwanted pesticide movement can be increased using nanopesticides (Bergeson 2010a). Many exclusive properties have been reported about nanomaterials and biocomposites. Stiffness permeability, crystallinity, thermal stability, solubility, and biodegradability (Bouwmeester et al. 2009; Bordes et al. 2009) needed for formulating nanopesticides are considered as some of these important advantages. High specific surface is another feature of the nanopesticides, and hence affinity to the target can be increased using these materials (Jianhui et al. 2005).

4.6 Developing New Nanopesticides

In the recent years, many efforts have been made to manage plague insects, for example, using biological control, which being time-consuming is considered as one of its disadvantages. Release systems in a controlled manner can be considered in this scenario as a very attractive alternative in this battle field. Controlled-release formulations associate the active compound with inert materials. Protecting and managing the rate of compound release into the target site in a defined period of time are considered as some important tasks of the last ones. Controlled-release systems are used, aimed to rule the (bio)availability of the active compound after the application (Wilkins 2004). Due to many progresses that have been reported in the nanotechnology, most of those controlled-release biopesticide applications were and still are successfully made. For decades, formulations based on nanomaterials have been able to obtain a special place. The first microcapsule-based formulation became commercially available in the 1970s (Fanger 1974).

4.7 Food Packaging

Degrading conventional food packaging materials is difficult, and it causes serious waste problems as solid waste material. Although there are many solutions for this problem including the use of biomass-based material in food packaging, their performance and cost-effectiveness are still raised as challenge (Siracusa et al. 2008; Farris et al. 2009). It has been reported that the use of nanomaterials in packaging biopolymers like cellulose and its derivative polyesters (Darder et al. 2007), plant oils and gelatins (Bordes et al. 2009; Sinha and Bousmina 2005), can provide necessary mechanical strength, better reinforcement, and barrier properties (Choudalakis and Gotsis 2009). According to the studies, lighter, stronger plastics with better heat resistance and barrier properties can be achieved using nanocomposites (Azeredo 2009). One of the materials which have been successfully used in food packaging is nanoclays and silicates and hydrated alumina-silicate layered clay (Weiss et al. 2006).

In food packaging, benefiting from nanotechnology can help make it stronger, lighter, or perform better. One of the materials which can be used in packaging to prevent spoilage of foods is antimicrobials such as nanoparticles of silver or titanium dioxide. Durethan is a transparent plastic film which can be produced by chemical giant Bayer produces. Durethan contains nanoparticles of clay. Durethan is considered as an engineering plastic based on polyamide 6 and polyamide 66; an excellent combination of properties is provided using these particles. These include high strength and toughness, abrasion resistance, chemical resistance, and resistance to cracking. Application of Durethan is not limited to special industries, and it can be widely used in various fields including packaging film for the medical field and food packaging. The nanoparticles are spread throughout the plastic and have the ability to block oxygen, carbon dioxide, and moisture from reaching fresh meats or other foods. One of positive features of use of nanoclay is it also makes the plastic lighter, stronger, and more heat resistant. Durethan film material with nanoparticles combines the advantages of polyamide 6 and ethylene vinyl alcohol to produce an inexpensive but still very airtight packaging material. The embedded nanoparticles prevent gases from penetrating the film and also keeping moisture from escaping (Lam and Smolander 2010).

4.7.1 Silver Nanoparticles and Nanocomposites as Antimicrobial Food Packaging Materials

For a long time, silver as an antimicrobial agent can be used in food and beverage storage applications. In the past, silver vessels were used to store wine and water. According to studies, silver has had many applications in various fields. There are many reliable reports of early settlers placing silver dollars or silver spoons at the bottom of milk and water bottles to prolong shelf-life and of seafaring ships or airliners lining their water tanks with silver to keep water potable for long periods of time. It has been reported that silver as the sterilization agent for water had many applications in the Russian MIR space station and on NASA space shuttles (Silver 2003). It should be noted that silver's broad-spectrum antimicrobial activity and relative low cost have made it a candidate as the active disinfecting agent for water in developing countries (Solsana and Mendez 2003; Corbett and River 2008; Aziz et al. 2015, 2016). The food additive regulations were modified by FDA in 2009 in order to allow adding silver nitrate as a disinfectant to commercially bottled water at concentrations less than 17 lg/kg (Fig.4.1).

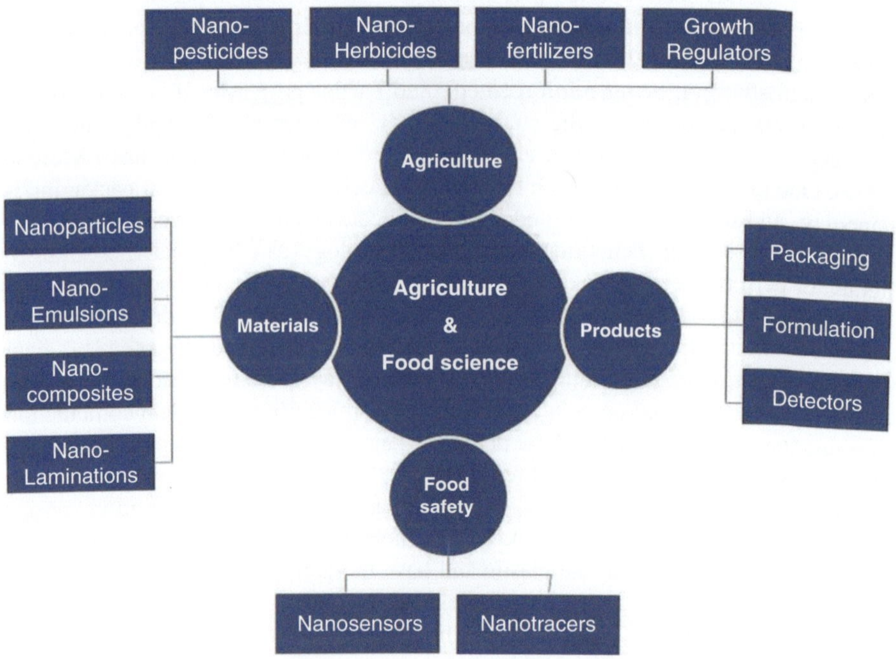

Fig. 4.1 A pictorial demonstration of applications of nanotechnology in agri-food sector

4.8 Conclusions

Nanotechnology is considered as an effective factor on the environment. Nanotechnology has many applications in various fields, including agriculture sector. The quality of life can be improved using nanotechnology, also it can improve quality of agricultural products and food for community. In the future, this technology will lead to dramatic changes around the world. Like any other new technology, this technology should be used with caution and probable and unexpected items be considered. Though, creating skilled prospect manpower for this novel technology is discussed as a significant issue for the future of a state. Therefore, the people must first be aware of the benefits of this technology, which will increase the awareness and innovation of novel applications in all fields.

References

Ali MA, Rehman I, Iqbal A, Din S, Rao AQ, Latif A, Samiullah TR, Azam S, Husnain T (2014) Nanotechnology, a new frontier in agriculture. Adv Life Sci 1(3):129–138

Azeredo H (2009) Nanocomposites for food packaging applications. Food Res Int 42(9):1240–1253

Aziz N, Faraz M, Pandey R, Sakir M, Fatma T, Varma A, Barman, Prasad R (2015) Facile algae-derived route to biogenic silver nanoparticles: synthesis, antibacterial and photocatalytic properties. Langmuir 31:11605–11612

Aziz N, Pandey R, Barman I, Prasad R (2016) Leveraging the attributes of *Mucor hiemalis*-derived silver nanoparticles for a synergistic broad-spectrum antimicrobial platform. Front Microbiol 7:1984. doi:10.3389/fmicb.2016.01984

Baruah S, Dutta J (2009) Nanotechnology applications in sensing and pollution degradation in agriculture. Env Chem Lett J 7:191–204

Bergeson LL (2010a) Nanosilver: US EPA's pesticide office considers how best to proceed. Environ Qual Manag 19(3):79e85

Bergeson LL (2010b) Nanosilver pesticide products: what does the future hold. Environ Qual Manag 19(4):73e82

Bickel A, Killorn R (2001) Spatial response of corn to banded zinc sulfate fertilizer in Iowa, 31st North central Extension Industry Soil Fertility Conference, Potash and Phosphate Institute, 605/692–6280, 17:14–15

Bordes P, Pollet E, Avérous L (2009) Nano-biocomposites: biodegradable polyester/nanoclay systems. Prog Polym Sci 34:125e155

Borm PJA, Robbins D, Haubold S, Kuhlbusch T, Fissan H, Donaldson K, Schins R, Stone V, Kreyling W, Lademann J, Krutmann J, Warheit D, Oberdörster E (2006) The potential risks of nanomaterials: a review carried out for ECETOC. Part Fibre Toxicol 3:11

Bouwmeester H, Dekkers S, Noordam MY, Hagens WI, Bulder AS, de Heer C, ten Voorde SECGS, Wijnhoven WP, Marvin HJP, Sips AJAM (2009) Review of health safety aspects of nanotechnologies in food production. Regul Toxicol Pharmacol 53:52e62

Brady NR, Weil RR (1999) In: Brady NR, Weil RR (eds) The nature and properties of soils. Prentice Hall, New Jersey, pp 415–473

Chinnamuthu CR, Boopathi PM (2009a) Nanotechnology and agroecosystem. Madras Agric J 96:17–31

Chinnamuthu CR, Boopathi PM (2009b) Review nanotechnology and agroecosystem. Madras Agric J 96(1–6):17–31

Choudalakis G, Gotsis A (2009) Permeability of polymer/clay nanocomposites: a review. Eur Polym J 45(4):967–984

Corbett S, Ron Rivera (2008) Solution in a pot. The New York Times.

Corredor E, Testillano PS, Coronado MJ, González-Melendi P, Fernández-Pacheco R, Marquina C, Ibarra R, de-la-Fuente JM, Rubiales D, Perez-de-Luque A, Risueno MC (2009) Nanoparticle penetration and transport living pumpkin plants: in situ subcellular identification. BMC Plant Biol 9:45

Cui HX, Sun CJ, Liu Q, Jiang J, Gu W (2010) Applications of nanotechnology in agrochemical formulation, perspectives, challenges and strategies. In: International conference on Nanoagri, Sao Pedro, Brazil, pp 28–33

Cui H, Jiang J, Liu Q (2011) On plant nutrition smart delivery systems and precision fertilization. Acta Metall Sin 17:494–499

Darder M, Aranda P, Ruiz-Hitzky E (2007) Bionanocomposites: a new concept of ecological, bioinspired and functional hybrid materials. Adv Mater 19(10):1309–1319

DeRosa MC, Monreal C, Schnitzer M, Walsh R, Sultan Y (2010) Nanotechnology in fertilizers. Nat Nanotechnol 5(2):91

Dhillon NK, Mukhopadhyay SS (2015) Nanotechnology and allelopathy: synergism in action. J Crop Weed 11(2):187–191

Ditta A (2012) How helpful is nanotechnology in agriculture? Adv Nat Sci Nanosci Nanotechnol 3:10

Doug R, Luvall JC, Shaw J, Mask P, Kissel D, Sullivan D (2009) Precision agriculture: changing the face of farming. http://www.agiweb.org/geotimes/nov03/feature_agric.html

Fanger GO (1974) Microencapsulation: a brief history and introduction. In: Vandegaer JE (ed) Microencapsulation: process and applications. Plenum Press, New York, pp 1–20

Farris S, Schaich KM, Liu L, Piergiovanni L, Yam KL (2009) Development of polyion-complex hydrogels as an alternative approach for the production of bio-based polymers for food packaging applications: a review. Trends Food Sci Technol 20(8):316–332

Ghormade V, Deshpande MV, Paknikar KM (2011) Perspectives for nano-biotechnology enabled protection and nutrition of plants. Biotechnol Adv 29:792–803

González-Melendi P, Fernández-Pacheco R, Coronado MJ, Corredor E, Testillano PS et al (2008) Nanoparticles as smart treatment-delivery systems in plants: assessment of different techniques of microscopy for their visualization in plant tissues. Ann Bot 101:187–195

Jianhui Y, Kelong H, Yuelong W, Suqin L (2005) Study on anti-pollution nanopreparation of dimethomorph and its performance. Chin Sci Bull 50(2):108e112

Khodakovskaya M, Dervishi E, Mahmood M, Xu Y, Li Z et al (2009) Carbon nanotubes are able to penetrate plant seed coat and dramatically affect seed germination and plant growth. ACS Nano 3(10):3221–3227

Lam D, Smolander M (2010) International forum on emerging technologies in food, pp 222–223

Leggo PJ (2000) An investigation of plant growth in an organo-zeolitic substrate and its ecological significance. Plant Soil 219:135–146

Milani N, McLaughlin M, Hettiaratchchi GM, Beak DK, Kirby JK, Stacey S (2010) Fate of nanoparticulate zinc oxide fertilisers in soil: solubility, diffusion and solid phase speciation. Soil Solutions Chang World:1–6

Mousavi SR, Rezaei M (2011) Nanotechnology in agriculture and food production. J Appl Environ Biol Sci 1(10):414–419

Naderi MR, Danesh-shahraki A (2013) Nanofertilizers and their roles in sustainable agriculture. J Agric Crop Sci JACS 5–9:2229–2232

Oberdörster G, Oberdörster E, Oberdörster J (2005) Nanotoxicology: an emerging discipline evolving from studies of ultrafine particles. Environ Health Perspect 113:823–839

Park HJ, Sung HK, Hwa JK, Seong HC (2006) A new composition of nanosized silica-silver for control of various plant diseases. Plant Pathol J 22:295–302

Pérez de Luque A, Rubiales D (2009) Nanotechnology for parasitic plant control. Pest Manag Sci 65:540–545

Prasad R (2014) Synthesis of silver nanoparticles in photosynthetic plants. J Nanopart Article ID 963961. http://dx.doi.org/10.1155/2014/963961
Prasad R, Kumar V, Prasad KS (2014) Nanotechnology in sustainable agriculture: present concerns and future aspects. Afr J Biotechnol 13(6):705–713
Prasad R, Pandey R, Barman I (2016) Engineering tailored nanoparticles with microbes: quo vadis. WIREs Nanomed Nanobiotechnol 8:316–330
Prasad R, Bhattacharyya A, Nguyen QD (2017) Nanotechnology in sustainable agriculture: recent developments, challenges, and perspectives. Front Microbiol 8:1014. doi:10.3389/fmicb.2017.01014
Rai V, Acharya S, Dey N (2012) Implications of nanobiosensors in agriculture. J Biomaterials Nanobiotechnol 3:315–324
Rai M, Ribeiro C, Mattaso L, Duran N (2016) Nanotechnology in food and agriculture. Springer, Cham. doi:10.1007/978-3-319-14024-7_4
Rameshaiah GN, Pallavi J, Shabnam S (2015) Nano fertilizer and nano sensor- an attempt for developing smart agriculture. J Eng Res Gen Sci 3:314–320
Shakhashiri (2010) Apply a fertilizer to the soil to keep cultivated plants healthy. 103-1. www.scifun.org
Sharon M, Choudhary AK, Kuma R (2010) Nanotechnology in agricultural diseases and food safety. J Phytology 2(4):83–92
Silver S (2003) Bacterial silver resistance: molecular biology and uses and misuses of silver compounds. FEMS Microbiol Rev 27:341–353
Sinha Ray S, Bousmina M (2005) Biodegradable polymers and their layered silicate nanocomposites: in greening the 21st century materials world. Prog Mater Sci 50(8):962–1079
Siracusa V, Rocculi P, Romani S, Rosa MD (2008) Biodegradable polymers for food packaging: a review. Trends Food Sci Technol 19(12):634–643
Solsana F, Mendez JP (2003) Water disinfection, Pan American Center for Sanitary Engineering and Environmental Sciences. Pan American Health Organization, Lima
Thakkar MN, Mhatre S, Parikh RY (2010) Biological synthesis of metallic nanoparticles. Nanotechol Biol Med 6:257–262
Vigani M, Rodríguez-Cerezo E (2014) Nanotechnology for the agricultural sector: from research to the field, Claudia Parisi, [work shop]
Warad HC, Dutta J (1995) Nanotechnology for agriculture and food systems-a view microelectronics. School of Advanced Technologies, Asian Institute of Technology, Thailand. http://www.nano.ait.ac.th
Weiss J, Takhistov P, McClements DJ (2006) Functional materials in food nanotechnology. J Food Sci 71(9):R107–R116
Wilkins RM (2004) Controlled release technology, agricultural. In: Seidel A (ed) Kirk-Othermer encyclopedia of chemical technology, 5th edn. Wiley, New Jersey
Wu G, Xiao M, Yang C, Yu YT (2011) U2 snRNA is inducibly pseudouridylated at novel sites by Pus7p and snR81 RNP. EMBO J 30(1):79–89
Yang F, Hong F, You W, Liu C, Gao F et al (2006) Influence of nano-anatase TiO_2 on the nitrogen metabolism of growing spinach. Biol Trace Elem Res 110(2):179–190

Nanoagrotechnology for Soil Quality, Crop Performance and Environmental Management

5

Jeyabalan Sangeetha, Devarajan Thangadurai, Ravichandra Hospet, Etigemane Ramappa Harish, Prathima Purushotham, Mohammed Abdul Mujeeb, Jadhav Shrinivas, Muniswamy David, Abhishek Channayya Mundaragi, Shivasharana Chandrabanda Thimmappa, Suresh Basavaraj Arakera, and Ram Prasad

Abstract

Nanotechnology is emerging as the key enabling technology that contributes to increased crop production with special emphasis on soil protection with environmental sustainability. Increasing worldwide food security and challenging climatic conditions are the key components for encouraging the scientific community to focus on accelerating the growth of nanoagrotechnology. Last few decades immensely contributed to the field of agriculture; technological innovations by

J. Sangeetha
Department of Environmental Science, Central University of Kerala,
Kasaragod, Kerala 671316, India

D. Thangadurai (✉) • R. Hospet • P. Purushotham • A.C. Mundaragi
Department of Botany, Karnatak University, Dharwad, Karnataka 580003, India
e-mail: drthanga.kud@gmail.com

E.R. Harish • J. Shrinivas • M. David
Department of Zoology, Karnatak University, Dharwad, Karnataka 580003, India

M.A. Mujeeb • S.C. Thimmappa
Department of Microbiology and Biotechnology, Karnatak University,
Dharwad, Karnataka 580003, India

S.B. Arakera
Department of Marine Biology, Karnatak University PG Centre,
Karwar, Karnataka 581303, India

R. Prasad
Amity Institute of Microbial Technology, Amity University,
Noida, Uttar Pradesh 201303, India

R. Prasad et al. (eds.), *Nanotechnology*, DOI 10.1007/978-981-10-4573-8_5

several hybrid varieties, synthetic chemical compounds and advanced techniques of biotechnology are an integral part of this achievement. The present decade emerged as the "decade of nanoagrotechnology", as a new origin of agricultural developments through most groundbreaking scientific finding in the field.

Keywords
Aptamers • Carbon nanotube • Pesticides • Fertilizers • Agrochemicals • Smart dust technology

5.1 Introduction

In recent years, the use of nanotechnology has created massive interest in the field of agriculture (Kuzma and Verhage 2006; Renton 2006; Chaudhry et al. 2008; Chinnamuthu and Boopathi 2009; Huyghebaert et al. 2010; Naidoo and Kistnasamy 2015). Although internationally there is no unified definition approved (Lövenstam et al. 2010), nanotechnology widely used to obtain products measures the scales less than 100 nm in one dimension. The scale having lesser range comprises larger surface area and volume ratio; the organic and inorganic properties of substance fundamentally differ due to larger substances present in corresponding materials (Aziz et al. 2015; Prasad et al. 2016). Oftenly, nanomaterials exhibit variations in thermodynamic, magnetic and optical properties in smaller quantity when compared to bulk materials (Schnettler et al. 2013). Literally, this property leads to development of new prospects in all sectors. A wide array of nanotechnology-based applications were developed in agriculture to overcome limitations such as packaging quality, food safety and processing technology (Doyle 2006; Garber 2006; Nord 2009; Yada 2009; Miller 2010; Neethirajan and Jayas 2010; Prasad et al. 2014; Prasad 2016) and also to promote sustainable agriculture to produce better-quality food products throughout the world (Larkins et al. 2008; Gruère 2011; Prasad et al. 2017a). Interesting applications include the use of nanoporous zeolites to improve fertilizers efficiency, nanosensors to measure soil quality and smart dust tools for fertilizer delivery (Chinnamuthu and Boopathi 2009; Guillaume 2012). For food and water safety, tremendous research projects are underway, nanosilver or nanoclay products have been developed for improved water filtration and nanosensors are being developed to detect and help track food pathogens (Chinnamuthu and Boopathi 2009; Gruère 2011; Prasad et al. 2014). In agricultural industries, certain nanomaterials such as nanoparticles (NPs), nanoclays (NCs) and nanoemulsions (NEs) are used. A number of methods are available for their synthesis and have many applications in the agrifood sector. NPs are unique in its nature, where the physicochemical properties depend on its surface characteristics. Thus, the chemical compound determines only the chemical composition and purity of the component, whereas, the nanomaterials demand comprehensive characterization of the compound.

One amongst the many challenges of agriculture is to optimize production and minimize losses in the field as well as during transport and storage (Khater 2011).

The main loss in the cultivation is due to the action of insects and pests that can be prevented by insecticides obtained naturally from plants or minerals and the use of nanotechnological tools for the production of new formulations (Khater 2011; Gogos et al. 2012; Forim et al. 2013). Control-release system is designed to enhance the target specificity, regulating the action of active ingredients and to reduce its residual criteria (Risch and Reineccius 1995; De Oliveira et al. 2014). Nanotechnology has shown tremendous progress in the formulation of nanocompounds to improve the stability and effectiveness (Ghormade et al. 2011; Perlatti et al. 2013). Hence, such byproducts help to release the active compound to the respective site, and it provides ability to release molecules to the site of action. They can also help to minimize undesirable toxic effects on nontarget organisms, as well as improve physicochemical stability and prevent degradation of the active agent by microorganisms (Gogos et al. 2012; Perlatti et al. 2013). The fertility of the soil is maintained by the release and carrier systems, designed to control the diffusion, erosion, swelling or mixture of these (Pothakamuri and Barbosa-Cánovas 1995; Arifin et al. 2006; Tramon 2014), depending upon the mass-transfer system involved. Many different matrices can be used to produce nanostructured systems, including biodegradable polymers, and a variety of preparation techniques have been reported (Gogos et al. 2012; Perlatti et al. 2013; Prasad et al. 2017b). Nanotechnology, the phenomenal development in terms of environmental protection and risk management, thus holds promise for cleanup of hazardous waste. Rather than this, nanotechnology implies smart device for detecting the location of pathogens and to apply fertilizers for prevention of diseases, which disturbs the yield (Bergeson 2010). Nanosensors are innovative technique for detecting bacterial, viral and fungal pathogens in plants (Baac et al. 2006; Boonham et al. 2008; Yao et al. 2009; Chartuprayoon et al. 2010). Fluorescent silica nanoparticles were used by Yao et al. (2009) in addition to that antibody for diagnosing *Xanthomonas axonopodis* pv. *vesicatoria*, which is responsible for the bacterial spot disease in a Solanaceae member; it is the significant nanoparticle for detection of pathogens. Similarly, nano-gold-based immunosensors were used by Singh et al. (2010) – surface plasmon resonance (SPR) which recognizes Karnal bunt (*Tilletia indica*) disease in wheat. In spite of this, researchers have designed the SPR sensor for detection of plant disease, and it helps in seed certification and obtaining plant quarantine in wheat crops. Nano-chips, type of nanosensors known for its rapid diagnosis of pathogens and prevention of diseases, includes fluorescent oligo-probes for hybridization detection, which is well known for its sensitivity and particularly in detecting variation in single nucleotide of several microbes (López et al. 2009).

Major impacts on the environment are due to the exploitation of nanoscience and nanotechnology. To overcome most of environmental cleanup problems, the newly emerged environmental remediation technology represented by nanoscale particles helps to provide low cost-effective solutions. For *in situ*, the modified Fe nanoparticles provide large surface density, reactivity and enormous flexibility. Nanoscale iron particles are used to transform or eliminate environmental pollutants such as chlorinated compounds, organochlorine pesticides and PCBs (Zhang 2003). Modified Fe nanoparticles are catalyzed for synthesizing and for the enhancement of speed and efficacy of remediation; hence, this modified Fe particles have several advantages such as effective transformation of bulk environmental pollutants, cost-effectiveness and less toxicity. Recently, researchers have developed nanoscale iron particles for reduction and catalyzation of the environmental pollutants including chlorinated organic compound and heavy metal ions. Chlorinated contaminants can be dechlorinated completely within the water and soil-water slurries rapidly. For instance, with a nanoscale Pd and Fe particle dose at 6.25 gL^{-1}, all chlorinated compounds were reduced to below detectable limits. Ethane was the major product in all tests. Greater than 99% removal was achieved with nanoscale iron particle in 24 h. Many pesticides that are persistent in aerobic environments are more readily degraded under reducing conditions. Zerovalent iron (ZVI) is used as a chemical reductant for the application of such techniques (Tratnyek and Johnson 2006; Garner and Keller 2014). For the removal of heavy metals ions from the polluted waters, "magnetic" bacteria seem to be useful (e.g. Ag, Hg, Pb, Cu, Zn, Sb, Mn, Fe, As, Ni, Al, Pt, Pd and Ru). With the presence of magnetic ions such as iron sulphide, heavy metals were precipitated onto bacterial cell walls, by making the bacterial cell sufficiently magnetized for removal from suspension by magnetic separation protocol. Some of the bacteria were able to synthesize iron sulphide, which may act as adsorbent for many metallic ions. Synthesis of mesoporous magnetic nanocomposite particle can be used for the removal of harmful agents present in the environment. This new technique employs molecular templates for coating nanoparticles of magnetite with that of mesoporous silica.

Currently, nanotechnology is widely used in agriculture and environmental cleanup; still it requires qualitative analysis for the assessment of toxicity and behavioural changes in the environmental systems, as well as to gain long-lasting bioavailability and durability.

5.2 Nanotechnology for Production and Protection of Crop Plants

Nanotechnology is one of the encouraging fields of interdisciplinary research. It promotes a wide array of possibilities in scientific areas like electronics, pharmaceuticals, medicine and agriculture. The recent approaches in the development of nanotechnology with biotechnology remarkably expanded the potential applications of nanoparticles in various fields of agriculture. Applications include insect pest control through the effective formulations of nanomaterial-based insecticides and pesticides, nanoparticle-mediated gene transfer in plants for the production of insect pest-resistant varieties, increase of agricultural outputs using bio-conjugated nanoelements in a steady release of water and nutrients and use of nanoparticles in making various kind of biosensors, which could help for remote sensing devices needed for site-specific crop management (Bhattacharyy et al. 2016a, b). Nanomaterials like metal-, metal-oxide- and carbon-based polymers and nanomaterials of biocomposites are being well developed (Nair et al. 2010). Their types include single-walled and multi-walled carbon nanotubes and silver, aluminium, gold, copper, silica, titanium dioxide, cerium oxide, zinc and zinc oxide nanoparticles. A wide application of these nanomaterials were observed in environmental remediation, water purification, water treatment, food processing, industrial and household needs, pharmaceutical purposes and smart sensor development (Jain 2005; Chau et al. 2007; Wei et al. 2007; Byrappa et al. 2008; Gao and Xu 2009; Qureshi et al. 2009; Zhang and Webster 2009; Lee et al. 2010; Bradley et al. 2011). The focused applications in these areas have well contributed to the improvement of agricultural production and protection (Bouwmeester et al. 2009; Emamifar et al. 2010; Nair et al. 2010; Sharon et al. 2010).

The demand for food and global population growth has led to optimizing agricultural production with minimizing losses (Khater 2011). The harmful effects of pests and insects are the major cause of crop losses, which can be minimized by way of applying natural botanicals as well as by implicating nanotechnology (Khater 2011; Gogos et al. 2012; Forim et al. 2013). Hence, there are a number of research works which are ongoing particularly on the effective and eco-friendly use of nanotechnologies in the agriculture field. Nanotechnology offers great promise in overcoming problems related to environmental impacts and target specificity of pest and insecticide and optimizing quality product yields. Nanotechnological formulation is applied to optimize the stability and effectiveness of various natural products (Ghormade et al. 2011; Perlatti et al. 2013). These nanoformulations offer the capacity to release the active compound to the particular organism and further provide controlled release of molecules at the specific site of action and also minimize the toxic effects on nontargeted organisms, as well as improve physicochemical stability and prevent degradation of the active compounds by microorganisms (Duran and Marcato 2013; Perlatti et al. 2013). Effective application of nanomaterials as fertilizer has been noticed (Raliya and Tarafdar 2013). Various nanoparticle compounds, for the most part, carbon-based nanomaterials and metal-based nanoparticles, have been utilized for their assimilation, accumulation, translocation and, more importantly, impacts on development and advancement of crop plants (Nair et al. 2010; Rico et al. 2011; Sekhon 2014). The acceptable morphological effects include improved germination rate and percentage, the vegetative biomass of seedlings and length of root and shoot. The enhanced physiological parameters observed as improved nitrogen metabolism and photosynthetic activity by metal-based nanoparticles in few crops, including soybean (Agrawal and Rathore 2014), peanut (Giraldo et al. 2014) and spinach (Zheng et al. 2005; Linglan et al. 2008), were reported. The existence of magnetic fluid in the maize seeds pertains a significant improvement in the nucleic acid level because of the regeneration reactions occurring in plant metabolism (Racuciu et al. 2009). Magnetic nanomaterials coated with that of tetramethylammonium hydroxide help to accelerate chlorophyll-a level in maize (Racuciu and Creanga 2006). In pumpkin application of iron oxide found to improve root elongation, and that ascribed to the iron cessation (Wang et al. 2011). Nanopesticides are one of the upcoming effective tools used to solve the problems of non-nanoparticles (Sasson et al. 2007). It covers a wide variety of quality nanoproducts; some of them are getting good marketing popularization also. Hence, some of the nanoformulations combine along with many polymers, the surfactants and certain nanoparticles in nanometer size. In the development of agro-compounds for crop protection, the scarcity of water solubility is observed to be one of the major limiting factors. Microencapsulation is a versatile technique for water-repellent pesticides for delivery of active components (Ragaei and Sabry 2014; Sekhon 2014). Organic polymers often used in the production of nanoparticle have been reported (Perlatti et al. 2013).

Nanomaterials serve similarly as additives and active constituents (Gogos et al. 2012). The controlled-release (CR) formulations of imidacloprid produced from polyethylene glycol and aliphatic acids through encapsulation show better control than that of commercial formulations against epidemic pests of soybean (whitefly and stem fly) (Sekhon 2014). The poly-amphiphilic polymer-based formulation exhibits potential performance for determining yellow mosaic virus transmitted through whitefly and stem fly incidence (Sekhon 2014). In addition to this, some of the improved CR formulations recorded increased yield over the control and commercial formulation (Adak et al. 2012a, b). The CR formulations along with imidacloprid and carbofuran examined as one of the efficient pests against the leafhopper and to the aphid when compared to any other conventional formulations. The residue of imidacloprid and carbofuran in the soil and potato tuber was not seen during the period of harvesting in any of the formulations (Kumar et al. 2011). Nanoparticles including iron oxide, gold, polymeric nanoparticles and silver particles are used as nanopesticides. Various concepts of nanoparticle formulation, effects, characterization and their applications in plant pest control are observed (Al-Samarrai 2012).

The potential use of nanoparticle in insect pest management has been successfully documented (Bhattacharyya et al. 2010, 2016a, b). In the control of polyphagous pest, *Helicoverpa armigera*, notable application of nanoparticles has been reported (Vinutha et al. 2013). Synthesized silver nanoparticles reveal excellent mosquito larvicidal and antilice activity (Jayaseelan et al. 2011; Ragaei and Sabry 2014; Sekhon 2014). Nanoencapsulation helps to promote the gradual release of chemical compounds for a distinct host for the management of insect pest by releasing some of the activities such as diffusion, biodegradation and osmotic pressure (Vidyalakshmi et al. 2009). The acceptable application of amorphous nanosilica as a pesticide has been found in several agricultural findings (Barik et al. 2008). Nanocopper is a modified nanoparticle which is suspended in water and used in a known compound, Bouisol, as fungicide for some grape varieties and other fruit crops. Because of mutagenesis, viral capsids could be changed to achieve several elements like the production of some enzymes and nucleic acids to act against parasites (Perez-de-Luque and Rubiales 2009). Nanoparticles with silver at 100 mg/kg restrain growth of mycelium and germination of conidia on cucurbits and inhibit the growth rate of powdery mildew in pumpkins (Lamsal et al. 2011). As one of the elite nanopesticides, silver nanoparticles show a significant application in agriculture practices (Afrasiabi et al. 2012). Treatment of mulberry leaves affected by grasserie disease is done by application of the ethanolic suspension of hydrophobic alumina-silicate nanoparticles, which has remarkably minimized the viral load (Goswami et al. 2010). DNA-tagged gold nanoparticles are powerful for *Spodoptera litura* and hence considered as a functional component for integrated pest management (Chakravarthy et al. 2012).

Antifungal activities of silica-silver nanoparticles against *Botrytis cinerea*, *Colletotrichum gloeosporioides* and *Rhizoctonia solani* (Park et al. 2006), silver nanoparticles against *Fusarium oxysporum* and *Aspergillus flavus* (Aziz et al. 2016) and polymer-based copper nanocomposites against pathogenic fungi have been well documented (Cioffi et al. 2004). Copper nanoparticles found in soda-lime glass powder expressed effective antimicrobial activity against gram-positive and negative bacteria (Esteban-Tejeda et al. 2009). Weed control is an essential part of enriching the productivity of any crop, and effective use of nanoherbicides seems to be an economically important substitute. The nano-silicon carrier consisting diatom frustules has been utilized for delivery of herbicides and pesticides in crop plants (Lodriche et al. 2013). The effective activity of fungicidal sulphur nanoparticles on two phytopathogens, *Venturia inaequalis* that is accountable for the apple scab disease and *Fusarium solani* responsible for early blight in tomato leaf, was reported (Rao and Paria 2013). For minimizing undesirable pest populations, the application of pheromones is one of the promising eco-friendly management to gain crop quality, for example, nanogel produced from pheromone called methyl eugenol. A detrimental pest for several of fruit crops, *Bactrocera dorsalis*, is effectively managed by the use of nanogelled pheromone (Bhagat et al. 2013). Scientists observed prominent effectiveness of nanoparticles with alumina as insecticide against two insect pests, *Rhyzopertha dominica* and *Sarocladium oryzae*. These pests are considered as major insect pests in preserved food supplies. Hence, compared to commercially available insecticides, nanostructured alumina can offer reliable and cost-effective substitute for control of insect pests (Stadler et al. 2012). The most notable use of nanofertilizers in crop production makes very innovative and effective initiation to boost the agricultural yields (Sekhon 2014).

As an alternative for conventional mode of fertilizer, nanofertilizer provides a new way for the release of nutrients into the soil by the slow and controlled way, thus minimizing autrification and water pollution (Naderi and Abedi 2012). Effect of titanium dioxide (TiO_2) nanoparticles on maize exhibits considerable change in growth (Moaveni and Kheiri 2011). In an experiment, titanium dioxide and silicon dioxide (SiO_2) nanoparticles enhance the nitrate reductase activity in soybeans to strengthen plant absorption ability (Lu et al. 2002). Researchers have found environmentally sustainable nano-organic iron-chelated fertilizer (Iran Nanotechnology Initiative Council 2009). Nanofertilizers exhibit various unique features like increased production yield, improved photosynthesis activity and an ability of ultra-high absorption which leads to significant expansion of leaves in crop plants (Singh et al. 2016). The moderate application of nanofertilizer in agricultural practices not only enhances the crop quality but also increases the efficacy of the soil compounds, thereby minimizing the utilization of chemical fertilizers (Naderi and Danesh 2013; Sekhon 2014).

Nanotechnology is considered to be one of the potential sources for crop protection and production system. Use of nanotechnology could be a promising way for enhancing agricultural production. With the results of the maximum output of crop yield by using the minimum amount of fertilizer, inputs make it more farmer friendly. There are several approaches for the development of improved nanoformulation of agrochemicals; in the meantime, issues related to biosafety and interaction with plant, soil and environment clearly need improvement.

5.3 Nanosensors for Monitoring Soil Conditions and Environmental Stresses

Nanotechnology provides details of compound in a nanoscale range, based on its physical, organic and inorganic properties (Sadik et al. 2009). Continuous use of chemical fertilizers adversely affects the soil microbes and microfauna, and the plants, which further leads toxicity. The cost of nanofertilizers is economically low in price, and it requires a lesser amount when compared to chemical fertilizers. The main cause for improper yield was due to uptake of nitrogen, which is identified by the farmers, recently. Currently, a sensing device is utilized to overcome certain environmental issues. To rectify the disturbances to soil microbes comprises few assays, but it holds some limitations, such as, time consuming with high price to perform analysis; to avoid such issues sensing device is used, which displays the exact image on conditions of the field (Tothill 2001). Sensing devices are used to monitor the variations or effects, which are caused by several pesticides, insecticides and inorganic fertilizers; it also monitors the physical properties of soil, such as pH of soil, humidity and the growth conditions of crop plant, stem, fruit and even root, and instantly it can monitor the toxicity. Typically, sensors are human friendly; it helps in detection and it cautions farmers to carry out proper measures, which have to be taken before rather than acting for an effect later (Rameshaia et al. 2015).

5.3.1 Carbon Nanotube

Nanotubes are composed of carbon molecules and are cylindrical in shape with slight variation in terms of wall construction. This kind of multi-layered carbon nanotubes has played a major role in agricultural sector to obtain maximized growth and to improve germination and water uptake and enhance the nutrient uptake from the soil. Along with this, the implementation of several ranges of carbon nanotubes exhibits better yield with the presence of an external Fe supplement, and Ca ion helps in maintaining the quality of yield (Tiwari et al. 2014). This multi-layered carbon nanotubes with the concentration of 50 μg/ml were used to some crops like maize, wheat, peanut and garlic and gave better result by increasing the length of root and shoot, to allow the seeds to imbibe at time of germination, to enhance the growth and to obtain a well-characterized root system (Srivastava and Rao 2014). Addition of C nanotubes helps in retaining water content in plants and increasing the production rate vigorously to that of lesser amount of nanomaterials like 50 μg/ml; researchers have stressed the use of fullerene to increase the productivity of tomato, which is a phenomenal work in the agricultural sector (Husen and Siddiqi 2014).

5.3.2 Nanoaptamers

Aptamers are single-stranded nucleic acids, which bind to the pre-targeted molecules with high affinity, that fix into the target molecules to form three-dimensional molecules to produce exact bonding of the substance in vitro selection method (Tai-Chia and Chih-Ching 2009). Those kinds of sensors give accurate measurement to identify plant pathogens and resistance of crops and to enhance productivity rate. Photoluminescence, present in sensor, helps in signalling, without disturbing the cell; it helps to obtain the proper regulation of the system, i.e. insulin-binding aptamer designed to assess light extinctions from a specific region to get the signal. Luminescent assay technique is one of the promising aptamer sensors for the assessment of toxic content in food (McKeague et al. 2011) specifically to identify herbicide and pesticidal properties (atrazine and malachite green).

5.3.3 Smart Dust Technology

For monitoring of environmental hazards and energy usage, smart dust technology is developed; it almost detects everything in the surroundings like monitoring the temperature and tracking the traffic. The technology gains popularity due to its unique way of regulating the system. This tool is usually monitored with the help of computer network wirelessly; they are distributed in the field to perform the tasks, and the device is undetectable due to small-size transducers regardless of location of the sensor (Bawankar et al. 2012). The devices consist of micro-sized electrochemical sensors in the system. However, this sensor has promising result and great potential for sensing the environmental variations, automation and computing, but it still has few limitations such as impact on environment, toxicity and how far this will be helpful in the field of agriculture, health and environment (Rameshaiah et al. 2015).

5.3.4 Wireless Sensors

Wireless sensors hold a strong proof that, to monitor the activities, there is no need to be at the location where the process takes place. A wireless technology is designed for the same purpose; it not only requires point-to-point arrangements but also the amount field trials estimated. Those types of sensors maintain an increase in the growth of plants crops by constant monitoring of soil and environmental conditions. Such kind of sensors maintains optimal growth of the crop plants by continuously monitoring the soil and environmental conditions. Closed-circuit television (CCTV) is installed and used in the agricultural fields to cattle monitoring, rainwater harvesting and water quality checking. The data obtained from the CCTV can be stored and analyzed for future purposes.

5.4 Nanocapsules for Efficient Delivery of Pesticides, Fertilizers and Agrochemicals

Agroforestry ecosystems are greatly governed by interactions between biotic and abiotic components (Mittler 2006). Biotic factors such as crop plants, weeds (34%), insect pests (18%), disease-causing pathogens and nematodes (16%) had negative impacts on agricultural production (Patterson et al. 1999; Oerke 2006). In order to overcome this issue, agrochemical was developed, and its uses were randomly increased in the wake of green revolution. Agrochemicals are the inorganic chemical substances, which include fertilizers, pesticides, hormones, and other chemical growth agents that are intended to increase productivity rate by preventing crop loss during or after harvesting (Aktar et al. 2009). The green revolution brought inorganic synthetic chemicals, organic and inorganic pesticides, hybrid seeds and new irrigation technique made much more impact on agriculture sector in terms of yield and resistant well. Now the times have been changed – as the green revolution is not as green as it was earlier – it has now become a curse to environment and nontarget organisms, due its improper delivery of agrochemicals and management (Pepper 2011). Since the agrochemicals that were found have broad-range toxics and have detrimental effect on nontarget organism that reside in both terrestrial and aquatic ecosystems, it was needed to replace these agrochemicals and to develop an effective and more environmentally friendly agrochemical delivery system with the help of precision farming practices and effective application of nanotechnology to agriculture.

5.4.1 Targeted Delivery of Agrochemicals Using Nanotechnology

In recent times, nanotechnology emerged as a potential tool in the field of material science, biological science, chemical science, engineering sciences and space science (Nair et al. 2010; Rai and Ingle 2012). From the last decade, its uses and benefits are enormous in the field of agricultural science (Scott et al. 2003) by enhancing the agricultural productivity with replacing conventional agricultural practices (Ghormade et al. 2011; Gogos et al. 2012; Khot et al. 2012; Rai and Ingle 2012). Safe application of conventional agrochemicals is a major concern, due to a number of problems associated with them. For example, >90% of applied agrochemicals are lost and unable to reach the target area; it may be influenced by a number of factors including techniques used, physicochemical properties of agrochemicals and environmental conditions (Mogul et al. 1996; Perlatti et al. 2013). The losses are due to emission, leaching, evaporation, degradation due to photolysis, hydrolysis and by microorganisms. In addition to their loss, it may cause pollution to the environment and toxicity to nontarget organism (van den Berg et al. 1999; Bedos et al. 2002; Nuruzzaman et al. 2016).

The conventional agrochemical practices are replaced by various nano-based formulations that are similar to conventional formulations developed by several scientists to overcome such issues with improved features. These include increased rate of solubility, stability, permeability, biodegradability, improved nutrient use efficiency and decreased rate of agrochemical spreading with uniform dispersion (Kah et al. 2013; Kah 2015). These nano-based formulations contain nanomaterials which can work as carrier material for agrochemicals and exhibited useful properties such as stability, solubility, stiffness and crystallinity and may release efficient dosage of water and nutrients for the purpose of pest detection; management resulted in better agricultural yield (Bordes et al. 2009).

For specific applications of agrochemicals, researchers have possessed several kinds of nano-based insecticides, and fertilizers, for improving agrochemical activities, simultaneously by retaining the environmental impact to a minimum. Most of these formulations include the structures having the nanometer range with minimum amount of pesticide ingredient, along with the precised nanopore network with surfactant. Such nano-based products have long desired goal to manage the agricultural inputs and to reduce the impact of modern agriculture (Kah 2015).

5.4.2 Nano-based Pesticides in Agriculture

Conventional pesticide practices are replaced by microemulsion formulation method for the first time by the method developed by Schulman et al. (1959) and commercially available in the 1970s (Fanger 1974). Nanocapsules disperse the agrochemicals in uniform spherical droplets of either oil or water in an appropriate continuous phase. These have very low surface tension, cost-effective approach, droplet in size and clear, transparent and thermodynamically stable dispersion of oil water and stabilized by interfacial film of surfactant frequently in combination with cosurfactant (Lawrence and Rees 2000; De et al. 2014). Later, several researchers encapsulate pesticide by using variety of cost-effective materials with larger surface area, higher stability and solubility and easily biodegradable nanomaterials such as polymer-based nanomaterials, block polymers, solid lipid nanoparticles, inorganic porous nanomaterials, nanoclays and layered double hydroxides forming different types of nanomaterials such as nanocapsules for herbicides, nanospheres, micelles, nanogels, liposomes and inorganic nanocages (Perez-de-Luque and Rubiales 2009; Kah et al. 2013; Nuruzzaman et al. 2016).

5.4.3 Nano-based Fertilizer Efficiency in Agriculture

Commonly used fertilizers may enhance the agricultural yield, but the availability of nutrients present in sprayed agrochemicals is not fully accessible to plants, of this about 40–70% of nitrogen, 80–90% of phosphorus and 50–90% of potassium contents were lost (Subramanian et al. 2015). The lost chemicals may reach environment through leaching, drift, runoff and evaporation and become fixed in soil and contribute to air, water and soil pollution and may disturb the soil mineral balance and to decrease soil fertility (Solanki et al. 2015). To minimize these pollution and nutrient losses, smarty delivery systems are developed by using nanomaterials (Mukal et al. 2009; Nair et al. 2010). Such nanostructured fertilizers could increase the efficacy of nutrients by using various mechanisms, such as targeted delivery and controlled release so that it could spray the respective constituents in response to environmental impacts and biological requirement. Solanki et al. (2015) revealed that they help to increase the crop productivity rate by increasing the germination in seed, growth of seedlings, net photosynthetic rate, nitrogen metabolism and carbohydrate and protein synthesis in the respective plant crop (Chen and Yadav 2011; Subramanian et al. 2015). The data reveals that the nanofertilizers used against the crops to reach the optimum requirement are just a ppm level per acre; this technology is economically reasonable, and it is safe for living beings and is eco-friendly in nature.

5.5 Improving Plant Traits Against Environmental Stresses Using Nanotechnology

Recent developments in plant biotechnology have revolutionized the agricultural sector to overcome environmental stresses, including cold, drought, diseases and salinity, due to increasing global population, and to meet the food requirements agricultural intensification has adopted with wide usage of harmful chemical pesticides. On contrary, the modern agriculture practices have been facing several challenges that need to be addressed immediately. The extensive use of inorganic pesticides and fertilizers has raised concerns over major populations that they have severe adverse effects on environment and health of animals. Though in recent past, several technologies such as nanotechnology and genetically engineering have been developed to overcome these constraints; however, they have been not implemented into the field clearly indicating that risk assessment is not yet been evaluated. Global warming is another major abiotic stress factor that has a negative impact on the plant growth and crop productivity. Due to which, soil microorganisms are adversely affected, wherein natural enzymes are destroyed. Subsequently, uptake of macro- and micronutrients affected hence more and more fertilizers poured into the agriculture fields for better crop yield (Nair et al. 2010). Several abiotic stresses such as drought, water deficit and high salinity result in decreased crop productivity; Nanotechnology is an emerging field of science that could solve these problems with ease. It also aids in monitoring and delivering deficient nutrients, including the fertilizers and pesticides at the specific affected part of the plant, thus increasing and improving the overall plant growth and crop yield. Several nanoparticles have been evaluated for its effect on germination, seed growth, nutrient efficiency and uptake of fertilizers. Recent advancements in instrumentation technology have aided in evaluation and monitoring the behaviour of nanoparticles in plants. Several studies indicate that nanoparticles enhance the biological activity (Khan et al. 2016).

The potential applications of nanotechnology in agricultural sciences are increasing day by day, and it has been studied in various streams including production, protection and improvement of crops. During the last two decades, several studies highlighted the potential applications of nanoparticles against plant stress;

nanoparticles such as zinc oxide (Torabian et al. 2016), silicon (Qados and Moftah 2015), titanium oxide (Hong et al. 2005; Gao et al. 2006; Jaberzadeh et al. 2013), copper (Adhikari et al. 2012), silver (Yin et al. 2012; Hojjat 2016) and carbon nanotubes (Pourkhaloee et al. 2011) have shown promising role in overcoming the plant environment stress-related problems (Monica and Cremonini 2009).

Soluble phosphates are extensively used in agriculture as fertilizers, and their uncontrolled usage has resulted in eutrophication, and on the contrary, other phosphates are not effective in fulfilling deficiency. Recent study suggests that synthetic apatite nanoparticle enables the availability of phosphorus to crops with limited or less adverse effect on environment. Furthermore, the use of apatite nanoparticles as a phosphorous fertilizer can be beneficial; also, their report indicated an up to 32.6% and 20.4% increase in soybean crop yield when compared to conventional phosphorous fertilizer usage (Liu and Lal 2014; Raliya et al. 2016). More recent studies on the zinc oxide NPs on mung bean indicated that there was a significant increase in uptake of phosphate utilization and biosynthesized zinc oxide improved plant phenology including stem height and root volume. Jaberzadeh et al. (2013) research group recommended that application of titanium dioxide NPs at 0.02% concentration on wheat has increased the plant growth including gluten and starch content under water-deficit stress.

Cold stress causes loss of water content and seepage of solutes from the cell leading to deprived growth and seed germination, subsequently affecting crop yield. Application of nanoparticles such as selenium, titanium oxide and silicon oxide in combination with short chilling treatment has known to be beneficial with increased growth and physiological activities (Hawrylak-Nowak et al. 2010; Mohammadi et al. 2013; Azimi et al. 2014; Hasanpour et al. 2015; Kohan-Baghkheirati and Geisler-Lee 2015).

In recent past, several research groups are evaluating the potential applications and risks of silicon NPs in plants. Silicon increases drought resistance in plants and improves water-deficit stress tolerance (Pei et al. 2010), and it stimulates the root elongation and the physiological activity (Currie and Perry 2007; Epstein 2009; Ashkavand et al. 2015). Its deficiency has been shown to have abnormalities in plant structure when compared with silicon-rich counterparts; furthermore, silicon-deficient plants are more susceptible to biotic and abiotic stresses (Rafi et al. 1997; Ma 2004; Gao et al. 2005). Suriyaprabha et al. (2012) analyzed maize seeds treated with naturally synthesized (from rice husk) silica NPs. They reported that there was significant increase in germination percentage and root growth upon nano-SiO_2 treatments. A recent report on hawthorn seedlings suggests that silicon NPs were tended to be critical for physiological and biochemical functions under drought stress conditions (Ashkavand et al. 2015). Iron, titanium oxide and silver NPs are few amongst others that have been studied against drought stress conditions in safflower (Zareii et al. 2014), flax (Aghdam et al. 2016) and ajwain (Seghatoleslami et al. 2015), respectively.

Another study demonstrated that application of analcite nanoparticles to soil successfully enhanced seed germination, seedling growth and photosynthetic activity against drought conditions (Zaimenko et al. 2014). To overcome the salinity stress, nanosilica was applied to tomato seeds; application of silicon NPs had minimized the potential harm to the seed germination and root length under high salinity (Haghighi et al. 2012; Haghighi and Pessarakli 2013). Application of nano-silicon on plants such as lentils, sunflower and safflower has been reported (Sabaghnia and Janmohammadi 2014, 2015). Foliar application of nano-silicon on safflower gave positive results (Janmohammadi et al. 2016); foliar application of zinc oxide nanoparticles increased plant growth, CO_2 assimilation rate and chlorophyll content of *Helianthus annuus* L., under salt stress (Torabian et al. 2016). Nevertheless, several previous studies have shown the negative impacts of metal oxide nanoparticles on plant growth, seed germination and environment, thus causing phytotoxicity, cytotoxicity and genotoxicity (Lin and Xing 2008; AshaRani et al. 2009; Musante and White 2010; Goix et al. 2014; Ko and Kong 2014).

The distinct properties such as the size and surface charge of the nanoparticles enhance their easy penetration into cell membranes and targeted delivery of the materials. Several reports suggest that lower concentrations of NPs have a positive effect. On contrary, NPs applied at higher concentration have shown negative effect and cause cytotoxicity and lead to generation of reactive oxygen species. Many reports have cautioned the use of certain nanoparticles that may pose risk to environment and animals.

In conclusion, nanoparticles improve tolerance in plants towards several abiotic stresses. The mechanisms involved in the regulation and resistance have not been well understood, or the reports are scarce. Hence, further systematic investigations are necessary at the molecular level to understand the role and significance of nanoparticles at subcellular level. Suitable biosensor techniques have to be adopted to control the activities of nanoparticles and to evaluate and characterize stress inhibitors and stress inducer.

5.6 Nanotechnology and Its Applications in Water Conservation

From the perspective of necessity, water is considered as next oil. Pure potable water is depleting rapidly across the globe, affecting as many as 1.2 billion people every year (Montgomery and Elimelech 2007). In principle, unlike oil, water is a renewable natural resource which only confines to the cycle of evaporation, condensation and precipitation. Given the fact that out of 10^{18} gallons of fresh water on earth, the rate of its consumption exceeds its replenishment back into the cycle (Schoen et al. 2010). Scientists have already concluded that demand may exceed supply in just another two decades (Kim et al. 2010). Two main causes of scarcity of potable water across the globe are technical boundaries in distribution of water from centralized purification facility to the site of its consumption, and pollution and contamination of soft water bodies through natural and anthropogenic activities, which makes water unusable. Although, natural water cycles are capable of purifying itself, these cycles are not fast enough to effectively clean up the contaminated sites. On the other hand, traditional water purification processes are however being used in purification and supply of clean drinking water and have their own limitations. These limitations are due to centralized purification units consuming a large amount of energy and space. Moreover, there are poor supply channels from the site of purification to the site of consumption.

Nanotechnology, on the other hand, has promising solutions for addressing the issue of water purification in an efficient manner. The strategy includes on-site purification (minimizing cost of transportation and technical limitations with centralized purification units). Due to their unique physical properties and large surface area, nanomaterials have peculiar functionalities. Such unique functional attributes of nanomaterials are useful to develop novel materials like membranes, adsorption materials and catalysts for water treatment processes (Gehrke et al. 2015).

Membrane filters are pressure-driven channels across a semi-permeable membrane where particle matters larger than 0.5 nm are rejected back, which means particle impurities lesser than 0.5 nm still pass through membrane barriers. Nanofiltration membranes, on the other hand, are characterized by their charge-based selection and repulsion property and allow only water molecules to pass through membranes (Jagadevan et al. 2012; Sharma and Sharma 2012; Qu et al. 2013). These nanofilters prove useful in purification of ground hard water. Nanofilters can also be used in desalination of seawater where conventional membrane filters are cost intensive.

Nanoparticles have a larger surface area and, hence, are appropriate candidates for adsorption of many organic compounds. Any solid material which is used to adsorb gases or dissolved impurities which are in close physical contact are adsorbents. Traditional adsorbents such as activated carbon particle system may require regular backwash because of chock formation within or amongst carbon particles (Pan and Xing 2008). Nano-adsorbents have high specific surface area for many organic and inorganic pollutants like micropollutants and heavy metals. Due to their nanoparticle size, these systems do not chock and require less maintenance regimes.

Traditional biological process of water treatment to nullify the effect of organic and microbial pollution through oxidation demands large amount of infrastructure and energy. However, the use of catalyst to breakdown complex organic molecules is an emerging field with larger scope of practical application. Photocatalysis, on the other hand, is an advanced oxidation process for mitigating the organic toxic pollution from wastewater (Friedmann et al. 2010). According to several studies published (Fujishima et al. 2008; Gaya and Abdullah 2008; Chong et al. 2010), suspended complex organic wastes in water can be degraded by making use of photocatalysts. It is due to their well-known particle properties, low toxicity and cost efficiency. Titanium oxide (TiO) nanoparticle is one of the most frequently used photocatalyst to date (Qu et al. 2013). TiO nanoparticles, when irradiated with UV light with appropriate wavelength (200–300 nm) range, become photo-excited. This will immediately follow formation of electron hole pair triggering a chain reaction of oxido-reduction reactions. Increase in rate of reaction will only make degradation of heavy decomposable substances easier.

Nanotechnology, although, newly emerging and reliable technology for application in wastewater treatment, has its unforeseen limitations. Commercialization of nanomaterials for water treatment procedures strongly depends on its eco-toxic potentials on aquatic fauna. Many studies to ascertain toxicological endpoints, pathways of biotransformation and impact on life cycle of these nanoparticles on aquatic ecosystem have been carried out (EPA 2010; Clemente et al. 2011; Asghari et al. 2012). Apart from toxicological point of view, efficiency of nanoparticles in certain processes like catalytic system poses question marks. Generally, during tertiary water treatment processes, micropollutants like antibiotics and other suspended organic particles are removed in polishing processes. Whereas, ultraviolet radiation is only efficient about 5% of that of sunlight making its efficiency low in terms of industrial scales (Asghari et al. 2012). Nevertheless, nanotechnology in terms of emerging field of science has many more favourable and cost-efficient applications in wastewater treatments such that limitations in this field can be oversighted for now.

5.7 Conclusion and Future Perspectives

Nanotechnology is an emerging trend to contribute increased crop production with special emphasis on soil protection with environmental sustainability. This includes important technologies such as nanobiotechnology which are giving rise to a number of applications with more environmentally efficient outcomes in the areas of energy production, consumer goods, agricultural crops and information and communication technologies, which have the potential to address major environmental concerns or help to adapt to changing environmental conditions (e.g. due to climate change).

Research works on agricultural nanotechnology development and its applications have been progressing with hope for solutions to several environmental and agricultural problems. Due to its extremely small size, nanomaterial with its uncommon physical, chemical and biological properties, which are very distinct from their bulk materials and individual compounds, was accepted to be much compelling and safe.

Nanomaterials are showing most promising results in crop protection, soil improvement, disease diagnosis, plant breeding, water purification and soil conservation by acting as smart delivery systems for nutrients and other agrochemicals, indicators of soil and plant health and chelators to remove toxic substances from the soil.

In spite of these tremendous potentials and claims from academic institutions, small entrepreneurs and large-scale industries for patents on nanomaterials, no new nanomaterial-based products have reached the market. High initial investment is one of the major barriers to the production of large-scale nanomaterials for field applications. Another difficulty is regulatory issues and public opinion to release these nanomaterials to be tested at open and multicentre agricultural field trials, pointing its nanosize and lack of insights for its biological risk and physiological target sites.

Future challenges for the further development and field level application of this potential technology are designing methods for large-scale and cost-effective production of nanomaterials and formulating integrated robust top-down and bottom-up procedures to assess the hazards of nanomaterials to humans and other non-target organisms in the environment.

References

Adak T, Kumar J, Dey D, Shakil NA, Walia S (2012a) Residue and bio-efficacy evaluation of controlled release formulations of imidacloprid against pests in soybean (*Glycine max*). J Environ Sci Health B 47(3):226–231

Adak T, Kumar J, Shakil NA, Walia S (2012b) Development of controlled release formulations of imidacloprid employing novel nano-ranged amphiphilic polymers. J Environ Sci Health B 47(3):217

Adhikari T, Kundu S, Biswas AK, Tarafdar JC, Rao AS (2012) Effect of copper oxide nano particle on seed germination of selected crops. J Agric Sci Technol 2:815–823

Afrasiabi Z, Eivazi F, Popham H, Stanley D, Upendran A, Kannan R (2012) Silver nanoparticles as pesticides. In: Capacity building grants program project director's meeting. National Institute of Food and Agriculture, Huntsville. September 16–19

Aghdam MTB, Mohammadi H, Ghorbanpour M (2016) Effects of nanoparticulate anatase titanium dioxide on physiological and biochemical performance of *Linum usitatissimum* (Linaceae) under well-watered and drought stress conditions. Braz J Bot 39(1):139–146

Agrawal S, Rathore P (2014) Nanotechnology pros and cons to agriculture: a review. Int J Curr Microbiol App Sci 3(3):43–55

Aktar MW, Sengupta D, Chowdhury A (2009) Impact of pesticides use in agriculture: their benefits and hazards. Interdiscip Toxicol 2(1):1–12

Al-Samarrai AM (2012) Nanoparticles as alternative to pesticides in management plant diseases – a review. Int J Sci Res Publ 2(4):1–4

Arifin DY, Lee LY, Wang CH (2006) Mathematical modeling and simulation of drug release from microspheres: implications to drug delivery systems. Adv Drug Deliv Rev 58:1274–1325

Asghari S, Johari SA, Lee JH, Kim YS, Jeon YB, Choi HJ, Moon MC, Yu IJ (2012) Toxicity of various silver nanoparticles compared with silver ions in *Daphnia magna*. J Nanobiotech 10:14. doi:10.1186/1477-3155-10-14

AshaRani PV, Mun GLK, Hande MP, Valiyaveettil S (2009) Cytotoxicity and genotoxicity of silver nanoparticles in human cells. ACS Nano 3:279–290

Ashkavand P, Tabari M, Zarafshar M, Tomášková I, Struve D (2015) Nanoparticles on drought resistance in hawthorn seedlings. Leśne Pr Badawcze 76(4):350–359

Azimi R, Borzelabad MJ, Feizi H, Azimi A (2014) Interaction of SiO_2 nanoparticles with seed prechilling on germination and early seedling growth of tall wheat grass (*Agropyron elongatum* L.) Pol J Chem Technol 16:25–29

Aziz N, Faraz M, Pandey R, Sakir M, Fatma T, Varma A, Barman I, Prasad R (2015) Facile algae-derived route to biogenic silver nanoparticles: synthesis, antibacterial and photocatalytic properties. Langmuir 31:11605–11612. doi:10.1021/acs.langmuir.5b03081

Aziz N, Pandey R, Barman I, Prasad R (2016) Leveraging the attributes of *Mucor hiemalis*-derived silver nanoparticles for a synergistic broad-spectrum antimicrobial platform. Front Microbiol 7:1984

Baac H, Hajós JP, Lee J, Kim D, Kim SJ, Shuler ML (2006) Antibody-based surface plasmon resonance detection of intact viral pathogen. Biotechnol Bioeng 94(4):815–819

Barik TK, Sahu B, Swain V (2008) Nanosilica – from medicine to pest control. Parasitol Res 103(2):253–258

Bawankar SD, Bhople SB, Jaiswal VD (2012) Mobile networking for smart dust with RFID sensor networks. Int J Smart Sensors Ad Hoc Netw 2(3):62–66

Bedos C, Cellier P, Calvet R, Barriuso E (2002) Occurrence of pesticides in the atmosphere in France. Agronomie 22:35–49

Bergeson LL (2010) Nanosilver: US EPA's pesticide office considers how best to proceed. Environ Qual Manag 19(3):79–85

Bhagat D, Samanta SK, Bhattacharya S (2013) Efficient management of fruit pests by pheromone nanogels. Sci Rep 3:1294. doi:10.1038/srep01294

Bhattacharyya A, Bhaumik A, Rani PU, Mandal S, Epidi TT (2010) Nanoparticles – a recent approach to insect pest control. Afr J Biotechnol 9(24):3489–3493

Bhattacharyya A, Duraisamy P, Govindarajan M, Buhroo AA, Prasad R (2016a) Nano-biofungicides: emerging trend in insect pest control. In: Prasad R (ed) Advances and applications through fungal nanobiotechnology. Springer International Publishing, Cham, pp 307–319

Bhattacharyya A, Prasad R, Buhroo AA, Duraisamy P, Yousuf I, Umadevi M, Bindhu MR, Govindarajan M, Khanday AL (2016b) One-pot fabrication and characterization of silver nanoparticles using *Solanum lycopersicum*: an eco-friendly and potent control tool against Rose Aphid, *Macrosiphum rosae*. J Nanosci Article ID 4679410, 7 pages, http://dx.doi.org/10.1155/2016/4679410

Boonham N, Glover R, Tomlinson J, Mumford R (2008) Exploiting generic platform technologies for the detection and identification of plant pathogens. Eur J Plant Pathol 121:355–363

Bordes P, Pollet E, Avérous L (2009) Nano-biocomposites: biodegradable polyester/nanoclay systems. Prog Polym Sci 34:125–155

Bouwmeester H, Dekkers S, Noordam MY, Hagens WI, Bulder AS, Voorde GT, Sips A (2009) Review of health safety aspects of nanotechnologies in food production. Regul Toxicol Pharmacol 53(1):52–62

Bradley EL, Castle L, Chaudhry Q (2011) Applications of nanomaterials in food packaging with a consideration of opportunities for developing countries. Trends Food Sci Technol 22:604–610

Byrappa K, Ohara S, Adschiri T (2008) Nanoparticle synthesis using supercritical fluid technology – towards biomedical applications. Adv Drug Deliv Rev 60:299–327

Chakravarthy AK, Chandrashekharaiah KSB, Bhattacharya A, Dhanabala K, Gurunatha K, Ramesh P (2012) Bio efficacy of inorganic nanoparticles CdS, Nano-Ag and Nano-TiO_2 against *Spodoptera litura* (Fabricius) (Lepidoptera: Noctuidae). Curr Biotica 6(3):271–281

Chartuprayoon N, Rheem Y, Chen W, Myung NV (2010) Detection of plant pathogen using LPNE grown single conducting polymer nanoribbon. In: Proceedings of the 218th ECS meeting. Las Vegas. October 10–15, p 2278

Chau CF, Wu SH, Yen GC (2007) The development of regulations for food nanotechnology. Trends Food Sci Technol 18(5):269–280

Chaudhry Q, Scotter M, Blackburn J, Ross B, Boxall A, Castle L, Aitken R, Watkins R (2008) Applications and implications of nanotechnologies for the food sector. Food Add Contam 25(3):241–258

Chen H, Yadav R (2011) Nanotechnologies in agriculture: new tools for sustainable development. Trends Food Sci Technol 22:585–594

Chinnamuthu CR, Boopathi PM (2009) Nanotechnology and agroecosystem. Madras Agric J 96(1–6):17–31

Chong MN, Jin B, Chow CW, Saint C (2010) Recent developments in photocatalytic water treatment technology: a review. Water Res 44:2997–3027

Cioffi N, Torsi L, Ditaranto N (2004) Antifungal activity of polymer-based copper nanocomposite coatings. Appl Phys Lett 85(12):2417–2419

Clemente Z, Castro VL, Jonsson CM, Fraceto LF (2011) Ecotoxicology of Nano-TiO – an evaluation of its toxicity to organisms of aquatic ecosystems. Int J Environ Res 6:33–50

Currie HA, Perry CC (2007) Silica in plants: biological, biochemical and chemical studies. Ann Bot 100:1383–1389

De A, Bose R, Kumar A, Mozumdar S (2014) Targeted delivery of pesticides using biodegradable polymeric nanoparticles. Springer India, New Delhi, pp 59–81

De Oliveira JL, Campos EVR, Bakshi M, Abhilash PC, Fraceto LF (2014) Application of nanotechnology for the encapsulation of botanical insecticides for sustainable agriculture: prospects and promises. Biotechnol Adv 32(8):1550–1561
Doyle M (2006) Nanotechnology: a brief literature review. Food Research Institute, University of Wisconsin, Madison
Duran N, Marcato PD (2013) Nanobiotechnology perspectives, role of nanotechnology in the food industry: a review. Int J Food Sci Technol 48(6):1127–1134
Emamifar A, Kadivar M, Shahedi M, Soleimanian-Zad S (2010) Evaluation of nanocomposite packaging containing Ag and ZnO on shelf life of fresh orange juice. Innovative Food Sci Emerg Technol 11:742–748
EPA (2010) Nanomaterial case studies: nanoscale titanium dioxide in water treatment and in topical sunscreen. RTP Division, European Protection Agency, Washington, DC. http://cfpub.epa.gov/ncea/cfm/recordisplay.cfm?deid=230972. Accessed 11 Aug 2016
Epstein E (2009) Silicon: its manifold roles in plants. Ann Appl Biol 155:155–160
Esteban-Tejeda L, Malpartida F, Esteban-Cubillo A, Pecharromán C, Moya JS (2009) Antibacterial and antifungal activity of a soda-lime glass containing copper nanoparticles. Nanotechnology 20(50):505–701
Fanger GO (1974) Microencapsulation: a brief history and introduction. In: Vandegaer JE (ed) Microencapsulation: process and applications. Plenum Press, New York, pp 1–20
Forim MR, Costa ES, da Silva MFGF, Fernandes JB, Mondego JM, Junior ALB (2013) Development of a new method to prepare nano-microparticles loaded with extracts of *Azadirachta indica*, their characterization and use in controlling *Plutella xylostella*. J Agric Food Chem 61(38):9131–9139
Friedmann D, Mendiveb C, Bahnemann D (2010) TiO for water treatment: parameters affecting the kinetics and mechanisms of photocatalysis. Appl Catal B Environ 99:398–406
Fujishima A, Zhang X, Tryk DA (2008) TiO_2 photocatalysis and related surface phenomena. Surf Sci Rep 63:515–582
Gao J, Xu B (2009) Applications of nanomaterials inside cells. Nano Today 4:37–51. doi:10.1016/j.nantod.2008.10.009
Gao X, Zou C, Wang L, Zhang F (2005) Silicon improves water use efficiency in maize plants. J Plant Nutr 27:1457–1470
Gao F, Hong F, Liu C, Zheng L, Su M, Wu X, Yang F, Wu C, Yang P (2006) Mechanism of nano-anatase TiO_2 on promoting photosynthetic carbon reaction of spinach. Biol Trace Elem Res 111(1–3):239–253
Garber C (2006) Nanotechnology food coming to a fridge near you. http://www.nanowerk.com/spotlight/spotid=1360.php. Accessed 11 Aug 2016
Garner KL, Keller AA (2014) Emerging patterns for engineered nanomaterials in the environment: a review of fate and toxicity studies. J Nanopart Res 16:250–253
Gaya UI, Abdullah AH (2008) Heterogeneous photocatalytic degradation of organic contaminants over titaniumdioxide: a review of fundamentals, process and problems. J Photochem Photobiol A Chem 9:1–12
Gehrke I, Geiser A, Somborn-Schulz A (2015) Innovations in nanotechnology for water treatment. Nanotechnol Sci Appl 8:1–17
Ghormade V, Deshpande MV, Paknikar KM (2011) Perspectives for nano-biotechnology enabled protection and nutrition of plants. Biotechnol Adv 29(6):792–803. doi:10.1016/j.biotechadv.2011.06.007
Giraldo JP, Landry MP, Faltermeier SM (2014) Plant nanobionics approach to augment photosynthesis and biochemical sensing. Nat Mater 13(4):400–408. doi:10.1038/nmat3890
Gogos A, Knauer K, Bucheli TD (2012) Nanomaterials in plant protection and fertilization: current state, foreseen applications, and research priorities. J Agric Food Chem 60(39):9781–9792. doi:10.1021/jf302154y
Goix S, Lévêque T, Xiong TT, Schreck E, Baeza-Squiban A, Geret F, Uzu G, Austruy A, Dumat C (2014) Environmental and health impacts of fine and ultrafine metallic particles: assessment of threat scores. Environ Res 133:185–194

Goswami A, Roy I, Sengupta S, Debnath N (2010) Novel applications of solid and liquid formulations of nanoparticles against insect pests and pathogens. Thin Solid Films 519(3):1252–1257. doi:10.1016/j.tsf.2010.08.079

Gruère GP (2011) Labeling nano-enabled consumer products. NanoToday 6(2):117–121

Haghighi M, Pessarakli M (2013) Influence of silicon and nano-silicon on salinity tolerance of cherry tomatoes (*Solanum lycopersicum* L.) at early growth stage. Sci Hortic 161:111–117

Haghighi M, Afifipour Z, Mozafarian M (2012) The effect of N-Si on tomato seed germination under salinity levels. J Biol Environ Sci 6:87–90

Hasanpour H, Maali-Amiri R, Zeinali H (2015) Effect of TiO_2 nanoparticles on metabolic limitations to photosynthesis under cold in chickpea. Russ J Plant Physiol 62:779–787

Hawrylak-Nowak B, Matraszek R, Szymańska M (2010) Selenium modifies the effect of short-term chilling stress on cucumber plants. Biol Trace Elem Res 138:307–315

Hojjat SS (2016) The effect of silver nanoparticle on lentil seed germination under drought stress. Int J Farm Allied Sci 5(3):208–212

Hong F, Zhou J, Liu C, Yang F, Wu C, Zheng L, Yang P (2005) Effect of nano titanium oxide on phytochemical reaction of chloroplast of spinach. Biol Trace Elem Res 105(1):269–279

http://www.iranreview.org/content/Documents/Iranians_Researchers_Produce_Nano_Organic_Fertilizer.htm. Accessed 11 Aug 2016

https://fri.wisc.edu/files/Briefs_File/FRIBrief_Nanotech_Lit_Rev.pdf. Accessed 11 Aug 2016

Husen A, Siddiqi KS (2014) Carbon and fullerene nanomaterials in plant system. J Nanobiotechnol 12:16. doi:10.1186/1477-3155-12-16

Huyghebaert A, Van Huffel X, Houins G (2010) Nanotechnology in the food chain: opportunities and risks. Springer, Berlin

Iran Nanotechnology Initiative Council (2009) First nano-organic iron chelated fertilizer invented in Iran

Jaberzadeh A, Moaveni P, Moghadam THR, Zahedi H (2013) Influence of bulk and nanoparticles titanium foliar application on some agronomic traits, seed gluten and starch contents of wheat subjected to water deficit stress. Not Bot Hortic Agrobot 41(1):201–207

Jagadevan S, Jayamurthy M, Dobson P, Thompson IPA (2012) Novel hybrid nanozerovalent iron initiated oxidation biological degradation approach for remediation of recalcitrant waste metal working fluids. Water Res 46:2395–2404

Jain KK (2005) The role of nanobiotechnology in drug discovery. Drug Discov Today 10(21):1435–1442. doi:10.1016/S1359-6446(05)03573-7

Janmohammadi M, Amanzadeh T, Sabaghnia N, Ion V (2016) Effect of nano-silicon foliar application on safflower growth under organic and inorganic fertilizer regimes. Bot Lithuanica 22(1):53–64

Jayaseelan C, Rahuman AA, Rajakumar G, Vishnu Kirthi A, Santhoshkumar T, Marimuthu S, Bagavan A, Kamaraj C, Zahir AA, Elango G (2011) Synthesis of pediculocidal and larvicidal silver nanoparticles by leaf extract from heart leaf moon seed plant, *Tinospora cordifolia* Miers. Parasitol Res 109(1):185–194. doi:10.1007/s00436-010-2242-y

Kah M (2015) Nanopesticides and nanofertilizers: emerging contaminants or opportunities for risk mitigation? Front Chem 3:64. doi:10.3389/fchem.2015.00064

Kah M, Beulke S, Tiede K, Hofmann T (2013) Nanopesticides: state of knowledge, environmental fate, and exposure modeling. Crit Rev Environ Sci Technol 43:1823–1867

Khan MN, Mobin M, Abbas ZK, AlMutairi KA, Siddiqui ZH (2016) Role of nanomaterials in plants under challenging environments. Plant Physiol Biochem. doi:10.1016/j.plaphy.2016.05.038

Khater HF (2011) Ecosmart biorational insecticides: alternative insect control strategies. In: Parveen F (ed) Insecticides: advances in integrated pest management. InTech, Croatia, pp 780–782

Khot LR, Sankaran S, Maja JM, Ehsani R, Schuster EW (2012) Applications of nanomaterials in agricultural production and crop protection: a review. Crop Prot 35:64–70

Kim SJ, Ko SH, Kang KH, Han J (2010) Direct seawater desalination by ion concentration polarisation. Nat Nanotechnol 5:297–301

Ko KS, Kong IC (2014) Toxic effects of nanoparticles on bioluminescence activity, seed germination, and gene mutation. Appl Microbiol Biotechnol 98:3295–3303

Kohan-Baghkheirati E, Geisler-Lee J (2015) Gene expression, protein function and pathways of *Arabidopsis thaliana* responding to silver nanoparticles in comparison to silver ions, cold, salt, drought, and heat. Nanomaterials 5:436–467

Kumar J, Shakil NA, Khan MA, Malik K, Walia S (2011) Development of controlled release formulations of carbofuran and imidacloprid and their bioefficacy evaluation against aphid, *Aphis gossypii* and leafhopper, *Amrasca biguttula* (Ishida) on potato crop. J Environ Sci Health B 46(8):678–682. doi:10.1080/03601234.2012.592066

Kuzma J, Verhage P (2006) Nanotechnology in agriculture and food production: anticipated applications. Woodrow Wilson International Center for Scholars, Washington, DC. http://www.nanotechproject.org/process/assets/files/2706/94_pen4_agfood.pdf. Accessed 11 Aug 2016

Lamsal K, Kim SW, Jung JH, Kim YS, Kim KS, Lee YS (2011) Inhibition effects of silver nanoparticles against powdery mildews on cucumber and pumpkin. Mycobiology 39(1):26–32. doi:10.4489/MYCO.2011.39.1.026

Larkins B, Bringgs S, Delmer D, Dick R, Flavell R, Gressel J, Habtemariam T, Lal R, Pell AN, St Leger R, Wall RJ (2008) Emerging technologies to benefit farmers in sub-Saharan Africa and South Asia. National Academies Press, Washington, DC

Lawrence MJ, Rees GD (2000) Microemulsion-based media as novel drug delivery systems. Adv Drug Deliv Rev 45(1):89–121. doi:10.1016/S0169-409X(00)00103-4

Lee J, Mahendra S, Alvarez PJJ (2010) Nanomaterials in the construction industry: a review of their applications and environmental health and safety considerations. ACS Nano 4(7):3580–3590. doi:10.1021/nn100866w

Lin D, Xing B (2008) Root uptake and phytotoxicity of ZnO nanoparticles. Environ Sci Technol 42:5580–5585. doi:10.1021/es800422x

Linglan M, Chao L, Chunxiang Q, Sitao Y, Jie L, Fengqing G, Fashui H (2008) Rubisco activase mRNA expression in spinach: modulation by nanoanatase treatment. Biol Trace Elem Res 122(2):168–178. doi:10.1007/s12011-007-8069-4

Liu R, Lal R (2014) Synthetic apatite nanoparticles as a phosphorus fertilizer for soybean (*Glycine max*). Sci Rep 4:5686. doi:10.1038/srep05686

Lodriche SS, Soltani S, Mirzazadeh R (2013) Silicon nanocarrier for delivery of drug, pesticides and herbicides, and for waste water treatment. United States Patent, US20130225412 A1

López MM, Llop P, Olmos A, Marco-Noales E, Cambra M, Bertolini E (2009) Are molecular tools solving the challenges posed by detection of plant pathogenic bacteria and viruses? Curr Issues Mol Biol 11:13–46

Lövenstam G, Rauscher H, Roebben G, Sokull Klüttgen B, Gibson N, Putaud JP, Stamm H (2010) Considerations on a definition of nanomaterial for regulatory purposes. Publication Office of the European Union, Luxembourg. https://ec.europa.eu/jrc/sites/jrcsh/files/jrc_reference_report_201007_nanomaterials.pdf. Accessed 11 Aug 2016

Lu CM, Zhang CY, Wen JQ, Wu GR, Tao MX (2002) Research on the effect of nanometer materials on germination and growth enhancement of *Glycine max* and its mechanism. Soybean Sci 21(3):168–171

Ma JF (2004) Role of silicon in enhancing the resistance of plants to biotic and abiotic stresses. Soil Sci Plant Nutr 50:11–18

McKeague MK, Giamberardino A, DeRosa MC (2011) Advances in aptamer based biosensors for food safety. In: Somerset V (ed) Environmental biosensors. InTech, Croatia, pp 17–42

Miller DD (2010) Food nanotechnology: new leverage against iron deficiency. Nat Nanotechnol 5(5):318–319

Mittler R (2006) Abiotic stress, the field environment and stress combination. Trends Plant Sci 11:15–19

Moaveni P, Kheiri T (2011) TiO2 nano particles affected on maize (*Zea mays* L). In: 2nd international conference on agricultural and animal science. Maldives, pp 160–163

Mogul MG, Akin H, Hasirci N, Trantolo DJ, Gresser JD, Wise DL (1996) Controlled release of biologically active agents for purposes of agricultural crop management. Resour Conserv Recycl 16:289–320

Mohammadi R, Maali-Amiri R, Abbasi A (2013) Effect of TiO_2 nanoparticles on chickpea response to cold stress. Biol Trace Elem Res 152:403–410

Monica RC, Cremonini R (2009) Nanoparticles and higher plants. Caryologia 62:161–165

Montgomery MA, Elimelech M (2007) Water and sanitation in developing countries: including health in the equation. Environ Sci Technol 44(1):17–24

Mukal D, Sexena N, Dwivedi PD (2009) Emerging trends of nanoparticles application in food technology: safety paradigms. Nanotoxicol 3:10–18

Musante C, White JC (2010) Toxicity of silver and copper to *Cucurbita pepo*: differential effects of nano and bulk-size particles. Environ Toxicol 27(9):510–517. doi:10.1002/tox.20667

Naderi MR, Abedi A (2012) Application of nanotechnology in agriculture and refinement of environmental pollutants. J Nanotechnol 11(1):18–26

Naderi MR, Danesh-Shahraki A (2013) Nanofertilizers and their roles in sustainable agriculture. Int J Agric Crop Sci 5(19):2229–2232

Naidoo L, Kistnasamy EJ (2015) A desktop evaluation of the potential impact of nanotechnology applications in the field of environmental health in a developing country. Am J Publ Health Res 3(5):182–186

Nair R, Varghese SH, Nair BG, Maekawa T, Yoshida Y, Kumar DS (2010) Nanoparticulate material delivery to plants. Plant Sci 179(3):154–163

Neethirajan S, Jayas DS (2010) Nanotechnology for the food and bioprocessing industries. Food Bioprocess Technol 4(1):39–47

Nord E (2009) Top 10 reasons for using nanotech in food. http://www.nanotech-now.com/news.cgi?story_id=32231. Accessed 11 Aug 2016

Nuruzzaman MD, Rahaman MM, Liu Y, Naidu R (2016) Nanoencapsulation, nano-guard for pesticides: a new window for safe application. J Agric Food Chem 64:1447–1487

Oerke EC (2006) Crop losses to pests. J Agric Sci 144:31–43

Pan B, Xing BS (2008) Adsorption mechanisms of organic chemicals on carbon nanotubes. Environ Sci Technol 42:9005–9013

Park HJ, Kim SH, Kim HJ, Choi SH (2006) A new composition of nanosized silica-silver for control of various plant diseases. Plant Pathol J 22(3):295–302

Patterson DT, Westbrook JK, Joyce RJV, Lingren PD, Rogasik J (1999) Weeds, insects and diseases. Clim Chang 43:711–727

Pei ZF, Ming DF, Liu D, Wan GL, Geng XX, Gong HJ, Zhou WJ (2010) Silicon improves the tolerance to water-deficit stress induced by polyethylene glycol in wheat (*Triticum aestivum* L.) seedlings. J Plant Growth Regul 29:106–115

Pepper D (2011) The toxic consequences of the green revolution. http://www.usnews.com/news/world/articles/2008/07/07/the-toxic-consequences-of-the-green-revolution. Accessed 11 Aug 2016

Perez-de-Luque A, Rubiales D (2009) Nanotechnology for parasitic plant control. Pest Manag Sci 65(5):540–545

Perlatti B, Bergo PLS, Fernandes da Silva MFG, Fernandes JB, Forim MR (2013) Polymeric nanoparticle-based insecticides: a controlled release purpose for agrochemicals. In: Trdan S (ed) Insecticides: development of safer and more effective technologies. InTech, Rijeka. http://www.intechopen.com/books/insecticides-development-of-safer-and-more-effective-technologies/polymeric-nanoparticle-based-insecticides-a-controlled-release-purpose-for-agrochemicals. Accessed 11 Aug 2016

Pothakamuri UR, Barbosa-Cánovas GV (1995) Fundamental aspects of controlled release in foods. Trends Food Sci Technol 6:397–406

Pourkhaloee A, Haghighi M, Saharkhiz MJ, Jouzi H, Doroodmand MM (2011) Investigation on the effects of carbon nanotubes (CNTs) on seed germination and seedling growth of salvia (*Salvia microsiphon*), pepper (*Capsicum annum*) and tall fescue (*Festuca arundinacea*). J Seed Technol 33:155–160

Prasad R (2016) Advances and applications through fungal nanobiotechnology. Springer, International Publishing, Cham. ISBN:978-3-319-42989-2

Prasad R, Kumar V, Prasad KS (2014) Nanotechnology in sustainable agriculture: present concerns and future aspects. Afr J Biotechnol 13(6):705–713

Prasad R, Pandey R, Barman I (2016) Engineering tailored nanoparticles with microbes: quo vadis. WIREs Nanomed Nanobiotechnol 8:316–330. doi:10.1002/wnan.1363

Prasad R, Bhattacharyya A, Nguyen QD (2017a) Nanotechnology in sustainable agriculture: recent developments, challenges, and perspectives. Front Microbiol 8:1014. doi:10.3389/fmicb.2017.01014

Prasad R, Pandey R, Varma A, Barman I (2017b) Polymer based nanoparticles for drug delivery systems and cancer therapeutics. In: Kharkwal H, Janaswamy S (eds) Natural polymers for drug delivery. CABI, Oxfordshire, pp 53–70

Qados AMSA, Moftah AE (2015) Influence of silicon and nano-silicon on germination, growth and yield of faba bean (*Vicia faba* L.) under salt stress conditions. Am J Exp Agric 5:509–524

Qu X, Alvarez PJ, Li Q (2013) Applications of nanotechnology in water and wastewater treatment. Water Res 47:3931–3946

Qureshi A, Kang WP, Davidson JL, Gurbuz Y (2009) Review on carbon-derived, solid-state, micro and nano sensors for electrochemical sensing applications. Diam Relat Mater 18:1401–1420

Racuciu M, Creanga DE (2006) TMA-OH coated magnetic nanoparticles internalized in vegetal tissue. Rom J Phys 52(3–4):395–402

Racuciu M, Miclauş S, Creanga DE (2009) The response of plant tissues to magnetic fluid and electromagnetic exposure. Rom J Biophys 19:73–82

Rafi MM, Epstein E, Falk RH (1997) Silicon deprivation causes abnormalities in wheat (*Triticum aestivum* L.) J Plant Physiol 152:497–501

Ragaei M, Sabry AH (2014) Nanotechnology for insect pest control. Int J Sci Environ Technol 3(2):528–545

Rai M, Ingle A (2012) Role of nanotechnology in agriculture with special reference to management of insect pests. Appl Microbiol Biotechnol 94(2):287–293

Raliya R, Tarafdar JC (2013) ZnO nanoparticle biosynthesis and its effect on phosphorous-mobilizing enzyme secretion and gum contents in Clusterbean (*Cyamopsis tetragonoloba* L). Agric Res 2(1):48–57

Raliya R, Tarafdar JC, Biswas P (2016) Enhancing the mobilization of native phosphorus in the mung bean rhizosphere using ZnO nanoparticles synthesized by soil fungi. J Agric Food Chem 64(16):3111–3118

Rameshaiah GN, Pallavi J, Shabnam S (2015) Nanofertilizers and nano sensors – an attempt for developing smart agriculture. Int J Eng Res Gen Sci 3(1):314–320

Rao KJ, Paria S (2013) Use of sulfur nanoparticles as a green pesticide on *Fusarium solani* and *Venturia inaequalis* phytopathogens. RSC Adv 3(26):10471–10478

Renton A (2006) Welcome to the world of nanofoods. http://observer.guardian.co.uk/foodmonthly/futureoffood/story/0,,1971266,00.html. Accessed 11 Aug 2016

Rico CM, Majumdar S, Duarte-Gardea M, Peralta-Videa JR, Gardea-Torresdey JL (2011) Interaction of nanoparticles with edible plants and their possible implications in the food chain. J Agric Food Chem 59(8):3485–3498

Risch SJ, Reineccius GA (1995) Encapsulation and controlled release of food ingredients. American Chemical Society, Washington, DC, p 590

Sabaghnia N, Janmohammadi M (2014) Graphic analysis of nano-silicon by salinity stress interaction on germination properties of lentil using the biplot method. Agric For 60(3):29–40

Sabaghnia N, Janmohammadi M (2015) Effect of nano-silicon particles application on salinity tolerance in early growth of some lentil genotypes. Ann UMCS Biol 69:39–55

Sadik OA, Zhou AL, Kikandi S, Du N, Wang Q, Varner K (2009) Sensors as tools for quantitation, nanotoxicity and nanomonitoring assessment of engineered nanomaterials. J Environ Monit 11(10):1782–1800

Sasson Y, Levy-Ruso G, Toledano O, Ishaaya I (2007) Nanosuspensions: emerging novel agrochemical formulations. In: Ishaaya I, Nauen R, Horowitz AR (eds) Insecticides design using advanced technologies. Springer, Berlin, pp 1–39

Schnettler B, Crisóstomo G, Mills N, Miranda H, Mora M, Lobos G, Grunert KG (2013) Preferences for sunflower oil produced conventionally, produced with nanotechnology or genetically modified in the Araucanía region of Chile. Cien Inv Agric 40(1):17–29

Schoen DT, Schoen AP, Hu L, Kim HS, Heilshorn SC, Cui Y (2010) High speed water sterilization using one dimensional nano structures. Nano Lett 10(9):3628–3632

Schulman JH, Stoeckenius W, Prince LM (1959) Mechanism of formation and structure of micro emulsions by electron microscopy. J Phys Chem 63(10):1677–1680

Scott N, Chen H, Rutzke CJ (2003) Nanoscale science and engineering for agriculture and food systems: a report submitted to cooperative state research, education and extension service. U.S. Department of Agriculture: National Planning Workshop, Washington, DC

Seghatoleslami MJ, Feizi H, Mousavi G, Berahmand A (2015) Effect of magnetic field and silver nanoparticles on yield and water use efficiency of *Carum copticum* under water stress conditions. Pol J Chem Technol 17:110–114

Sekhon BS (2014) Nanotechnology in agri-food production: an overview. Nanotechnol Sci Appl 7:31–53

Sharma V, Sharma A (2012) Nanotechnology: an emerging future trend in wastewater treatment with its innovative products and processes. Int J Enhanc Res Sci Tech Eng 1:121–128

Sharon M, Choudhary A, Kumar R (2010) Nanotechnology in agricultural diseases and food safety. J Phytol 2(4):83–92

Singh A, Singh S, Prasad SM (2016) Scope of nanotechnology in crop science: profit or loss. Res Rev J Bot Sci 5(1):1–4

Singh D, Singh SC, Kumar S, Lal B, Singh NB (2010) Effect of titanium dioxide nanoparticles on the growth and biochemical parameters of *Brassica oleracea*. In: Riberio C, de Assis OBG, Mattoso LHC, Mascarenas S (eds) International conference on food and agricultural applications of nanotechnologies. São Pedro

Solanki P, Bhargava A, Chhipa H, Jain N, Panwar J (2015) Nano-fertilizers and their smart delivery system. In: Rai M, Ribeiro C, Mattoso L, Duran N (eds) Nanotechnologies in food and agriculture. Springer, New York, pp 81–102

Srivastava A, Rao DP (2014) Enhancement of seed germination and plant growth of wheat, maize, peanut and garlic using multiwalled carbon nanotubes. Eur Chem Bull 3(5):502–504

Stadler T, Buteler M, Weaver DK, Sofie S (2012) Comparative toxicity of nanostructured alumina and a commercial inert dust for *Sitophilus oryzae* (L.) and *Rhyzopertha dominica* (F.) at varying ambient humidity levels. J Stored Prod Res 48:81–90

Subramanian KS, Manikanda A, Thirunavukkarasu M, Rahale CS (2015) Nano-fertilizers for balanced crop nutrition. In: Rai M, Ribeiro C, Mattoso L, Duran N (eds) Nanotechnologies in food and agriculture. Springer, New York. doi:10.1007/978-3-319-14024-7_3

Suriyaprabha R, Karunakaran G, Yuvakkumar R, Rajendran V, Kannan N (2012) Silica nanoparticles for increased silica availability in maize (*Zea mays* L.) seeds under hydroponic conditions. Curr Nanosci 8:902–908

Tai-Chia C, Chih-Ching H (2009) Aptamer-functionalized nano-biosensors. Sensors 9(12):10356–10388

Tiwari DK, Dasgupta-Schubert N, Cendejas LMJV, Villegas J, Montoya LC, Garcia SEB (2014) Interfacing carbon nanotubes (CNT) with plants: enhancement of growth, water and ionic nutrient uptake in maize (*Zea mays*) and implications for nanoagriculture. Appl Nanosci 4(5):577–591

Torabian S, Zahedi M, Khoshgoftar AH (2016) Effects of foliar spray of two kinds of zinc oxide on the growth and ion concentration of sunflower cultivars under salt stress. J Plant Nutr 39:172–180

Tothill IE (2001) Biosensors developments and potential applications in the agricultural diagnosis sector. Comput Electron Agric 30:205–218

Tramon C (2014) Modelling the controlled release of essential oils from a polymer matrix – a special case. Ind Crop Prod 61:23–30

Tratnyek PG, Johnson RL (2006) Nanotechnologies for environmental cleanup. NanoToday 1:44–48

van den Berg F, Kubiak R, Benjey WG, Majewski MS, Yates SR, Reeves GL, Smelt JH, van der Linden AMA (1999) Emission of pesticides into the air. In: Van Dijk HFG, Van Pul WAJ, De Voogt P (eds) Fate of pesticides in the atmosphere: implications for environmental risk assessment. Springer, Netherlands, pp 195–218

Vidyalakshmi R, Bhakyaraj R, Subhasree RS (2009) Encapsulation "the future of probiotics" – a review. Adv Biol Res 3(3–4):96–103

Vinutha JS, Bhagat D, Bakthavatsalam N (2013) Nanotechnology in the management of polyphagous pest *Helicoverpa armigera*. J Acad Ind Res 1(10):606–608

Wang H, Kou X, Pei Z, Xiao JQ, Shan X, Xing B (2011) Physiological effects of magnetite (Fe_3O_4) nanoparticles on perennial ryegrass (*Lolium perenne* L.) and pumpkin (*Cucurbita mixta*) plants. Nanotoxicol 5(1):30–42

Wei C, Yamato M, Wei W, Zhao X, Tsumoto K, Yoshimura T, Ozawa T, Chen YJ (2007) Genetic nanomedicine and tissue engineering. Med Clin N Am 91:889–898

Yada R (2009) Nanotechnology: a new frontier in foods, food packaging, and nutrient delivery. In: Pray L, Yaktine A (eds) Nanotechnology in food products. National Academies Press, Washington, DC

Yao KS, Li SJ, Tzeng KC, Cheng TC, Chang CY, Chiu CY, Liao CY, Hsu JJ, Lin ZP (2009) Fluorescence silica nanoprobe as a biomarker for rapid detection of plant pathogens. Adv Mater Res 79(82):513–516

Yin L, Colman BP, McGill BM, Wright JP, Bernhardt ES (2012) Effects of silver nanoparticle exposure on germination and early growth of eleven wetland plants. PLoS One 7(10):476–474

Zaimenko NV, Didyk NP, Dzyuba OI, Zakrasov OV, Rositska NV, Viter AV (2014) Enhancement of drought resistance in wheat and corn by nanoparticles of natural mineral analcite. Ecol Balkanica 6(1):1–10

Zareii FD, Roozbahani A, Hosnamidi A (2014) Evaluation the effect of water stress and foliar application of Fe nanoparticles on yield, yield components and oil percentage of safflower (*Carthamus tinctorious* L.) Int J Adv Biol Biomed Res 2:1150–1159

Zhang W-X (2003) Nanoscale iron particles for environmental remediation: an overview. J Nanopart Res 5(3):323–332

Zhang L, Webster TJ (2009) Nanotechnology and nanomaterials: promises for improved tissue regeneration. NanoToday 4:66–80

Zheng L, Hong F, Lu S, Liu C (2005) Effect of nano-TiO_2 on strength of naturally aged seeds and growth of spinach. Biol Trace Elem Res 104(1):83–92

Nanoengineering Superabsorbent Materials: Agricultural Applications

6

Majid Peyravi, Peyman Pouresmaeel Selakjani, and Soodabeh Khalili

Abstract

The superabsorbent polymers (SAP) are able to absorb large amounts of water. Superabsorbent materials are known as hydrophilic complexes which have the capacity to absorb a large volume of aqueous fluids in a short time and desorb the absorbed water under stress condition. The absorption capacity of SAP is one of the important parameters of that could limit the application of this material. The use of superabsorbent polymers for water managing and the renewal of arid as well as desert environment have been considered greatly. The encouraging results show that superabsorbent materials can help agriculture and environment by irrigation water consumption reduction, fertilizer retention time improvement in soil, lowering the death rate of plants, and plant growth rate increment. Overall, the modification of superabsorbent polymers using nanotechnology could include employment of nanomaterials for preparation of superabsorbent nanocomposite materials. One of the most convenient and useful superabsorbent nanocomposites is superabsorbent/clay nanocomposites which are introduced in this chapter.

Keywords

Superabsorbent polymers • Agricultural nanotechnology • Nanocomposites • Nano clay • Absorption

M. Peyravi (✉) • P.P. Selakjani
Nanotechnology Research Institute, Faculty of Chemical Engineering,
Babol Noshirvani University of Technology, Babol, Iran
e-mail: majidpeyravi@nit.ac.ir

S. Khalili
Oil, Gas and Chemical Engineering Department, Semnan University, Semnan, Iran

R. Prasad et al. (eds.), *Nanotechnology*, DOI 10.1007/978-981-10-4573-8_6

6.1 Introduction

Another class of polymers called superabsorbent polymers (SAP) can retain a lot of water, normally more than general sponges (Esposito et al. 1996). By and large, they comprise of polymeric organize that are cross-connected together. In many sorts of these materials, there are ionic gatherings alongside the polymer chains to encourage dispersion of water inside the system (Raju et al. 2003). In other words, superabsorbent materials are known as hydrophilic edifices which have the ability to ingest vast volume of watery liquids in a brief timeframe and desorb the assimilated water under anxiety condition. As indicated by Bowman and Evans (1991) and Woodhouse and Johnson (1991), superabsorbent polymer can hold 400–1500 g of water for each dry gram of its network. Johnson and Veltkamp (1985) have recommended that when the hydrogel is legitimately utilized, 95% of the put-away water is accessible for plant. Three classifications of SAP were presented by Woodhouse and Johnson (1991): starch-polyacrylonitrile joint polymers (starch copolymers), vinyl liquor acrylic corrosive copolymers (polyvinyl alcohols), and acrylamide sodium acrylate copolymers (cross-connected polyacrylamides). At the point when the absorbents are added to soil, the following mechanism suggests. Amid hydration handle, a shapeless coagulated material is framed. Because of their ability of assimilation and desorption for a drawn-out stretch of time, these sponges go about as a moderate discharge lattice for water and broke down supplements in it. The survival of sapling is extended by addition of the shrinking time amid precipitation periods and may prompt to higher yields in specific conditions. The dry season stretch impacts can be lessened by the utilization of super spongy polymer and enhances plant yield and farming generation soundness (Vundavalli et al. 2015; Khadem et al. 2010; Ali et al. 2014).

6.2 Formation and Structure of Cross-Linked Polyacrylates

Lately, there has been a critical consideration regarding SAP materials that are equipped for assimilation and holding a lot of water. These polymers have discovered variable business applications as retentive in individual care items, for example, newborn child diapers, female cleanliness items, and incontinence items (Brannon-Peppas and Harland 2012) and additionally have gotten many enthusiasm for an assortment of more expert applications, for example, base grids used in chemical immobilization (Samsonov and Kuznetsova 1992), biosorbents in chromatography (Samsonov and Kuznetsova 1992), materials for horticultural items (Kazanskii and Dubrovskii 1992), and networks for controlled discharge vehicles (Colombo 1993). Water-retentive polymers are typically shaped amid free radical cross-linking polymerization response of hydrophilic acrylate or methacrylate monomers with little amounts of cross-linking operators that incorporate (at least two) polymerizable twofold securities (Colombo 1993). A few cases of cross-linking specialists are N,N′-methylenebisacrylamide, triallylamine, ethylene glycol diacrylate, tetraethylene glycol diacrylate, trimethylolpropane triacrylate, and the methacrylate analogs of the previously mentioned acrylates. Hydrophilic esters of acrylic or methacrylic corrosive (e.g., 2-hydroxyethylmethacrylate and its analogs) have been broadly utilized to shape hydrogels which nearly display a most extreme swelling of 40–50 wt% water. The one-of-a-kind arrangement of attributes displayed by the hydroxyethyl (meth)acrylate hydrogels have prompted to their wide application in biomedical materials and gadgets (Peppas 1987; Langer and Peppas 1981; Cooper and Peppas 1982). Hydrogels showing much higher swelling limits might be created via doing the free radical cross-linking polymerizations with ionogenic monomers, for example, acrylic and methacrylic corrosive (or their sodium salts). For example, poly(acrylic corrosive) hydrogels may indicate the greatest swelling of more than 99 wt% water. Models which have been exhibited to clarify free radical cross-linking polymerizations might be named (1) measurable models, (2) active portrayals, and (3) dynamic gelation models based upon PC recreated arbitrary strolls. At the point when a monomer containing one polymerizable twofold bond is copolymerized with little estimations of a divinyl cross-linker, simultaneous copolymerization and cross-linking responses happen. A developing radical which responds with one of the twofold obligations of a cross-linker to frame a polymer chain with a pendant twofold bond is a case for this wonder. This twofold bond may respond with a moment developing chain and go about as the cross-linker between chains. The nearness of the cross-links between polymers binds prompts to the most essential normal for these sorts of polymerizations: the likelihood of three-dimensional polymer arrangement development.

6.2.1 Statistical Models

Measurable models expect that polymer structures incorporate monomer units as per probabilistic guidelines for security development. Some factual computational models have been accounted for in the literary works, for example, the recursive strategy (Dotson et al. 1988; Macosko 1985), the piece technique (Williams and Vallo 1988), and the production capacity strategy (Scranton and Peppas 1990; Scranton et al. 1991). Free radical cross-linking polymerization, where identical measurable depictions depend on the suppositions of perfect arbitrary cross-linking, contains (1) equivalent and autonomous reactivity of all carbon twofold bonds, (2) no intramolecular cyclization, (3) end by chain exchange or disproportionation, and (4) change in free energy. This approach has been accounted for to incorporate transformation subordinate energy, inexact treatment of cyclization, and different end components (Kinney and Scranton 1994).

Figure 6.1 represents the connection likelihood design for producing capacities for a free radical cross-linking polymerization response. As it appeared in the figure, there are six conceivable fortified conditions of the monomer in a free radical cross-linking polymerization. Note that the reinforced conditions of a monomer unit are presented with two properties: (1) the quantity of shaped bonds with different monomers and (2) the kind of fortified unit(s). Thus, as illustrated in the figure, the monovinyl acrylate monomer has six possible bonded states as follows: unreacted, no bonds with different units; one bond shaped with another unit, for example,

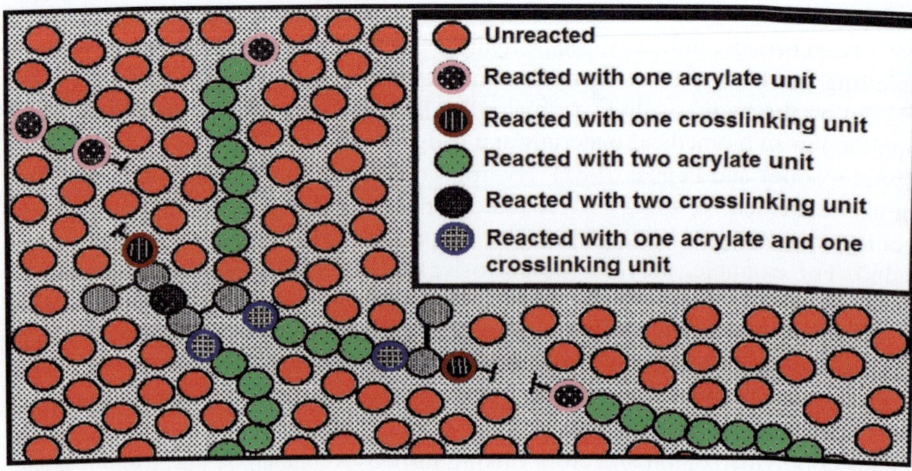

Fig. 6.1 Monomer reacted states of the acrylate monomer during a free radical cross-linking polymerization (Modified from reference, Kinney and Scranton 1994)

another acrylate monomer; or a cross-linker; two framed bonds with different units, which might be two other acrylate monomers; an acrylate and a cross-linker; or two cross-linkers.

6.3 Mechanisms of Swelling in Superabsorbent Polymers

It is important to comprehend the reasons why the SAP materials swell. There are a few components for the way toward swelling, all of which add to the last swelling limit (or centrifuge retention capacity CRC – which is the amount of 0.9 wt% saline solution that a SAP can hold under free swelling conditions when surface water has been expelled in a centrifuge) (Elliott 2004).

Figure 6.2 is a diagrammatic representation of the SAP polymer arrange. The polymer spine in SAP is hydrophilic in light of the fact that it incorporates water cherishing carboxylic corrosive gatherings (–COOH). At the point when watery arrangement is added to SAP, there is a polymer/dissolvable association; hydration and hydrogen holdings are two of these collaborations.

6.3.1 Hydration

The communication of particles of a solute with atoms of a dissolvable, i.e., COO- and Na+ particles that pulls in the polar water atoms is called hydration (Fig. 6.3).

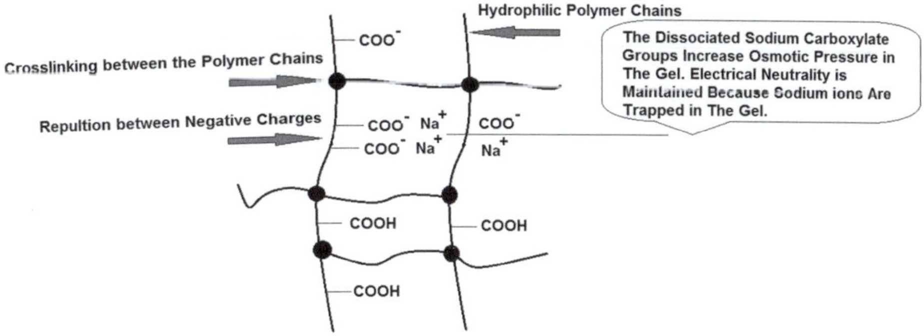

Fig. 6.2 Schematic of SAP network during swelling process

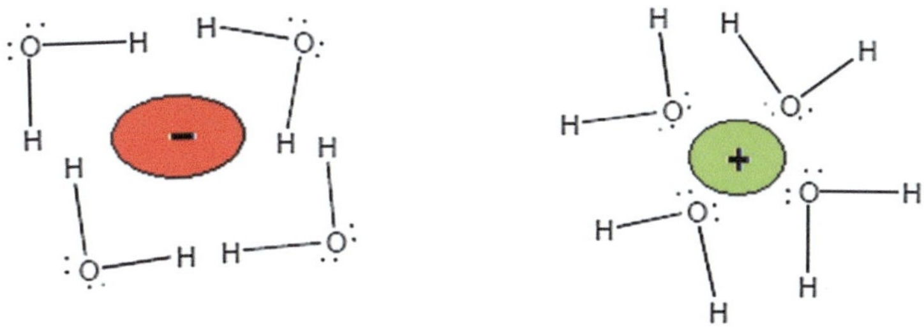

Fig. 6.3 Schematic of hydration

6.3.2 Hydrogen Bonds

Hydrogen bond is a concoction bond where a hydrogen iota of one particle is pulled into an electronegative molecule of another atom particularly N, O, or F iotas. The hydrogen particles are pulled into the non-holding electron sets (solitary sets) on other neighboring electronegative iotas (Fig. 6.4).

In watery conditions the electronegative iota is oxygen which pulls the electrons of hydrogen toward itself and a dipole sets up in the atom. The emphatically charged hydrogen particles are pulled into the oxygen solitary sets of electrons on other water atoms. Oxygen has two solitary sets of electrons and each is fit for hydrogen clinging to two other water particles (Elliott 2004).

These impact the increment of the entropy of the framework. As indicated by the hydrophilic way of SAP system, the polymer affixes have a slant to scatter in the water (i.e., they are attempting to break down in the water), which prompts to expand the entropy. The reason that keeps disintegration of polymer from water is cross-interfaces between polymer chains that shape a solid three-dimensional system. This is caused by the flexible powers of the system and is joined by abatement in entropy of the chains, as they get to be distinctly stiffer from their typical curled state.

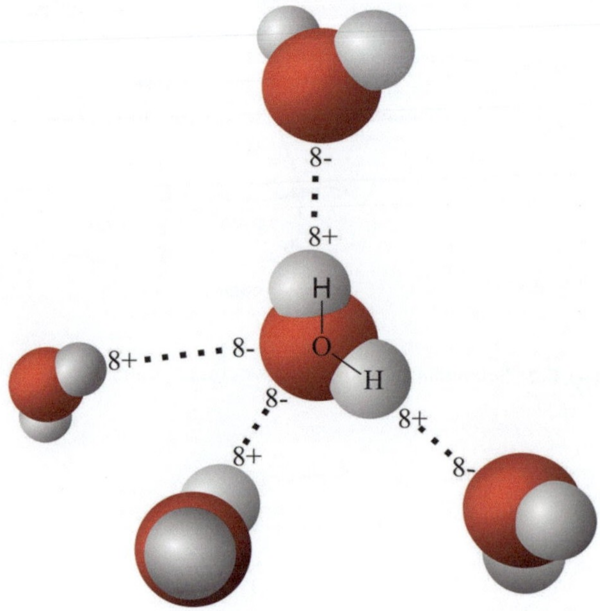

Fig. 6.4 Hydrogen bonding

6.4 Important Basic Properties of Superabsorbent Polymers

6.4.1 Absorption of Aqueous Solution

The assimilation limit of SAP must be measured in different conditions, since this trademark could restrict the use of this material. For instance, poly(acrylic corrosive) SAP can't be utilized for ocean water ingestion on the grounds that the receptiveness of poly(acrylic corrosive) SAP is significantly diminished because of its poor sponginess soundness against polyvalent cations. SAP absorbance could be figured by Eq. 6.1 as the swelling proportion of the cross-linked polyelectrolyte organizes framework (*qm*).

$$q_m^{5/3} = \frac{\left(i / 2v_u S^{*1/2}\right)^2 + \left(1/2 - x_1\right)/v_1}{V_0 / v_e} \tag{6.1}$$

where *qm* is the swelling proportion of the system at balance; *i* is the ionization degree increased by the system settled charge valency and the structure unit molar volume, S^* is the outer arrangement ionic quality; x_1 is the system and the dissolvable cooperation parameter, the dissolvable molar volume, and the chains' compelling number in the system; and V_0 is the unswollen organized volume. In Eq. 6.1, the first and second terms in the top of the right side represent the ionic charges in the polymer network and in the solution, and the affinity contribution between the SAP network and the solvent, respectively. The term speaks to the cross-link thickness of

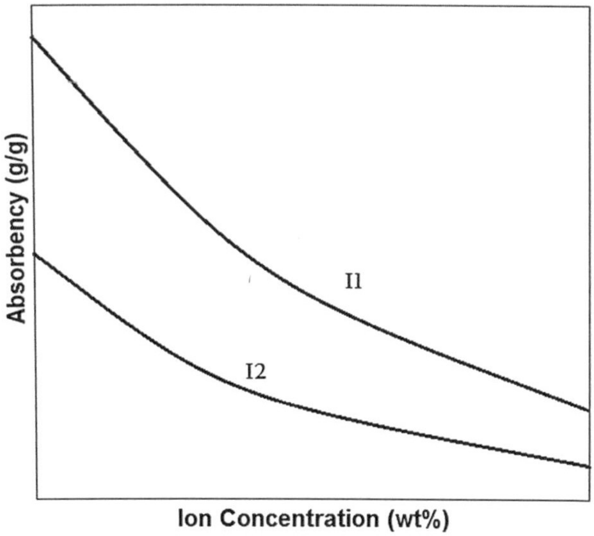

Fig. 6.5 Absorbency dependence of the poly (acrylic acid) SAP on the concentration of I1 and I2 solution

the system. Really, Eq. 6.1 presents the retaining conduct of a superabsorbent polymer basically and can without much of a stretch be utilized for outlining a SAP since it gives major thoughts regarding what the fundamental chain structure and the SAP cross-link thickness ought to be utilized for a particular application. Figure 6.5 demonstrates the retentiveness reliance of a poly(acrylic corrosive) SAP on the convergence of two unique particle watery arrangements (I1 and I2) schematically. The receptiveness diminishes by augmentation of salt focus; this is clarified by the expansion of S*1/2 in Eq. 6.1. On account of I2 arrangement, the absorbance of poly(acrylic corrosive) SAP is likewise subject to the splashing time. The retentiveness of the poly(acrylic corrosive) SAP is plainly diminished lastly; it is totally halted. This is a result of the extra cross-links made by the irreversible particle trade on carboxyl gatherings by calcium and sodium particles, as it were a result of the expansion of in Eq. 6.1.

6.4.2 Moisture Absorption

With an examination between SAP material and helpful permeable materials like silica gel, the dampness assimilation of SAP is much higher than that of silica gel at high relative mugginess. Figure 6.6 shows the retentiveness of SAP and silica gel schematically. The more extreme incline at high relative moistness demonstrates that the SAP is more delicate to the stickiness change and more adequately ingests and discharges dampness than silica gel. In this manner, the SAP could be a superior dampness-controlling material than silica gel (Shimomura and Namba 1994).

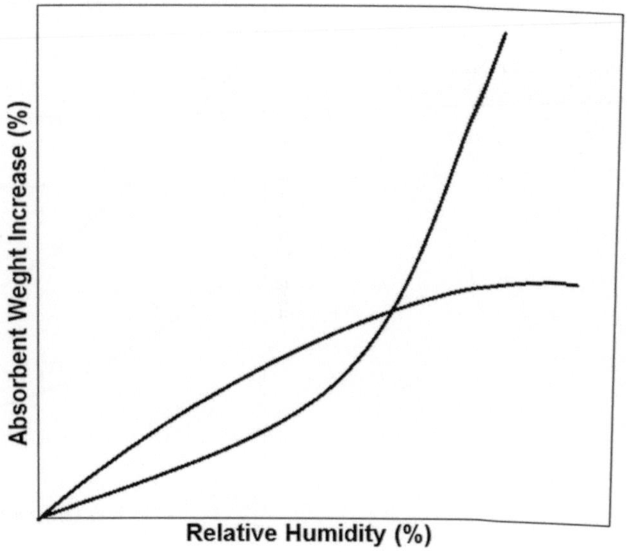

Fig. 6.6 Moisture absorption of the poly (acrylic acid) SAP and silica gel

6.5 Superabsorbent Polymers Application in Agriculture

Farming and cultivation stand out among the most essential utilizations of SAP materials in light of the fact that these materials could transform dry deserts into green-rich terrains. The SAP powder utilized as a part of cleanliness applications has a distance across around 300 μm. In any case, for agribusiness and agriculture, it is ideal to utilize the particles with bigger distance across (ca. 1–3 mm) and higher gel quality since little or delicate gels fill spaces in soil and keep roots from breathing and moving or from water depletion. The measure of SAPs blended in soil is essential, and around 0.1 wt% of SAP is normally a reasonable sum for adding to the dirt.

Another preferred standpoint of SAP materials also of water maintenance is augmentation of the air content in the dirt. By and large, the accompanying favorable circumstances can be acquired by including the SAP in soil:

(a) Longer interim between watering
(b) Less watering
(c) Lower demise rate of plants
(d) Higher developing rate of plants
(e) Higher manure maintenance

Less watering diminishes the salt gathering contained in soil or water, and this might be an incredible help in the greening of deserts and furthermore in human well-being because of the less substantial metal particles in the farming items. From a report by Shimomura and Namba (1994), yields of *Brassica rapa*, a sort of Chinese vegetable usually known as "Komatsuna," expanded up to the third gathering in similar pots utilizing the water-sparing condition, and this is represented in Fig. 6.7.

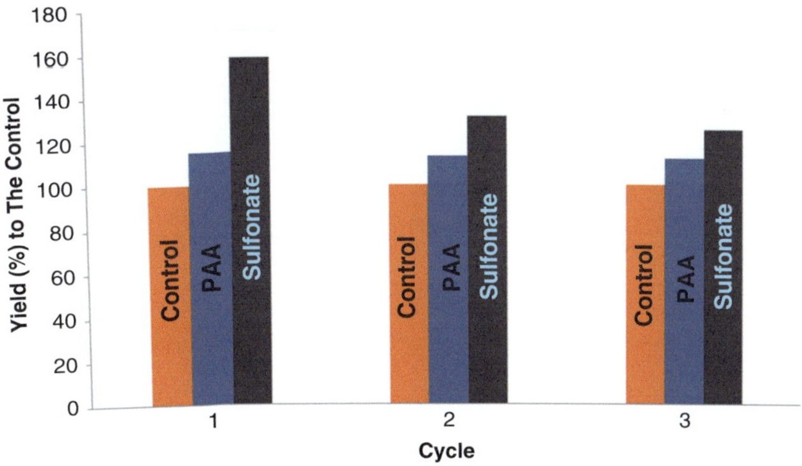

Fig. 6.7 Persistence test for the poly(acrylic acid) SAP and the sulfonate SAP as water retainer

"In this investigation, 0.15% of SAP powder was blended in the dirt with composts, and Komatsuna was planted and refined in pots for 30 days. In the wake of gathering, new Komatsuna was planted and refined again in a similar pot. The entire method was rehashed up to the third reap, and Komatsuna gathered every time was weighed. The poly (acrylic corrosive) SAP expanded the yield around 10% contrasted with the control. The changes are significantly more prominent when sulfonate sort SAP is utilized. This is on account of the sulfonate SAP is less touchy to polyvalent cations from soil, water, or manures and can keep up its execution for quite a while" (Shimomura and Namba 1994).

Another use of SAP materials in the farming is overwhelming metal particle evacuation amid the planting. Lately, arsenic and cadmium metals tainting in drinking water and in characteristic items have turned into an imperative natural issue and require consideration everywhere throughout the world. Furthermore, arsenic and cadmium cause neural issues, renal unsettling influences, lung issues, stomach-related issues, bone injuries, and hypertension in people. Likewise, both metal particles are nondegradable and in charge of serious issues like malignancy in people and also in creatures (Rivas and Muñoz 2010; Rao et al. 2010).

In a matter of seconds, adsorption is the most reasonable process for the expulsion of focused metal particles from wastewater in light of its solid proclivity, high proficiency, and monetary advantages (Zhang et al. 2012). Adsorption by utilizing superabsorbent materials is a key strategy for expelling dangerous species from regular assets and wastewater (Meng et al. 2012). Superabsorbent hydrogels have sought properties and potential applications in this field (Patel and Patel 2013). In a study by Mahida and Patel (2014), the superabsorbent poly(NIPAAm/AA/N--allylisatin) nanohydrogels were combined, portrayed, and utilized as an adsorbent material for the evacuation of As (V) and Cd (II) metal particles from fluid arrangements. As indicated by the aftereffects of this study, the chosen SAP material has a

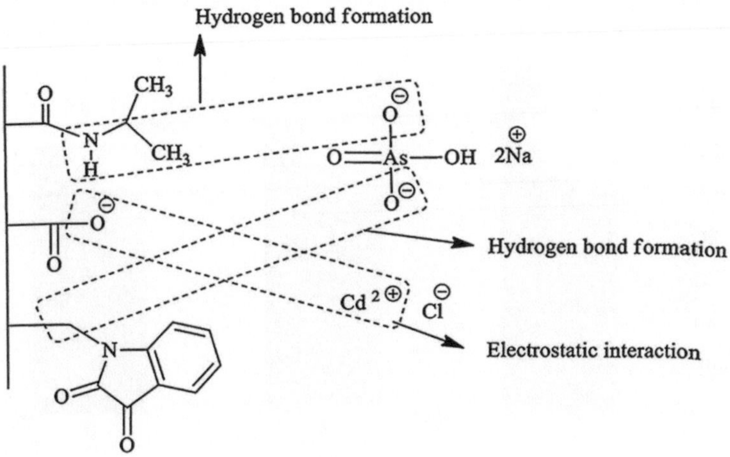

Fig. 6.8 Possible complexation process between nanohydrogel and metal ions (Adopted from reference Mahida and Patel 2014)

decent potential to adsorb overwhelming metal particles. Utilization of the SAPs in the dirt could assimilate the overwhelming metal particles and trap the particles in the cross-connected system of the polymer. Figure 6.8 has been embraced from the reference of Mahida and Patel (2014) and delineates a schematic of conceivable complexation prepared among nanohydrogel and metal particles.

6.6 Nanotechnology Application in Superabsorbent Polymers

Lately, the utilization of superabsorbent polymers for water overseeing and the restoration of parched and in addition abandoned environment have been considered enormously. The empowerment comes to demonstrate that superabsorbent materials can help farming and environment water system water utilization diminishment, compost maintenance time change in soil, bringing down the demise rate of plants, and plant development rate increase. Generally speaking, the adjustment of superabsorbent polymers utilizing nanotechnology could incorporate work of nanomaterials to planning of superabsorbent nanocomposite materials. Because of the better execution, high particular surface region utilization of the items, nanosized superabsorbent materials are utilized. Likewise, these materials diminish the material use and increment the execution yield. Nanotechnology has numerous applications in the field of horticulture. In dry land territories, variety in sum and conveyance of precipitation straightforwardly influence the yield creation. As another illustration, soil dampness is the most basic element for the second harvest of rice after the first under rainfed condition (Jatav et al. 2013). Superabsorbent hydrogels may offer the fancied properties and potential applications in this examination territory. A few attempts have been made to adjust the properties of superabsorbent polymeric materials, from full scale to miniaturized scale to nanoscales. Considerably more consideration has been centered around the change of the swelling capacity, gel quality, and mechanical and warm soundness of superabsorbent utilizing nanotechnology as another perspective (Mahida and Patel 2014). Many studies were centered around utilization of nanotechnology in superabsorbent polymer change. In the present part, the fixation is on the nanocomposite superabsorbent polymer and the impact of nanoparticles and unique impact of nanoclay on adjustment of the SAP materials.

6.7 Superabsorbent/Clay Nanocomposites

As per a definition by Sigma-Aldrich Co (http://www.sigmaaldrich.com/), nanoclays are dirt minerals which can be adjusted for use in polymer-clay nanocomposite multiuseful material frameworks with a lot of property improvements focused for an extraordinary application. Nanoclays are an expansive class of characteristic inorganic minerals, and platelike montmorillonite is the most normally utilized as a part of material applications. Montmorillonite incorporates ~1-nm-thick aluminosilicate layers substituted in the surface with metal cations and stacked in ~10-μm multilayer stacks (Fig 6.9a). Polymer-earth nanocomposite is normally shaped by scattering of the stacks in the polymer framework (Fig 6.9b). Inside the nanocomposite, singular nm-thick clay layers are completely isolated to frame platelike nanoparticles with high (nm × μm) viewpoint proportion. Indeed, even at low nanoclay stacking, the clay surface dwelling in close contact with the whole nanocomposite comprises of interfacial polymer, with lion's share of polymer chains. This can drastically influence the properties of a nanocomposite contrasted with the unadulterated polymer (Table 6.1). Potential advantages incorporate unrivaled fire resistance, diminished gas porousness, expanded mechanical quality, and even upgraded straightforwardness when scattered nanoclay plates stifle polymer crystallization (Giannelis 1996; Vaia et al. 1996; LeBaron et al. 1999; Morgan and Gilman 2007).

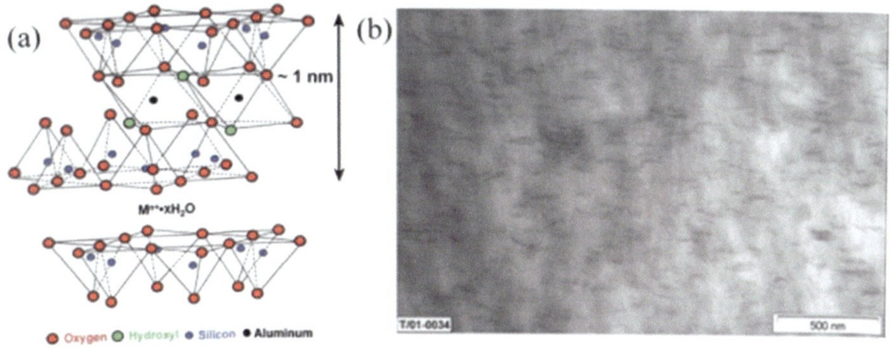

Fig. 6.9 (**a**) Schematic of nm-thick montmorillonite clay aluminosilicate layers. (**b**) TEM micrograph of 2% Nanoclay, Nanomer® I.34TCN-Nylon six nanocomposite showing complete dispersion of clay layers into distinct platelike nanoparticles (Adopted from Sigma-Aldrich Co. http://www.sigmaaldrich.com/)

Table 6.1 Example of property improvements in a polymer-clay nanocomposite

Material	Tensile strength (MPa)	Tensile modulus (MPa)	Flexural strength (MPa)	Flectural modulus (MPa)
Nylon 6. Control	75	3140	114	3112
5% I.34TCN composite	80	4200	142	4223

Data adopted from Sigma-Aldrich Co. (http://www.sigmaaldrich.com/)

Montmorillonite was found in 1847 in Montmorillon in the Vienne prefecture of France. A critical wellspring of montmorillonite in nature is bentonite. A stone is framed from exceptionally colloidal and plastic dirt for the most part made out of montmorillonite (Organization 2005). Bentonite particles are indistinct from kaolin clay minerals; notwithstanding, the thickness is the fundamental distinction. Hypothetically, the thickness of sodium or potassium salts of bentonite that were peeled into thin plates could be 1 nm (Runcy et al. 2013). Notwithstanding montmorillonite, bentonite may incorporate crystalline quartz, feldspar, and cristobalite. For the most part, the earth minerals in view of bentonite may show the properties of high water retention, thixotropic-gel arrangement with water, and high cation-trade limit. These properties could be distinctive in dirt minerals in light of the way of interstitial water and interchangeable cations in the interlayer (Uddin 2008).

The montmorillonite (MMT) has been broadly utilized as a part of the superabsorbent polymer alteration. In a late study by Rashidzadeh et al. (2014), the MMT nanocomposite with NaAlg-g-poly(AA-co-AAm) has been utilized as a superabsorbent for compost-controlled discharge. The creators displayed a presentation about the advantages of utilizing MMT and its nanocomposites as superabsorbent in the horticultural applications particularly in compost-controlled discharge as underneath: "The high cost of the superabsorbents in the market disregarding the upsides of superabsorbent application as manure bearer makes its application unfeasible. Organization of superabsorbent polymers with earth minerals, for example, montmorillonite (Wu et al. 2003), attapulgite (Liang and Liu 2006), and kaolin (Hussien et al. 2012) in high extents is an option answer for its high cost issue." This presentation not just lessens the last cost of the item firmly but additionally enhances gel qualities and mechanical and warm soundness and swelling property (Schexnailder and Schmidt 2009). In addition, it has been demonstrated that the consideration of inorganic dirt into hydrogels diminishes the discharge rate of manure from moderate discharge definition. Controlled arrival of urea manure has been finished by poly acrylamide/methyl cellulose/montmorillonite nanocomposite hydrogel (Bortolin et al. 2013). The study was endeavored to blend a novel moderate discharge manure (SRF) superabsorbent nanocomposite and researches its water permeableness and compost moderate discharge properties. From the aftereffects of the study, the nearness of the montmorillonite in the framework brought about to free the supplement in a more controlled way than that the superabsorbent without montmorillonite. The best possible controlled discharge compost property and also great water maintenance time (retention and desorption limit) demonstrated that this plan is conceivably reasonable for application in farming as a manure transporter

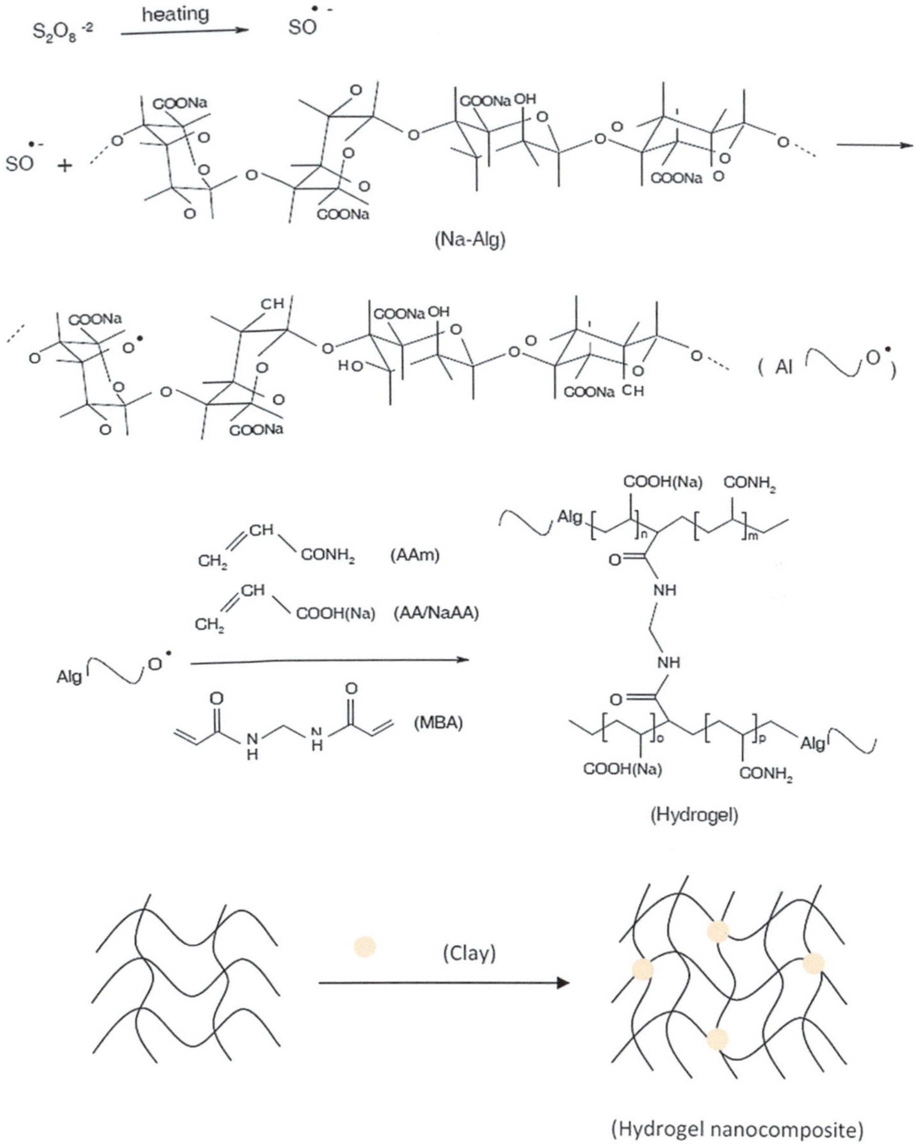

Fig. 6.10 Proposed reaction mechanism for synthesis of superabsorbent nanocomposite (Adopted from reference Rashidzadeh and Olad 2014)

(Rashidzadeh and Olad 2014). Alginate-g-poly (acrylic corrosive co-acrylamide)/ MMT superabsorbent nanocomposite was readied utilizing unite copolymerization of acrylamide and acrylic corrosive onto sodium alginate by utilizing of MBA as cross-linking specialist and montmorillonite clay.

Figure 6.10 Illustrates the response component for amalgamation of the NaAlg-g-poly(AA-co-AAm)/MMT superabsorbent nanocomposite.

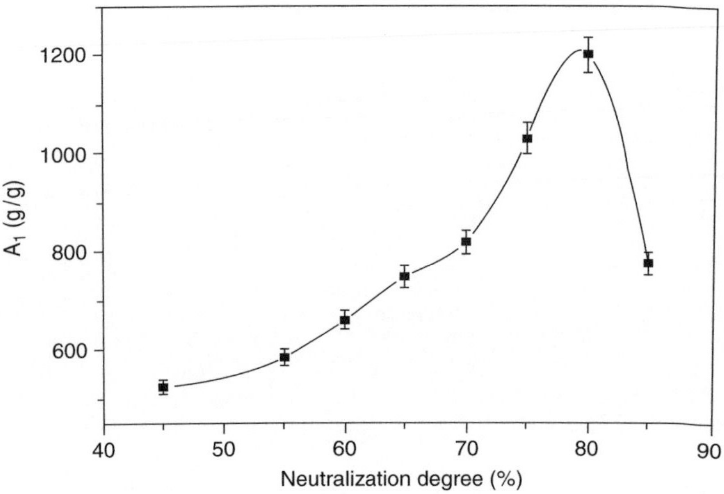

Fig. 6.11 Effect of the neutralization degree on the water absorbency of the nanocomposite (5% MMT, 0.15% cross-linker, and 0.3% initiator) (Adopted from reference Liu et al. 2006)

In a study by Liu et al. (2006), a superabsorbent nanocomposite was blended by the intercalation polymerization of incompletely killed acrylic corrosive and a sodium-sort montmorillonite powder with N,N0-methylenebisacrylamide cross-linker and ammonium persulfate and sodium sulfite as a kind of blended redox initiator. The impacts of components, for example, the measures of the sodium-sort montmorillonite, cross-linker, and initiator and balance degree, were examined on the water retentiveness of the nanocomposite. The ideal sponginess of the nanocomposite was 1201 and 83 g/g in refined water and saline water (NaCl fixation = 0.9%) individually.

Figure 6.11 (from reference Liu et al. 2006) demonstrates the impact of the balance degree on the water permeableness of the superabsorbents. It appears that the balance builds water receptiveness from 45 to 80% and encourage augmentation of balance degree diminishes water sponginess. This pattern might be because of synergetic impact of carboxyl gatherings and carboxylate bunches. Commitment in the water fondness of the system is ascribed to carboxyl gatherings, while carboxylate bunches add to the penetration weight when the superabsorbent is coasted in water environment.

Also, one of the imperative variables that influence the polymer water sponginess as indicated by Flory's system hypothesis is the charge thickness of the polymer (Flory 1953). The immensity of balance specifically decides the charge thickness: a deficient balance degree prompts to less penetration weight, though a high balance degree causes repugnance among carboxylate aggregates and diminished water permeableness. Subsequently, the helpful impact will achieve its greatest when carboxylate and carboxyl gatherings collaborate in a reasonable proportion. The got ideal permeableness of the nanocomposite was appeared at 80% balance degree.

6.8 Special Characteristics

Reswelling capacity (reswellability) is an imperative parameter for horticultural superabsorbents, since these smaller-than-expected water supplies ought to hold their water ingestion capacity for quite a long while in conditions with no normal swelling-deswelling cycle. Liu et al. considered reswellability of poly(AA-SA)-diatomite composite hydrogels. From the outcomes, reswelling capacity of the composite hydrogel is higher than the flawless examples (Qi et al. 2007). Wang and colleagues examined on the reswellability of AAm-based superabsorbent hydrogels stacked with various muds including mica, vermiculite, kaolin, MMT, and ATA (Zhang and Wang 2007). It was accounted for that the most elevated reswellability was for the composite hydrogels stacked with mica (Kabiri et al. 2011).

One of the fundamental targets of get ready SAP nanocomposites is to enhance the mechanical quality of the hydrogel. SAPs ought to have legitimate gel quality amid their administration in rural applications due to brutal condition that the material has in the dirt. Many studies have demonstrated that the mud or organ-modified dirt are reasonable fillers and can influence the hydrogels emphatically. Utilitarian gatherings at the surface of dirt particles and the polymer can respond with each other to frame new cross-links, or mud particles can be shed to go about as a receptive multifunctional added substance (Nie et al. 2005; Haraguchi and Takehisa 2002). Mechanical properties are in any case enhanced through these extra communications/cross-links.

References

Ali MA, Rehman I, Iqbal A, Din S, Rao AQ, Latif A, Samiullah TR, Azam S, Husnain T (2014) Nanotechnology, a new frontier in agriculture. Adv Life Sci 1(3):129–138

Bortolin A, Aouada FA, Mattoso LH, Ribeiro C (2013) Nanocomposite PAAm/methyl cellulose/montmorillonite hydrogel: evidence of synergistic effects for the slow release of fertilizers. J Agric Food Chem 61(31):7431–7439

Bowman DC, Evans RY (1991) Calcium inhibition of polyacrylamide gel hydration is partially reversible by potassium. Hortscience 26(8):1063–1065

Brannon-Peppas L, Harland RS (2012) Absorbent polymer technology. Elsevier, New York

Colombo P (1993) Swelling-controlled release in hydrogel matrices for oral route. Adv Drug Deliv Rev 11(1):37–57

Cooper SL, Peppas NA (1982) Biomaterials, interfacial phenomena and applications. American Chemical Society, Washington, DC

Dotson NA, Galvan R, Macosko CW (1988) Structural development during nonlinear free-radical polymerizations. Macromolecules 21(8):2560–2568

Elliott M (2004) Superabsorbent polymers. Product development scentist for SAP BASF Aktiengesellschaft ss 13

Esposito F, Del Nobile M, Mensitieri G, Nicolais L (1996) Water sorption in cellulose-based hydrogels. J Appl Polym Sci 60(13):2403–2407

Flory PJ (1953) Principles of polymer chemistry. Cornell University Press, Ithaca

Giannelis EP (1996) Polymer layered silicate nanocomposites. Adv Mater 8(1):29–35

Haraguchi K, Takehisa T (2002) Nanocomposite hydrogels: a unique organic-inorganic network structure with extraordinary mechanical, optical, and swelling/de-swelling properties. Adv Mater 14(16):1120

Hussien RA, Donia AM, Atia AA, El-Sedfy OF, El-Hamid ARA, Rashad RT (2012) Studying some hydro-physical properties of two soils amended with kaolinite-modified cross-linked poly-acrylamides. Catena 92:172–178

Jatav GK, Mukhopadhyay R, De N (2013) Characterization of swelling behaviour of nanoclay composite. Int J Innov Res Sci Eng Technol 2:1560–1563

Johnson MS, Veltkamp CJ (1985) Structure and functioning of water-storing agricultural polyacrylamides. J Sci Food Agric 36(9):789–793

Kabiri K, Omidian H, Zohuriaan-Mehr M, Doroudiani S (2011) Superabsorbent hydrogel composites and nanocomposites: a review. Polym Compos 32(2):277–289

Kazanskii K, Dubrovskii S (1992) Chemistry and physics of "agricultural" hydrogels. In: Polyelectrolytes hydrogels chromatographic materials. Springer, pp 97–133

Khadem SA, Galavi M, Ramrodi M, Mousavi SR, Rousta MJ, Rezvani-Moghadam P (2010) Effect of animal manure and superabsorbent polymer on corn leaf relative water content, cell membrane stability and leaf chlorophyll content under dry condition

Kinney A, Scranton AB (1994) Formation and structure of cross-linked polyacrylates: methods for modeling network formation. In: ACS symposium series (USA)

Langer R, Peppas N (1981) Present and future applications of biomaterials in controlled drug delivery systems. Biomaterials 2(4):201–214

LeBaron PC, Wang Z, Pinnavaia TJ (1999) Polymer-layered silicate nanocomposites: an overview. Appl Clay Sci 15(1):11–29

Liang R, Liu M (2006) Preparation and properties of coated nitrogen fertilizer with slow release and water retention. Ind Eng Chem Res 45(25):8610–8616

Liu PS, Li L, Zhou NL, Zhang J, Wei SH, Shen J (2006) Synthesis and properties of a poly (acrylic acid)/montmorillonite superabsorbent nanocomposite. J Appl Polym Sci 102(6):5725–5730

Macosko CW (1985) Rheological changes during crosslinking. Br Polym J 17(2):239–245

Mahida VP, Patel MP (2014) Synthesis of new superabsorbent poly (NIPAAm/AA/N-allylisatin) nanohydrogel for effective removal of As (V) and Cd (II) toxic metal ions. Chin Chem Lett 25(4):601–604

Meng Y, Wang JN, Xu L, Li AM (2012) Fast removal of Pb 2+ from water using new chelating fiber modified with acylamino and amino groups. Chin Chem Lett 23(4):496–499

Morgan AB, Gilman J (2007) Polymer-clay nanocomposites: design and application of multifunctional materials. Mater Matter 2:20–25

Nie J, Du B, Oppermann W (2005) Swelling, elasticity, and spatial inhomogeneity of poly (N-isopropylacrylamide)/clay nanocomposite hydrogels. Macromolecules 38(13):5729–5736

Organization WH (2005) Bentonite, kaolin, and selected clay minerals. WHO, Geneva

Patel YN, Patel MP (2013) A new fast swelling poly [DAPB-co-DMAAm-co-AASS] superabsorbent hydrogel for removal of anionic dyes from water. Chin Chem Lett 24(11):1005–1007

Peppas NA (1987) Hydrogels in medicine and pharmacy, vol 3. CRC press, Boca Raton

Qi X, Liu M, Chen Z, Liang R (2007) Preparation and properties of diatomite composite superabsorbent. Polym Adv Technol 18(3):184–193

Raju KM, Raju MP, Mohan YM (2003) Synthesis of superabsorbent copolymers as water manageable materials. Polym Int 52(5):768–772

Rao K, Mohapatra M, Anand S, Venkateswarlu P (2010) Review on cadmium removal from aqueous solutions. Int J Eng Sci Technol 2(7):81–103

Rashidzadeh A, Olad A (2014) Slow-released NPK fertilizer encapsulated by NaAlg-g-poly (AA-co-AAm)/MMT superabsorbent nanocomposite. Carbohydr Polym 114:269–278

Rivas BL, Muñoz C (2010) Functional water-insoluble polymers with ability to remove arsenic (V). Polym Bull 65(1):1–11

Runcy W, Nithin C, Sabu T (2013) Layered clay rubber composites. In: Key engineering materials. Trans Tech Publ, pp 197–213

Samsonov GV, Kuznetsova NP (1992) Crosslinked polyelectrolytes in biology. In: Polyelectrolytes hydrogels chromatographic materials. Springer, Berlin, pp 1–50

Schexnailder P, Schmidt G (2009) Nanocomposite polymer hydrogels. Colloid Polym Sci 287(1):1–11

Scranton AB, Peppas NA (1990) A statistical model of free-radical copolymerization/crosslinking reactions. J Polym Sci A Polym Chem 28(1):39–57

Scranton AB, Klier J, Peppas NA (1991) Statistical analysis of free-radical copolymerization/crosslinking reactions using probability generating functions: reaction directionality and general termination. Macromolecules 24(6):1412–1415

Shimomura T, Namba T (1994) Preparation and application of high-performance superabsorbent polymers. In: ACS symposium series

Uddin F (2008) Clays, nanoclays, and montmorillonite minerals. Metall Mater Trans A 39(12):2804–2814

Vaia RA, Jandt KD, Kramer EJ, Giannelis EP (1996) Microstructural evolution of melt intercalated polymer-organically modified layered silicates nanocomposites. Chem Mater 8(11):2628–2635

Vundavalli R, Vundavalli S, Nakka M, Rao DS (2015) Biodegradable nano-hydrogels in agricultural farming-alternative source for water resources. Proc Mater Sci 10:548–554

Williams RJ, Vallo CI (1988) Statistics of free-radical polymerizations revisited using a fragment approach. 2. Polyfunctional monomers. Macromolecules 21(8):2571–2575

Woodhouse J, Johnson M (1991) The effect of gel-forming polymers on seed germination and establishment. J Arid Environ 20(3):375–380

Wu J, Wei Y, Lin J, Lin S (2003) Study on starch-graft-acrylamide/mineral powder superabsorbent composite. Polymer 44(21):6513–6520

Zhang J, Wang A (2007) Study on superabsorbent composites. IX: synthesis, characterization and swelling behaviors of polyacrylamide/clay composites based on various clays. React Funct Polym 67(8):737–745

Zhang MC, Zhou Q, Zhou Y, Li AM, Shuang CD (2012) Efficient adsorption and desorption of Cu 2+ by a novel acid-resistant magnetic weak acid resin. Chin Chem Lett 23(11):1267–1270

Nanotechnology in Agriculture, Food Process Product, and Food Packaging

7

A. Allwyn Sundarraj

Abstract

Nanotechnology is a new boundary, and food engineering is a rising technology. Recently, "nanobiotechnology" has become a contemporary technological field with a broad space of analysis with relation to agriculture, food processing, protection, packing, dairy, enlargement of useful foods, and soured food products. Food and dairy farm makers, agrarian manufacturers, and customers may gain additional aggressive situation through technology. The grocery store requirement technology, that is crucial to stay smart management within the food processing business to supply contemporary genuine, appropriate, and flavorful food products. Nanotechnology has the possibility to transform the food business. Nanopacking solution can focus additionally on food security by dominant microorganism growth, detain reaction, visibility, and satisfaction. Nano has prospective nano-developed agrochemicals, enrich biological process values, and generation of novel product through bioactive encapsulation. The subject of this chapter includes the appliance of nanotechnology in agriculture, food processing, packaging, food safety, and security also as regulation of nanotechnologies to confirm safety.

Keywords

Nanobiotechnology • Food • Agriculture • Processing • Preservation • Packaging • Safety

A.A. Sundarraj, B.Tech., M.Tech., Ph.D. (✉)
Department of Food Processing and Engineering, Karunya University,
Karunya Nagar, Coimbatore, 641114 Tamil Nadu, India
e-mail: asrthegreat@gmail.com

R. Prasad et al. (eds.), *Nanotechnology*, DOI 10.1007/978-981-10-4573-8_7

7.1 Introduction

In today's aggressive store, new boundary technology is important to stay within the agriculture, food processing, and packaging trade (Sundarraj et al. 2013). Nanotechnology could be a new bough of technology, which has high expectations to change the world radically (Leinonen and kivisaari 2010). At one-billionth of a meter, a nanometer is minuscule much too small for the human eye to see. For most humans, anything 100 nm or less may be intolerable to understand as noteworthy (Sundarraj et al. 2014a).

Nanotechnology uses different abilities, processing, and various parts at the supramolecular level, roughly in a scale between 1 and 100 nm (Tony 2009; Suman et al. 2010; Aziz et al. 2015; Prasad et al. 2016) to generate new belongings and to restore desired advantage (IRGC 2008). However, researchers have commenced a field of study that could exactly fit on a person's fingernail: nanoscience and technology, engineering, physics, an imitative of chemistry, and microfabrication capabilities (Ravichandran 2009, 2010).

The current nanomaterials which can offers many new chances for the food processing and packing industries, which includes food coloring, flavoring, biological process food supplements, and antimicrobial constituents for food packing (Prasad et al. 2017). Furthermore, nanotechnologies are used in completely dissimilar sectors, such as medicines, veterinary medicines, animal feeds, food cleanliness, and biotechnology—the area referred to as nanobiotechnology (Roco and Bainbridge 2002; Alfadul and Elnesway 2010; Prasad 2014; Prasad et al. 2016).

Food security means that all food should be shielded from physicochemical, biological (Sundarraj et al. 2013), and radiation pollution through processing, handling, and issuing (Alfadul and Elnesway 2010). The current chapter is based on the applications of technology within the agriculture, food processing, preservation, and food packing with special observation to demonstration of quality and security.

7.2 Nanotechnology

Nanotechnology can possibly affect many features of food and farming management. Food safety, illness investigation, new implements for granular and cellular biology, new parts for microorganism observation, and preservation of the surroundings area unit samples are the vital links of nanotechnology of farming and food networks (Ozimek et al. 2010; Kaur et al. 2012; Sundarraj et al. 2013). Examples of nanoscience as implements to attain any promotion within the food trade are:

- Enlarged safety of production, process, and shipping of food merchandise between sensors for microorganism and contamination (Kaur et al. 2012).

Table 7.1 Kinds of nanoparticles, specific functions, and their targeted applications

Type of nanoparticles	Function	Application
1. Metal nanoparticles (Silver, ZnO)	Food ingredients	Enhanced gastrointestinal uptake of metal
	Packing parts	Increase barrier properties
	Food production devices	
2. Sprays	Refrigerators, storage bottles	Anti-bacterial coating
3. Complicated nanostructures	Nanosensors in packaging	Observation of food decaying
4. Hand-held devices	Depot ordering investigation	Observing of impurities
5. Includes energetic nanoparticles	Moving out of packing materials	Oxygen forage
6. Filters with nano-holes contaminants	Water decontaminate	Transfer of microorganism
7. Nano-sized foods	Food supplements	Professed increased uptake
8. Conveyance systems	Food supplements	Protecting and carriage of content

Source: www.food.actapol.net, Kaur et al. (2012)

- Networks that offer incorporation of observing, localization, and device control of food merchandise and safety of food process and transportation (Ozimek et al. 2010).
- Enclosing and transportation arrangements that carry and secure useful food additives to their specific site of action.

Nanomaterials and nanostructures can offer the particular targeted applications. Examples are listed in Table 7.1.

The applications of nanoscience and technology in food and agriculture are new compared with its use in drugs and prescription drugs. The second approaches of nanotechnology—bottom-up and top-down—build the fundamentals of this century's frontier technology (Ozimek et al. 2010; Sundarraj et al. 2013).

7.3 Nanotechnology in Food Sector

Nanotechnology has the possibility to reorganize the novel agricultural and food safety systems, disease-treatment delivery ways, cellular biology, sensors for infectious agent, chemical substance, packing materials, atmospheric condition, and education of the general public and thereafter are examples of the necessary contact that nanoscience and technology might have on the science and technology of farming and food networks (Moraru 2003; Bouwmeester 2009; Garcia et al. 2010; Prasad 2014; Prasad et al. 2014, 2016; Aziz et al. 2015; Cristina et al. 2015).

The inclusion of merchandise in Table 7.2 relies on stamping info on the merchandise as understood by the producer (Bouwmeester 2009). The claims that

Table 7.2 Different approaches of nanoscience and technology in agriculture, food processing, and packaging sector

Chain phase	Application	Nanotechnology	Function
1. Farming manufacture	Nanosensors	Nanospray on food items	Binds and colours organisms
	Pesticides	Nano-emulsions, capsules	Observation of mycotoxins and organism
2. Manufacturing and process of food	Food manufacture	Nanoceramic appliances	Large reactive area, anti-bacterial coating
3. Preservation	Food product	Nano-sized silver sprays	Anti-bacterial coating
			Detection of food deterioration
4. Useful food, consumption	Ingredients	Colloidal metal nanopores	Claimed increased fascinating uptake of metal

Bouwmeester (2009) and Garcia et al. (2010)

these merchandise obtain nanoscience and technology cannot be verified from the data given. Yet, Table 7.2 offers direction for the sort of merchandise that might be found within the market, as well as areas of use. Researchers and controllers ought to bear in mind this, and use the data within the prioritization of analysis (Bouwmeester 2009).

7.4 Nanotechnology Within the Food Trade

The term "nanofood" narrates the food that has been cultured, manufactured, or prepacked. Exploitation nanotechnology expertise nanomaterials are value-added (Morris 2007; Momin et al. 2013; Sundarraj et al. 2013). The applications of nanoscience and technology for the food sections make up the subsequent main classifications:

- Where nano-sized, nano-enclosing have dimensions of only a few nanoparticles ingredients are used;
- Where food additives are developed to make nanostructures (Momin et al. 2013);
- Where nanomaterials are included to expand enhanced, vigorous materials for food packaging (Momin et al. 2013);
- Where nanotechnology-based implements and substances are used, e.g., for nanostraining, water investigation (Momin et al. 2013);
- Where nanosensors are used for food security and attributable and material perception (McCall 2007; Chaudhry et al. 2008; Momin et al. 2013).

7.4.1 Nano Foodstuff

The worldwide sales of nanoscience and technology merchandise within the food process merchandise and potable packaging sector increased from U.S. $150 million in 2002 to U.S. $860 million in 2004 (Momin et al. 2013) and are expected to increase to U.S. $20.4 billion by 2010 (Helmut Kaiser Consultancy 2004). Cientifica has calculated the (2006) food approaches of nanotechnologies at approximately $410 million (food process U.S. $100 million, food additives U.S. $100 million, and food wrapping U.S. $201 million) (Momin et al. 2013).

7.5 Nanotechnology in Agriculture

The Institute of Food Science and Technology (IFST) initiated its "statistical affirmation on nanoscience and technology 2006" with the subsequent stress on industrial applications (Meetoo 2011). With the lack of dependable information, the Nanowerk website issued an outline of contemporary fields of approaches to farming, food processing, food wrapping, and food additives, which is shown in Table 7.3 (Nanowerk Food 2007).

Many reports establish agriculture as a serious field of potential for nanoscience and technology applications. In the near future, nanostructured catalysts are obtainable, which will expand the potency of rodents and herbicides, permitting lower quantities to be used (IRGC 2008).

7.6 Nanotechnology in Food Processing

Nanoscience and technology appealed to food manufacturing equipment. Examples of this kind of nanotechnologies are coating of mechanisms or the employment of nanosieves (e.g., to separate out bacteria) (Source: www.scielo.br). Whereas straight food connection is clear, this approach of nanoscience and technology is predicted to own negligible extra safety considerations compared with typical capabilities—over to food is predicted to be insignificant (Garcia et al. 2010). The kind of matter (and wear off as a result of the use) of coatings would possibly need some study; however, this cannot be solely associated with the protection of nanosciences (Garcia et al. 2010).

7.6.1 Nanodispersions and Nanocapsules

Functional ingredients are essential parts of a good variety of business merchandise, together with prescription drugs, tending merchandise, agrochemicals, and foods process merchandise. These practical ingredients are available in a range of various molecular and physical forms, such as polarities, molecular weights, and solid, liquid, and gaseous states. Functional additives are infrequently used in their pure

Table 7.3 Future fields of nanoscience and technology applications in farming, food processing, food wrapping, and food additives

Agriculture	Food process	Food wrapping	Supplements
Single molecule diversion to work out enzyme/substrate interactions	Nanocapsule to improve bioavailability of nutraceuticals	Antibodies hooked up to fluorescent nanoparticles to detect chemicals	Nanosize powders to extend immersion of nutritional supplements
Transportation of advance hormones during a controlled fashion	Nanotubes and nanoparticles as solidification and viscosifying representatives	Nanoclays and nanofilms as fence substance to stop decomposition	Nanoencapsulation of nutraceuticals for higher absorption
Nanosensors for watching soil conditions	Nanocapsule essence of plant-based steroids to interchange a meat's cholesterin	Electrochemical nanosensors to observe ethylene	Nanochelates to transport nutrients additional expeditiously to cells while not moving color or taste of food
Nanochips for distinctive preservation and tracking	Nanoparticles by selection bind and take away chemicals	Active against microbes and antifungal exterior layering with nanoparticles	Vitamin sprays distribute vital molecules into nanodroplets for higher incorporation
Nanocapsules to transport vaccines	Nanoemulsions and particles for higher convenience and dissemination of nutrients	Bright, powerful and additional heat-impenetrable films with silicate nanoparticles	–
Nanoparticles to deliver Deoxyribonucleic acid to plants	–	Modified permeation behavior of foils	–

Nanowerk Food (2007), Meetoo (2011), and Sunandan and Dutta (2009)

form. Alternatively, they often are included in some form of transportation network (Weisis et al. 2006).

The attributes of the transport network are one of the most dominant parts determining the ability of useful foods in much commercial merchandise (Weisis et al. 2006; Mishra 2015).

7.6.2 Association Colloids

Alliance colloids are thermodynamically approving networks whose generation is usually operated by the hydrophobic result—that is, the depletion of the communication area linking the nonpolar categories of the surfactant that contain the alliance colloid and water (Golding and Sein 2004; Wesis et al. 2006). The type of alliance colloid established and the creation of the resultant formations determined on the absorptions and molecular predictable of the surfactant and co-surfactant used as well as the conquer atmospheric orders.

7.6.3 Nanoemulsions

Nanoemulsions have been made and studied for several years; therefore, an outsized body of documentation of their extraction, isolation, creation, and implementation exists (McClements 2004). Encapsulating utility parts among the droplets usually allows a delay of chemical mortification action by technology—the belongings of the surface sheet encompass them (McClements and Decker 2000; Wesis et al. 2006).

7.6.4 Nanolaminates

Nanotechnology provides food scientists (Sundarraj et al. 2013) with variety of ways to make novel laminate films appropriate to be used within the food business (Wesis et al. 2006). One of the foremost powerful ways is predictable on the LBL perishing capabilities, within which the imposed surfaces are covered with surface films made up of multiple nanolayering of various substances (Decher and Schlenoff 2003).

7.6.5 Nanofrying

The U.S.-based Oil-fresh Agency has retailed a replacement nanoceramic product that decreases oil use in hotels and sustenance retailers by 0.5 as a result of its massive extent (Joseph and Morrison 2006; Pehanich 2006; Momin et al. 2013).

7.6.6 Novel Foods

Wrapping paper foods and NanoteK council have arrangements to include the electronic tongue into foods to unleash correctly managed quantities of the appropriate molecules for the custom-made tailor foods (Shelke 2006; Sozer and Kokini 2009; Momin et al. 2013).

7.6.7 Nanofiltration

Nanofiltration dividing substances of less than 0.001 μ in size decline powerfulness and multivalent ions (Cuartas et al. 2007; Momin et al. 2013). It is an alternative to reverse diffusion and is cost-effective (Sangamithra and Thirupathi 2009; Momin et al. 2013).

7.6.8 Nanocomposites

Chemical compound clay nanocomposites are collected from natural polymer compound matrices and organophilic clay fillers (Kim et al. 2003). Exfoliation may be a state within which the silicate layering are utterly divided and spread in a very continual polymeric compound matrix. The formation and belongings of the ensuring nanocomposites will be modified by dominant, delicate polymer–clay interactions (Oya et al. 2000; Wesis et al. 2006).

7.6.9 Nanotubes

Carbon nanotubes are used widely as a non-food application of nanotechnology. Circular proteins from milk can be created—gathering into equally structured nanotubes below applicable atmospheric orders (Wesis et al. 2006).

7.6.10 Nanoceuticals

"Nanoceuticals" to obtain quality and business dairy and food ingredients obtaining nanoparticles is possible (Chen et al. 2006; Sinha et al. 2008; Sundarraj et al. 2014a, b).

Examples of food-connected nanoproducts:

- Carotenoids nanoparticles is unfold in H_2O and should be superimposed to fruit drinks for better systemic availability (Sinha et al. 2008);
- Canola oil primarily based totally nanosized micellar network is declared to produce transport of various substances, such as vitamins and phytochemicals (Sundarraj et al. 2014a, b)
- To forestall the buildup of cholester in a variety of nutraceuticals incorporated among the conveyors embrace antioxidant, beta-carotenes, and phytosterols (Sundarraj et al. 2014a, b)

7.6.11 Inactivation of Enzymes

Nanoporous media, nanofibers, and carbon nanotubes are exploitation for catalyst inactivation (Mao et al. 2006; Kim and Wang 2008; Kosseva et al. 2009; Momin et al. 2013).

7.7 Nanoscience and Technology in Food Packaging and Preservation

Food could be an advanced biomaterial that is issue to several biological and physicochemical changes that have an effect on its amount and time period. Organic altered are created by organisms, insects, and rodents (Sharma and Singh 2000; Giueppe and Monica 2010). The main goals of food wrapping are to scale back the speed of quality loss and to extend merchandise time period to the extent needed by the administration system (Han 2003; Giueppe and Monica 2010). Packing techniques are invented to fulfill these goals, such as high-barrier packing and MAP.

Nano wrapping approaches as Food Contact Materials (FCMs) are expected to grow from a $66 million business in 2003 to more than $360 million by 2008 (Scrinis and Lyons 2007). Approaches for FCMs victimization nanoscience and technology are as follows:

- FCMs integrating nanomaterials to enhance packing (Mommin et al. 2013)
- Active FCMs that assimilate nanoparticles with antimicrobial or gas scavenging belongings (Mommin et al. 2013)
- Intelligent or sensible food packing includes nanosensors for sensing and communication of microorganism and organic chemistry modifies, unleash of antimicrobials, antioxidants, enzymes, and nutraceuticals to enlarge time period (Mommin et al. 2013)
- Biodegradable polymer-nanomaterial composites by initiation of inanimate spots, like clay, into the biopolymeric matrix and may even be managed with surfactants that are victimization for the moderation of superimposed silicate (Doyle 2006; Joseph and Morrison 2006; Lopez et al. 2006; Brody 2007; Choudhary et al. 2008; Miller and Sejnon 2008; Sozer and Kokini 2009; Momin et al. 2013)

7.7.1 Improved Packing

A range of nanoparticle strengthened polymers, additionally entitled as "nanocomposites" have evolved (Mommin et al. 2013), which usually contain up to 5 % w/w nanoparticles with clay nanoparticle composites with better railing barrier belongings (80–90% reduction) for the production of bottles for brew, cooking oils, and effervescent drinks and films (Brody 2007; McCall 2007; Chaudhry et al. 2008; Momin et al. 2013). The United States Food and Drug Administration (USFDA) has accepted the use of nanocomposite in foods (Sozer and Kokini 2009; Momin et al. 2013).

7.7.2 Active Packing

"Active" FCMs that incorporate nanoparticles with antimicrobial or oxygen scavenging properties (Aziz et al. 2016).

7.7.3 Smart/Intelligent Packaging

Mommin et al. (2013) have developed a smart food packaging incorporating nanosensors for sensing and sgnaling of microbial and biochemical changes, release of antimicrobials, antioxidants and enzymes.

7.7.4 Edible Nano Coating

Waxy coating is used widely for some foods such as apples and cheeses. Recently, nanotechnology has enabled the development of nanoscale edible coatings as thin as 5 nm wide, which are invisible to the human eye. Edible coatings and films are currently used on a wide variety of foods, including fruits, vegetables, meats, cheese and bakery products (Momin et al. 2013).

7.8 Nanoscience and Technology in Food Safety

Identification of a small quantity of a chemical microorganism within the food system is another possible use of technology (Source: www.bioline.org.br). The exciting risk of mixing biology and nanotechnology into sensors is promising, because it can take a considerably reduced response time to sense a possible drawback. This may lead to additional safety for the food processing system (Alfadul and Elneshwy 2010).

7.8.1 Role of the International Scientific Community on Food Safety

The vision for future food manufactures was summarized as "safe, property, and ethical" (Giuseppe and Monica 2010). Although abundant progress has been made for future decades and a few claim that "our food has never been safer," those concerned with the security of food offered ought to acknowledge that there is still far to go before that goal is reached (FAO 2000, 2003, 2007; USDA 2002; Giuseppe and Monica 2010).

The work of the World Health Organization (WHO) incorporate becomes stronger food security systems, encouraging sensible producing practices, and educating retailers and shoppers concerning acceptable food handling. Education of shoppers and coaching of food handlers in safe food holding is one of the foremost crucial intercessions for the interference of food-borne ill health (Source: qou-au.com).

WHO is promoting the utilization of all food technologies that can improve public health, such as sterilization, food irradiation, and fermentation (Giuseppe and Monica 2010). WHO has initiated work to determine a content that specializes in a border analysis of risks, advantages, and different issues associated with the assembly and consumption of foods derived from biotechnology (Giuseppe and Monica 2010).

7.8.2 Pharmacological Medicine and the Safety Side of Nanoparticles

Nanoparticles also may be taken up by damaged skin or brain cells (Miller and Sejnon 2008). Molecules smaller than 70 nm will enter cell nuclei and even cause disability of polymer replication and transcription (Chaudhry et al. 2008; Momin et al. 2013).

7.9 Laws

The EU Union laws for food and food packing have counseled that for the initiation of recent technology, specific safety orders and analyzing procedures are needed (Halliday 2007). The National Nanotechnology initiative (NNI) includes 26 savage agencies, as well as Atmospheric Protection agency, the Department of Health and Human Services, and the Client Product Safety Commission (Dingman 2008). In the United States, nanofoods and most of the food packing are controlled by the USFDA (Badgley et al. 2007), whereas in Australia, nanofood supplements are balanced by Food Standards Australia and New Zealand (FSANZ), underneath the Food Standards Code (Bowman and Hodge 2006; Momin et al. 2013).

Many organizations are concerned with technology analysis, laws, and suggestions. I debate their actions below. The USFDA has issued guidance on nanotechnology on their website (http://www.fda.gov/nanotechnology/). One vital truth is that the office controls merchandise, not innovations, and the website makes the subsequent observations regarding problems that the USFDA anticipate within the balancing of technology products:

- The chance that a lot of the technology merchandise that the Agency regulates is variation merchandise.
- Because USFDA regulates merchandise supported their statutory classification instead of technology they use, restrictive thought that an approach involves a technology merchandise might not occur until the beginning enlargement of that technology (USFDA 2006; Sundarraj et al. 2014a, b).

7.10 Future for Nanotechnology

Nanoscience and technology is slowly entering into well-liked culture. There is promise that the long run of technology is incredibly bright. It will confer distinctive blessings on processed foods in many ways. Nano-based polymers with silicon oxide—primarily based nanoparticles sandwiched will amplify the belongings of pressure—delicate adhesive labels and make perishable belongings in them (Ravichandran 2010).

The Govt. of Republic of India has created a National Nanotechnology Institute to develop nanofoods by victimization nanotechnology throughout cultivation, process, preservation, and packing of food; however, it is not clear whether the communication has any helping of the Govt. of Republic of India or whether any serious design has gone into the proposal (Kaynak and Tasan 2006; Ravichandran 2009). Nanotechnology in agriculture and food does not mean atomically changed food process (or) food made by nanomachines (Sorrentino et al. 2007; Ravichandran 2010).

7.11 Conclusions

Nanoscience and technology has had an incredible impact on food merchandise, making them tastier and more nutritious, and has created new food process merchandise, new food wrapping, and security. Agricultural producers, consumers, food makers, and researchers may gain through the application of food and agricultural nanotechnology. The new cohort of wrapping substances is able to match the requirements of protective fruits, vegetables, beverages, meat, cereals, and alternative food and food process merchandise. Using additional, acceptable nanoparticles, it will be possible to provide packages with a stringer, barrier, and thermal presentation.

Nanostructured matters can forestall the initiation of microorganism as a priority for food security. The nanosensors fixed within the packing might alert the buyer if a food has declined. It has authorized food security and food grade features. However, shopper acceptance of nanoscience and technology will ultimately dictate its success in food applications (Mommin et al. 2013). Nanotechnology has become more and more necessary for the food and agriculture sector. The integration of nanomaterials into food packing can improve the barrier belongings of wrapping substances and thereby will facilitate to scale back the employment of precious raw materials and the cohort of residue. It is widely expected that nanoscience and technology-derived food merchandise will become more accessible to shoppers worldwide within the coming years.

References

Alfadul SM, Elneshwy AA (2010) Use of nanotechnology in food processing, packaging and safety – review. Afr J Food Agric Nutr Dev 10:6

Aziz N, Faraz M, Pandey R, Sakir M, Fatma T, Varma A, Barman I, Prasad R (2015) Facile algae-derived route to biogenic silver nanoparticles: Synthesis, antibacterial and photocatalytic properties. Langmuir 31:11605–11612

Aziz N, Pandey R, Barman I, Prasad R (2016) Leveraging the attributes of *Mucor hiemalis*-derived silver nanoparticles for a synergistic broad-spectrum antimicrobial platform. Front Microbiol 7:1984. doi:10.3389/fmicb.2016.01984

Badgley C, Moghtader J, Quintero E, Zakem E, Chappel LM, Vazquez KA, Salon A, Perfecto I (2007) Organic agriculture and the global food supply. Renew Agr Food Syst 22(2):86–108

Bouwmeester H (2009) Review of health safety aspects of nanotechnologies in food production. Regular Toxicol Pharm 6(1):52–62

Bowman D, Hodge G (2006) Nanotechnology: mapping the wild regulatory frontier. Future 28:1060–1073
Brody AL (2007) Case studies on nanotechnologies for food packaging. Food Technol 7:102–107
Chaudhry Q, Scotter M, Blackburn J, Ross B, Boxall A, Castle L, Aitken R, Watkins R (2008) Applications and implications of nanotechnologies for the food sector. Food Addit Cont 25(3):241–258
Chen H, Wesis J, Shahidi F (2006) Nanotechnology in nutraceuticals and functional foods. Food Technol 3:30–36
Cristina M, Sabilo, Hongda C, Rickey YY (2015) Polymeric nanoparticles for food applications. Nanotechnology and functional foods. Wiley, Chichester
Cuartas U, Miranda BMIA, Pia AB (2007) Comparison of the behaviour of two nanofiltration membranes for sweet whey demineralization. J Dairy Sci 90:1094–11101
Decher G, Schenoff JB (2003) Multilayer thin films: sequential assembly of nanocomposite materials. Wiley – VCH, Weinheim, p 543
Dingman J (2008) Nanotechnology: its impact on food safety. J Environ Health 70(6):47–50
Doyle ME (2006) Nanotechnology: a brief literature review. http://www.wisc.edu/fri/briefs/FRIBrief_Nanotech_Lit_Rev.pdf
FAO (2000) Twenty second regional conference for Europe: food safety and quality as affected by organic farming. Porto, Portugal
FAO (2003) Agricultural biotechnology meeting the needs of the poor?
FAO (2007) Countries urged to be more vigilant about food safety. ROME/GENEVA
FDA – Food and Drug Administration (2006) FDA and nanotechnology products. Available at: http://www.fda.gov/nanotechnology/faqs.html. Accessed 28 Aug 2006
Garcia M, Forbe T, Gonzalez E (2010) Potential applications of nanotechnology in the agro – food sector. Cienc Technol Aliment Campinas 30(3):573–581
Giuseppe E, Monica S (2010) Science for food safety, security and quality: a review – part 2. Quality of Life 1(1):41–54
Golding M, Sein A (2004) Surface rheology of aqueous casein – monoglyceride dispersions. Food Hydrocoll 18(3):451–461
Halliday J (2007) EU Parliament votes for tougher additives regulation. FoodNavigator.com
Han JH (2003) Antimicrobial food packaging. In: Ahevenainen R (ed) Novel food PackTechniq. Woodhead Publishing Ltd, Cambridge, pp 50–70
Helmut Kaiser Consultancy (2004) Nanotechnology in food and food processing industry worldwide 2008–2010–2015 Report. Zurich, Switzerland
IRGC (2008) International risk governance council, towards an Integrative Approach, Geneva. www.irgc.org
JosephT,MorrisonM(2006)Nanotechnologyinagriculturefood:ananoforumreportpublishedin2006. http://www.worldcat.org/title/nanotechnology-in-agriculture-and-food-a-nanoforum-report/oclc/71212728
Kaur G, Singh T, Kumar A (2012) Nanotechnology: a review. Inter J Edu and Appli Res 2(1):50–53
Kaynak C, Tasan C (2006) Effects of production parameters on the structure of resol type phenolic resin/layered silicate nanocomposites. Ear Polym J 42:1908–1921
Kim JJWG, Wang P (2008) Nanobiocatalysis and its potential applications. Trends Bio 26(11):639–646
Kim KY, Lim HJ, Park SM, Lee SJ (2003) Synthesis and characterization of high impact polystyrene / organically modified layered silicate nanocomposites. Polymer (Korea) 27(4):377–384
Kosseva MRPSP, Kaur G, Kennedy JF (2009) Use of immobilized biocatalysts in the processing of cheese whey. Int J Biol Macromol 45:437–337
Leinonen A, Kivisaari S (2010) Nanotechnology perceptions, the literature review on media coverage, public opinion and NGO perspectives. Espo, VTT Tiedotteita – Research notes 2559:55
Lopez RA, Gavara R, Lagaron JM (2006) Bioactive packaging: turning foods into healthier foods through biomaterials. Trends Food Sci Technol 17:567–575

Mao X, Huang J, Leung MF, Du Z, Ma L, Huang Z, Li P, Gu L (2006) Novel core shell nanoparticles and their application in high capacity immobilization of enzymes. Appl Biochem Biotechnol 135:229–239

McCall, J (2007) On some factors that affect the 'feel' of molecules. Thesis submitted in School of Applied Science, Cranfield University

McClements DJ (2004) Food emulsions: principles, practice and techniques, 2nd edn. CRC Press, Boca Raton

McClements DJ, Decker EA (2000) Lipid oxidation in oil -in -water emulsions: impact of molecular environment on chemical reactions in heterogeneous food systems. J Food Sci 65(8):1270–1282

Meetoo DD (2011) Nanotechnology and the food sector: from the farm to the table. Emirates J Food Agric (EJFA) 23(5):387–403

Miller G, Senjen R (2008) Out of the laboratory and onto our plates – nanotechnology in food and agriculture. http://www.foeeurope.org/activities/nanotechnology/Documents/Nano_food_report.pdf

Mishra M (2015) Encapsulation technologies for modifying food performance. In: Handbook of encapsulation and controlled release. CRC Press, November 11, 2015, p 1512

Momin J, Jayakumar C, Prajapati JB (2013) Potential of nanotechnology in functional foods. Emir J Food Agri 25(1):10–19

Moraru CI (2003) Nanotechnology: a few frontiers in food science. Food Technol 57(12):24–29

Morris V (2007) Nanotechnology and food. Available: http://www.iufost.org/reports_resources/bulletins/documents/IUF.SIB.Nanotechnology.pdf

Nanowerk Food (2007) Nanowerk, the promises of food nanotechnology. http://www.nanowerk.com/spotlight/spotid=1846.php. Accessed 28 Sept 2016

Oya A, Kurokawa Y, Yasuda H (2000) Factors controlling mechanical properties of clay mineral/polypropylene nanocomposites. J Mater Sci 35(5):1045–1050

Ozimek L, Pospiech E, Narine S (2010) Nanotechnology in food and meat processing. Acta Sci Pol Technol Aliment 9(4):401–412

Pehanich M (2006) Small gains in processing and packaging. Food Process 11:46–68

Prasad R (2014) Synthesis of silver nanoparticles in photosynthetic plants. J Nanoparticles, Article ID 963961. http://dx.doi.org/10.1155/2014/963961

Prasad R, Kumar V, Prasad KS (2014) Nanotechnology in sustainable agriculture: present concerns and future aspects. Afr J Biotechnol 13(6):705–713

Prasad R, Pandey R, Barman I (2016) Engineering tailored nanoparticles with microbes: quo vadis. WIREs Nanomed Nanobiotechnol 8:316–330

Prasad R, Bhattacharyya A, Nguyen QD (2017) Nanotechnology in sustainable agriculture: recent developments, challenges, and perspectives. Front Microbiol 8:1014. doi:10.3389/fmicb.2017.01014

Ravichandran R (2009) Nanoparticles in drug delivery: potential green-nano-biomedicine applications. Int J Nanotechnol Biomed 1:108–130

Ravichandran R (2010) Nanotechnology applications in food and food processing: innovative green approaches, opportunities, and uncertainties for the global market. Inter J Green Nanotech: Phys Chem 1(2):72–P96. doi:10.1080/19430871003684440

Roco MC, Bainbridge WS (2002) Converging technologies for improving human performance: integrating from the nanoscale. J Nanopart Res 4:281. doi:10.1023/A:1021152023349

Sangamithra A, Thiruppathi V (2009) Nanotechnology in food. http://www.technoprenur.net/infromationdesk/sciencetechmagazine/2009/jan09/nanotechnology.pdf

Scrinis G, Lyons K (2007) The emerging nanocorporated paradigm: nanotechnology and the transformation of nature, food and agro – food systems. Int J Socio Food Agri 15(2):22–44

Sharma RM, Singh RR (2000) Harvesting, postharvest handling and physiology of fruits and vegetables. In: Verma LR, Joshi VK (e) (eds) Postharvest technology of fruits and vegetables. Handling, processing, fermentation and waste management. Indus Publishing Co., New Delhi, pp 94–147

Shelke K (2006) Tiny invisible ingredients. http://www.foodprocessing.com/articles/2006/227.html
Sinha VK, Vinay A, Bhinge JR (2008) Nanocochleates: a novel drug delivery technology. Pharma info Net 6:5
Sorrentino A, Gorrasi G, Vittoria V (2007) Potential perspectives of bio- nanocomposites for food packaging applications. Trends Food Sci Technol 18:84–95
Sozer N, Kokini JL (2009) Nanotechnology and its applications in the food sector. Trends Bio 27(2):82–89
Suman PR, Jain VK, Varma A (2010) Role of nanomaterials in symbiotic fungus growth enhancement. Curr Sci 99:1189–1191
Sunandan B, Dutta J (2009) Nanotechnology applications in pollution sensing and degradation in agriculture: a review. J Environ Chem Lett 7:191–204
Sundarraj AA, Ragavi J, Rubila S, Tirouthchelvamae D, Ranganathan TV (2013) Recent trends in nanotechnology applications in foods. Inter J Eng Res and Tech 2(10):956–961
Sundarraj AA, GuhanNath S, Aaron SI, Ranganathan TV (2014a) Recent innovations in nanotechnology in food processing and its various applications – a review. Int J Pharm Sci Rev Res 29(2):116–124
Sundarraj AA, Wesley SJ, Raja P, Tiroutchelvamae D (2014b) Review on nanotechnology applications in food packaging and safety. Inter J Eng Res 3(11):645–651
Tony J (2009) Application of zinc oxide quantum dots in food safety. Nanotoxicity
USDA (United States Department of Agriculture) (2002) Traceability for food marketing and food safety: what's the next step? Agricultural outlook/January–February, 21–25
Wesis J, Takhistow P, McClements DJ (2006) Functional materials in food nanotechnology. J Food Sci 71(9):107–116

Website References

www.scielo.br
qol-au.com
www.bioline.org.br
www.food.actapol.net

Green Nanotechnology: Biomimetic Synthesis of Metal Nanoparticles Using Plants and Their Application in Agriculture and Forestry

8

Mohammadhassan Gholami-Shabani, Zeynab Gholami-Shabani, Masoomeh Shams-Ghahfarokhi, Fatemehsadat Jamzivar, and Mehdi Razzaghi-Abyaneh

Abstract

Biomimetic nanotechnology is an outstanding investigation area at the meeting place of life sciences with physics and engineering. It is an uninterrupted emerging field that deals with knowledge transfer from biology to nanotechnology. Biomimetic nanotechnology is a scope that has the potential to support extensively successful mastering of major global challenges and solve the problems. Metallic nanoparticles are being utilized in every phase of science along with engineering, including agriculture fields, and are still charming the scientists to explore new dimensions for their respective worth, which is generally credited to their corresponding small sizes. The up-and-coming researches have proven their antimicrobial significance. The present chapter is devoted to the possibility of metal nanoparticle synthesis using plant extracts and microorganisms. This approach has been actively pursued in recent years as an alternative, efficient, low-cost, and environmentally safe technique for producing nanoparticles with specified properties. The main attention is on the role of the natural plant biomolecules involved in the bioreduction of metal salts during the nanoparticle synthe-

M. Gholami-Shabani
Department of Mycology, Pasteur Institute of Iran, Tehran 13164, Iran

Department of Nanobiotechnology, Pasteur Institute of Iran, Tehran 13164, Iran

Z. Gholami-Shabani
Faculty of Aerospace, Science and Research Campus, Islamic Azad University, Tehran, Iran

M. Shams-Ghahfarokhi
Department of Mycology, Faculty of Medical Sciences, Tarbiat Modares University, Tehran 14115-331, Iran

F. Jamzivar • M. Razzaghi-Abyaneh (✉)
Department of Mycology, Pasteur Institute of Iran, Tehran 13164, Iran
e-mail: mrab442@yahoo.com; mrab442@pasteur.ac.ir

R. Prasad et al. (eds.), *Nanotechnology*, DOI 10.1007/978-981-10-4573-8_8

sis. Moreover, attempts to apply nanotechnology in agriculture began with the growing realization that conventional agriculture technologies would neither be able to grow productivity any further nor restore ecosystems damaged by existing technologies back to their pristine situation, in particular because the long-term effects of farming with "miracle seeds," in conjunction with pesticides, irrigation, and fertilizers, have been questioned both at the scientific and policy levels and must be steadily phased out. Nanotechnology in agricultural science has gained momentum in the past decade with a plenty of public funding, but the pace of growth is modest, even though many disciplines come under the umbrella of agriculture. This could be credited to a unique nature of farm production, which functions as an open system whereby energy and material are swapped freely; the scale of request of input materials is gigantic in contrast with industrial metal nano-products; an absence of control over the input nanomaterials in contrast with industrial nano-products (e.g., the cell phone) and because their fate has to be conceived on the geosphere (pedosphere), hydrosphere, biosphere, and atmosphere continuum; the time delay of emerging technologies reaching the farmers' field, specifically given that many emerging economies are reluctant to spend on innovation; and the lack of foresight subsequent from agricultural education not having attracted an enough number of clear minds the world over, whereas personnel from kindred disciplines might absence an understanding of agricultural production methods. If these issues are taken care of, nanotechnological impact in farming has bright views for improving the efficiency of nutrient use through nano-formulations of fertilizers, breaking yield obstacles through bionanotechnology, surveillance, and control of "pests and diseases," apprehension mechanisms of host-parasite interactions at the cellular and molecular levels, growth of new-generation pesticides and their carriers, packaging and preservation of foodstuff and food additives, strengthening of natural fibers, removal of contaminants from water and soil, improving the shelf life of flowers and vegetables, clay-based nanoresources for reclamation of salt-affected soils, precision water management, and stabilization of erosion-prone surfaces to name a few.

Keywords
Biomimetics • Nanobiotechnology • Nanoparticles • Agriculture • Natural products • Plant extracts

8.1 Introduction

Bionanotechnology and nanobiotechnology are important disciplines that discuss the intersection of biology and nanotechnology (Amin et al. 2011; Gholami-Shabani et al. 2014; Nalawade et al. 2014). Bionanotechnology as complementary part for various related technologies usually refers to the survey of how the purposes of nanotechnology can be guided by investigating how biological "systems" work and acclimatize these biological subjects into improving existing nanotechnologies or creating new ones (Jayaseelan et al. 2013). Nanobiotechnology refers to the methods that nanotechnology is used to make devices to study biological systems (Steinmetz and Evans 2007; Baharara et al. 2015; Prasad et al. 2016). In the past decade, the evolution of extraordinary "green chemistry" approaches for synthesis of metal nanoparticles has grown a major attention for researchers (Karthick et al. 2012; Prasad 2014; Gholami-Shabani et al. 2015, 2016). These approaches were used to discover an eco-friendly creation of well-characterized nanoparticles. One of the most discussed approaches is synthesis of metal nanoparticles via organisms of which plants are considered to be the best candidates, and they are appropriate for large-scale bioproduction of nanoparticles. Metal nanoparticle synthesis by plants is more stable and the frequency of production is faster than microorganisms. Also, metal nanoparticles are more different in size and shape compared with those synthesized by other organisms. The increasing advantages of using plants and herbs to synthesize metal nanoparticles encourage the researchers to search mechanisms of metal ions' bio-reduction and to discover the probable mechanism of metal nanoparticle synthesis in plants.

Historically, humans have tried to use organisms and plant to develop their performance and beneficial properties for thousands of years. A long time ago, any technical information of heredity and the methods of breeding, crossing, and choosing plants were extensively experienced with the goal to realize superior harvests, extra beautiful flowers, and advantageous medicinal treatments (Nalawade et al. 2014; Vijayakumar et al. 2013; Baharara et al. 2015). During the past century, technical labors were focused toward refining plants, and in current period usage, the implements of molecular biology have been carried. Currently, our considerate of the genomics of plants and their multifaceted gene expression composition is exciting. Genetically engineered plant productions with developed agriculture characters have complete the move from research laboratory benches and greenhouses to grounds all over the world, where their existence grown up commercially. The fusion of biotechnology and nanotechnology on the past period has managed to creative cooperation and making of the novel field of "bionanotechnology"; nonetheless, only very newly has it been known that this knowledge could be useful to crops (Karthick et al. 2012; Kumar et al. 2014). Crops and agriculture are essential to trade worldwide, principally food products.

The variation of nanomaterials usable, together with lilting surface attributes, creates a great platform for a wide range of biomedical and biological applications (Roopan et al. 2012; Ganaie et al. 2014; Huang et al. 2015; Aziz et al. 2015). The worldwide activity in bionanotechnology expanded to the agriculture research. Printed work is increasing in the range of study and usage of artificial metal nanoparticles for "gene delivery" to transport plant safety materials and to improve other beneficial plant attributes.

There are a lot of significant applications for metal nanoparticles in pharmacy and medicine. Gold and silver nanoparticles are the highest ones used for biomedical applications and in emerging interdisciplinary field of nanobiotechnology (Gade et al. 2010; Abbasi et al. 2015; Prasad et al. 2014, 2016). For example, oligonucleotide-covered gold nanoparticles have been used for polynucleotide or protein finding using various tracing/characterization approaches by gel electrophoresis, atomic force microscopy, Raman spectroscopy, scanometric assay, amplified voltametric detection, surface plasmon resonance imaging, and chronocoulometry (Iravani and Zolfaghari 2013). Also, gold nanoparticles have been working in capillary electrophoresis, protein analysis, immunoassay, and cancer cell detection (Iravani 2011; Sathishkumar et al. 2014; Gholami-Shabani et al. 2016).

In the field of medicine, gold nanoparticles are used for several goals. They may be used as signs for biologic tests. After uptake of nanoparticles, they can act as exact and strong thermal knifes to kill cancer (Kathiravan et al. 2014). Furthermore, gold nanoparticles are able to persuade apoptosis in chronic lymphoid leukemia (B-cell chronic lymphocytic leukemia). Also researchers used silver nanoparticles in their wide applications, for example, bio-labeling, sensors, and filters combined circuits, antimicrobial spray fibers, cell conductors, and antimicrobials (Roopan et al. 2012; Vasanth et al. 2014; Premanand et al. 2016).

Antimicrobial effects of silver nanoparticles reason the use of these nanoparticles in different industries, livestock, various fields of medicine (such as metal sensing, drug delivery imaging, tissue repair and disinfection), accessories, agriculture, packaging, makeups, military, and health. Silver nanoparticles show potential antimicrobial effects against infectious organisms (Aziz et al. 2015, 2016). This chapter discusses about nanomaterial interplay with plants and the starts of money-making utilization. Special attention will be made to the action and reaction of nanomaterials with plants, with specific focus on the possibility for useful application in agronomy and connected profitable parts (Fig. 8.1). In this chapter, we highlighted the role of plants in metal nanoparticle synthesis and green-nanotechnology applications.

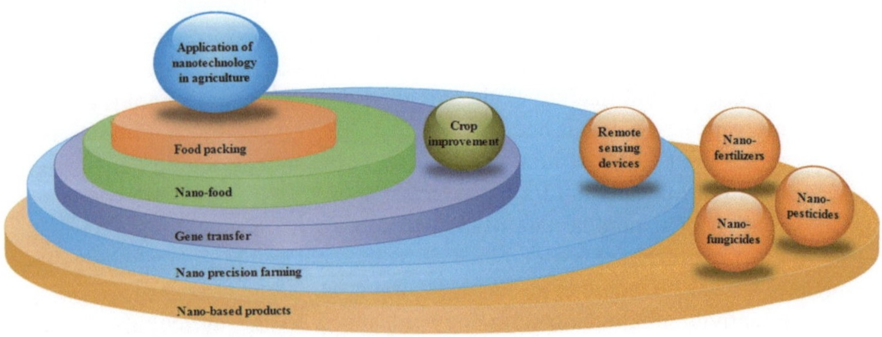

Fig. 8.1 Various applications of nanotechnology in agriculture

8.2 Role of Plants, Animals, and Microbial Cells in Nanobiotechnology

8.2.1 Plant and Animal Cells

There are 15 major differences in biotechnological applications of plant and animal cells (Table 8.1). Plant cells in two important subjects are different from animal cells. The plant cells are including photosynthetic organizations and a cell wall (Fig. 8.2). Photosynthesis process occurs in organelles called chloroplast, in the masses of thylakoids, the membranous sub-organelles, which are ready in masses called grana, and contain two compound protein units, photosystem I and photosystem II (Showalter 1993; Duca 2015). Photosystem I (PSI) has a response center and 14 protein subunits plus 4 membrane-related feeler compound proteins that take light and leader its energy to the response center. This assembled ordinance of

Table 8.1 Comparative discrepancy between animal and plant cells

	Plant cells	Animal cells
1	Plant cell is usually larger in size	Animal cell is comparatively smaller in size
2	It is enclosed by a rigid cellulose cell wall in addition to plasma membrane	It is enclosed by a thin, flexible plasma membrane only
3	It cannot change its shape	Animal cells often can change shape
4	Plastids are present. Plant cells exposed to sunlight contain chloroplast	Plastids are usually absent
5	A mature plant cell contains a large central vacuole	Animals cells often possess many small vacuoles
6	Nucleus lies on one side in the peripheral cytoplasm	Nucleus usually lies in the center
7	Centrioles are usually absent except in motile cells of lower plants	Centrioles are practically present in animal cells
8	Lysosomes are rare	Lysosomes are always present in animal cells
9	Glyoxysomes may be present	They are absent
10	Tight junctions and desmosomes are lacking. Plasmodesmata is present	Tight junctions and desmosomes are present between cells. Plasmodesmata are usually absent
11	Reserve food is generally in the form of starch	Reserve food is usually glycogen
12	Plant cell synthesize all amino acids, coenzymes and vitamins required by them	Animal cells cannot synthesize all amino acids, co-enzymes and vitamins required by them
13	Spindles formed during cell divisions in anastral i.e. without asters at opposite poles	Spindle formed during cell division is amphiastral, i.e., has an ester at each pole
14	Cytokinesis occurs by cell plate method	Cytokinesis occurs by construction or furrowing
15	Plant cell does not burst if placed in hypotonic solution due to the presence of the cell wall	Animal cell lacking contractile vacuoles usually burst, if placed in hypertonic solution

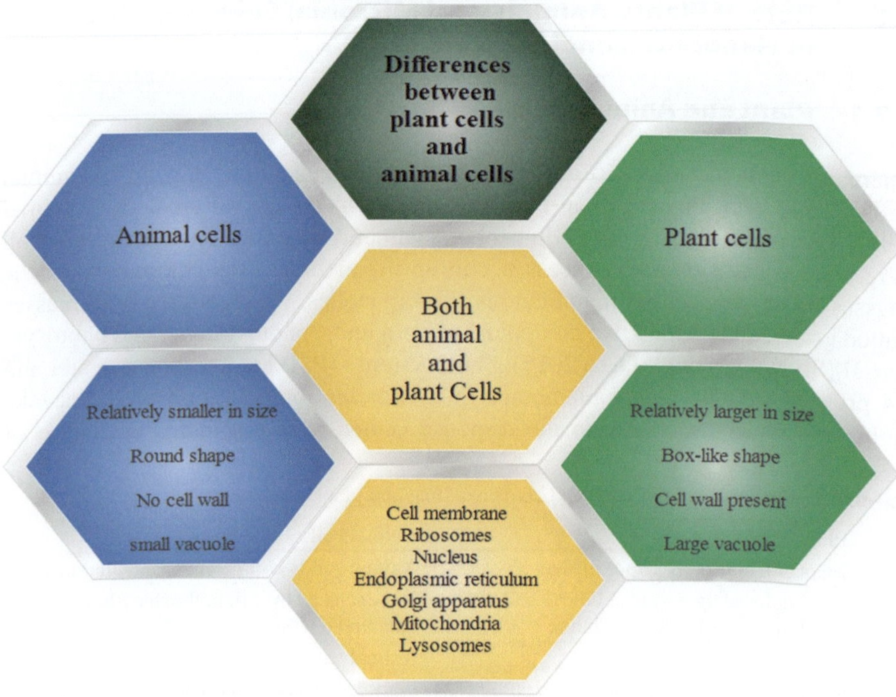

Fig. 8.2 Role of animal and plant cells in nanobiotechnology

protein treillage and chlorophyll molecules enable the light-collecting feeler and electron carriage path for transformation of photons into electric possible. The all structure of the PSI protein compound from a pea was certain by x-ray crystallography in 2003 (Heinemeyer et al. 2004). Although its convolution, PSI is very effective, and almost each photon engrossed effects in stimulation of the specific chlorophyll couple that transforms the photon energy to "±" bar differentiation. This ability is used to produce ATP. Development has been completed to use from photosynthesis to make combination nanoplans and electric potential.

8.2.2 Plants and Fungi

The most important difference between plants and fungi is that plants can make their own food; however, fungi cannot. Plants use carbon dioxide, sunlight, and water to create their own food. This process is recognized as photosynthesis. Fungi, on the other hand, are incapable of making their own food. They generally eat off their host as parasites or decompose material and take it as their food. This is the most important difference you essential to remember about plants and fungi. This brings us to the second difference. Fungi do not have chlorophyll—that green matter that gives plants their beautiful green color and supports in photosynthesis. The next difference between plants and fungi is related to their process of reproduction. Reproduction is one of the main things that differentiate a living thing from a dead one. Plants reproduce through pollen and seeds. However, fungi reproduce through frequent spores. They do not have pollen, fruit, or seeds. Another important difference between them relates to the way they are involved. All plants have a structure of roots that attaches the plant to the ground and helps it to soak moisture. If you look at fungi very closely, you will find them spreading a sort of net of filaments on the external of the plant or wherever they attach. This helps them to attach to their host. There are not any complex root systems, stems, or leaves in fungi. Plants and fungi also have different roles to play in the total ecological system. Plants are mainly considered to be producers, because they produce food. They create biomass through the development of photosynthesis. The role of fungi is just the opposite. They are the decomposers who disintegrate biomass. Imagine what this ground would be without these busy cleaners – just a large dustbin that was never cleaned out! Finally, the cell walls on a plant are covered with cellulose, whereas those of the fungi are made of chitin and other polysaccharides (Fig. 8.3).

8.2.3 Plants and Bacteria

Bacteria are the smallest and most primitive unicellular organism. The building of bacterial cell is actually simple demonstrating the prokaryotic organization. The protoplasm of the cell is enclosed by a cell wall. The cell wall is made up of mostly polysaccharide (glucans, mannans, and galactans), chitin, muramic acid, and diaminopimelic acid. However, plant cells are more complex and advanced. Plant cells

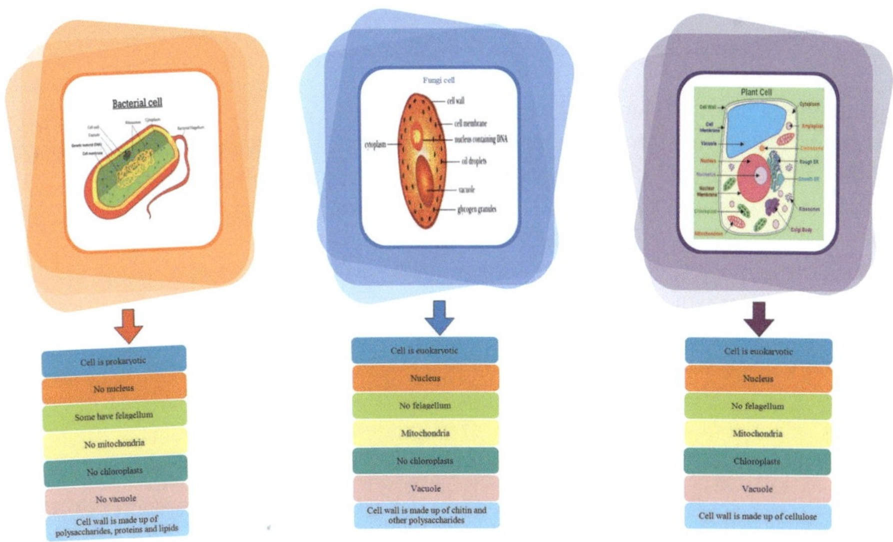

Fig. 8.3 Comparative applications of microbial and plant cells in nanobiotechnology

are enclosed by a definite rigid, protective envelope called the cell wall. The wall is made up of chiefly cellulose and frequently with some other resources (suberin, lignin, pectin, etc.) in minor amounts (Fig. 8.3).

8.3 Plant Cell Wall

Plant cell walls are generally separated in textbooks into two types: primary walls that envelopment growing cells or cells capable of growth, and secondary walls that are thickened structures that include lignin and circumambient specialized cells, for example, vessel elements or fiber cells (Keegstra et al. 1973). The cell wall contains mainly polysaccharides and cellulose microfibrils representing the primary structural units, entrenched in a medium of hemicellulose and pectin, and also secondary wall lignin, with a tinier, but functionally appropriate, supplement of proteins. This hard obstacle often causes a problem for transport of materials toward the plant cell. Some inimitable features of "nanomaterials" can interpenetrate this obstacle. Different biochemical knowledge for studying plant biochemistry includes eliminating or processing this cell wall with pectinase enzymes, cellulose, and hemicellulose sendoff a cell membrane more evocative of an animal cell, which covers the cytoplasm, chloroplasts, mitochondria, nucleus, and other organelles. The resultant package is named "a protoplast," and in this situation, the plant cell biochemical system can still work. When there is a single cell membrane, most of the normal animal cell methods for material transition athwart the membrane. Then, experimental gene insert, drug delivery, or other protocols can cause regrowth of the cell wall to a complete plant cell. Protoplasts are whole plant cells, seeds, and complete plants.

8.4 Preparation of Plant Extracts

Extraction is the important first step for nanoparticle synthesis by plants, because it is required to extract the chosen chemical components from the plant materials for further separation and characterization. The basic process involved steps, such as prewashing, drying of plant materials or freeze-drying, grinding to achieve a homogenous sample and frequently improving the kinetics of analytic extraction and also growing the interaction of sample surface with the solvent system. Suitable actions must be taken to avouch that potential active components are not lost, partial or completely destroyed during the research of the extract from plant samples. Numerous methods can be working to extract the plant material. While water is used as an extracting in several traditional protocols, organic solvents of variable polarities are usually selected in modern approaches of extraction to use the different solubility of plant components. Various solvent systems are available to extract the bioactive compound from natural crops. The extraction of hydrophilic compounds uses polar solvents, such as ethanol or ethyl-acetate and methanol. For extraction of more lipophilic complexes, dichloromethane or a mixture of methanol/dichloromethane in fraction of 1:1 are used. In some examples, extraction with hexane is used to remove chlorophyll (Cos et al. 2006; Sasidharan et al. 2011; Dey 2012).

8.4.1 Maceration

This simple generally used technique includes separating the triturated plant by soaking in a convenient solvent in a closed flask. Simple soaking is achieved at room temperature via mixing the powdered plant with the solvent (triturated plant solvent ratio: 1:5 or 1:10) and separating the mixture for some days with infrequent shaking or stirring. The extract is then separated from the plant constituent part by straining. The action is repeated once or twice with fresh solvent. Lastly, the final residue of extract is compressed of the plant particles by a mechanical mangle or a centrifuge. Dynamic maceration varies from simple soaking to stable stirring. This technique is convenient for both primary and bulk extraction. The major difficulty of maceration is that the method can be time-consuming, from hours to several days or weeks.

8.4.2 Ultrasound-Aided Solvent Extraction

This is an improved maceration process where the extraction is assisted by the use of ultrasound. The plant triturate is placed in a flask. The flask is placed in an ultrasonic device, and ultrasound is used to make a mechanical pressure on the cells through cavitation in the sample. The cellular separation increases the solubilization of metabolites in the solvent and better extraction yields. It is typically used for the primary extraction of a small rate of material.

8.4.3 Percolation Solvent Extraction

The triturated plant material is macerated first in a solvent in a percolator device. Added solvent is then transferred on top of the plant material and allowed to pass slowly (drop wise) out of the end of the percolator. Then, filtration of the extract is not necessary, because there is a filter at the vent of the percolator. Percolation is suitable for both primary and large-scale extraction. The main difficulties are materials and powders plant, such as resins that swell (e.g., plant materials containing mucilage), can block the percolator. Also, if the plant materials are not isolated homogenously in the flask, the solvent may not influence all parts, and the extraction will be unfinished.

8.4.4 Soxhlet Solvent Extraction

This process is suitable for both primary and bulk extractions. The plant powders are placed in a cellulose cover in an extraction space, which is placed on top of a collecting container under a reversal condenser. An adequate solvent is extra to the flask, and the setup is thermal under reversal. When a determined level of condensate solvent has collected in the pipe, it falls into the flask below. The major advantage of Soxhlet extraction is that it is a consecutive process.

8.4.5 Pressurized Solvent Extraction

The plant triturated material is loaded into an extraction cell, which is located in an oven. The solvent is then pumped from a tank to fill the cell, which is thermal and pressed at planned levels for a set period of time. The cell is excited with nitrogen gas, and the extract, which is mechanically filtered, is collected in a container. Fresh solvent is used to rinse the cell and to solubilize the residual components. A final removal with nitrogen gas is achieved to dry the material. This method suggests a more economical and environment-friendly alternative to conventional approaches.

8.4.6 Extraction by Reflux and Steam Distillation

Plant material is immersed in a solvent in a round-bottomed container, which is linked to a condenser. The solvent is heated until it achieves its boiling point. As the steam is condensed, the solvent is recycled to the container. It is usually practical to the extraction of plant essential oils. The major difficulty is that thermo-labile components risk being degraded.

8.4.7 Extraction with Supercritical Liquid

Supercritical fluids (SCFs) are increasingly replacing organic solvents. For example, chloroform, dichloromethane, *n*-hexane, and other organic solvents are usually used in industrial extraction processes because of regulatory and environmental pressures on hydrocarbon and ozone-depleting emissions. Most of the presently offered solvent-free extraction systems use CO_2, which is commonly considered safe for solvent-free extraction methods. The essential steps involved in solvent-free extraction are as follows: liquid CO_2 is required into supercritical state by adaptable its temperature and pressure. Supercritical CO_2 has solvent power and extracts mostly lipophilic and volatile compounds. Gassy CO_2 returns to CO_2 tank after a complete round, the new extraction starts with circulating CO_2.

8.4.8 Countercurrent Extraction

This is a continuous operation in which the plant material transfers against the solvent. It is convenient method for making of large amounts of extracts on an industrial scale. Various types of extractors are available. In the screw and bolt extractor, the plant material is transported by a screw through a pipe and meets the solvent, which is pumped in a different direction.

8.5 Plants as Green Factories for Synthesis Metal Nanoparticles

Nanoparticles have been synthesized by physicochemical methods for a long time, but in recent decades, microorganisms and biological systems considerably progressed to use in creation of metal nanoparticles. Metal nanoparticles can be produced by pure chemical complexes, but this way has a lot of restriction, such as being expensive and dangerous to health. The usage of organisms in this scope is quickly emerging in due to their achievement and simplicity of creation of nanoparticles. Furthermore, biosynthesis of metal nanoparticles is a naturally and co-friendly technique called "green chemistry" that does not use dangerous, toxic, or expensive chemicals (Suman et al. 2014; Zuas et al. 2014; Premanand et al. 2016). The biological synthesis of nanoparticles is compatible with the environment—a very important field in chemistry, biochemistry, biology, and material sciences. The biosynthesis of particles at nanoscale via plant extracts has been recognized since the early 1900s. The operation of plant extracts for the biological production of nanoparticles has grown significantly due to the increase of the chemicals, physical, and biological properties of practices created by this way. Nanoparticles biosynthesis, by microorganisms and plant extracts or natural combination, has been used for multiple reasons, including a relatively easy, inexpensive, and environmentally friendly compared with conventional chemical methods of production and thus gains an upper hand. Furthermore, it is being appropriately setup for large-scale nanoparticles production (Arunachalam et al. 2013; Hassan et al. 2016). There have been recent reports on the biosynthesis of metal nanoparticles using plant extracts. Several reports imply the biological production of metal nanoparticles using plant extracts (Fig. 8.4). Metal nanoparticles synthesized by using various natural plants are summarized in Table 8.2. The most former studies active biomolecules (such as vitamins, sugars, carbohydrates, enzymes, proteins, and amino acid, etc.), various kind of complete cells of different microorganisms (such as fungi, bacteria, algae), or various plant resources (such as leaves, flowers, roots, seeds, fruit, and bark powders) for the production of metal nanoparticles. Specifically, naturally developed plant species serve as biotic sources of phytochemicals, which may function as an environmentally friendly source for the synthesis of metal nanoparticles. Also, it reduces hectic works, various purification steps, and retention of microbial cell cultures (Philip et al. 2011; Sivaraman et al. 2013; Gupta et al. 2014). A major part of the biological synthesis of metal nanoparticles is the potential of different plant parts (leaves, root, fruit, flower, and bark) in the green synthesis reaction; for instance, the production of nanoparticles using purified apiin extracted from plants, such as henna leaves at ambient situations (Kasthuri et al. 2009). Black and green tea extracts are used as reducing agents. It is possible to produce stable synthesis gold, silver, and copper nanoparticles in watery solution at ambient condition (Begum et al. 2009; Kora et al. 2010; Boruah et al. 2012; Brumbaugh et al. 2014). Table 8.2 shows plant species that are used for biosynthesis of metal nanoparticles. Plants have reducing agents as the chemicals which perform reduction reactions.

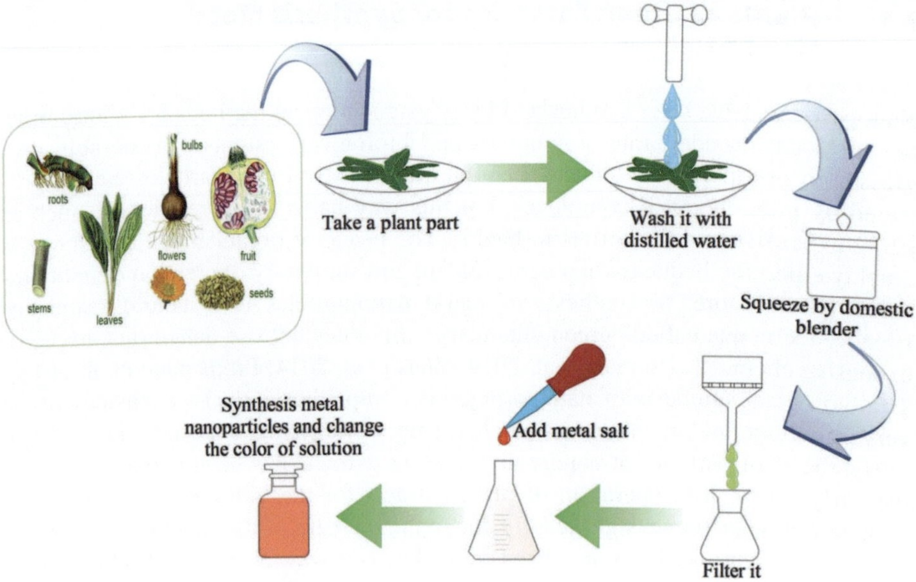

Fig. 8.4 Schematic feature of metallic nanoparticle synthesis using plant extracts

Naturally reducing agents are used in the green synthesis of metal nanostructures. They are generally reducing metal salts into pure metals (Fig. 8.5).

8.6 Environmental Toxicity of Nanoparticles on the Plants

Various studies have down on the environmental toxicity of nanoparticles, mostly in the last period, as engineered nanomaterial come in the marketplace. Unusual novel materials alike carbon nanotubes and fullerenes, also other single nanoparticles, for example quantum dots (naturally prepared from cadmium selenide (CdSe)), taken the public attention. Nano forms of recognized inorganic materials similar to non-nanotitanium dioxide (TiO_2) and other nanometals finally developed as the major production materials of anxiety in the environment. Several studies on public part of environmental nano-toxicology are existing (Navarro et al. 2008; Stampoulis et al. 2009; Ma et al. 2010; Li et al. 2011). Although mechanisms for nanomaterial to transmission across cell walls and membranes are not understood, when inside cells, nanomaterial can modify transformation in membranes and other cell structures (Lin and Xing 2007; Chomoucká et al. 2010). The surface characterization of engineered nanomaterial is of important significance for their collection performance and resulting aqueous movement and transport. Ecotoxicological study of engineered nanomaterials revealed toxic effects on invertebrates and fish (Handy et al. 2008). Little information on worldly plant species are accessible (Panda et al. 2011). The past decade saw developments in nanotechnology and a consistent growth in the usage of nanoparticles in crops. Another study by Klaine et al. (2008)

Table 8.2 Synthesis of nanoparticles using plants

Plant source	Parts/extracts	Metal NPS	Size (nm)	Morphology	Pharmacological applications	References
Abelmoschus esculentus	Seed extract	Gold	45–75	Spherical	Antifungal activity	Jayaseelan et al. (2013)
Acacia auriculiformis	Pods extract	Silver	20–150	Spherical	Antibacterial activity	Nalawade et al. (2014)
Achillea biebersteinii	Flower extract	Silver	5–35	Irregular	Anticancer activity	Baharara et al. (2014, 2015)
Adhathoda vasica	Leaf extract	Silver	21–29	Spherical	Antimicrobial activity	Bhumi et al. (2015)
Adhathoda vasica	Leaf extract	Gold	22–47	Spherical	ND	Karthick et al. (2012)
Abutilon indicum	Leaf extract	Silver	15–20	Spherical	Antimicrobial, antidiabetic activity	Bandi and Vasundhara (2012)
Arbutus unedo	Leaf extract	Silver	3–20	Spherical	ND	Kouvaris et al. (2012)
Artocarpus heterophyllus Lam.	seed extract	Silver	3–25	Irregular	Antibacterial activity	Jagtap and Bapat (2013)
Artemisia nilagirica	Leaf extract	Silver	70–90	Irregular	Antibacterial activity	Vijayakumar et al. (2013)
Aloe vera	Leaf extract	Gold and silver	10–20	Spherical	ND	Chandran et al. (2006)
Aloe vera (*Aloe barbadensis* Miller)	Leaf extract	Indium oxide	5–50	Cubic	ND	Maensiri et al. (2008)
Alternanthera dentata	Leaf extract	Silver	10–80	Spherical	Antibacterial activity	Kumar et al. (2014a)
Acalypha indica	Leaf extract	Silver	20–30	Spherical	Antibacterial activity	Krishnaraj et al. (2010)
Amphibious weed ipomoea	Leaf, stems, roots extract	Silver	Different size	Spherical	ND	Ganaie et al. (2014)
Amphibious weed ipomoea	Leaf, stems, roots extract	Gold	30–40	Irregular	ND	Abbasi et al. (2015)
Annona squamosa L peel	Fruit extract	Palladium	100	Spherical	Acaricidal, insecticidal, larvicidal activity	Roopan et al. (2012)

(continued)

Table 8.2 (continued)

Plant source	Parts/extracts	Metal NPS	Size (nm)	Morphology	Pharmacological applications	References
Ananas comosus (Pineapple)	Fruit extract	Gold	10	Spherical	Antimicrobial activity	Asavegowda et al. (2013)
Apple	Fruit extract	Silver	30	Spherical	Antimicrobial activity	Ali et al. (2016)
Henna leaves	Apiin extract	Silver	39	Quasi-spherical	ND	Kasthuri et al. (2009)
Henna leaves	Apiin extract	Gold	21	Anisotropic	ND	Kasthuri et al. (2009)
Avena sativa (oat)	Biomass	Gold	5–85	Irregular	ND	Armendariz et al. (2004)
Azadirachta indica (neem)	Leaf extract	Gold, silver and Au core–Ag shell	15–100	Spherical	ND	Shankar et al. (2004a)
Azadirachta indica	Leaf extract	Silver	10–100	Spherical	ND	Tripathy et al. (2010)
Azadirachta indica	Leaf extract	Zinc oxide	9.6–25.5	Spherical	Antibacterial and photocatalytic	Bhuyan et al. (2015)
Bacopa monnieri (Linn.) Wettst.	Leaf, stems, roots extract	Silver	2–50	Spherical, cubical	ND	Krishnaraj et al. (2012)
Blackberry	Fruit extract	Silver and gold	25–150	Spherical	ND	Nadagouda et al. (2014)
Black tea	Leaf extract	Gold and silver	20	Spherical	ND	Begum et al. (2009)
Black pepper	Leaf extract	Silver	5–50	Spherical	Antibacterial activity	Augustine et al. (2014)
Blueberry	Fruit extract	Silver and gold	50–150	Spherical, triangular	ND	Nadagouda et al. (2014)
Boerhaavia diffusa	Leaf extract	Silver	25	Spherical	Antibacterial activity	Kumar et al. (2014b)
Brassica juncea (mustard)	Seed extract	Silver	ND	ND	ND	Shekhawat and Arya (2009)
Bryophyllum pinnatum	Leaf extract	Silver	70–90	Spherical	Antibacterial activity	Baishya et al. (2012)

Plant source	Parts/extracts	Metal NPS	Size (nm)	Morphology	Pharmacological applications	References
Catharanthus roseus Linn	Leaf extract	Silver	27–50	Spherical	Antibacterial activity	Kotakadi et al. (2013)
Camellia sinensis (green tea)	Leaf extract	Gold	2–10	Spherical	ND	Boruah et al. (2012)
Carob	Leaf extract	Silver	5–40	Spherical	Antibacterial activity	Awwad et al. (2013)
Carica papaya	Fruit extract	Silver	10–20	Cubic	Antibacterial activity	Jain et al. (2009)
Cassia fistula	Leaf extract	Zinc oxide	5–15	Irregular	Antibacterial activity	Suresh et al. (2015)
Centella asiatica	Callus extract	Silver	5–40	Spherical	Antibacterial activity	Netala et al. (2015)
Chenopodium album	Leaf extract	Gold and silver	4–30	Spherical, triangular	ND	Dwivedi and Gopal (2010)
Chenopodium murale	Leaf extract	Silver	30–50	Spherical	Antibacterial activity	Abdel-Aziz et al. (2014)
Citrus sinensis	Fruit extract	Silver	10–35	Spherical	Antibacterial activity	Kaviya et al. (2011)
Citrus sp.	Fruit extract	Gold	15–80	Irregular	ND	Sujitha and Kannan (2013)
Crataegus douglasii	Fruit extract	Silver	40–60	Spherical	Antibacterial activity	Ghaffari-Moghaddam and Hadi-Dabanlou (2014)
Cinnamomum camphora	Leaf extract	Gold and silver	55–80	Triangular, spherical	ND	Huang et al. (2007)
Cinnamomum camphora	Leaf extract	Palladium	3–6	Spherical	ND	Yang et al. (2010)
Cinnamon zeylanicum (cinnamon)	Bark extract	Palladium	15–20	Spherical	ND	Sathishkumar et al. (2009a)
Cinnamon zeylanicum (cinnamon)	Bark extract	Silver	31–100	Irregular	Antibacterial activity	Sathishkumar et al. (2009b)
Citrus limon (lemon)	Fruit extract	Silver	50	Spherical	ND	Prathna et al. (2011)

(continued)

Table 8.2 (continued)

Plant source	Parts/extracts	Metal NPS	Size (nm)	Morphology	Pharmacological applications	References
Citrus limon (lemon)	Leaf extract	Silver	15–30	Heterogeneously	Antifungal activity	Vankar and Shukla (2012)
Cocos nucifera	Coir extract	Silver	23	Spherical	Larvicidal activity	Roopan et al. (2013)
Cocos nucifera	Inflorescences extract	Silver	22	Spherical	Antibacterial activity	Mariselvam et al. (2014)
Cochlospermum gossypium	Gum tears of grade	Silver	3	Spherical	Antibacterial activity	Kora et al. (2010)
Cochlospermum gossypium	Gum tears of grade	Silver, gold, platinum	2–10	Spherical	ND	Vinod et al. (2011)
Coriandrum sativum (coriander)	Leaf extract	Gold	5–70	Irregular	ND	Rao (2011)
Coleus aromaticus	Leaf extract	Silver	40–50	Spherical	Antibacterial activity	Vanaja and Annadurai (2013)
Cranberry	Fruit extract	Silver	6–60	Irregular	Antimicrobial activity	Puišo et al. (2014)
Cymbopogon flexuosus (lemongrass)	Leaf extract	Gold	200–500	triangular	ND	Shankar et al. (2004b)
Cyperus sp.	Rotundus grass extracts	Silver	1–100	Spherical	ND	Siva et al. (2014)
Cycas sp. (cycas)	Leaf extract	Silver	2–6	Spherical	ND	Jha and Prasad (2010)
Cypress leaves	Leaf extract	Gold	5–80	Spherical	ND	Noruzi et al. (2012)
Datura metel	Leaf extract	Silver	40–60	Irregular	*Anax Immaculifrons, Anopheles Stephensi*	Murugan et al. (2015)
Dendrophthoe falcata	Leaf extract	Silver	5–45	Spherical	Anticancer activity	Sathishkumar et al. (2014)
Dalbergia spinosa	Leaf extract	Silver	18 ± 4	Spherical	Antibacterial activity	Muniyappan and Nagarajan (2014)

Plant source	Parts/extracts	Metal NPS	Size (nm)	Morphology	Pharmacological applications	References
Desmodium triflorum	Leaf extract	Silver	5–20	Spherical	Antibacterial activity	Ahmad et al. (2010)
Delphinium denudatum	Root extract	Silver	58	Spherical	Antibacterial, mosquito larvicidal activity	Suresh et al. (2014)
Diospyros kaki (persimmon)	Leaf extract	bimetallic gold/silver	Different size	Irregular	Antibacterial, *anticancer activity*	Kuppusamy et al. (2016)
Dioscorea bulbifera	Tuber extract	Silver	8–20	Irregular	Antibacterial activity	Ghosh et al. (2012)
Eclipta sp.	Leaf extract	Silver	60	ND	Antibacterial activity	Saminathan (2015)
Emblica officinalis (indian gooseberry)	Fruit extract	Gold, silver	10–25	Irregular	ND	Ankamwar et al. (2005)
Eucalyptus camaldulensis	Leaf extract	Gold	2–100	Spherical	ND	Golmoraj et al. (2011)
Eucalyptus leucoxylon	Leaf extract	Silver	50	Spherical	Antioxidant activities	Rahimi-Nasrabadi et al. (2014)
Eucalyptus hybrida (safeda)	Leaf extract	Silver	5–150	Spherical	ND	Dubey et al. (2009)
Euphorbia hirta	Leaf extract	Silver	30–60	Spherical, cubic	ND	Devi et al. (2014)
Euphorbia condylocarpa M. bieb	Root extract	Au/Pd	79	Cubic	ND	Nasrollahzadeh et al. (2014)
Euphorbia condylocarpa M. bieb	Root extract	Pd/Fe_3O_4	38	Irregular	ND	Nasrollahzadeh et al. (2015a)
Ficus bengalensis (marri)	Leaf extract	Silver	10–50	Spherical	ND	Saware et al. (2014)
Ficus benghalensis	Leaf extract	Silver	16	Spherical	Antibacterial activity	Saxena et al. (2012)

(continued)

Table 8.2 (continued)

Plant source	Parts/extracts	Metal NPS	Size (nm)	Morphology	Pharmacological applications	References
Feronia elephantum (Rutaceae)	Leaf extract	Silver	18–45	Irregular	Filariasis, malaria dengue vector mosquitoes	Veerakumar and Govindarajan (2014)
Garcinia mangostana (mangosteen)	Leaf extract	Silver	6–57	Spherical	Antibacterial activity	Veerasamy et al. (2011)
Garcinia kola	Leaf extract	Silver	10	Spherical	Antibacterial activity	Hassan et al. (2016)
Gardenia jasminoides Ellis	Fruit extract	Palladium	3–5	Spherical	ND	Jia et al. (2009)
Gliricidia sepium	Leaf extract	Silver	10–50	Cubic	ND	Raut-Rajesh et al. (2009)
Gloriosa superba	Leaf extract	CuO	5–10	Spherical	Antibacterial activity	Naika et al. (2015)
Hibiscus rosa sinensis	Leaf extract	Silver, gold	13, 14	Spherical	ND	Philip (2010)
Hibiscus cannabinus	Leaf extract	Silver	7–25	Spherical	Antibacterial activity	Bindhu and Umadevi (2013)
Hippophae rhamnoides	Leaf extract	Palladium	2–14	Spherical	ND	Nasrollahzadeh et al. (2015b)
Hovenia dulcis	Fruit extract	Gold	20	Spherical, hexagonal	Antibacterial activity	Basavegowda et al. (2014)
Hordeum vulgare	Leaf extract	Iron oxide	100–200	Spherical	ND	Makarov et al. (2014)
Hydrilla verticilata	Leaf extract	Silver	65	Spherical	ND	Sable et al. (2012)
Ipomoea aquatica	Leaf extract	Silver	77–624	ND	Antibacterial activity	Sivaraman et al. (2013)
Iresine herbstii	Leaf extract	Silver	44–64	Cubic	Antibacterial activity	Dipankar and Murugan (2012)
Jatropha curcas	latex extract	Silver	10–20	Irregular	ND	Bar et al. (2009a)
Jatropha curcas	Seed extract	Silver	15–50	Spherical	ND	Bar et al. (2009b)
Kalopanax pictus	Leaf extract	Silver	10–30	Spherical	Antibacterial activity	Salunke et al. (2014a)

Plant source	Parts/extracts	Metal NPS	Size (nm)	Morphology	Pharmacological applications	References
Kalopanax septemlobus	Leaf extract	Silver	5–100	Irregular	Antibacterial activity	Salunke et al. (2014b)
Lawsonia inermis (Henna)	Leaf extract	Silver	5–45	Spherical	Antimicrobial activity	Gupta et al. (2014)
Leptadenia reticulata	Leaf extract	Silver	5–70	Spherical	Antibacterial, antioxidant, cytotoxic activity	Swamy et al. (2015)
Lemon peels	Fruit extract	Silver	17–61	Spherical	Antidermatophytic activity	Nisha et al. (2014)
Lingonberry	Fruit extract	Silver	6–60	Irregular	Antimicrobial activity	Puišo et al. (2014)
Magnolia kobus	Leaf extract	Copper	50–250	Spherical	Antibacterial activity	Lee et al. (2013)
Memecylon umbellatum	Leaf extract	Silver, gold	15–25	Spherical	Antibacterial activity	Arunachalam et al. (2013)
Melia dubia	Leaf extract	Silver	3–11	Spherical	Anticancer activity	Kathiravan et al. (2014)
Moringa oleifera	Bark extract	Silver	40	Spherical, pentagon	Anticancer activity	Vasanth et al. (2014)
Mimusops elengi	Seed extract	Silver	12–30	Spherical	Antibactrial and antioxidant activities	Kumar et al. (2014c, d)
Murraya koenigii	Leaf extract	Silver	10	Spherical	ND	Philip (2010)
Murraya koenigii		Gold	20	Spherical	ND	Philip et al. (2011)
Morinda citrifolia	Root extract	Silver	30–55	Spherical	Anticancer activity	Suman et al. (2013)
Morinda citrifolia	Root extract	Gold	12–38	Triangle, spherical	ND	Suman et al. (2014)
Musa paradisiaca	Fruit extract	Silver	23	Spherical	Antimicrobial activity	Ibrahim (2015)
Mulberry	Leaf extract	Silver	20	Spherical	Antibacterial activity	Awwad and Salem (2012)
Myrmecodia pendan (Sarang Semut plant)	Fruit extract	Silver	10–20	Cubic	ND	Zuas et al. (2014)

(continued)

Table 8.2 (continued)

Plant source	Parts/extracts	Metal NPS	Size (nm)	Morphology	Pharmacological applications	References
Nelumbo nucifera (lotus)	Leaf extract	Silver	30–40	Spherical	Antibacterial activity	Premanand et al. (2016)
Ocimum sanctum (Tulsi)	Leaf extract	Silver	4–30	Spherical	Antibacterial activity	Singhal et al. (2011)
Ocimum basilicum L. var. *purpurascens* Benth.	Leaf extract	Zinc oxide	50	Hexagona	ND	Salam et al. (2014)
Olive	Leaf extract	Silver	90	Spherical	Cytotoxic activity	Rashidipour and Heydari (2014)
Pear	Fruit extract	Gold	12–20	Hexagonal	ND	Ghodake et al. (2010)
Plectranthus amboinicus	Leaf extract	Silver	16	Spherical	Antimicrobial activity	Ajitha et al. (2014)
Physalis alkekengi	Aerial parts extract	Zinc oxide	50–200	Triangular, elongated	ND	Qu et al. (2011a)
Phoenix dactylifera L.	Leaf extract	Gold	32–45	Spherical	Catalytic activity	Zayed and Eisa (2014)
Pomegranate	Fruit extract	Silver, gold	5–50	Spherical	ND	Nadagouda et al. (2014)
Piper longum	Leaf extract	Silver	17–41	Spherical	Cytotoxic activity	Jacob et al. (2012)
Piper longum	Fruit extract	Silver	46	Spherical	Antioxidant, antibacterial and cytotoxic activity	Reddy et al. (2014)
Plumeria rubra	Latex extract	Silver	32–200	Spherical	Larvicidal activity	Patil et al. (2012)
Psidium guajava (guava)	Leaf extract	Gold	25–30	Spherical	ND	Raghunandan et al. (2009)
Punica granatum	Fruit extract	Gold	23–35	Irregular	Catalytic activity	Dash and Bag (2014)
Rumex acetosa	Leaf extract	Iron oxide	100–200	Spherical	ND	Makarov et al. (2014)
Trigonella foenum-graecum	Seed extract	Gold	15–25	Spherical	Catalytic activity	Aromal and Philip (2012)
Vitex negundo	Leaf extract	Silver	20–100	Spherical	Anticancer activity	Prabhu et al. (2013)

Plant source	Parts/extracts	Metal NPS	Size (nm)	Morphology	Pharmacological applications	References
Sacha inchi (*Plukenetia volubilis*)	Leaf extract	Silver	4–25	Spherical	Antioxidant activity	Kumar et al. (2014)
Saraca indica	Bark extract	Gold	15–23	Irregular	Catalytic reduction	Dash et al. (2014)
Santalum album	Leaf extract	Silver	80–200	Spherical	Antimicrobial activity	Swamy and Prasad (2012)
Scutellaria barbata D. Don (Barbated skullcup)	Herb extract	Gold	5–30	Spherical	ND	Wang et al. (2009)
Sedum alfredii Hance	Shoots extract	Zinc oxide	53	Spherical	ND	Qu et al. (2011b)
Sesbania drummondii (leguminous shrub)	Seed extract	Gold	6–20	Spherical	ND	Sharma et al. (2007)
Sesbania grandiflora	Leaf extract	Silver	10–25	Spherical	Antibacterial activity	Das et al. (2013)
Solanum tricobatum	Fruit extract	Silver	12–42	Spherical	Antibacterial, anticancer activity	Ramar et al. (2015)
Solanum lycopersicum	Fruit extract	Silver	13	Spherical	Insecticidal activity	Bhattacharyya et al. (2016)
Sorbus aucuparia (rowan)	Leaf extract	Silver, gold	16–18	Irregular	ND	Dubey et al. (2010a)
Syzygium aromaticum (clove)	Flower extract	Gold	5–100	Irregular	ND	Raghunandan et al. (2010)
Syzygium cumini (jambul)	Bark extract	Silver	20–60	Spherical	Antibacterial activity	Prasad and Swamy (2013)
Syzygium cumini (jambul)	Leaf extract	Silver	100–160	Spherical	Antibacterial activity	Prasad et al. (2012)

(continued)

Table 8.2 (continued)

Plant source	Parts/extracts	Metal NPS	Size (nm)	Morphology	Pharmacological applications	References
Tabernaemontana divaricate	Leaf extract	Copper oxide	48	Spherical	Antibacterial activity	Sivaraj et al. (2014)
Tanacetum vulgare (tansy fruit)	Fruit extract	Gold, silver	10–40	Spherical	ND	Dubey et al. (2010b)
Tea	Leaf extract	Copper	3	Spherical	ND	Brumbaugh et al. (2014)
Terminalia catappa (almond)	Leaf extract	Gold	10–35	Spherical	ND	Ankamwar (2010)
Terminalia chebula	Fruit extract	Silver	100	Spherical and triangular	Antibacterial activity	Kumar et al. (2012)
Thevetia peruviana	Latex extract	Silver	10–30	Spherical	ND	Rupiasih et al. (2015)
Trianthema decandra	Root extract	Gold, silver	36–74	Irregular	Antimicrobial activity	Geethalakshmi and Sarada (2012)
Tribulus terrestris	Leaf extract	Silver	15–40	Spherical	ND	Ashokkumar et al. (2014)
Turmeric	Fruit extract	Silver, gold	5–100	Spherical	ND	Nadagouda et al. (2014)
Withania somnifera	Leaf extract	Silver	5–30	Spherical	Antimicrobial activity	Raut et al. (2014)
Ziziphora tenuior	Leaf extract	Silver	8–40	Spherical	ND	Sadeghi and Gholamhoseinpoor (2015)
Zingiber officinale	Root extract	Silver, gold	5–20	Irregular	Antibacterial activity	Velmurugan et al. (2014)

ND not determined

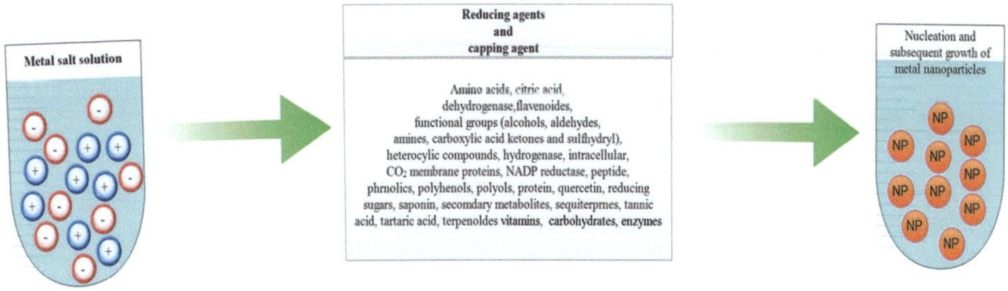

Fig. 8.5 Reducing and capping agents for the synthesis of metallic nanostructures

disproved the absence of information base for performance, disposition, and toxicity of synthetic nanomaterials in plants. Lin and Xing (2007) reported a study of five kinds of nanoparticles in six dissimilar plant organisms. Phytotoxicity studies of multi-walled carbon-nanotubes (MWNTs), also nanoparticle zinc, zinc oxide, aluminum and alumina, on seed pullulating and root growing of six plant type (corn, lettuce, rape radish, ryegrass, and cucumber) were performed. Most nanoparticles indicated low result in these organisms, except for Zn NPs and ZnO NPs (Lin and Xing 2007). Seed pullulation was not pretentious, excluding for the inhibition by Zn NPs on ryegrass and ZnO nanoparticles on corn at 2,000 mg/L. The 50% inhibitory concentrations (IC_{50}) of Zn NPs and ZnO nanoparticles were predictable to be approximately 20 mg/L for rape and ryegrass and 50 mg/L for radish. These conclusions indicated that considerable environmental agents could effect from unsuitable use and removal of such engineered nanoparticles. A recent survey introduced a wide study of a number of several nanoparticles in comparison with its bulk material (Nel et al. 2006; Roduner 2006; Lin and Xing 2007). This study evaluated some information from five generic nanoparticles using three various experiment systems to measure toxicity. The effects on zucchini seed germination of the attendance or lack attendance of experiment combinations, via both the nanoparticles and the resultant bulk materials, were evaluated (Stampoulis et al. 2009). A high variation was detected when the nanomaterials were isolated via a surfactant. There appeared to be more of an efficacy on the sodium dodecyl sulfate surfactant than on the complexes tested and trouble in sample water-insoluble materials. The second experiment contained effects of test materials on root growth, with little effects detected for any material. The last assay used general biomass to test effects on plant growth. These studies were to some extent problematic quantitatively because of the problem of defining the concentrations of nanoparticles compared with metal ions. One exciting result—the effect of multiwall carbon nanotubes carbon on aggregate biomass of zucchini seedlings—was assessed. The information appeared to show that carbon nanotubes delay zucchini seedling growth (De La Torre-Roche et al. 2013).

8.7 Effects of Metal Nanoparticles on the Plant Growth and Development

8.7.1 Titanium Dioxide Nanoparticles (TiO_2 NPs)

The effects of TiO_2 NPs and non-nano-TiO_2 in different forms were considered with old spinach seed, and considerable effects on plant growth were detected. It was indicated that the biological effects are dependent on the size of particles at a 0.25–4.0% concentration. The mechanism by which TiO_2 NPs develop the growth of spinach seeds requires more investigation (Navarro et al. 2008; Asli and Neumann 2009). In an independent study, it has been shown that willow trees were not sensitive to TiO_2 NPs. Properties were recorded for zinc oxide and zinc particles but were due to heavy metal toxicity and were not nanoparticle-size-specific (Ghosh et al. 2010; Feizi et al. 2012). Another study surveyed the effects of nanoscale titanium dioxide particles on plant growth and development. Due to the widespread cultivation of canola in Iran and other parts of the world and in view of the potential effect of titanium on its growth, this plant was selected as the model system. Canola seeds were separately treated with dissimilar concentrations of nanoscale titanium, and the influence this treatment was considered on seed germination and seedling vitality. Treatment of nanoscale TiO_2 (20-nm average particle size) at 2000 mg/L concentration promoted both seed germination and seedling vigor. The lowest and the highest germination rate were found in 1500 and 2000 mg/L treatments, respectively. Higher concentrations of nanoscale TiO_2 (1200 and 1500 mg/L) indicated large radicle and plumule growth of seedling (Mahmoodzadeh et al. 2013).

8.7.2 Iron Nanoparticles (Fe_3O_4 NPs)

Phytotoxicity study on the *Cucurbita maxima* (pumpkin) was performed by Wang et al. (2011). The authors indicated that plants grown in an aqueous medium, including magnetite nanoparticles (Fe_3O_4 NPs), can absorb, translocate, and huddle nanoparticles in the plant tissues. Particular toxic effects were not indicating, but the efficient transport of the iron NPs led researchers to develop their studies and use magnetite nanoparticles as a possible carrier to deliver combination to plants (Wang et al. 2011). Also, Zhu et al. (2008) indicated that pumpkin plants (*Cucurbita maxima*), grown in an aqueous medium, including magnetite (Fe_3O_4) nanoparticles, are able to absorb, translocate, and accumulate the particles in the plant tissues. These effects recommend that plants, as a vital component of the environmental and ecological methods, are essential when evaluating the overall fate, transport, and exposure pathways of nanoparticles in the environment.

8.7.3 Zinc Oxide Nanoparticle (ZnO NPs)

Lin and Xing reported the internalization and growing translocation of ZnO NPs by *Lolium perenne* (rye grass) and compared the effects with Zn^{+2} salts. Characterization and phytotoxicity of nanoparticles were imagined by light and electron microscopy. In this experiment of ZnO NPs, ryegrass biomass was meaningfully reduced, root ends shrank, and root epidermal and cortical cells collapsed. Effects of Zn^{+2} salts at same concentrations were lesser. So, the phytotoxicity of ZnO NPs was due to the nanoparticles and not their resolution in the bulk nutrient solution. The authors reported that ZnO NPs were observed at the root surface, and separate ZnO NPs were seen in apoplast and protoplast tissues of the root endodermis and stele. Finally, their inferences showed that less ZnO NPs would transmit up the ryegrass stems. In another study, Kumari et al. (2011) reported cytogenetic and genotoxic effects of ZnO NPs on the root cells of *Allium cepa*. The special effects of ZnO NPs on the mitotic index (MI), chromosomal aberration index, micronuclei index (MN index), and lipid peroxidation were specified through the hydroponic culturing of *A. cepa*. *A. cepa* roots were treated with the distributions of ZnO NPs at four dissimilar concentrations (25, 50, 75, and 100 μg/L). With the increasing concentrations of ZnO NPs, MI reduced with the growth of pyknotic cells. Conversely, MN and chromosomal aberration index were increased. The frequency of micronucleated cells was higher in ZnO NPs-treated cells compared with control. The number of cells in each mitotic phase differs upon ZnO NPs treatment. The effect of ZnO NPs on lipid peroxidation was evident compared with bulk ZnO. Internalization of ZnO NPs-like particles was confirmed by TEM analysis. SEM image of treated *A. cepa* showed that the internalized nanoparticles agglomerated depending on the physicochemical environment inside the cell. Results demonstrated that ZnO NPs can be a clastogenic/genotoxic and cytotoxic agent. Also, Alharby et al. (2016) studied the effect of ZnO NPs on some callus growth traits, plant regeneration rate, mineral element contents, and changes in the activity of superoxide dismutase (SOD) and glutathione peroxidase (GPX) in five tomato cultivars, which were investigated in a callus culture exposed to elevated concentrations of salt (3.0 and 6.0 g/L NaCl) and in the presence of zinc oxide nanoparticles (15 and 30 mg/L). The relative callus growth rate was inhibited by 3.0 g/L NaCl; this was increased dramatically at 6.0 g/L. Increasing exposure to NaCl was related to significantly higher sodium content and SOD and GPX activities. Zinc oxide nanoparticles mitigated the effects of NaCl, at concentrations of 15–30 mg/L. This finding shows that zinc oxide nanoparticles should be investigated further as a potential anti-stress agent in crop fabrication. Tomato cultivars showed different degrees of tolerance to salinity in the existence of ZnO-NP.

8.7.4 Copper Nanoparticles (Cu NPs)

In another study, bioaccumulation of nonwater-soluble copper nanoparticle was observed (Lee et al. 2008). The plants tested were mung bean (*Phaseolus radiatus*) and wheat (*Triticum aestivum*). Growth inhibition of seedlings by various concentrations of Cu NPs surveyed in the plant agar media, and information indicated that all concentrations examined were toxic to both plants. Research indicated that cupric ion released from Cu NPs had small effects in the concentration ranges of the present survey and that the obvious toxicity resulted from Cu NP. The authors recommended that the plant agar tests are a nice protocol for analysis of the phytotoxicity of nonwater-soluble nanoparticles. Moreover, Adhikari et al. (2016) reported that copper nanoparticle were able to enter the plant cell and manage the growth of the maize plant. A solution culture test was conducted to investigate the effect of Cu nanoparticles (<50 nm) on the growth and enzymatic activity of maize (*Zea mays* L.) plant. Bioaccumulation of Cu NPs in plant also was investigated. Results showed that Cu NPs can enter the plant cell through roots and leaves. Cu NPs in solution and spray forms significantly enhanced the growth of maize. The different enzymatic activities, such as glucose-6-phosphate dehydrogenase (G6PD), succinate dehydrogenase, superoxide dismutase, catalase, and guaiacol peroxidase, were studied to find a possible mechanism of action of NPs. Among the enzymes, the activity of G6PD was highly influenced by copper oxide (CuO) nanoparticles application by spray as well as in solution. Experimental results revealed that CuO NPs affected the pentose phosphate pathway of maize plant. The obtained experimental results indicated that the nanoparticles considered under this study could enter the plant cell and also improve its growth by regulating the different enzyme activities.

8.7.5 Silver Nanoparticles (Ag NPs)

A study of the application and separating of silver nanoparticles was reported (Harris and Bali 2008). The authors studied the uptake of Ag NPs by two usual metallophytes: Indian *Brassica juncea* (coward) and *Medicago sativa* (alfalfa). Indian coward adsorbs up to 12 wt% silver in an aqueous substrate, including 1,000 ppm of $AgNO_3$ for 72 h, but alfalfa collects up to 14 wt% silver in an aqueous substrate, including 10,000 ppm $AgNO_3$ for 24 h. In both plants, the silver was stowed as separate nanoparticles, with an average size of 50 nm. Sosan et al. (2016) determined the early stages of interactions between Ag NPs and plant cells and investigated their physiological roles. They showed that the addition of Ag NPs to cultivation medium, at levels above 300 mg/L, inhibited *Arabidopsis thaliana* root elongation and leaf development, while it decreased photosynthetic efficiency and the extreme accumulation of Ag in tissues. Ag NPs slightly inhibited plasma membrane K^+ efflux and Ca^{2+} influx currents, and activated Gd^{3+}-sensitive Ca^{2+} influx channels. Ag NPs were not able to catalyze hydroxyl radical generation, but they directly oxidized the major plant antioxidant, l-ascorbic acid. Overall, the data presented show that these include the induction of classical stress signaling reactions (mediated by $[Ca^{2+}]_{cyt}$ and ROS) and a specific effect on the plasma membrane conductance and the reduced ascorbate.

8.7.6 Carbon Nanotubes (CNTs)

Probable toxic effects of multiwall carbon nanotubes (MWNTs) on *Oryza sativa* cells (rice) were studied (Tan et al. 2009). When a plant cell suspension was cultured with MWNTs, a rise in reactive oxygen species (ROS) and a reduction in cell viability were shown. When the antioxidant ascorbic acid was presented into the medium, the ROS amount reduced and cell viability increased. Transmission electron microscopy (TEM) shown separate tubes in interaction with cell walls. The rice suspension cells with separate MWNTs linked with their cell walls appeared to undergo an oversensitive response identified as the ROS protection response force, which led to reduced viability. Also Cañas et al. (2008) conducted a pilot study on the effects of functionalized (with poly-3-aminoben-zenesulfonic acid) and non-functionalized carbon nanotubes on root elongation of six crop species (cabbage, carrot, cucumber, lettuce, onion, and tomato). Root growth was measured at 0, 24, and 48 h after exposure. Scanning-electron microscopy was used to evaluate potential uptake of carbon nanotubes and to observe the interaction of nanotubes with the root surface. Nonfunctionalized nanotubes inhibited root elongation in tomato and enhanced root elongation in onion and cucumber. Functionalized nanotubes inhibited root elongation in lettuce. Cabbage and carrots were not affected. Microscopy images showed the presence of nanotube sheets on the root surfaces. Khodakovskaya et al. (2009) reported that carbon nanotubes (CNTs) were able to go through tomato seeds and support water uptake inside seeds, a process that eventually affects seed germination and growth of tomato seedlings. Khodakovskaya et al. (2012) demonstrated that multi-walled carbon nanotubes (MWCNTs) have the ability to enhance the growth of tobacco cell culture (55–64 %) in concentrations of 5–500 μg/mL. Activated carbon (AC) stimulated cell growth at low concentration of 5 μg/mL while dramatically inhibiting cellular growth at higher concentrations of 100–500 μg/mL. The authors showed that the expression of the tobacco aquaporin (*NtPIP1*) gene and production of corresponding protein, as well as the expression of marker genes for cell division (*CycB*) and cell wall extension (*NtLRX1*), significantly increased in cells exposed to MWCNTs compared with control cells or those exposed to AC.

8.7.7 Alumina Nanoparticles (Al_2O_3 NPs)

The phytotoxicity of alumina nanoparticles overloaded with and without the chemical phenanthrene was examined by way of root elongation tests (Yang and Watts 2005). Five plant species, cucumber (*Cucumis sativus*), corn (*Zea mays*), cabbage (*Brassica oleracea*), soybean (*Glycine max*), and carrot (*Daucus carota*), were used in the research of phytotoxicity. The surface specifications of phenanthrene with and without nanoparticles were examined by Fourier transform infrared spectroscopy. It was established that when the phenanthrene monolayer concentration increased, the degree of the root elongation inhibition caused using the particles to decrease. When mixed with dimethylsulfoxide as a hydroxyl radical at concentrations from 0.5% to

1.0%, the nonloaded particles also indicated reduced inhibition of root elongation. The authors suggest that the surface features of the particles play a significant role in the phytotoxicity of alumina nanoparticles. In a proteomic study by Yasmeen et al. (2016), the effects of Al_2O_3 NPs on the recovery of soybean from flooding were examined. The authors showed that survival percentage and weight/length of roots counting hypocotyl were improved after 2 and 4 days of flooding with 50 ppm Al_2O_3 NPs. Overall, a total of 211 common proteins were changed in abundance during the recovery period after treatment with or without Al_2O_3 NPs. These proteins have been related to protein synthesis, stress, cell wall, and signaling. Among the known stress-related proteins, *S*-adenosyl-l-methionine–dependent methyltransferases were recovered from flooding with Al_2O_3 NPs. Hierarchical clustering divided the recognized proteins into three clusters. Cluster II exhibited the greatest change in proteins associated to protein synthesis, transport, and development during the recovery from flooding with Al_2O_3 NPs. However, enolase activity remained unaffected during flooding leading to subsequent recovery with Al_2O_3 NPs. These consequences suggest that *S*-adenosyl-l-methionine–dependent methyltransferases and enolase might be involved in mediating recovery responses by Al_2O_3 NPs.

8.8 Use of Metal Nanoparticles for Compound Delivery to Plants

The effective application of different nanoplatforms in medicine under in vitro conditions led to some important issues in agriculture nanotechnology. This knowledge holds the potential of controlled release of agrochemicals and site-targeted delivery of different macromolecules required for enhanced plant disease resistance, enhanced plant growth, and efficient nutrient utilization. Processes for example nano-encapsulation indicate the advantage of more efficient use and nontoxic managing of pesticides with less spotlight to the environment that guarantees eco-protection. The uptake effectiveness and properties of different nanoparticles on the growth and metabolic functions vary differently between plants. Metal nanoparticles mediated plant transformation possesses the potential for genetic modification of plants for further improvement. Specifically, application of metal nanoparticle technology in plant pathology targets exact agricultural problems in plant–pathogen interactions and offer new methods for crop protection.

8.8.1 Silica Nanoparticles as Delivery Systems in Plants

Although all plants work on the environmental toxicity of nanoparticles, cognition of the real effect of nanotechnology on plant science was not documented until 2007, when the delivery of DNA to plants by nanoparticle transporters was reported. The article explained the use of engineered mesoporous silica nanoparticles (MSNPs) to deliver DNA into tobacco protoplasts (Torney et al. 2007). MSNPs were covered with plasmid DNA coding for GFP (green fluorescent protein).

MSNPs and DNA formed a stable compound, as concluded by the finding of no free DNA in solution. Action of tobacco protoplasts with the above MSNPs/DNA nanoparticles gave up to 7% temporary expression of the GFP. Also, the MSNPs/DNA nanoparticles were covered on gold nanoparticles and presented into corn using biolistic machinery. Torney et al. (2007) was widely cited and could be said to have hurled the notice in nanoparticles and application of green nanotechnology in the agriculture area. Several more articles have studied the use of nanoparticles as delivery systems to plants (Nair et al. 2010; Vivero-Escoto et al. 2013).

8.8.2 Magnetic Nanoparticles as Delivery Systems in Plants

Magnetic nanoparticles were used to transfer materials into plants and really to use their magnetic characterization to guide carriage and localization. Marrow plants (*Cucurbita pepo*) were cultured *in vitro* and treated with carbon-coated iron nanoparticles. Localization of the magnetic nanoparticles was specified by light, confocal, and electron microscopy. Influence and transposition of magnetic nanoparticles in complete living plants could be observed (González-Melendi et al. 2008).

8.9 Potentials of Nanotechnology in Forest Sector

The forest product industry relies on a huge renewable resource base to produce an extensive array of products that are crucial to our modern society. Wood and paper product companies produce a million tons of products each year that touch every part of our lives, contribute over billion dollars per year to the gross domestic product, and employ a million people in the world. Developing nanotechnologies offers the potential to improve entirely new methods for manufacturing engineered wood and fiber-based materials. They also can enable the growth of a wide range of new or improved wood-based materials and products that offer cost-effective substitutes for nonrenewable materials exploited in the manufacture of metallic, plastic, or ceramic products. Nanotechnology could change the forest product industry in effectively all aspects, ranging from production of raw materials to innovative applications for paper products and composite to new generations of efficient nanoscale lignocellulosics. Investigation and development in nanotechnology is critically vital to the inexpensive and sustainable manufacture of these new generations of forest-based resources, materials that will meet societal requirements while improving forest health and contributing to the further development of the biomass-based economy. Nanotechnology can be used to tap the vast undeveloped potential that trees possess as photochemical "factories"—those rich sources of renewable raw materials via sunlight, water, and carbon dioxide. The utilization of carbon dioxide in the production of these raw materials provides a carbon sink for this significant greenhouse gas. Through harnessing this potential, nanotechnology can provide benefits that extend well beyond fiber manufacture and new materials progress and into the areas of sustainable energy production, storage, and utilization. Nanotechnology for example may provide new methods for obtaining and utilizing energy from sunlight based on the process of the plant cell. Novel new methods to produce energy, chemicals, and other innovative products and methods from this renewable, domestic resource base will help to address main issues facing our nation, counting national energy security, global climate change, air and water quality, and global industrial competitiveness (Moon et al. 2006; Wegner and Jones 2007).

Potential uses for nanotechnology comprise developing intelligent wood and paper produces with an array of nanosensors built in to measure forces, loads, temperature, moisture levels, chemical emissions, pressure, attack by wood decaying bacteria, fungi, etc. Structure functionality onto lignocellulosic surfaces at the nanoscale possibly will open new opportunities for such things as pharmaceutical products, electronic lignocellulosic devices, and self-sterilizing surfaces. The use of nanodimensional building blocks will enable the synthesis of functional materials and substrates by substantially higher strength properties, which will allow the manufacture of lighter-weight products from less material as well as with less energy requirements. Important improvements in surface properties and functionality will be possible, manufacture of existing products extremely effective, and enable the development of many newer products. Nanotechnology can be used to progress processing of wood-based materials into a myriad of wood and paper products by improving water elimination and removing rewetting: reducing energy utilization in drying and tagging fibers, flakes, and particles to allow modified property enhancement in processing. Various challenges stand in the way of developing the potential benefits of nanotechnology science in the forest products industry, and much investigation will be needed to move forward in this arena. Investigators will need to address technical challenges, such as the absence of essential understanding of lignocellulosic material establishment at the nanoscale and the lack of adequate technology for measuring and characterizing these materials at the nanoscale. Applicants in this effort will be essential to come from not only academia but from industry and government as well; they will need to come together to form an organization and move forward as a cohesive unit working together toward a single aim of the advancement of nanotechnology into the forest and wood products industry. Advancing the nanotechnology study agenda efficiently and effectively will need consensus on investigation needs and priorities among the forest products industry, institution of higher education with forest products investigation, and education departments and programs, technology inventers and suppliers, research organizations and laboratories helping the forest products industry, and mission-oriented agencies with supportive goals, such as Department of Agriculture, National Science Foundation, and Department of Energy. In building consensus, the forest products part can capitalize on the good working interactions that the forest products industry has with its university investigation community and with agencies, such as the Department of Agriculture Forest Service, Cooperative State Research, Education, and Extension Service, Industries of the Future Program, and Biomass Program. In addition, the forest products part can take advantage of the links it has with investigation communities across the globe. As the industry's operation and marketplaces become more universal in nature, international cooperation and collaboration are imperative (Moon et al. 2006; Wegner and Jones 2007).

8.10 Conclusions

Chemical synthesis approaches lead to existence of some toxic chemical absorbed on the surface that may have harmful results in the medical and agricultural applications. Green synthesis makes available innovation over chemical and physical systems as it is totally environmental-friendly, cost-effective, easily scaled up for large-scale synthesis, and no necessity to use energy, temperature, high pressure, and toxic chemicals (Dubey et al. 2009; Gholami-Shabani et al. 2013). The prospect for bionanotechnology application in agriculture is amazing. Investigation of the applications of nanotechnology in agricultural science is less than a decade old. However, as conventional agricultural practices develop and requirements exceed the carrying ability of the terrestrial ecosystem, we have the option to explore nanotechnology in all parts of agriculture. It is well recognized that approval of new technology is vital to growth of national wealth (Peters et al. 2016). Nanotechnology promises an innovation to improve our currently abysmal nutrient use productivity through nano-formulation of fertilizers, nutritional quality, and breaking yield barriers through green nanotechnology, investigation, and control of diseases and pests, understanding the mechanism of host-parasite interactions at the molecular scale, improvement of new production pesticides and safe carriers, preservation and packaging of foodstuff and food additives, removal of contaminants from soil and water bodies, strengthening of natural fiber, improving the shelf life of flowers and vegetables, and usage of clay minerals as receptacles for nanoresources, including nutrient ion receptors, precision water management, reclamation of salt-affected soils, regenerating soil fertility, checking acidification of stabilization of erosion-prone surfaces, and irrigated lands (Prasad et al. 2017). Revisiting our perception of the theoretical foundations of the agricultural production method along the geosphere (pedosphere)-biosphere-atmosphere continuum combined with application of advanced theories, such as the theory of chaos and string, may open up new avenues. Nanotechnology requires a detailed understanding of science, in addition to fabrication and material technology, in conjunction with knowledge of the agricultural production method. The care of this challenge might interest brilliant minds to choose agricultural science as a career. To achieve success in the field, human resources require sophisticated training, for which new education programs, especially at the graduate level, are immediately needed. Research estimates that any technology takes approximately 20 years to emerge from the research laboratory and be commercialized (Knauer and Bucheli 2009). Nanotechnology in agriculture might take a few decades to move from research laboratory to land, especially since it has to avoid the difficulties experienced with biotechnology. For this to happen, sustained funding on the part of policy planners and science administrators, along with reasonable expectations, would be crucial for this nascent field to blossom.

References

Abbasi T, Anuradha J, Ganaie SU, Abbasi SA (2015) Gainful utilization of the highly intransigent weed ipomoea in the synthesis of gold nanoparticles. J King Saud University-Sci 27:15–22. doi:10.1016/j.jksus.2014.04.001

Abdel-Aziz MS, Shaheen MS, El-Nekeety AA, Abdel-Wahhab MA (2014) Antioxidant and antibacterial activity of silver nanoparticles biosynthesized using *Chenopodium murale* leaf extract. J Saudi Chem Soc 18:356–363. doi:10.1016/j.jscs.2013.09.011

Adhikari T, Sarkar D, Mashayekhi H, Xing B (2016) Growth and enzymatic activity of maize (*Zea mays* L.) plant: solution culture test for copper dioxide nano particles. J Plant Nutr 39:99–115. doi:10.1080/01904167.2015.1044012

Ahmad N, Sharma S, Singh VN, Shamsi SF, Fatma A, Mehta BR (2010) Biosynthesis of silver nanoparticles from *Desmodium triflorum*: a novel approach towards weed utilization. Biotechnol Res Int 2011:1–8. doi:10.4061/2011/454090

Ajitha B, Reddy YA, Reddy PS (2014) Biosynthesis of silver nanoparticles using *Plectranthus amboinicus* leaf extract and its antimicrobial activity. Spectrochim Acta Mol Biomol Spectrosc 128:257–262. doi:10.1016/j.saa.2014.02.105

Alharby HF, Metwali EM, Fuller MP, Aldhebiani AY (2016) Impact of application of zinc oxide nanoparticles on callus induction, plant regeneration, element content and antioxidant enzyme activity in tomato (*Solanum lycopersicum* Mill.) under salt stress. Arch Biol Sci 00:17. doi:10.2298/ABS151105017A

Ali ZA, Yahya R, Sekaran SD, Puteh R (2016) Green synthesis of silver nanoparticles using apple extract and its antibacterial properties. Adv Mater Sci Eng:1–6. doi:10.1155/2016/4102196

Amin R, Hwang S, Park SH (2011) Nanobiotechnology: an interface between nanotechnology and biotechnology. Nano 6:101–111. doi:10.1142/s1793292011002548

Ankamwar B (2010) Biosynthesis of gold nanoparticles (green-gold) using leaf extract of *Terminalia catappa*. E-J Chem:71334–71339. doi:10.1155/2010/745120

Ankamwar B, Damle C, Ahmad A, Sastry M (2005) Biosynthesis of gold and silver nanoparticles using *Emblica officinalis* fruit extract, their phase transfer and transmetallation in an organic solution. J Nanosci Nanotechnol 5(10):1665–1671

Armendariz V, Herrera I, Jose-yacaman M, Troiani H, Santiago P, Gardea-Torresdey JL (2004) Size controlled gold nanoparticle formation by *Avena sativa* biomass: use of plants in nanobiotechnology. J Nanopart Res 6:377–382. doi:10.1007/s11051-004-0741-4

Aromal SA, Philip D (2012) Green synthesis of gold nanoparticles using *Trigonella foenum-graecum* and its size-dependent catalytic activity. Spectrochim Acta Mol Biomol Spectrosc 97:1–5. doi:10.1016/j.saa.2012.05.083

Arunachalam KD, Annamalai SK, Hari S (2013) One-step green synthesis and characterization of leaf extract-mediated biocompatible silver and gold nanoparticles from *Memecylon umbellatum*. Int J Nanomedicine 8:1307–1315. doi:10.2147/IJN.S36670

Asavegowda N, Sobczak-Kupiec A, Malina D, Yathirajan HS, Keerthi VR, Chandrashekar N, Liny P (2013) Plant mediated synthesis of gold nanoparticles using fruit extracts of *Ananas comosus* (L.) (pineapple) and evaluation of biological activities. Adv Mater Lett 4:332–337

Ashokkumar S, Ravi S, Kathiravan V, Velmurugan S (2014) Synthesis, characterization and catalytic activity of silver nanoparticles using *Tribulus terrestris* leaf extract. Spectrochim Acta Mol Biomol Spectrosc 121:88–93. doi:10.1016/j.saa.2013.10.073

Asli S, Neumann PM (2009) Colloidal suspensions of clay or titanium dioxide nanoparticles can inhibit leaf growth and transpiration via physical effects on root water transport. Plant Cell Environ 32:577–584. doi:10.1111/j.1365-3040.2009.01952.x

Augustine R, Kalarikkal N, Thomas S (2014) A facile and rapid method for the black pepper leaf mediated green synthesis of silver nanoparticles and the antimicrobial study. Appl Nanosci 4:809–818. doi:10.1007/s13204-013-0260-7

Awwad AM, Salem NM (2012) Green synthesis of silver nanoparticles by mulberry leaves extract. Nanosci Nanotechnol 2:125–128. doi:10.5923/j.nn.20120204.06

Awwad AM, Salem NM, Abdeen AO (2013) Green synthesis of silver nanoparticles using carob leaf extract and its antibacterial activity. Int J Ind chem 4:1–6. doi:10.1186/2228-5547-4-29

Aziz N, Faraz M, Pandey R, Sakir M, Fatma T, Varma A, Barman I, Prasad R (2015) Facile algae-derived route to biogenic silver nanoparticles: synthesis, antibacterial and photocatalytic properties. Langmuir 31:11605–11612

Aziz N, Pandey R, Barman I, Prasad R (2016) Leveraging the attributes of *Mucor hiemalis*-derived silver nanoparticles for a synergistic broad-spectrum antimicrobial platform. Front Microbiol 7:1984. doi:10.3389/fmicb.2016.01984

Baharara J, Namvar F, Ramezani T, Hosseini N, Mohamad R (2014) Green synthesis of silver nanoparticles using *Achillea Biebersteinii* flower extract and its anti-angiogenic properties in the rat aortic ring model. Molecules 19:4624–4634. doi:10.3390/molecules19044624

Baharara J, Namvar F, Ramezani T, Mousavi M, Mohamad R (2015) Silver nanoparticles biosynthesized using *Achillea Biebersteinii* flower extract: apoptosis induction in MCF-7 cells via caspase activation and regulation of Bax and Bcl-2 gene expression. Molecules 20:2693–2706. doi:10.3390/molecules20022693

Baishya D, Sharma N, Bora R (2012) Green synthesis of silver nanoparticle using *Bryophyllum pinnatum* (Lam.) and monitoring their antibacterial activities. Arch Appl Sci Res 4:2098–2104

Bandi S, Vasundhara K (2012) Green synthesis of silver nanoparticles using *Adhatoda vasica* methanolic extract and its biological activities. J Atom Mol 2:282

Bar H, Bhui DK, Sahoo GP, Sarkar P, De SP, Misra A (2009a) Green synthesis of silver nanoparticles using latex of *Jatropha Curcas*. Colloids Surf A Physicochem Eng Aspect 339:134–139. doi:10.1016/j.colsurfa.2009.02.008

Bar H, Bhui DK, Sahoo GP, Sarkar P, Pyne S, Misra A (2009b) Green synthesis of silver nanoparticles using seed extract of *Jatropha curcas*. Colloids Surf A Physicochem Eng Asp 348:212–216. doi:10.1016/j.colsurfa.2009.07.021

Basavegowda N, Idhayadhulla A, Lee YR (2014) Phyto-synthesis of gold nanoparticles using fruit extract of *Hovenia dulcis* and their biological activities. Ind Crop Prod 52:745–751. doi:10.1016/j.indcrop.2013.12.006

Begum NA, Mondal S, Basu S, Laskar RA, Mandal D (2009) Biogenic synthesis of Au and Ag nanoparticles using aqueous solutions of black tea leaf extracts. Colloids Surf B: Biointerfaces 71:113–118. doi:10.1016/j.colsurfb.2009.01.012

Bhattacharyya A, Prasad R, Buhroo AA, Duraisamy P, Yousuf I, Umadevi M, Bindhu MR, Govindarajan M, Khanday AL (2016) One-pot fabrication and characterization of silver nanoparticles using *Solanum lycopersicum*: an eco-friendly and potent control tool against Rose Aphid, Macrosiphum rosae. J Nanosci:4679410. 7 p. http://dx.doi.org/10.1155/2016/4679410

Bhumi G, Rao ML, Savithramma N (2015) Green synthesis of silver nanoparticles from the leaf extract of *Adhatoda vasica* nees. and assessment of its antibacterial activity. Asian J Pharm Clin Res 8:62–67

Bhuyan T, Mishra K, Khanuja M, Prasad R, Varma A (2015) Biosynthesis of zinc oxide nanoparticles from *Azadirachta indica* for antibacterial and photocatalytic applications. Mater Sci Semicond Process 32:55–61

Bindhu MR, Umadevi M (2013) Synthesis of monodispersed silver nanoparticles using *Hibiscus Cannabinus* leaf extract and its antimicrobial activity. Spectrochim Acta A Mol Biomol Spectrosc 101:184–190. doi:10.1016/j.saa.2012.09.031

Boruah SK, Boruah PK, Sarma P, Medhi C, Medhi OK (2012) Green synthesis of gold nanoparticles using *Camellia Sinensis* and kinetics of the reaction. Adv Mater Lett 3:481

Brumbaugh AD, Cohen KA, St. Angelo SK (2014) Ultra small copper nanoparticles synthesized with a plant tea reducing agent. ACS Sus Chem Engin 2:1933–1939. doi:10.1021/sc500393t

Cañas JE, Long M, Nations S, Vadan R, Dai L, Luo M, Ambikapathi R, Lee EH, Olszyk D (2008) Effects of functionalized and nonfunctionalized single-walled carbon nanotubes on root elongation of select crop species. Environ Toxicol Chem 27:1922–1931

Chandran SP, Chaudhary M, Pasricha R, Ahmad A, Sastry M (2006) Synthesis of gold nanotriangles and silver nanoparticles using *Aloe vera* plant extract. Biotechnol Prog 22:577–583. doi:10.1021/bp0501423

Chomoucká J, Drbohlavová J, Hubálek J, Babula P, Adam V, Kizek R (2010) Toxicity of nanoparticles for plants. List Cukrova Repa. 2010 126:400

Cos P, Vlietinck AJ, Berghe DV, Maes L (2006) Anti-infective potential of natural products: how to develop a stronger in vitro 'proof-of-concept'. J Ethnopharmacol 106:290–302. doi:10.1016/j.jep.2006.04.003

Das J, Das MP, Velusamy P (2013) *Sesbania grandiflora* leaf extract mediated green synthesis of antibacterial silver nanoparticles against selected human pathogens. Spectrochim Acta Mol Biomol Spectrosc 104:265–270. doi:10.1016/j.saa.2012.11.075

Dash SS, Bag BG (2014) Synthesis of gold nanoparticles using renewable *Punica granatum* juice and study of its catalytic activity. Appl Nanosci 4:55–59. doi:10.1007/s13204-012-0179-4

Dash SS, Majumdar R, Sikder AK, Bag BG, Patra BK (2014) *Saraca indica* bark extract mediated green synthesis of polyshaped gold nanoparticles and its application in catalytic reduction. Appl Nanosci 4:485–490. doi:10.1007/s13204-013-0223-z

De La Torre-Roche R, Hawthorne J, Deng Y, Xing B, Cai W, Newman LA, Wang Q, Ma X, Hamdi H, White JC (2013) Multiwalled carbon nanotubes and C60 fullerenes differentially impact the accumulation of weathered pesticides in four agricultural plants. Environ Sci Technol 47:12539–12547. doi:10.1021/es4034809

Devi GD, Murugan K, Selvam CP (2014) Green synthesis of silver nanoparticles using *Euphorbia hirta* (Euphorbiaceae) leaf extract against crop pest of cotton bollworm, *Helicoverpa armigera* (Lepidoptera: Noctuidae). J Biopest 7:54

Dey PM (2012) Methods in plant biochemistry, vol 1 APL. Academic, London. ISBN:978-0-12-461020-0

Dipankar C, Murugan S (2012) The green synthesis, characterization and evaluation of the biological activities of silver nanoparticles synthesized from *Iresine Herbstii* leaf aqueous extracts. Colloids Surf B Biointerfaces 98:112–119. doi:10.1016/j.colsurfb.2012.04.006

Dubey M, Bhadauria S, Kushwah BS (2009) Green synthesis of nanosilver particles from extract of *Eucalyptus Hybrida* (safeda) leaf. Dig J Nanomater Biostruct 4:537–543

Dubey SP, Lahtinen M, Särkkä H, Sillanpää M (2010a) Bioprospective of *Sorbus aucuparia* leaf extract in development of silver and gold nanocolloids. Colloids Surf B: Biointerfaces 80:26–33. doi:10.1016/j.colsurfb.2010.05.024

Dubey SP, Lahtinen M, Sillanpää M (2010b) Tansy fruit mediated greener synthesis of silver and gold nanoparticles. Process Biochem 45:1065–1071. doi:10.1016/j.procbio.2010.03.024

Duca M (2015) Plant cell physiology. In: Plant physiology. Springer, Cham, pp 13–37. ISBN-13: 978-3319179087

Dwivedi AD, Gopal K (2010) Biosynthesis of silver and gold nanoparticles using *Chenopodium Album* leaf extract. Colloid Surf A: Physicochem Engin Asp 369:27–33. doi:10.1016/j.colsurfa.2010.07.020

Feizi H, Moghaddam PR, Shahtahmassebi N, Fotovat A (2012) Impact of bulk and nanosized titanium dioxide (TiO_2) on wheat seed germination and seedling growth. Biol Trace Elem Res 146:101–106. doi:10.1007/s12011-011-9222-7

Gade A, Ingle A, Whiteley C, Rai M (2010) Mycogenic metal nanoparticles: progress and applications. Biotechnol Lett 32:593–600. doi:10.1007/s10529-009-0197-9

Ganaie SU, Abbasi T, Anuradha J, Abbasi SA (2014) Biomimetic synthesis of silver nanoparticles using the amphibious weed ipomoea and their application in pollution control. J King Saud University-Sci 26:222–229. doi:10.1016/j.jksus.2014.02.004

Geethalakshmi R, Sarada DV (2012) Gold and silver nanoparticles from *Trianthema decandra*: synthesis, characterization, and antimicrobial properties. Int J Nanomedicine 7:5375–5384. doi:10.2147/ijn.s36516

Ghaffari-Moghaddam M, Hadi-Dabanlou R (2014) Plant mediated green synthesis and antibacterial activity of silver nanoparticles using *Crataegus douglasii* fruit extract. J Ind Eng Chem 20:739–744. doi:10.1016/j.jiec.2013.09.005

Ghodake GS, Deshpande NG, Lee YP, Jin ES (2010) Pear fruit extract-assisted room-temperature biosynthesis of gold nanoplates. Colloids Surf B: Biointerfaces 75:584–589. doi:10.1016/j.colsurfb.2009.09.040

Gholami-Shabani MH, Akbarzadeh A, Mortazavi M, Emadzadeh MK (2013) Evaluation of the antibacterial properties of silver nanoparticles synthesized with *Fusarium oxysporum* and *Escherichia coli*. Int J Lifesc Bt Pharm Res 2:342–348

Gholami-Shabani M, Akbarzadeh A, Norouzian D, Amini A, Gholami-Shabani Z, Imani A, Chiani M, Riazi G, Shams-Ghahfarokhi M, Razzaghi-Abyaneh M (2014) Antimicrobial activity and physical characterization of silver nanoparticles green synthesized using nitrate reductase from *Fusarium oxysporum*. Appl Biochem Biotechnol 172:4084–4098. doi:10.1007/s12010-014-0809-2

Gholami-Shabani M, Shams-Ghahfarokhi M, Gholami-Shabani Z, Akbarzadeh A, Riazi G, Ajdari S, Amani A, Razzaghi-Abyaneh M (2015) Enzymatic synthesis of gold nanoparticles using sulfite reductase purified from *Escherichia coli*: a green eco-friendly approach. Process Biochem 50:1076–1085. doi:10.1016/j.procbio.2015.04.004

Gholami-Shabani M, Imani A, Shams-Ghahfarokhi M, Gholami-Shabani Z, Pazooki A, Akbarzadeh A, Riazi G, Razzaghi-Abyaneh M (2016) Bioinspired synthesis, characterization and antifungal activity of enzyme-mediated gold nanoparticles using a fungal oxidoreductase. J Iran Chem Soc 9:1–10. doi:10.1007/s13738-016-0923-x

Ghosh M, Bandyopadhyay M, Mukherjee A (2010) Genotoxicity of titanium dioxide (TiO_2) nanoparticles at two trophic level: plant and human lymphocytes. Chemosphere 81:1253–1262. doi:10.1016/j.chemosphere.2010.09.022

Ghosh S, Patil S, Ahire M, Kitture R, Kale S, Pardesi K, Cameotra SS, Bellare J, Dhavale DD, Jabgunde A, Chopade BA (2012) Synthesis of silver nanoparticles using *Dioscorea bulbifera* tuber extract and evaluation of its synergistic potential in combination with antimicrobial agents. Int J Nanomedicine 7:483–496

Golmoraj VE, Reza Khoshayand M, Amini M, Mollazadeh Moghadamd K, Amin G, Reza Shahverdi A (2011) The surface chemistry and stability of gold nanoparticles prepared using methanol extract of *Eucalyptus camaldulensis*. J Exp Nanosci 6:200–208. doi:10.1080/17458080.2010.489581

González-Melendi P, Fernández-Pacheco R, Coronado MJ, Corredor E, Testillano PS, Risueño MC, Marquina C, Ibarra MR, Rubiales D, Pérez-de-Luque A (2008) Nanoparticles as smart treatment-delivery systems in plants: assessment of different techniques of microscopy for their visualization in plant tissues. Ann Bot 101:187–195. doi:10.1093/aob/mcm283

Gupta A, Bonde SR, Gaikwad S, Ingle A, Gade AK, Rai M (2014) *Lawsonia inermis*-mediated synthesis of silver nanoparticles: activity against human pathogenic fungi and bacteria with special reference to formulation of an antimicrobial nanogel. IET Nanobiotechnol 8:172–178. doi:10.1049/iet-nbt.2013.0015

Handy RD, Owen R, Valsami-Jones E (2008) The ecotoxicology of nanoparticles and nanomaterials: current status, knowledge gaps, challenges, and future needs. Ecotoxicology 17:315–325. doi:10.1007/s10646-008-0206-0

Harris AT, Bali R (2008) On the formation and extent of uptake of silver nanoparticles by live plants. J Nanopart Res 10:691–695. doi:10.1007/s11051-007-9288-5

Hassan LA, Elijah AT, Ojiefoh OC, Joseph O, Sunday BO, Olugbenga DE, Anuoluwapo AA (2016) Biosynthesis of silver nanoparticles using *Garcinia kola* and its antimicrobial potential. Afr J Pure Appl Chem 10:1–7. doi:10.5897/ajpac2015.0650

Heinemeyer J, Eubel H, Wehmhöner D, Jänsch L, Braun HP (2004) Proteomic approach to characterize the supramolecular organization of photosystems in higher plants. Phytochemistry 65:1683–1692. doi:10.1016/j.phytochem.2004.04.022

Huang J, Li Q, Sun D, Lu Y, Su Y, Yang X, Wang H, Wang Y, Shao W, He N, Hong J, Chen C (2007) Biosynthesis of silver and gold nanoparticles by novel sundried *Cinnamomum camphora* leaf. Nanotechnology 18:105104

Huang J, Lin L, Sun D, Chen H, Yang D, Li Q (2015) Bio-inspired synthesis of metal nanoparticles and application. Chem Soc Rev 44:6330–6374. doi:10.1039/C5CS00133A

Ibrahim HM (2015) Green synthesis and characterization of silver nanoparticles using banana peel extract and their antimicrobial activity against representative microorganisms. J Radiat Res Appl Sci 8:265–275. doi:10.1016/j.jrras.2015.01.007

Iravani S (2011) Green synthesis of metal nanoparticles using plants. Green Chem 13:2638–2650. doi:10.1039/C1GC15386B

Iravani S, Zolfaghari B (2013) Green synthesis of silver nanoparticles using *Pinus eldarica* bark extract. Bio Med Res Int 2013:639725. doi:10.1155/2013/639725

Jacob SJ, Finub JS, Narayanan A (2012) Synthesis of silver nanoparticles using *Piper longum* leaf extracts and its cytotoxic activity against Hep-2 cell line. Colloids Surf B: Biointerfaces 91:212–214. doi:10.1016/j.colsurfb.2011.11.001

Jagtap UB, Bapat VA (2013) Green synthesis of silver nanoparticles using *Artocarpus heterophyllus* Lam. seed extract and its antibacterial activity. Ind Crop Prod 46:132–137

Jain D, Daima HK, Kachhwaha S, Kothari SL (2009) Synthesis of plant-mediated silver nanoparticles using papaya fruit extract and evaluation of their antimicrobial activities. Dig J Nanomater Biostruct 4:557–563

Jayaseelan C, Ramkumar R, Rahuman AA, Perumal P (2013) Green synthesis of gold nanoparticles using seed aqueous extract of *Abelmoschus esculentus* and its antifungal activity. Ind Crop Prod 45:423–429. doi:10.1016/j.indcrop.2012.12.019

Jha AK, Prasad K (2010) Green synthesis of silver nanoparticles using Cycas leaf. Int J Green Nanotechnol: Phys Chem 1:110–117. doi:10.1080/19430871003684572

Jia L, Zhang Q, Li Q, Song H (2009) The biosynthesis of palladium nanoparticles by antioxidants in *Gardenia jasminoides* Ellis: long lifetime nanocatalysts for p-nitrotoluene hydrogenation. Nanotechnology 20:385601. doi:10.1088/0957-4484/20/38/385601

Karthick V, Ganesh-Kumar V, Maiyalagan T, Deepa R, Govindaraju K, Rajeswari A, Stalin-Dhas T (2012) Green synthesis of well dispersed nanoparticles using leaf extract of medicinally useful *Adhatoda vasica* nees. Micro Nanosys 4:192–198

Kasthuri J, Veerapandian S, Rajendiran N (2009) Biological synthesis of silver and gold nanoparticles using apiin as reducing agent. Colloids Surf B: Biointerfaces 68:55–60

Kathiravan V, Ravi S, Ashokkumar S (2014) Synthesis of silver nanoparticles from *Melia dubia* leaf extract and their in vitro anticancer activity. Spectrochim Acta Part A: Mol Biomol Spectrosc 130:116–121

Kaviya S, Santhanalakshmi J, Viswanathan B, Muthumary J, Srinivasan K (2011) Biosynthesis of silver nanoparticles using *Citrus sinensis* peel extract and its antibacterial activity. Spectrochim Acta Mol Biomol Spectrosc 79:594–598. doi:10.1016/j.saa.2011.03.040

Keegstra K, Talmadge KW, Bauer WD, Albersheim P (1973) The structure of plant cell walls III. A model of the walls of suspension-cultured sycamore cells based on the interconnections of the macromolecular components. Plant Physiol 51:188–197

Khodakovskaya M, Dervishi E, Mahmood M, Xu Y, Li Z, Watanabe F, Biris AS (2009) Carbon nanotubes are able to penetrate plant seed coat and dramatically affect seed germination and plant growth. ACS Nano 3:3221–3227. doi:10.1021/nn900887m

Khodakovskaya MV, de Silva K, Biris AS, Dervishi E, Villagarcia H (2012) Carbon nanotubes induce growth enhancement of tobacco cells. ACS Nano 6:2128–2135. doi:10.1021/nn204643g

Klaine SJ, Alvarez PJ, Batley GE, Fernandes TF, Handy RD, Lyon DY, Mahendra S, McLaughlin MJ, Lead JR (2008) Nanomaterials in the environment: behavior, fate, bioavailability, and effects. Environ Toxicol Chem 27:1825–1851. doi:10.1897/08-090.1

Knauer K, Bucheli T (2009) Nano-materials-the need for research in agriculture. Agrarforschung 16:390–395

Kora AJ, Sashidhar RB, Arunachalam J (2010) Gum kondagogu (*Cochlospermum gossypium*): a template for the green synthesis and stabilization of silver nanoparticles with antibacterial application. Carbohydr Polym 82:670–679. doi:10.1016/j.carbpol.2010.05.034

Kotakadi VS, Rao YS, Gaddam SA, Prasad TNVKV, Reddy AV, Gopal DS (2013) Simple and rapid biosynthesis of stable silver nanoparticles using dried leaves of *Catharanthus roseus*. Linn. *G. Donn* and its antimicrobial activity. Colloids Surf B: Biointerfaces 105:194–198. doi:10.1016/j.colsurfb.2013.01.003

Kouvaris P, Delimitis A, Zaspalis V, Papadopoulos D, Tsipas SA, Michailidis N (2012) Green synthesis and characterization of silver nanoparticles produced using *Arbutus Unedo* leaf extract. Mater Lett 76:18–20. doi:10.1016/j.matlet.2012.02.025

Krishnaraj C, Jagan EG, Rajasekar S, Selvakumar P, Kalaichelvan PT, Mohan N (2010) Synthesis of silver nanoparticles using *Acalypha indica* leaf extracts and its antibacterial activity against water borne pathogens. Colloids Surf B: Biointerfaces 76:50–56. doi:10.1016/j.colsurfb.2009.10.008

Krishnaraj C, Jagan EG, Ramachandran R, Abirami SM, Mohan N, Kalaichelvan PT (2012) Effect of biologically synthesized silver nanoparticles on *Bacopa monnieri* (Linn.) Wettst. plant growth metabolism. Process Biochem 47:651–658. doi:10.1016/j.procbio.2012.01.006

Kumar KM, Sinha M, Mandal BK, Ghosh AR, Kumar KS, Reddy PS (2012) Green synthesis of silver nanoparticles using *Terminalia chebula* extract at room temperature and their antimicrobial studies. Spectrochim Acta Part A: Mol Biomol Spectrosc 91:228–233

Kumar B, Smita K, Cumbal L, Debut A (2014a) Synthesis of silver nanoparticles using Sacha inchi (*Plukenetia volubilis* L.) leaf extracts. Saudi J Biol Sci 21:605–609. doi:10.1016/j.sjbs.2014.07.004

Kumar DA, Palanichamy V, Roopan SM (2014b) Green synthesis of silver nanoparticles using *Alternanthera dentata* leaf extract at room temperature and their antimicrobial activity. Spectrochim Acta Part A: Mol Biomol Spectrosc 127:168–171. doi:10.1016/j.saa.2014.02.058

Kumar HA, Mandal BK, Kumar KM, Babu Maddinedi S, Kumar TS, Madhiyazhagan P, Ghosh AR (2014c) Antimicrobial and antioxidant activities of *Mimusops elengi* seed extract mediated isotropic silver nanoparticles. Spectrochimica Acta Part A: Molecul Biomolecul Spectroscopy 130:13–18. doi:10.1016/j.saa.2014.03.024

Kumar PV, Pammi SVN, Kollu P, Satyanarayana KVV, Shameem U (2014d) Green synthesis and characterization of silver nanoparticles using *Boerhaavia diffusa* plant extract and their antibacterial activity. Ind Crop Prod 52:562–566. doi:10.1016/j.indcrop.2013.10.050

Kumari M, Khan SS, Pakrashi S, Mukherjee A, Chandrasekaran N (2011) Cytogenetic and genotoxic effects of zinc oxide nanoparticles on root cells of *Allium cepa*. J Hazard Mater 190:613–621

Kuppusamy P, Yusoff MM, Maniam GP, Govindan N (2016) Biosynthesis of metallic nanoparticles using plant derivatives and their new avenues in pharmacological applications–an updated report. Saudi Pharm J 24:473–484. doi:10.1016/j.jsps.2014.11.013

Lee WM, An YJ, Yoon H, Kweon HS (2008) Toxicity and bioavailability of copper nanoparticles to the terrestrial plants mung bean (*Phaseolus radiatus*) and wheat (*Triticum aestivum*): plant agar test for water-insoluble nanoparticles. Environ Toxicol Chem 27:1915–1921. doi:10.1897/07-481.1

Lee HJ, Song JY, Kim BS (2013) Biological synthesis of copper nanoparticles using *Magnolia kobus* leaf extract and their antibacterial activity. J Chem Technol Biotechnol 88:1971–1977. doi:10.1002/jctb.4052

Li M, Zhu L, Lin D (2011) Toxicity of ZnO nanoparticles to *Escherichia coli*: mechanism and the influence of medium components. Environ Sci Technol 45:1977–1983. doi:10.1021/es102624t

Lin D, Xing B (2007) Phytotoxicity of nanoparticles: inhibition of seed germination and root growth. Environ Pollut 150:243–250. doi:10.1016/j.envpol.2007.01.016

Ma X, Geiser-Lee J, Deng Y, Kolmakov A (2010) Interactions between engineered nanoparticles (ENPs) and plants: phytotoxicity, uptake and accumulation. Sci Total Environ 408:3053–3061. doi:10.1016/j.scitotenv.2010.03.031

Maensiri S, Laokul P, Klinkaewnarong J, Phokha S, Promarak V, Seraphin S (2008) Indium oxide (In2O3) nanoparticles using *Aloe vera* plant extract: synthesis and optical properties. J Optoelectron Adv Mater 10:161–165

Mahmoodzadeh H, Nabavi M, Kashefi H (2013) Effect of nanoscale titanium dioxide particles on the germination and growth of canola (*Brassica napus*). J Ornamental Hortic Plants 3:25–32

Makarov VV, Makarova SS, Love AJ, Sinitsyna OV, Dudnik AO, Yaminsky IV, Taliansky ME, Kalinina NO (2014) Biosynthesis of stable iron oxide nanoparticles in aqueous extracts of *Hordeum vulgare* and *Rumex acetosa* plants. Langmuir 30:5982–5988. doi:10.1021/la5011924

Mariselvam R, Ranjitsingh AJA, Nanthini AUR, Kalirajan K, Padmalatha C, Selvakumar PM (2014) Green synthesis of silver nanoparticles from the extract of the inflorescence of *Cocos nucifera* (family: Arecaceae) for enhanced antibacterial activity. Spectrochim Acta A Mol Biomol Spectrosc 129:537–541. doi:10.1016/j.saa.2014.03.066

Moon RJ, Frihart CR, Wegner T (2006) Nanotechnology applications in the forest products industry. Forest Prod J 56:4

Muniyappan N, Nagarajan NS (2014) Green synthesis of silver nanoparticles with *Dalbergia spinosa* leaves and their applications in biological and catalytic activities. Process Biochem 49:1054–1061. doi:10.1016/j.procbio.2014.03.015

Murugan K, Dinesh D, Kumar PJ, Panneerselvam C, Subramaniam J, Madhiyazhagan P, Suresh U, Nicoletti M, Alarfaj AA, Munusamy MA, Higuchi A (2015) *Datura metel*-synthesized silver nanoparticles magnify predation of dragonfly nymphs against the malaria vector Anopheles stephensi. Parasitol Res 114:4645–4654. doi:10.1007/s00436-015-4710-x

Nadagouda MN, Iyanna N, Lalley J, Han C, Dionysiou DD, Varma RS (2014) Synthesis of silver and gold nanoparticles using antioxidants from blackberry, blueberry, pomegranate, and turmeric extracts. ACS Sus Chem Engin 2:1717–1723. doi:10.1021/sc500237k

Naika HR, Lingaraju K, Manjunath K, Kumar D, Nagaraju G, Suresh D, Nagabhushana H (2015) Green synthesis of CuO nanoparticles using *Gloriosa superba* L. extract and their antibacterial activity. J Taibah University Sci 9:7–12. doi:10.1016/j.jtusci.2014.04.006

Nair R, Varghese SH, Nair BG, Maekawa T, Yoshida Y, Kumar DS (2010) Nanoparticulate material delivery to plants. Plant Sci 179:154–163. doi:10.1016/j.plantsci.2010.04.012

Nalawade P, Mukherjee P, Kapoor S (2014) Biosynthesis, characterization and antibacterial studies of silver nanoparticles using pods extract of *Acacia auriculiformis*. Spectrochim Acta A Mol Biomol Spectrosc 129:121–124. doi:10.1016/j.saa.2014.03.032

Nasrollahzadeh M, Sajadi SM, Rostami-Vartooni A, Khalaj M (2014) Journey on greener pathways: use of *Euphorbia condylocarpa* M. bieb as reductant and stabilizer for green synthesis of Au/Pd bimetallic nanoparticles as reusable catalysts in the Suzuki and Heck coupling reactions in water. RSC Adv 4:43477–43484. doi:10.1039/c4ra07173e

Nasrollahzadeh M, Sajadi SM, Rostami-Vartooni A, Khalaj M (2015a) Green synthesis of Pd/Fe_3O_4 nanoparticles using *Euphorbia condylocarpa* M. bieb root extract and their catalytic applications as magnetically recoverable and stable recyclable catalysts for the phosphine-free Sonogashira and Suzuki coupling reactions. J Molecul Catal A: Chem 396:31–39. doi:10.1016/j.molcata.2014.09.029

Nasrollahzadeh M, Sajadi SM, Maham M (2015b) Green synthesis of palladium nanoparticles using *Hippophae rhamnoides* Linn leaf extract and their catalytic activity for the Suzuki–Miyaura coupling in water. J Molecul Catal A: Chem 396:297–303. doi:10.1016/j.molcata.2014.10.019

Navarro E, Baun A, Behra R, Hartmann NB, Filser J, Miao AJ, Quigg A, Santschi PH, Sigg L (2008) Environmental behavior and ecotoxicity of engineered nanoparticles to algae, plants, and fungi. Ecotoxicology 17:372–386. doi:10.1007/s10646-008-0214-0

Nel A, Xia T, Mädler L, Li N (2006) Toxic potential of materials at the nanolevel. Science 311:622–627. doi:10.1126/science.1114397

Netala VR, Kotakadi VS, Nagam V, Bobbu P, Ghosh SB, Tartte V (2015) First report of biomimetic synthesis of silver nanoparticles using aqueous callus extract of *Centella asiatica* and their antimicrobial activity. Appl Nanosci 5:801–807. doi:10.1007/s13204-014-0374-6

Nisha SN, Aysha OS, Rahaman JS, Kumar PV, Valli S, Nirmala P, Reena A (2014) Lemon peels mediated synthesis of silver nanoparticles and its antidermatophytic activity. Spectrochim Acta A Mol Biomol Spectrosc 124:194–198. doi:10.1016/j.saa.2013.12.019

Noruzi M, Zare D, Davoodi D (2012) A rapid biosynthesis route for the preparation of gold nanoparticles by aqueous extract of cypress leaves at room temperature. Spectrochim Acta A Mol Biomol Spectrosc 94:84–88. doi:10.1016/j.saa.2012.03.041

Panda KK, Achary VM, Krishnaveni R, Padhi BK, Sarangi SN, Sahu SN, Panda BB (2011) In vitro biosynthesis and genotoxicity bioassay of silver nanoparticles using plants. Toxicol in Vitro 25:1097–1105. doi:10.1016/j.tiv.2011.03.008

Patil CD, Patil SV, Borase HP, Salunke BK, Salunkhe RB (2012) Larvicidal activity of silver nanoparticles synthesized using *Plumeria rubra* plant latex against *Aedes aegypti* and *Anopheles stephensi*. Parasitol Res 110:1815–1822. doi:10.1007/s00436-011-2704-x

Peters RJ, Bouwmeester H, Gottardo S, Amenta V, Arena M, Brandhoff P, Marvin HJ, Mech A, Moniz FB, Pesudo LQ, Rauscher H (2016) Nanomaterials for products and application in agriculture, feed and food. Trends Food Sci Technol 54:155–164. doi:10.1016/j.tifs.2016.06.008

Philip D (2010) Green synthesis of gold and silver nanoparticles using *Hibiscus rosa sinensis*. Phys E Low-Dimensional Syst Nanostruct 42:1417–1424. doi:10.1016/j.physe.2009.11.081

Philip D, Unni C, Aromal SA, Vidhu VK (2011) *Murraya koenigii* leaf-assisted rapid green synthesis of silver and gold nanoparticles. Spectrochim Acta A Mol Biomol Spectrosc 78:899–904. doi:10.1016/j.saa.2010.12.060

Prabhu D, Arulvasu C, Babu G, Manikandan R, Srinivasan P (2013) Biologically synthesized green silver nanoparticles from leaf extract of *Vitex negundo* L. induce growth-inhibitory effect on human colon cancer cell line HCT15. Process Biochem 48:317–324. doi:10.1016/j.procbio.2012.12.013

Prasad R (2014) Synthesis of silver nanoparticles in photosynthetic plants. J Nanoparticle:963961. http://dx.doi.org/10.1155/2014/963961

Prasad R, Swamy VS (2013) Antibacterial activity of silver nanoparticles synthesized by bark extract of *Syzygium cumini*. J Nanopart 2013:1–6. doi:10.1155/2013/431218

Prasad R, Swamy VS, Prasad KS, Varma A (2012) Biogenic synthesis of silver nanoparticles from the leaf extract of *Syzygium cumini* (L.) and its antibacterial activity. Int J Pharm Bio Sci 3(4):745–752

Prasad R, Kumar V, Prasad KS (2014) Nanotechnology in sustainable agriculture: present concerns and future aspects. Afr J Biotechnol 13(6):705–713

Prasad R, Pandey R, Barman I (2016) Engineering tailored nanoparticles with microbes: quo vadis. WIREs Nanomed Nanobiotechnol 8:316–330. doi:10.1002/wnan.1363

Prasad R, Bhattacharyya A, Nguyen QD (2017) Nanotechnology in sustainable agriculture: recent developments, challenges, and perspectives. Front Microbiol 8:1014. doi:10.3389/fmicb.2017.01014

Prathna TC, Chandrasekaran N, Raichur AM, Mukherjee A (2011) Biomimetic synthesis of silver nanoparticles by *Citrus limon* (lemon) aqueous extract and theoretical prediction of particle size. Colloids Surf B: Biointerfaces 82(1):152–159. doi:10.1016/j.colsurfb.2010.08.036

Premanand G, Shanmugam N, Kannadasan N, Sathishkumar K, Viruthagiri G (2016) *Nelumbo nucifera* leaf extract mediated synthesis of silver nanoparticles and their antimicrobial properties against some human pathogens. Appl Nanosci 6:409–415. doi:10.1007/s13204-015-0442-6

Puišo J, Jonkuvienė D, Mačionienė I, Šalomskienė J, Jasutienė I, Kondrotas R (2014) Biosynthesis of silver nanoparticles using lingonberry and cranberry juices and their antimicrobial activity. Colloids Surf B: Biointerfaces 121:214–221. doi:10.1016/j.colsurfb.2014.05.001

Qu J, Yuan X, Wang X, Shao P (2011a) Zinc accumulation and synthesis of ZnO nanoparticles using *Physalis alkekengi* L. Environ Pollut 159:1783–1788. doi:10.1016/j.envpol.2011.04.016

Qu J, Luo C, Hou J (2011b) Synthesis of ZnO nanoparticles from Zn-hyperaccumulator (*Sedum alfredii* Hance) plants. IET Micro Nano Lett 6:174–176. doi:10.1049/mnl.2011.0004

Raghunandan D, Basavaraja S, Mahesh B, Balaji S, Manjunath SY, Venkataraman A (2009) Biosynthesis of stable polyshaped gold nanoparticles from microwave-exposed aqueous extracellular anti-malignant guava (*Psidium guajava*) leaf extract. NanoBiotechnology 5:34–41. doi:10.1007/s12030-009-9030-8

Raghunandan D, Bedre MD, Basavaraja S, Sawle B, Manjunath SY, Venkataraman A (2010) Rapid biosynthesis of irregular shaped gold nanoparticles from macerated aqueous extracellular dried clove buds (*Syzygium aromaticum*) solution. Colloids Surf B: Biointerfaces 79:235–240

Rahimi-Nasrabadi M, Pourmortazavi SM, Shandiz SA, Ahmadi F, Batooli H (2014) Green synthesis of silver nanoparticles using *Eucalyptus leucoxylon* leaves extract and evaluating the antioxidant activities of extract. Nat Prod Res 28:1964–1969

Ramar M, Manikandan B, Marimuthu PN, Raman T, Mahalingam A, Subramanian P, Karthick S, Munusamy A (2015) Synthesis of silver nanoparticles using *Solanum trilobatum* fruits extract and its antibacterial, cytotoxic activity against human breast cancer cell line MCF 7. Spectrochim Acta Mol Biomol Spectrosc 140:223–228

Rao SV (2011) Picosecond nonlinear optical studies of gold nanoparticles synthesised using coriander leaves (*Coriandrum sativum*). J Mod Opt 58:1024–1029

Rashidipour M, Heydari R (2014) Biosynthesis of silver nanoparticles using extract of olive leaf: synthesis and in vitro cytotoxic effect on MCF-7 cells. J Nanostruct Chem 4:1–6

Raut RW, Mendhulkar VD, Kashid SB (2014) Photosensitized synthesis of silver nanoparticles using *Withania somnifera* leaf powder and silver nitrate. J Photochem Photobiol B Biol 132:45–55

Raut-Rajesh W, Lakkakula Jaya R, Kolekar Niranjan S, Mendhulkar Vijay D, Kashid Sahebrao B (2009) Phytosynthesis of silver nanoparticle using *Gliricidia sepium* (Jacq.) Curr Nanosci 5:117–122

Reddy NJ, Vali DN, Rani M, Rani SS (2014) Evaluation of antioxidant, antibacterial and cytotoxic effects of green synthesized silver nanoparticles by *Piper longum* fruit. Mat Sci Engin: C 34:115–122. doi:10.1016/j.msec.2013.08.039

Roduner E (2006) Size matters: why nanomaterials are different. Chem Soc Rev 35:583–592

Roopan SM, Bharathi A, Kumar R, Khanna VG, Prabhakarn A (2012) Acaricidal, insecticidal, and larvicidal efficacy of aqueous extract of *Annona squamosa* L peel as biomaterial for the reduction of palladium salts into nanoparticles. Colloids Surf B: Biointerfaces 92:209–212

Roopan SM, Madhumitha G, Rahuman AA, Kamaraj C, Bharathi A, Surendra TV (2013) Low-cost and eco-friendly phyto-synthesis of silver nanoparticles using *Cocos nucifera* coir extract and its larvicidal activity. Ind Crop Prod 43:631–635. doi:10.1016/j.indcrop.2012.08.013

Rupiasih NN, Aher A, Gosavi S, Vidyasagar PB (2015) Green synthesis of silver nanoparticles using latex extract of *Thevetia peruviana*: a novel approach towards poisonous plant utilization. In Recent Trend Physic Mat Sci Technol 204:1–10. doi:10.1007/978-981-287-128-2_1

Sable NE, Gaikwad SW, Bonde SH, Gade AN, Rai M (2012) Phytofabrication of silver nanoparticles by using aquatic plant *Hydrilla verticilata*. Nusantara Bio sci 4:45–49

Sadeghi B, Gholamhoseinpoor F (2015) A study on the stability and green synthesis of silver nanoparticles using *Ziziphora tenuior* (Zt) extract at room temperature. Spectrochim Acta Mol Biomol Spectrosc 134:310–315. doi:10.1016/j.saa.2014.06.046

Salam HA, Sivaraj R, Venckatesh R (2014) Green synthesis and characterization of zinc oxide nanoparticles from *Ocimum basilicum* L. var. *purpurascens Benth.-Lamiaceae* leaf extract. Mater Lett 131:16–18. doi:10.1016/j.matlet.2014.05.033

Salunke BK, Sawant SS, Kim BS (2014a) Potential of *Kalopanax septemlobus* leaf extract in synthesis of silver nanoparticles for selective inhibition of specific bacterial strain in mixed culture. Appl Biochem Biotechnol 174:587–601. doi:10.1007/s12010-014-1077-x

Salunke BK, Shin J, Sawant SS, Alkotaini B, Lee S, Kim BS (2014b) Rapid biological synthesis of silver nanoparticles using *Kalopanax pictus* plant extract and their antimicrobial activity. Korean J Chem Eng 31:2035–2040. doi:10.1007/s11814-014-0149-5

Saminathan K (2015) Herbal synthesis of silver nanoparticles using *Eclipta alba* and its antimicrobial activity. Int J Curr Microbiol App Sci 4(3):1092–1097

Sasidharan S, Chen Y, Saravanan D, Sundram KM, Latha LY (2011) Extraction, isolation and characterization of bioactive compounds from plants' extracts. African J Tradition Complement Alt Med 8:1–10

Sathishkumar M, Sneha K, Kwak IS, Mao J, Tripathy SJ, Yun YS (2009a) Phyto-crystallization of palladium through reduction process using *Cinnamom zeylanicum* bark extract. J Hazard Mater 171:400–404. doi:10.1016/j.jhazmat.2009.06.014

Sathishkumar M, Sneha K, Won SW, Cho CW, Kim S, Yun YS (2009b) *Cinnamon zeylanicum* bark extract and powder mediated green synthesis of nano-crystalline silver particles and its bactericidal activity. Colloids Surf B: Biointerfaces 73:332–338. doi:10.1016/j.colsurfb.2009.06.005

Sathishkumar G, Gobinath C, Wilson A, Sivaramakrishnan S (2014) *Dendrophthoe falcata* (Lf) Ettingsh (Neem mistletoe): a potent bioresource to fabricate silver nanoparticles for anticancer effect against human breast cancer cells (MCF-7). Spectrochim Acta A Mol Biomol Spectrosc 128:285–290. doi:10.1016/j.saa.2014.02.096

Saware K, Sawle B, Salimath B, Jayanthi K, Abbaraju V (2014) Biosynthesis and characterization of silver nanoparticles using *Ficus benghalensis* leaf extract. Int J Res Eng Technol 3:867–874. doi:10.1007/s10876-014-0697-1

Saxena A, Tripathi RM, Zafar F, Singh P (2012) Green synthesis of silver nanoparticles using aqueous solution of *Ficus benghalensis* leaf extract and characterization of their antibacterial activity. Mater Lett 67:91–94

Shankar SS, Rai A, Ahmad A, Sastry M (2004a) Rapid synthesis of Au, Ag, and bimetallic Au core–Ag shell nanoparticles using Neem (*Azadirachta indica*) leaf broth. J Colloid Interface Sci 275:496–502. doi:10.1016/j.jcis.2004.03.003

Shankar SS, Rai A, Ankamwar B, Singh A, Ahmad A, Sastry M (2004b) Biological synthesis of triangular gold nanoprisms. Nature Mat 3:482–488. doi:10.1038/nmat1152

Sharma NC, Sahi SV, Nath S, Parsons JG, Gardea-Torresde JL, Pal T (2007) Synthesis of plant-mediated gold nanoparticles and catalytic role of biomatrix-embedded nanomaterials. Environ Sci Technol 41:5137–5142

Shekhawat GS, Arya V (2009) Biological synthesis of Ag nanoparticles through in vitro cultures of *Brassica juncea* C. zern. In Adv Mat Res 67:295–299

Showalter AM (1993) Structure and function of plant cell wall proteins. Plant Cell 5:9. doi:10.1105/tpc.5.1.9

Singhal G, Bhavesh R, Kasariya K, Sharma AR, Singh RP (2011) Biosynthesis of silver nanoparticles using *Ocimum sanctum* (Tulsi) leaf extract and screening its antimicrobial activity. J Nanopart Res 13:2981–2988. doi:10.1007/s11051-010-0193-y

Siva S, Sameem SM, Sudharsan S, Kannan RS (2014) Green, effective biological route for the synthesis of silver nanoparticles using *Cyperus rotundus* grass extracts. Int J Curr Res 6:4532–4538

Sivaraj R, Rahman PK, Rajiv P, Salam HA, Venckatesh R (2014) Biogenic copper oxide nanoparticles synthesis using *Tabernaemontana divaricate* leaf extract and its antibacterial activity against urinary tract pathogen. Spectrochim Acta Mol Biomol Spectrosc 133:178–181. doi:10.1016/j.saa.2014.05.048

Sivaraman D, Panneerselvam P, Muralidharan P, Prabhu TP, Kumar RV (2013) Green synthesis, characterization and anti-microbial activity of silver nanoparticles produced using *Ipomoea aquatica forsk* leaf extract. Int J Pharm Sci Res 4:2280

Sosan A, Svistunenko D, Straltsova D, Tsiurkina K, Smolich I, Lawson T, Subramaniam S, Golovko V, Anderson D, Sokolik A, Colbeck I (2016) Engineered silver nanoparticles are sensed at the plasma membrane and dramatically modify the physiology of *Arabidopsis thaliana* plants. Plant J 85:245–257. doi:10.1111/tpj.13105

Stampoulis D, Sinha SK, White JC (2009) Assay-dependent phytotoxicity of nanoparticles to plants. Environ Sci Technol 43:9473–9479. doi:10.1021/es901695c

Steinmetz NF, Evans DJ (2007) Utilisation of plant viruses in bionanotechnology. Org Biomol Chem 5:2891–2902. doi:10.1039/b708175h

Sujitha MV, Kannan S (2013) Green synthesis of gold nanoparticles using Citrus fruits (*Citrus limon*, *Citrus reticulata* and *Citrus sinensis*) aqueous extract and its characterization. Spectrochim Acta A Mol Biomol Spectrosc 102:15–23. doi:10.1016/j.saa.2012.09.042

Suman TY, Rajasree SR, Kanchana A, Elizabeth SB (2013) Biosynthesis, characterization and cytotoxic effect of plant mediated silver nanoparticles using *Morinda citrifolia* root extract. Colloids Surf B: Biointerfaces 106:74–78. doi:10.1016/j.colsurfb.2013.01.037

Suman TY, Rajasree SR, Ramkumar R, Rajthilak C, Perumal P (2014) The green synthesis of gold nanoparticles using an aqueous root extract of *Morinda citrifolia* L. Spectrochim Acta Mol Biomol Spectrosc 118:11–16. doi:10.1016/j.saa.2013.08.066

Suresh G, Gunasekar PH, Kokila D, Prabhu D, Dinesh D, Ravichandran N, Ramesh B, Koodalingam A, Siva GV (2014) Green synthesis of silver nanoparticles using *Delphinium denudatum* root extract exhibits antibacterial and mosquito larvicidal activities. Spectrochim Acta Mol Biomol Spectrosc 127:61–66. doi:10.1016/j.saa.2014.02.030

Suresh D, Nethravathi PC, Rajanaika H, Nagabhushana H, Sharma SC (2015) Green synthesis of multifunctional zinc oxide (ZnO) nanoparticles using *Cassia fistula* plant extract and their photodegradative, antioxidant and antibacterial activities. Mat Sci Semicond Proc 31:446–454. doi:10.1016/j.mssp.2014.12.023

Swamy VS, Prasad R (2012) Green synthesis of silver nanoparticles from the leaf extract of *Santalum album* and its antimicrobial activity. J Optoelectron Biomed Mater 4(3):53–59

Swamy MK, Sudipta KM, Jayanta K, Balasubramanya S (2015) The green synthesis, characterization, and evaluation of the biological activities of silver nanoparticles synthesized from *Leptadenia reticulata* leaf extract. Appl Nanosci 5:73–81. doi:10.1007/s13204-014-0293-6

Tan XM, Lin C, Fugetsu B (2009) Studies on toxicity of multi-walled carbon nanotubes on suspension rice cells. Carbon 47:3479–3487. doi:10.1016/j.carbon.2009.08.018

Torney F, Trewyn BG, Lin VS, Wang K (2007) Mesoporous silica nanoparticles deliver DNA and chemicals into plants. Nat Nanotechnol 2:295–300. doi:10.1038/nnano.2007.108

Tripathy A, Raichur AM, Chandrasekaran N, Prathna TC, Mukherjee A (2010) Process variables in biomimetic synthesis of silver nanoparticles by aqueous extract of *Azadirachta indica* (Neem) leaves. J Nanopart Res 12:237–246. doi:10.1007/s11051-009-9602-5

Vanaja M, Annadurai G (2013) *Coleus aromaticus* leaf extract mediated synthesis of silver nanoparticles and its bactericidal activity. Appl Nanosci 3:217–223. doi:10.1007/s13204-012-0121-9

Vankar PS, Shukla D (2012) Biosynthesis of silver nanoparticles using lemon leaves extract and its application for antimicrobial finish on fabric. Appl Nanosci 2:163–168. doi:10.1007/s13204-011-0051-y

Vasanth K, Ilango K, MohanKumar R, Agrawal A, Dubey GP (2014) Anticancer activity of *Moringa oleifera* mediated silver nanoparticles on human cervical carcinoma cells by apoptosis induction. Colloid Surfaces B: Biointer 117:354–359. doi:10.1016/j.colsurfb.2014.02.052

Veerakumar K, Govindarajan M (2014) Adulticidal properties of synthesized silver nanoparticles using leaf extracts of *Feronia elephantum* (Rutaceae) against filariasis, malaria, and dengue vector mosquitoes. Parasitol Res 113:4085–4096. doi:10.1007/s00436-014-4077-4

Veerasamy R, Xin TZ, Gunasagaran S, Xiang TF, Yang EF, Jeyakumar N, Dhanaraj SA (2011) Biosynthesis of silver nanoparticles using mangosteen leaf extract and evaluation of their antimicrobial activities. J S Chem Soc 15:113–120. doi:10.1016/j.jscs.2010.06.004

Velmurugan P, Anbalagan K, Manosathyadevan M, Lee KJ, Cho M, Lee SM, Park JH, Oh SG, Bang KS, Oh BT (2014) Green synthesis of silver and gold nanoparticles using *Zingiber officinale* root extract and antibacterial activity of silver nanoparticles against food pathogens. Bioproc Biosys Engin 37:1935–1943. doi:10.1007/s00449-014-1169-6

Vijayakumar M, Priya K, Nancy FT, Noorlidah A, Ahmed ABA (2013) Biosynthesis, characterisation and anti-bacterial effect of plant-mediated silver nanoparticles using *Artemisia nilagirica*. Ind Crop Prod 41:235–240. doi:10.1016/j.indcrop.2012.04.017

Vinod VTP, Saravanan P, Sreedhar B, Devi DK, Sashidhar RB (2011) A facile synthesis and characterization of Ag, Au and Pt nanoparticles using a natural hydrocolloid gum kondagogu (*Cochlospermum gossypium*). Colloids Surf B: Biointerfaces 83:291–298. doi:10.1016/j.colsurfb.2010.11.035

Vivero-Escoto JL, Rieter WJ, Lau H, Huxford-Phillips RC, Lin W (2013) Biodegradable polysilsesquioxane nanoparticles as efficient contrast agents for magnetic resonance imaging. Small 9:3523–3531. doi:10.1002/smll.201300198

Wang Y, He X, Wang K, Zhang X, Tan W (2009) *Barbated Skullcup* herb extract-mediated biosynthesis of gold nanoparticles and its primary application in electrochemistry. Colloids Surf B: Biointerfaces 73:75–79. doi:10.1016/j.colsurfb.2009.04.027

Wang H, Kou X, Pei Z, Xiao JQ, Shan X, Xing B (2011) Physiological effects of magnetite (Fe3O4) nanoparticles on perennial ryegrass (*Lolium perenne* L.) and pumpkin (*Cucurbita mixta*) plants. Nanotoxicology 5:30–42. doi:10.3109/17435390.2010.489206

Wegner T, Jones P (2007) Nanotechnology for the forest products industry. Wood Fiber Sci 37:549–551

Yang L, Watts DJ (2005) Particle surface characteristics may play an important role in phytotoxicity of alumina nanoparticles. Toxicol Lett 158:122–132

Yang X, Li Q, Wang H, Huang J, Lin L, Wang W, Jia L (2010) Green synthesis of palladium nanoparticles using broth of *Cinnamomum camphora* leaf. J Nanopart Res 12:1589–1598. doi:10.1007/s11051-009-9675-1

Yasmeen F, Raja NI, Mustafa G, Sakata K, Komatsu S (2016) Quantitative proteomic analysis of post-flooding recovery in soybean root exposed to aluminum oxide nanoparticles. J Proteome 30:136–150. doi:10.1016/j.jprot.2016.03.014

Zayed MF, Eisa WH (2014) *Phoenix dactylifera* L. leaf extract phytosynthesized gold nanoparticles; controlled synthesis and catalytic activity. Spectrochimica Acta Part A: Molecul Biomolecul Spectroscopy 121:238–244. doi:10.1016/j.saa.2013.10.092

Zhu H, Han J, Xiao JQ, Jin Y (2008) Uptake, translocation, and accumulation of manufactured iron oxide nanoparticles by pumpkin plants. J Environ Monit 10:713–717. doi:10.1039/b805998e

Zuas O, Hamim N, Sampora Y (2014) Bio-synthesis of silver nanoparticles using water extract of *Myrmecodia pendan* (Sarang Semut plant). Mat Lett May 123:156–159. doi:10.1016/j.matlet.2014.03.026

Nanomaterials for Delivery of Nutrients and Growth-Promoting Compounds to Plants

9

Josef Jampílek and Katarína Kráľová

Abstract

Nanotechnology is a rapidly expanding field that affords the development of materials in nanoscale dimensions that have unique properties and a wide spectrum of applications. Nanomaterials can be found more frequently in agriculture and the food sector. The application of nanomaterials for delivery of nutrients and growth-promoting compounds to plants has become more and more popular, and their utilization at the proper place, at the proper time, in the proper amount and of the proper composition emends the efficacy of fertilizers. This contribution reviews the potential application of various nanocarriers used for delivery of N, P and K macronutrients and plant growth-stimulating nanoscale essential metals nutrients (Fe, Zn, Cu, Mn, Co) as well as carbon-based (single- and multiwalled carbon nanotubes) and non-essential metal (Ti, Ag, Au, Ce, Al)- and metalloid (Si, Se)-based nanomaterials showing beneficial effects on plant growth that could be used in agricultural practice.

Keywords

Carbon nanomaterials • Clays • Fertilizers • Metalloids • Metals • Nanoparticles • Nutrients

J. Jampílek (✉)
Department of Pharmaceutical Chemistry, Faculty of Pharmacy, Comenius University, Odbojárov 10, 832 32 Bratislava, Slovakia
e-mail: josef.jampilek@gmail.com

K. Kráľová
Institute of Chemistry, Faculty of Natural Sciences, Comenius University, Ilkovičova 6, 842 15 Bratislava, Slovakia
e-mail: kata.kralova@gmail.com

R. Prasad et al. (eds.), *Nanotechnology*, DOI 10.1007/978-981-10-4573-8_9

9.1 Introduction

Nowadays a significant increase of nanotechnology use could be observed. The application of nanotechnology/nanomaterials/nanoparticles represents indisputably modern trends, for example, not only in industry (e.g. various construction composites, optoelectronic and electronic materials or materials with high-energy storage capacity) (Luther et al. 2004; Loeffler et al. 2005; Lee et al. 2010a, b; Charitidis et al. 2014) and biomedical sciences (e.g. nanomaterials for tissue engineering, nanodiagnostics, nanopharmaceutics) (Ventola 2012a, b, c; Jampílek et al. 2013, 2014, 2015; Jampílek and Kráľová 2017a, b; Opatřilová et al. 2013; Černíková et al. 2014, 2015; Vaculíková et al. 2012a, b, 2014, 2016; Weissig et al. 2014; Weissig and Guzman-Villanueva 2015; Prasad 2014; Aziz et al. 2015; Prasad et al. 2016) but also in agriculture and the food sector (Chaudhry and Castle 2011; Rashidi and Khosravi-Darani 2011; Chellaram et al. 2014; Sekhon 2014; Parisi et al. 2015; Pradhan et al. 2015; Unsworth et al. 2016; Jampílek and Kráľová 2015). The use of nanotechnologies can significantly contribute to sustainable intensification of agricultural production, because they cannot only facilitate the protection of plants against pesticides but also contribute to enhanced plant growth, secure rising of global food production, guarantee enhanced food quality and minimize the waste (Garcia et al. 2010; Sonkaria et al. 2012; Pérez-de-Luque and Hermosín 2013; Prasad et al. 2014; Sekhon 2014; Jampílek and Kráľová 2015); thus, the agricultural production and food industry belong to important areas of nanotechnology utilization (Kuzma and VerHage 2006; Ghormade et al. 2011; Coles and Frewer 2013; Raliya et al. 2013; Chen et al. 2014).

The exact definition of the term "nanomaterial" by the European Commission is as follows: "Nanomaterial" is a natural, incidental or manufactured material containing particles, in an unbound state or as an aggregate or as an agglomerate and where, for 50% or more of the particles in the number size distribution, one or more external dimensions are in the size range 1–100 nm. In specific cases and where warranted by concerns for the environment, health, safety or competitiveness, the number size distribution threshold of 50% may be replaced by a threshold between 1% and 50%. By derogation from the above, fullerenes, graphene flakes and single-wall carbon nanotubes with one or more external dimensions below 1 nm should be considered as nanomaterials (European Commission 2008). US National Nanotechnology Initiative defines nanoparticles (NPs) in the range 1–100 nm, as well (National Nanotechnology Initiative 2008).

Generally, nanoscale materials differ from their bulk form in physicochemical properties (Borm et al. 2006; Buzea et al. 2007; Fröhlich 2013; Dolez 2015), and permeation through cell walls/membranes to cells and tissues is probably the most affected parameter in this respect. In this context, mainly adverse effects of NPs accumulated in the cell leading to intracellular changes, such as disruption of organelle integrity, gene alterations, etc., or cytotoxic effects by generation of reactive oxygen species (ROS) as well as reactive nitrogen species resulting in the damage of plasma membrane, cell organelles and intracellular proteins have been studied. Especially particles with particle size <100 nm are critical, because they can practically unlimitedly permeate through biomembranes. NPs may be more easily taken up by any organism, which could result in their longer persistence in the environmental systems. The small size (an extrinsic property) of NPs influences these effects more significantly than a unique nanoscale property representing an intrinsic property (Buzea et al. 2007; De Jong and Borm 2008; Auffan et al. 2009; Kumar et al. 2012; Brayner et al. 2013; Janrao et al. 2014). Considering the potential toxicity of NPs and different nanomaterials on living organisms and also on human health, it is indispensable to minimize their entry into environment (Ventola 2012c; Berkner et al. 2016; Vestel et al. 2016; Jampílek and Kráľová 2017c).

In agricultural sector, increasing attention is paid to beneficial impacts of NPs applied in low doses on crops connected, for example, with induction of enhanced biosynthesis of secondary metabolites (e.g. Syu et al. 2014; Ghorbanpour and Hadian 2015), increased activity of antioxidant enzymes (e.g. Ghanati et al. 2005; Ghorbanpour and Hadian 2015), more effective absorption of water and fertilizers (e.g. Li et al. 2015a), positive effects on photosynthetic apparatus (e.g. Ze et al. 2011; Zheng et al. 2007), etc., resulting finally in stimulation of plant growth and increased yield and bringing economic profit. Therefore, in this chapter, advantageous effects of nanomaterials on plants focused not only on plant growth-promoting nanosized essential metal nutrients (Fe, Zn, Cu, Mn, Co) but also on other nanomaterials exhibiting positive effects on plant growth, including carbon-based (single- and multiwalled carbon nanotubes) and non-essential metal (Ti, Ag, Au, Ce, Al)- and metalloid (Si, Se)-based nanoformulations, are discussed. Special attention is devoted also to application of various nanocarriers used for delivery of N, P and K macronutrients.

9.2 Clays and Other Materials as Nanocarriers of N, P and K Macronutrient Fertilizers

For the retention and persistence of inorganic and organic compounds in soil, an important role is attributed to clay minerals representing natural materials, and modification of surface properties and charge of clay minerals enable their use for many specific purposes. Moreover, it could be mentioned that polymer nanocomposites composed of clay minerals are structural hybrids with modified properties and thus they can be used for slow release of fertilizers and enhancement of the water-holding capacity of soil (Basak et al. 2012). Zou et al. (2015) who evaluated the solubility characteristics and slow-release mechanism of nitrogen from organic-inorganic compound coated urea found that the solubility process of coated urea could be described by the first-order reaction kinetic equation, where the rate constantly increases with temperature increasing. With increasing temperature, the microstructure of the coating layer changed into a flocculent structure, and a considerable enhancement of the number of tiny pores and holes on the membrane surface with increasing temperature resulted in an increased N solubility rate.

Properly applied mineral fertilizers, containing mainly nitrogen, phosphorus and potassium considering the required amount and composition as well as the right time and the right place of application, greatly contribute to enhancement of fertilizer efficiency (Bindraban et al. 2015).

Borges et al. (2015) developed multi-element slow-release fertilizers by mechanochemical activation of mixtures of kaolinite and ammonium or potassium monohydrogen phosphates using milling the materials in a high-energy ball mill. The slow-release behaviour of phosphates could be connected with more thorough interaction of the aluminium ions in the mineral structure with phosphate than with potassium or ammonium, although all of the nutrients were released slowly. Sarkar et al. (2015) prepared nanoclay/superabsorbent polymer composites (NCPCs) using nanoclays containing mainly smectite, illite and kaolinite, loaded them with $(NH_4)_2HPO_4$ and urea solution and tested the release of phosphorus and total mineral nitrogen from these nanocomposites in alfisol, inceptisol and vertisol. Compared to the traditional fertilizers, both the cumulative P and total mineral N recovery showed significantly higher effectiveness for NCPC application, reaching even +88 and +27% for smectite-doped NCPC, +57 and +16% for kaolinite-doped NCPC and +55 and +15% for illite-doped NCPC, whereby the best slow-release property was estimated with smectite-doped NCPC. Slow release of urea was observed also from nanocomposites prepared by urea intercalation into montmorillonite (MMT) clay, in which urea represented 50–80 wt.% and clay lamellae were exfoliated into urea matrix, already at small amounts of clay (20% w/w) (Pereira et al. 2012).

Treatment with phosphorus NPs significantly increased P uptake, photosynthetic activity and plant weight of salt-stressed basil seedlings, and their application as P source promoted *Ocimum basilicum* growth at a rate of 20%, being more effective than the regular P fertilizer (Alipour 2016). By dispersion of hydroxyapatite within urea and thermoplastic starch/urea, representing two water-soluble matrices, hydroxyapatite solubility can be effectively increased, and prepared nanocomposite fertilizer could be considered as a smart nanostructured fertilizer enabling faster release of phosphate phases (characterized by insufficient solubility) and, on the other hand, slower release of urea showing excellent aqueous solubility (Giroto et al. 2015). Sufficient supply of P to crops could be provided also by NPs of synthetic apatite that exhibit lower mobility in the environment resulting in reduced risks of water eutrophication. Stimulation of the growth rate and improvement in seed yield of soybean compared to the treatment with conventional fertilizer with the application of nano-apatite fertilizer was observed by Liu and Lal (2014). The use of nano-P improved the nutrient uptake of cotton plants under water stress conditions as well as regular irrigation. Treatment with 0.5 g/L nano-P promoted the nutrient uptake when irrigation was missing at budding, while at lacking of irrigation at flowering stages, the dose 1.0 g/L was more efficient (Hussien et al. 2015). NPs of hydroxyapatite modified by urea that were entrapped into cavities of soft wood of quick-stick (*Gliricidia sepium*) and used as nitrogen nanofertilizer in soils of different acidity (pH 4.2, 5.2 and 7.0) released nitrogen after an initial burst slowly up to day 60. On the other hand, commercial preparation was characterized with intensive nitrogen release at the start and subsequent releasing of low and uneven amounts up to ca. 30th day (Kottegoda et al. 2011).

Investigation of natural zeolite of Iranian provenience with different particle sizes ranging from nanometres to millimetres focused on the removal of NH_4^+ ions from aqueous solution with 0.03, 0.1 and 0.3 mol/L Na^+ concentrations showed that by decreasing Na^+ concentration, the efficiency of removal increased due to enhanced amounts of exchanged NH_4^+ and following the rapid release during the first hour slowed down; the release was found to increase with the rise of the ionic strength of the solution. The applicability of the above-mentioned zeolite to remove NH_4^+ ions as an ion-exchanger pointed out also its suitability for the use as NH_4^+ fertilizer with controlled release (Malekian et al. 2011). The nanosized zeolite fortified with anionic SO_4^{2-} nutrient represents a long-lasting nanofertilizer securing slow release of sulphur nutrient (Selva Preetha et al. 2014). The application of zeolite-based nitrogen nanofertilizers resulted consistently in higher growth, yield, quality and nutrient uptake of maize plants, and increased grain nitrogen content was determined in treated plants compared to control with higher efficiency in alfisol than inceptisol (Manikandan and Subramanian 2016). Also surface modification of zeolite NPs with hexadecyltrimethylammonium and dioctadecyldimethylammonium greatly enhanced the sorption and slow release of nitrate. In addition, fertilizer showing slow-release properties prepared from a nanocomposite of zeolite NaP1 and hydroxyapatite A containing 17 mg/g NH_4^+, 55 mg/g K^+ and 24 mg/g of PO_4^{3-} ions, in which surfaces of zeolite were covered with hydroxyapatite needle-like crystals (ca. 200 nm diameter) by a hydrothermal treatment of the calcium-exchanged zeolite NaP1 in a mixed solution of potassium and ammonium phosphates at 40 °C, slowly released phosphate ions by the dissolution of hydroxyapatite in water, while NH_4^+ and K^+ ions were released by cation exchange with Ca^{2+} co-dissolved with phosphate (Watanabe et al. 2014).

As a promising slow-release NH_4^+ fertilizer, a palygorskite nanocomposite with the content of N 13% was proposed. Thus modified nanocomposite released in soil 60% of N for 10 days, which had beneficial effects on soil moisture content by reducing its evaporation rate. The advantage of this palygorskite nanocomposite designed primarily as an adsorbent for removal and recovery of NH_4^+ ions from wastewaters and applicable in pH range from four to eight consists just in the possibility of its agronomic reuse as a fertilizer (Wang et al. 2014). Adding of palygorskite sodium polyacrylate-polyacrylamide (PAM) complex to conventional fertilizer resulted in a fertilizer that is able to control water and nutrient loss and showed beneficial effects on maize growth, which were reflected in significantly elevated concentration of nutrients contained in the fertilizer in the maize stem. In the aqueous phase, the above-mentioned complex could self-assemble forming three-dimensional (3D) micro-/nano-networks, in which water and nutrient content could be optimized and retained and the filtering effect of soil secures that these networks will be preserved in the soil. Consequently, these properties of the above-discussed water and nutrient loss control fertilizer could contribute to reduction of serious environmental pollution caused by the migration of nitrogen fertilizers into the environment by volatilization, runoff or leaching (Zhou et al. 2015).

For controlled release of fertilizers also hydrogels containing PAM, methylcellulose and calcium MMT could be considered; they are characterized by effective loading of fertilizer, and the components incorporated in the formulation can exhibit synergetic effects. Incorporation of the clay into hydrogel polymer matrix resulted in enhancement of water absorption rate, and MMT component in the hydrogel contributed to more controlled nutrient release compared to pure hydrogel in different pH ranges. Larger amounts of released nutrient as well as its ca. 200-fold slower release from this system when compared to the application of pure urea indicate that this hydrogel formulation is characterized by very good desorption properties (Bortolin et al. 2013). Surface modification of clinoptilolite and MMT improved adsorption potential for PO_4^{3-} as compared to the corresponding unmodified form, and the increased amount of silicate-loaded surfactant resulted in the enhancement of PO_4^{3-} removal ability of the surface-modified material, where phosphate release could be observed even after 15 days of leaching (Bhardwaj et al. 2014). An eco-friendly slow-release urea fertilizer employing mulberry branch-g-poly(acrylic acid-co-acrylamide) superabsorbent blended with sodium alginate, urea and $CaCl_2$ solutions that was characterized with 420 g/g water absorbency, 60 (g/g)/min water absorbency rate in deionized water and water retention of 7.2% wt after 25 days exhibited typical slow-release behaviour for urea release in both deionized water and soil, where it was buried for 90 days (Zhang et al. 2014).

Pereira et al. (2015) used various concentrations of polymers (hydrophilic or hydrophobic) to prepare nanocomposite nitrogen fertilizers associated with an exfoliated clay mineral showing slow-release properties. The nanocomposites that acted as a structural matrix could load urea in a high extent (75% w/w), and the addition of PAM hydrogel and polycaprolactone (<4% w/w) to this nanocomposite affected mechanical properties as well as urea release profiles. The advantage of such nanocomposites containing hydrophilic or hydrophobic polymers consisted in substantially reduced N_2O emissions in the field. A formulation consisting of a NO_3^--layered hydroxide material matrix designed by Berber et al. (2014) was found to be suitable for delivery of nitrogen into soil. At 15 °C, it exhibited sustained controlled NO_3^- release into acidic soil for 16 days and into neutral soil for 20 days, and a soil temperature increase resulted in a little enhancement of NO_3^- release.

Core-shell fibres with two biodegradable polymers, polyhydroxybutyrate as the shell and polylactic acid mixed with fertilizer as the core, prepared by coaxial electrospinning can control the manner and timing of fertilizer delivery, because at a fixed flow rate of shell solution, the core-shell electrospun fibre mats exhibited a lower flow rate of core solution causing a lower release rate of fertilizer, and it was found that an electrospun mat can release fertilizer for 1 month without degradation (Kampeerapappun and Phanomkate 2013). Using free radical polymerization of sodium alginate, acrylic acid, acrylamide and clinoptilolite, in which *N,N′*-methylenebisacrylamide was applied as a cross-linker and $(NH_4)_2S_2O_8$ as an initiator, a hydrogel nanocomposite was fabricated showing both favourable water adsorption capacity and slow fertilizer release. The clinoptilolite zeolite contained in this formulation ensured better controlled release of nutrient compared to pure hydrogel, suggesting that the formulation could represent a suitable nutrient carrier for agricultural application (Rashidzadeh et al. 2014). The PAM/graphite oxide (GO) superabsorbent nanocomposites prepared by PAM polymerization using the same cross-linker and initiator as Rashidzadeh et al. (2014) exhibited remarkable improvement of the comprehensive swelling performance, which was connected with the satisfactory dispersion of GO nanoplatelets in the polymeric network, and the superabsorbent nanocomposites could absorb water twice as compared with cross-linked PAM superabsorbent showing a weight gain of 400 g/g at low loading of GO, and embedding of ammonium salt into superabsorbent nanocomposite resulted in its slow release from the network when swelling in water (Zhu et al. 2013).

Chitosan (CS)-poly(methacrylic acid) (PMAA) NPs prepared by MAA polymerization in CS solution and applied to investigate the incorporation of NPK fertilizer into these NPs showed spherical size of ca. 78 nm. The presence of the separated species from urea, KCl and $Ca(H_2PO_4)_2$ affected the stability of CS-PMAA suspension, the stability being higher for nitrogen and potassium than for P, which could be connected with the higher anion charge from $Ca(H_2PO_4)_2$ compared to that of KCl and urea. CS NPs were found to interact with N, P and K present in added fertilizers. However, the addition of different compounds resulted in the increase of the average diameter of the CS-PMAA NPs suggesting aggregation of the elements on the surface of CS NPs (Corradini et al. 2010). The application of nano-CS-NPK fertilizer more effectively increased the yield of wheat plants grown on sandy soil compared to the treatment with normal bulk NPK fertilizer, and compared to normal-fertilized *Triticum aestivum* plants showing life cycle of 170 days for yield production from the date of sowing, the life cycle for plants treated with CS-NPK fertilizer was shortened to 130 days corresponding to ca. 24% reduction (Abdel-Aziz et al. 2016).

9.3 Non-metal Nanomaterials Beneficial for Plant Growth

9.3.1 Carbon-Based Nanomaterials

The application of biochar (charcoal) that is fabricated by pyrolysis of plant biomass and crop residues is suitable for carbon sequestration, soil fertilization and improvement of soil structure (Sigua et al. 2015). Based on monitoring of microbial dynamics, it was concluded that co-location of various resources in and around biochars is possibly connected with better resource use (Lehmann et al. 2011). Biochar can also alleviate the toxic effects of trace elements in crops, what is reflected in increased plant growth and plant biomass, higher photosynthetic pigment concentrations and improvement of grain yield and quality (Rizwan et al. 2016). In addition, biochar obtained from anaerobically digested tailings of *Beta vulgaris* var. *altissima* plants could be considered as a perspective alternative adsorbent suitable for removing PO_4^{3-} ions from water or diminish PO_4^{3-} leaching from fertilized soils, whereby the PO_4^{3-} loaded biochar containing excess of beneficial nutrients can be subsequently used as a slow-release fertilizer for better soil fertility and carbon sequestration (Yao et al. 2011). The positive effects of biochar addition to agricultural soils were summarized by Biederman and Harpole (2013).

Carbon-based nanomaterials are widely used in various fields of human activity, including agriculture. When such nanomaterials with appropriate size applied at optimum concentrations penetrate into the plant system, they can affect the metabolic functions, which finally results in beneficial effects on plant growth, and therefore they could represent a suitable solution for enhanced crop yield and fruit manifold requiring low costs (Mukherjee et al. 2016; Liu and Lal 2015; Husen and Siddiqi 2014). Fan et al. (2012) reported beneficial effects on yield and nitrogen use of *Oryza sativa* plants cultivated on saline-alkali soil after combined application of N fertilizer and nanoscale carbon ensuring the increased utilization rate of the fertilizer and maintenance of the N fertilizer in agricultural production. Sun et al. (2012) investigated the effects of nitrogen regulators on the transformation of nitrogen of urea and $(NH_4)HCO_3$ fertilizers in meadow cinnamon soil and on *Brassica chinensis* growth. They found that co-application of dicyandiamide and nanoscale carbon exhibited a synergistic effect with respect to inhibition of soil NH_4^+ oxidation, whereby significant stimulation of the *B. chinensis* growth and N utilization at early growth stages and reduction of the plant NO_3^- level at harvesting stage was observed as well. On the other hand, higher concentrations of multiwalled carbon nanotubes (MWCNTs) showed phytotoxic effects connected mainly with the oxidative stress (Lin and Xing 2007; Ghodake et al. 2010; Begum and Fugetsu 2012; Begum et al. 2012). A comprehensive review focused on the application of carbon nanomaterials in agricultural practice and evaluation of the published observations concerning beneficial and adverse effects of different carbon nanomaterials (e.g. single-walled carbon nanotubes (SWCNTs), MWCNTs, fullerenes, carbon NPs and carbon nanohorns) on terrestrial plants and associated soil-dwelling microbes was presented by Mukherjee et al. (2016). Liu and Lal (2015) summarized findings concerning nanomaterials that can improve growth of plants in definite concentration ranges resulting in increased yields of crops and simultaneously minimizing the environmental pollution and consequently can be used in agriculture as macronutrient nanofertilizers, micronutrient nanofertilizers, nutrient-loaded nanofertilizers and plant growth-promoting nanomaterials. Effects of carbon nanotubes (CNTs), fullerenes and fullerol on germination rate, biomass increase, absorption and translocation in different plant organs were outlined by Husen and Siddiqi (2014).

Aerial oxidization of the raw carbon nanoparticles (rCNPs) occurring in biochar during ageing results in the incorporation of hydrophilic COOH and OH groups on the surface of biochar causing the increase of its spongy structure, and consequently improvement of its absorptivity to water and ionic nutrients could be observed. By introduction of COOH and OH groups following chemical oxidation of rCNPs in biochar, water-soluble carbon NPs (wsCNPs) could be prepared, which when applied in solution at concentrations 10–150 mg/L enhanced the growth rate of *T. aestivum* plants. From experiments using NH_4NO_3, it was evident that both rCNPs and wsCNPs retain cationic (NH_4^+) and anionic (NO_3^-) nutrients. The release of positively and negatively charged nutrients from these carbon NPs was slow over time, and the application of wsCNPs instead of manure or pure fertilizer could contribute to improved assimilation by plants specially due to controlled and slow release of nutrients (Saxena et al. 2014).

Carbon nano-onions (wsCNOs) isolated from wood wool (a wood-based pyrolysis waste product of wood) showing aqueous solubility promoted the overall growth rate of chickpea plants. At treatment with *Cicer arietinum* plants in laboratory conditions using a dose up to 30 μg/mL during 10 days, an increase in plant biomass was detected, while cultivation of plants pretreated with wsCNOs in soil resulted in better plant productivity, which was reflected in a larger number of grammes. In the wsCNOs-treated plants, also slightly increased content of C and H elements in shoots was estimated reflecting that in these plants, the synthesis of organic biomass increased, suggesting that this material acting as effective growth promoter can be considered as a suitable plant growth stimulant (Sonkar et al. 2012).

Single-walled carbon nanohorns (SWCNHs) are single graphitic tubules with diameter of 2–5 nm and lengths of 40–50 nm (Iijima et al. 1999; Kasuya et al. 2002), which exist in spherical aggregates reaching a diameter of 50–100 nm and have a cylindrical inner nanospace (capable to store a considerable number of molecules) and interstitial channels (Lahiani et al. 2015). SWCNHs were found to act as a beneficial regulator during crop germination and development. Beside activation of seed germination, enhanced growth of different organs of *Zea mays*, *Solanum lycopersicum*, *O. sativa* and *Glycine max* plants was observed after SWCNHs treatment. For example, a considerable increase of fresh weight (FW) of shoots of SWCNHs-treated maize, rice and soybean seedlings compared to control plants (70%, 76% and 11%, respectively) was estimated. SWCNHs also accelerated the growth of tobacco cell culture causing a 78% enhancement in growth of tobacco cells with respect to the control. The observed beneficial effect of SWCNHs on FW of callus but not on dry weight (DW) could be connected with greater water uptake of SWCNHs-treated cells in comparison with control cells. Besides, SWCNHs can influence stress signalling in plants and expression of genes that are associated with cell growth channels (Lahiani et al. 2015).

Significantly faster germination of graphene-treated seeds compared to control seeds was observed by Zhang et al. (2015), and it was shown that graphene penetrated seed husks and breaking of the husks due to penetration facilitated water uptake, which led to faster germination as well as higher germination rates. At the stage of seedling growth, penetration of graphene to root tip cells was estimated. Although significantly longer stems and roots of seedlings germinated from seeds treated with graphene compared to control plants were measured, treated plants were characterized by slightly lower biomass accumulation. Treatment of coriander and garlic seeds with 0.2 mg/mL of graphene quantum dots (QDs) for 3 h before planting resulted in enhanced growth rate of plant organs, including flowers and fruits, indicating that graphene QDs could be used as plant growth regulators in food plants to the benefit of high yield (Chakravarty et al. 2015). Water-soluble carbon nanodots enhanced the growth of *T. aestivum* plant organs both in light and dark and were found to enter inside the plant (Tripathi and Sarkar 2015).

Experiments with bitter melon (*Momordica charantia*) showed that fullerol [$C_{60}(OH)_{20}$] treatment caused increases in biomass yield (up to 54%) and in water content (up to 24%) and also in fruit length (up to 20%), fruit number (59%) and fruit weight (70%), which resulted in an enhancement in fruit yield up to 128%.

Moreover, fullerol treatment also significantly enhanced the content of several secondary metabolites (cucurbitacin-B, lycopene, charantin and insulin) exhibiting healing effects (Kole et al. 2013).

Tripathi et al. (2016) studied the effects of the morphology of carbon nanostructures on the growth stimulation of gramme plants and found that the <10 nm diameter SWCNTs were better promoters of plant growth than the ca. 180 nm straight open-ended MWCNTs post 7 days. Less effective growth stimulation was observed with 150–200 nm thick close-ended functionalized and annealed carbon nanowhiskers. The researchers suggested that the studied carbon-based nanomaterials probably prefer the apoplastic route to reach the root's interior and therefore cause extension of cell membrane pores due to their high potential gradient, whereby carbon NPs with smaller diameters are better growth stimulators.

CNTs significantly increased the rate of germination of rice seeds, whereby increased water content in treated seeds was estimated compared to controls and seedlings treated with CNTs were characterized by superior root and shoot systems and healthy appearance to a greater extent than control seedlings, indicating that CNTs acted as promoters of the *O. sativa* seedling growth (Nair et al. 2012). Water-soluble CNTs applied at concentrations up to 6.0 μg/mL caused increased root and shoot growth as well as branching in the common gramme plants indicating improved water absorption and retention compared to control plants (Tripathi et al. 2011).

SWCNTs accelerated the growth of corn seminal roots due to increased expression of genes associated with seminal root, while reduction of the root hair associated gene expression led to inhibition of root hair growth, whereby their effect on the primary root growth was not significant. Moreover, they selectively induced the upregulation of epigenetic modification enzyme genes causing overall deacetylation of histone H3 (likewise as plant response to other stresses), and it can be supposed that the change in gene expression due to NPs-root cell interaction affects relative root growth and development (Yan et al. 2013). Pourkhaloee et al. (2011) described beneficial effects of SWCNTs on seed germination and seedling growth of salvia, pepper and tall fescue under laboratory and greenhouse conditions. The application of 10, 20 and 30 mg/L of SWCNTs had the most pronounced effects on seedling FW and DW as compared to the control, and the best seed germination and seedling growth for salvia and tall fescue were achieved at 30 mg/L of SWCNTs and at 10 mg/L for pepper. The study of the effects of non-functionalized SWCNTs as well as those functionalized with poly-3-aminobenzenesulfonic acid on root growth of six crops showed that root length was affected by non-functionalized SWCNTs to a greater extent than by functionalized SWCNTs. Treatment with non-functionalized SWCNTs resulted in root growth inhibition of *S. lycopersicum* plants, while it caused root growth stimulation in *Allium cepa* and *Cucumis sativus* plants and did not influence root growth in *Brassica oleracea* and *Daucus carota* plants. On the other hand, root growth inhibition due to application of functionalized SWCNTs was observed in *Lactuca sativa* plants, while *B. oleracea* and *D. carota* roots were not affected (Canas et al. 2008).

Khodakovskaya et al. (2012) found that MWCNTs applied in a wide concentration range from 5 to 500 µg/mL effectively stimulated the growth of tobacco cell culture that reached 55–64% increase over the control, while treatment with a low concentration of activated carbon (AC) (5 µg/mL) alone caused only 16% increase of cell growth, and the application of higher AC concentrations (100–500 µg/mL) resulted in significant inhibition of the cellular growth. Tobacco cells exposed to MWCNTs were characterized with significantly increased expression of the tobacco aquaporin (NtPIP1) gene and production of the NtPIP1 protein compared to control cells or tobacco cells treated with AC, and upregulation of the expression of the matter genes for cell division (CycB) and cell wall extension (NtLRX1) in cells treated with MWCNTs was estimated as well. Khodakovskaya et al. (2013) in another paper introduced that plant phenotype and the composition of soil microbiota can be affected by MWCNTs. For example, the number of flowers and fruits in *S. lycopersicum* cultivated in soil supplemented with MWCNTs was doubled in comparison to plants cultivated in control soil, and an increase of MWCNTs concentration resulted in increased relative abundances of Bacteroidetes and Firmicutes and reduced abundance of Proteobacteria and Verrucomicrobia. In addition, MWCNTs showed a positive effect on the seed germination and plant growth of *S. lycopersicum* when compared to the normal medium, the most effective concentration being 40 µg/mL (Morla et al. 2011). Seedlings of pepper and tomato cultivated on ground supplemented with carbon composites, in which the portion of MWCNTs varied in the range from 25% to 100%, showed enhanced plant growth: the shoots of tomatoes and peppers were 1.6–1.8 and 2.0–2.3-fold longer than in the control group, respectively, and the area of the root system was 2–2.5-fold larger (Onishchenko et al. 2015). The germinations of alfalfa and wheat seed were tolerant up to 2560 mg/L MWCNTs, and in seedlings of both species exposed to MWCNTs, enhanced root elongation was estimated (Miralles et al. 2012). Industrial material Taunit containing MWCNTs stimulated the growth of roots and stems of *Onobrychis arenaria* seedlings and induced an increase in peroxidase activity. The researchers stated that changed physiological parameters are connected with the adsorption of MCWNTs on root surface as well as their penetration and accumulation in plant cells and tissues (Smirnova et al. 2011). Oxidized MWCNTs effectively stimulated cell elongation in roots of *T. aestivum* plants and increased dehydrogenase activity, which led to faster root growth and higher biomass production (Wang et al. 2012a). Ghorbanpour and Hadian (2015) reported about stimulation of callus induction, biosynthesis of secondary metabolites and antioxidant capacity in *Satureja khuzestanica* cultivated in vitro due to treatment with MWCNTs. At lower MWCNTs concentration, calli growth increased significantly, culminating at 50 µg/mL, but a rapid decrease was estimated at 500 µg/mL. Moreover, at treatment with 100 µg/L MWCNTs, a considerable increase of antioxidant activity reflected in the lowest IC_{50} value was estimated, and the application of suitable MWCNTs concentrations was assumed to act as an effective elicitor for in vitro biosynthesis of worthful secondary metabolites and antioxidant drugs such as rosmarinic acid, caffeic acid, phenolics and flavonoids.

Zhai et al. (2015) studied the effect of differently charged MWCNTs, namely, neutral pristine MWCNT (p-MWCNT), positively charged MWCNT-NH_2 and negatively charged MWCNT-COOH, applied in the concentration range from 10.0 to 50.0 mg/L on hydroponically cultured *Z. mays* and *G. max* plants during 18-day exposures and found that MWCNTs stimulated the growth of *Z. mays* and enhanced its dry biomass while inhibited the growth of *G. max* at applied doses. Moreover, the exposure of *Z. mays* plants to MWCNT-COOHs (50 mg/L) resulted in nearly two-fold higher cumulative transpiration of water compared to control plants. Guan et al. (2015) found that the binding of MWCNTs-COOH to copper-zinc superoxide dismutase (Cu/ZnSOD) was a weak endothermic process, indicating that hydrophobic interaction is the prevalent force of this binding. While only little changes in the microenvironment of the amino acid residues of the enzyme were detected, conformational changes in Cu/ZnSOD were noticeable; however, enzyme activity was not significantly affected by MWCNTs-COOH. Mondal et al. (2011) reported that treatment of *Brassica juncea* seeds with low concentration of oxidized MWCNTs (o-MWCNTs) could result in enhanced biomass production or increased growth rate of plants, while for advancing the germination rate of *B. juncea* seeds, pretreatment (soaking-drying treatment) of the seeds with 2.3×10^{-3} mg/mL of o-MWCNTs is suitable.

According to Chen et al. (2015), MWCNTs can act as contaminant carriers and be transported to the edible parts of crops. Addition of MWCNTs to sterile agar medium or their deposition on seed surfaces resulted in early seed germination and activation of plant growth in *Hordeum vulgare*, *G. max* and *Z. mays* seedlings. Nanotube agglomerates were detected inside seeds treated with MWCNTs, and increased expression of genes encoding several types of water channel proteins was observed in seeds coated with MWCNTs in comparison with uncoated seeds. The above results suggest the potential of CNTs as regulators of germination and plant growth (Lahiani et al. 2013). On the other hand, interesting results concerning an alleviation of Cd and Pb toxicity by increasing total chlorophyll (Chl) content and improving growth parameters of canola plants caused by MWCNTs treatment were published by Oloumi et al. (2014).

9.4 Nanomaterials Based on Metals Essential for Plant Growth

9.4.1 Iron-Based Nanomaterials

Iron (Fe), an essential micronutrient for plants, plays crucial role in several metabolic processes, including photosynthesis, respiration and DNA synthesis. Fe is also involved in the synthesis of Chl and required for the maintenance of the structure and function of chloroplasts, and thus it is indispensable for healthy plant development (e.g. Marschner 1995; Rout and Sahoo 2015). However, in calcareous soils, in which Fe occurs in insoluble oxidized form, its availability is low (Guerinot and Yi 1994), and Fe-deprived plants are usually chlorotic due to the inhibition of Chl synthesis, and their growth is inhibited. Because iron deficiency in humans belongs to the most prevalent nutritional problems in the world today (Scrimshaw 1991; Stoltzfus and Dreyfuss 1998), fertilization with suitable formulations containing Fe in available form is desirable and helpful for healthy plant growth; however for the important staple crop such as rice, it is also necessary to develop Fe-rich plants using conventional breeding or directed genetic modification (Sperotto et al. 2012; Bashir et al. 2013). However, application of high concentrations of Fe ions is phytotoxic resulting in the inhibition of photosynthetic electron transport (Kráľová et al. 2008) and ROS formation causing damage of vital cellular constituents (e.g. membranes) due to lipid peroxidation (Hendry and Brocklebank 1985; Becana et al. 1998). The role of Fe in plant growth and metabolism was comprehensively reviewed by Rout and Sahoo (2015).

Both soil and foliar application of FeNPs to *Solanum tuberosum* plants exhibited superior effect on the quantitative and qualitative traits of potato than the treatment with Fe chelate; however, the combined treatment using foliar application of amino acids and soil application of FeNPs showed the best results in the percentage of protein, the percentage of Fe, the mean weight of a single tuber and the economic yield (Pourali and Roozbahani 2015).

Joseph et al. (2015) investigated the effects of artificially aged enriched biochar-mineral complexes with higher mineral content, surface functionality, exchangeable cations, high concentration of magnetic FeNPs and higher water-extractable organic compounds on the mycorrhizal colonization, *T. aestivum* growth and nutrient uptake and soil quality improvement. The application of 100 kg/ha of such formulations resulted in significantly greater uptake of P and N by wheat shoots and consequently in enhanced growth of plants. Enhanced growth could be connected with increased phosphorus uptake by plants, which could be connected with an increase in mycorrhizal colonization as well as with the properties of the biochar-mineral complexes. Also iron oxides (FeO_x) NPs were reported as effective nanofertilizers that significantly stimulated the growth of lettuce seedlings when applied at concentrations <50 ppm (Liu et al. 2016). Libralato et al. (2016) evaluated the effects of ionic $FeCl_3$ (1.29–1570 mg/L), as well as micro- (1.71–10.78 mg/L) and nanosized zerovalent iron (nZVI) (4.81–33.560 mg/L) on seed germination, seedling elongation, germination index and biomass of *Lepidium sativum*, *Sinapis alba* and *Sorghum saccharatum* and observed stimulation with application of the highest micro-Fe and nano-Fe concentrations, indicating that these Fe forms are more bioavailable than ionic Fe at these concentrations and act as micronutrient. According to Li et al. (2013a), penetration of the peanut seed coats by nZVI results in increased water uptake, which stimulates seed germination. The application of nZVI at concentrations <320 μmol/L promoted the growth and root development of *Arachis hypogaea*, while the concentration of 320 μmol/L was already phytotoxic. It could be noted that treatment with 40 and 80 μmol/L of nZVI resulted in more effective plant growth stimulation than application of EDTA-Fe. Based on TEM analyses, it could be concluded that the beneficial effect was most likely connected with the uptake of nZVI by the plants. Phosphate sorbed by nZVI was found to be bioavailable to both algae *Selenastrum capricornutum* and spinach. The concentration of phosphate of algae increased by sevenfold and spinach by two- to fourfold in the presence of spent nZVI (where NPs were the only source of phosphate) compared to algae and spinach plants grown in standard nutrient media (including phosphate). Moreover, iron content in spinach plant organs treated with spent nZVI was 21 (roots), 11 (leaves) and 7 (stems) times higher compared to the controls (Almeelbi and Bezbaruah 2014). Foliar application of nano- and common forms of Fe using doses of 0.25 and 0.5 g/L and Mg (using a dose of 0.5 g/L for nanoform and 0.5 g/L for common form) applied 56 and 72 days after sowing to black-eyed pea generally improved some production and physiological characteristics such as yield, leaf Fe content, stem Mg content, plasma membrane stability and Chl content as estimated after day 85, which can be connected with more efficient photosynthesis (Delfani et al. 2014). Kim et al. (2015) observed that treatment of *Arabidopsis thaliana* plants with nZVI induced high plasma membrane H^+-ATPase activity, resulting in the decrease in apoplastic pH, while the leaf area increased and stomatal aperture enlarged. Exposure of plants to nZVI led to fivefold higher levels of the H^+-ATPase isoform responsible for stomatal opening, AHA2, compared to control plants. This indicates a possibility of increased CO_2 uptake by nZVI-treated plants. Enhanced plant growth at treatment of *Typha latifolia* with low nZVI concentrations (<200 mg/L) with minimal upwards transport of nZVI to shoots was observed by Ma et al. (2013a).

The Fe_2O_3 NPs (6 nm) and citrate-coated Fe_2O_3 NPs showed beneficial effect on root growth of *G. max*, compared to the bulk Fe_2O_3 suspensions of concentrations >500 mg/L. Moreover, the foliar application of citrate-coated Fe_2O_3 NPs in the form of spray at the eight-trifoliate leaf stage resulted in significant enhancement of photosynthetic rates, which was attributed to increases in stomatal opening rather than increased CO_2 uptake activity at the chloroplast level, and foliar application of Fe_2O_3 NPs was found to produce more pronounced positive effects than soil treatment (Alidoust and Isoda 2013). A positive correlation between root phytohormone inhibition and the concentration of Fe_2O_3 NPs in transgenic and non-transgenic *O. sativa* plants was observed by Gui et al. (2015), whereby Fe_2O_3 NPs applied in low concentrations resulted in markedly higher activities of antioxidant enzymes. Jeyasubramanian et al. (2016) observed dose-dependent enhancement in the stem and root length of spinach plants cultivated in hydroponics in the presence of Fe_2O_3 NPs, where also enhanced saturation magnetization of spinach plants was estimated through Fe uptake. In watermelon seedlings, Fe_2O_3 NPs were found to increase seed germination and plant growth, and partial enhancement of physiological function was estimated as well; the beneficial effects increased rapidly but with increasing Fe_2O_3 NPs concentrations were reduced. The application of 20 mg/L Fe_2O_3 NPs increased root activity most effectively, and ferric reductase activity, root apoplastic iron content and watermelon biomass were considerably affected as well, indicating that exposure to Fe_2O_3 NPs ultimately improved also the resistance of watermelon to environmental stresses (Li et al. 2013a). The application of 0.75 g/L of nano-Fe oxide increased leaf + pod DW and pod DW of soybean, while the highest increase in grain yield (48%) was estimated with lower (0.5 g/L) nano-Fe oxide concentration (Sheykhbaglou et al. 2010).

Lower concentrations of Fe_3O_4 NPs with particle size of 20 nm stimulated algal growth in stationary and decline phases compared to the control; however treatment with 200 mg/L significantly reduced Chl*a* content and the viable cell concentration in the exponential growth phase of *Picochlorum* sp. (Chlorophyta). The adverse effect of Fe_3O_4 NPs was manifested only at the early stages, when Fe_3O_4 NPs adhered to the cell and blocked critical pores and membrane function (Hazeem et al. 2015). The exposure of hydroponically cultivated soybean plants to superparamagnetic iron oxide NPs (SPIONs) increased Chl concentrations with no signs of toxicity, and the increase of Chl content in subapical leaves of *G. max* plants was closely connected with physicochemical characteristics of SPIONs. On the other hand, photosynthesis efficiency did not show considerable differences due to application of iron chelate and SPIONs used as Fe source (Ghafariyan et al. 2013). The combined application of compost and nano-magnetite stimulated the growth of *Helianthus annuus* (>25%) cultivated in the contaminated soil, which could be connected mainly with the immobilization of zinc from the soil pore water that is available to plants. Consequently, this resulted in its reduced uptake by sunflower, and therefore it exhibited less toxicity to the roots and aerial parts of plants; during water stress, accumulation of proline and total amino acids in plants did not increase (Martinez-Fernandez et al. 2015). Also a positive effect on *Vigna radiata* biomass was estimated due to treatment with low concentrations of nano-FeO as well as nano-ZnCuFe-oxide (Dhoke et al. 2013). Fe_3O_4 NPs stimulated the root growth of *S. lycopersicum* plants (153%) and caused an imbalance of K^+ translocation in plants (Antisari et al. 2015).

Foliar spraying of chickpea plants with nano-Fe chelate fertilizer increased seed number per pod (17%), pod number per plant (48%), 100 seed weight (13%) and grain yield (65%) compared to control (Valadkhan et al. 2015). Application of biocompatible fertilizer for plant development prepared by grafting EDTA on the surface of Fe_3O_4 NPs applied as spray (1 mg/L) or soil amendment (1 g/kg) four times at four to six leaf stage and within 15 days after the first treatment improved aerial organ dry biomass, number of leaves, plant height, Chl content and Fe content of sunflower plants in comparison with the control and Fe-EDTA application. In general, more pronounced positive effects were obtained through soil amendment of Fe_3O_4-EDTA NPs than by foliar application, although higher aerial organ biomasses and Fe content were estimated for the foliar treatment (Shahrekizad et al. 2015). Foliar and soil application of Fe_3O_4 NPs (13 nm) applied at concentrations 1, 2 and 3 mg/L significantly enhanced plant height, branches/plant, leaves/plant, FW and DW of *O. basilicum* plants compared to control plants, foliar application being more effective (Elfeky et al. 2013). Spraying of *C. sativus* leaves with liquid fertilizers Nanonat and Ferbanat produced using nanotechnology until becoming wet at 10-day intervals for three times during plant growth significantly improved the yield compared to the control. The highest average fruit weight (149 g) and fruit length (17 cm) were obtained with Nanonat treatment (4.0 L/ha), while the highest fruit diameter was reached with application of Ferbanat at 3.0 L/ha (Ekinci et al. 2014).

9.4.2 Zn-/ZnO-Based Nanomaterials

Zinc, an essential element for plants, has been found to be an integral constituent of nearly 300 metalloenzymes and a cofactor for several enzymes; it is also involved in regulating the nitrogen metabolism and auxin synthesis and in the synthesis of nucleic acid and proteins (Vallee and Auld 1990). Enhanced growth of plant organs and plant biomass was estimated due to treatment with low Zn concentrations (e.g. Peralta et al. 2001; Aydinalp and Marinova 2009; Tuhy et al. 2015); however, the application of high zinc concentrations showed adverse effects (Kalyanaraman and Sivagurunathan 1993; Vijayarengan 2012; Vijayarengan and Mahalakshmi 2013). Root growth inhibition by high Zn concentrations is due to inhibition of cell division or cell elongation (eventually due to combination of both), and reduced dry matter yield could be connected with the fact that heavy metal-stressed plants spent more energy for their survival, which otherwise would be available for their overall growth processes (Vijayarengan 2013). Together with nitrogen, phosphorus and potassium, zinc belongs to the most important yield-limiting nutrients in some regions (Adhikari et al. 2016a).

The use of nanoscale Zn fertilizer considerably increased the height and root length of rice plants, where Zn-loaded nanofertilizer released nutrients slowly and steadily during critical growth period thereby improving growth of rice (Yuvaraj and Subramanian 2014). Zn encapsulated into a Mn hollow core shell showed improved Zn efficiency in *O. sativa*, which was reflected in reduced loss of nutrients and significant reduction of environmental pollution. While Zn-fortified core shell released Zn for more than 696 h, in $ZnSO_4$-fertilized soil, the release of zinc stopped after 408 h, and plants treated with Zn-fortified core shell showed also higher Zn uptake representing 5.6 mg/hill under submerged and 3.5 mg/hill under aerobic moisture regimes (Yuvaraj and Subramanian 2015). Treatment with foliar zinc nanofertilizer (15 and 25 nm) significantly improved shoot length (15%), root length (4%) and root area (24%) of pearl millet as well as plant dry biomass (12%) and grain yield (37%) as compared to the control. Moreover, nano-Zn fertilizer applied at concentration 10 mg/L also increased Chl content and total soluble leaf protein of the plant and caused enhancement of the activity of some enzymes (acid phosphatase, alkaline phosphatase, phytase, dehydrogenase) in the rhizosphere of 6-week-old plants (Tarafdar et al. 2014).

Zinc-rich ZnO NPs showing a spherical shape and particle size of 20–30 nm enhanced the level of indole acetic acid in roots (sprouts) of *C. arietinum*, which stimulated growth of plants (Pandey et al. 2010). ZnO NPs enhanced germination, seedling growth, pigments, sugar and protein contents along with increased activities of antioxidant enzymes in cabbage, cauliflower and tomato vegetable crops, while bulk ZnO particles were phytotoxic and adversely affected germination, seedling growth and biochemical parameters (Singh et al. 2013). Mukherjee et al. (2014) applied ZnO NPs and bulk ZnO in organic matter-enriched soil at doses 125–500 mg/kg on pea plants and found that treatment with all ZnO NPs concentrations markedly increased root growth compared to the control plants but did not affect the stem, while all treatments with bulk ZnO caused a significant increase of both root and stem length.

Torabian et al. (2016) studied the effects of foliar treatment using bulk and nano-sized ZnO in the form of spray on the growth and ion concentration of *H. annuus* varieties under salt stress and observed beneficial impact resulting in increased shoot DW, leaf area, net CO_2 assimilation rate, sub-stomatal CO_2 concentration, Chl content and the Chl fluorescence parameter F_v/F_m reflecting the maximum quantum efficiency of photosystem II (PSII) and Zn content, while leaf Na content showed a decrease. However the enhancement in shoot weight, Chl content and F_v/F_m parameter was stronger after ZnO NPs application. ZnO NPs applied at concentration 1.5 ppm in the form of aqueous foliar spray stimulated shoot DW of 10-day-old seedlings of chickpea (*C. arietinum* var. HC-1) plants compared to control plants, which could be attributed to low ROS levels in the presence of ZnO NPs resulting in reduced lipid peroxidation (Burman et al. 2013). El-Kereti et al. (2013) used combined foliar spray application of ZnO NPs (10–15 nm) with pre-sowing He-Ne laser irradiation on sweet basil plants and found that from the aspect of plant height, branches/plant and FW as well as total Chl, total carbohydrate, essential oil levels and Zn content, this treatment was more beneficial than the application of ZnO nanofertilizer alone, and treatment with 20 mg/L concentration was the most effective for all measured traits. Spinach plants that were sprayed with 500 and 1000 ppm ZnO NPs 14 days after sowing and then cultivated for 45–50 days showed a significant increase in leaf length, protein, fat and fibre content compared to control plants, and the obtained results suggested that by application of 1000 ppm ZnO NPs, the nutritive value of spinach can be enhanced for vegetarian diet by providing protein, fibre and the required amount of vegetarian fat (Kisan et al. 2015).

Prasad et al. (2012) observed that treatment with ZnO NPs with average particle size 25 nm using a dose of 1000 ppm stimulated peanut seed germination and plant vigour, and the plants were characterized by early flowering and higher leaf Chl content. Applications of ZnO NPs effectively increased stem and root growth of peanut plants, giving 34% higher pod yield per plant than bulk $ZnSO_4$. Moreover, at field experiments, foliar application of 15-fold lower concentration of ZnO NPs as the recommended $ZnSO_4$ dose resulted in 30% and 26% higher pod yield. Adhikari et al. (2016a) coated the seeds of *Z. mays*, *G. max*, *Cajanus cajan* and *Abelmoschus esculentus* with micron-scale (<3 μm) and nanosized (<100 nm) ZnO powder at 25 or 50 mg Zn/g seed, respectively, and observed that the seed coating with ZnO caused enhancement of the germination percentage (93–100%) in comparison with uncoated seeds (80%), while the crop growth was similar to that observed with $ZnSO_4.7H_2O$. Due to application of zinc, also enhanced production of auxin indole-3-acetic acid in roots causing improvement of the overall plant growth was observed. However, because at the time of seed germination the seed coating with micro- and nanoscale ZnO did not exert any osmotic potential, the use of such seeds enabled to load effectively the total Zn amount required by the plant.

Treatment with ZnO NPs (20 nm) at concentrations 10 and 20 μg/mL improved germination of onion seeds and caused enhancement in root and shoot lengths, FW and DW of seedlings; however higher concentrations of ZnO NPs exhibited negative effects on these characteristics (Raskar and Laware 2014). Xiang et al. (2015) reported that application of ZnO NPs in the concentration range 1–80 mg/L did not affect the germination rate of *Brassica rapa* ssp. *pekinensis* but significant inhibition in elongation of plant roots and shoots was observed, whereby the roots were found to be more sensitive.

9.4.3 Copper-Based Nanomaterials

Copper is an essential micronutrient; it is indispensable for the plant growth and development of plants; it is involved in many physiological processes and is an essential cofactor for many metalloproteins. On the other hand, treatment of plants with excess Cu concentrations results in plant growth inhibition and impairment of important cellular processes (e.g. Maksymiec 1997; Yurela 2005). Cu toxicity in plants is closely connected with the production of highly toxic hydroxyl radicals formed as a result of the redox cycling between Cu^{2+} and Cu^{+} and also with inhibition of photosynthetic electron transport due to interaction with some constituents of PSII and due to the formation of Cu-chlorophylls which are unsuitable for photosynthesis, resulting in a damage of photosynthetic apparatus, including the damage of oxygen evolving complex on the donor side of PSII (Maksymiec 1997; Yurela 2005; Küpper et al. 1996; Šeršeň et al. 1997; Kráľová et al. 1994; Masarovičová et al. 2011). On the other hand, treatment with low Cu^{2+} concentrations showed beneficial effects on growth of plant organs of *H. annuus* (Lin et al. 2003) and *Z. mays* (Jiang et al. 2001; Bashmakov et al. 2005) and biomass of *Brassica napus* (Peško and Kráľová 2013). The application of low CuO NPs (Adhikari et al. 2012, 2016b) or CuNPs (Hafeez et al. 2015) concentrations was also reflected in beneficial effects on plant growth. CuO NPs (<50 nm) supplied in solution culture as well as in the form of spray improved the growth of *Z. mays* plant by 51% compared to the control and significantly affected the activity of glucose-6-phosphate dehydrogenase and the pentose phosphate pathway of maize plant (Adhikari et al. 2016b). CuO NPs affected the growth of *G. max* and *C. arietinum* plants at different concentrations. The best growth was observed after treatment with 100 ppm in soybean and 60 ppm in chickpea plants, and beyond these concentrations, plan growth was inhibited (Adhikari et al. 2012). The application of CuNPs (10–50 ppm) to soil in pots caused a considerable increase of growth and yield of wheat, whereby treatment with 30 ppm CuNPs led also to significantly higher Chl content, leaf area, number of spikes/pot, number of grains/spike, 100 grain weight and grain yield as compared with the control (Hafeez et al. 2015).

9.4.4 Manganese and Cobalt Nanoparticles

Manganese is one of the nine essential nutrients indispensable for plant growth that is necessary in processes of chloroplast formation, photosynthesis, nitrogen metabolism and synthesis of some enzymes (Marschner 1995). MnNPs were found to affect the assimilatory process in *V. radiata* supplemented with macro- and micronutrients and plants cultivated under controlled conditions by enhancing the net flux of nitrogen assimilation through nitrate reductase-nitrite reductase and glutamine synthetase-glutamate synthetase pathways indicating that NPs could contribute to safe nano-/micronutrient-based crop management (Pradhan et al. 2014).

The essential element cobalt belongs to transition metals, and low concentrations of Co^{2+} ions in medium were found to stimulate growth of photosynthetic organisms, while high concentrations exhibited phytotoxic effects (Tiwari and Mohanty 1993; Czerpak et al. 1994; El-Sheekh et al. 2003; Palit et al. 1994). Retardation of leaf senescence, increase in drought resistance in seeds, inhibition of ethylene biosynthesis and regulation of alkaloid accumulation in medicinal plants could be mentioned as beneficial effects of cobalt (Palit et al. 1994; Jiao et al. 2006a, b; Koval'skii et al. 1971; Jain and Nainawatee 2000). From cobalt oxides NPs, only Co_3O_4 NPs effects on radish seedlings were studied, and it was found that treatment with concentration 5 g/L resulted in improved root elongation (Wu et al. 2012).

9.5 Nanomaterials Based on Non-essential Metals and Metalloids Beneficial for Plant Growth

9.5.1 TiO_2 Nanoparticles

Titanium applied at low doses can be considered as beneficial element that improves various physiological characteristics such as plant biomass and yield (Pais 1983; Carvajal et al. 1994), Chl contents (Hrubý et al. 2002; Carvajal et al. 1994) and essential element contents (Carvajal et al. 1994), although at higher concentrations it is phytotoxic (Hrubý et al. 2002). It is also assumed that N deficiency could be compensated by Ti. Haghighi et al. (2012) cultivated tomato plants in the presence of 1 or 2 mg/L in a nutrient solution showing reduced N content by 25% and by 50%, respectively, and investigated the nutrient uptake by plants. The decreased concentration of N in nutrient solution was reflected in a decline of Ca and Ti concentrations in tomato leaves and an increase of K, Mn, Fe, Cu and Zn concentrations. Moreover, the application of 1 mg/L Ti to nutrient solution, the N concentration of which was decreased by 50%, could compensate the decrease of all photosynthetic parameters and physiological characteristics, including flower induction. In addition, low concentrations of TiNPs promoted plant growth of *Lemna minor*, while high concentrations inhibited it (Song et al. 2012). Moreover, TiO_2 NPs exhibited negative effect on the growth rate or Chl*a* content of duckweed, even at the application of 5 mg/L and the prolonged exposure time of 14 days (Li et al. 2013b).

Similarly to Fe_3O_4 NPs, TiO_2 NPs were found to affect rhizosphere P availability and uptake by lettuce, whereby the effects on P accumulation for roots ($TiO_2 > Fe_3O_4 >$ control) and shoots ($Fe_3O_4 > TiO_2 >$ control) differed each from other. Improved P availability and uptake by the plants using NPs can be utilized for better nutrient management to guarantee food security (Zahra et al. 2015). The enhanced adhesion of beneficial bacteria on oilseed rape roots by TiO_2 NPs resulted in enhanced crop growth, and TiO_2 NPs protected the plants against the fungal pathogen *Alternaria brassicae* (Palmqvist et al. 2015). The shoot and root lengths of *L. sativa* were increased up to 49% and 62%, respectively, at treatment of soil with TiO_2 NPs of sizes less than 65 nm at concentration 100 mg TiO_2 NPs/kg soil, as compared to the control treatment. It could be noted that the treatment with the above-mentioned concentration of TiO_2 NPs increased the concentration of phytoavailable P in soil up to 56% after 72 h incubation at 25 °C in Petri dishes; shoot and root P concentrations were increased up to 36% and 175%, respectively, and P uptake per plant was five-fold as compared to the control, and *L. sativa* significantly acidified its rhizosphere (Hanif et al. 2015). Burke et al. (2014) reported that even low concentrations of TiO_2 NPs may influence some important groups of soil microbes, such as mycorrhizal fungi, but changes in the composition of microbial communities may not affect plant growth under conditions of adequate moisture and nutrients.

Dehkourdi and Mosavi (2013) investigated the impact of different concentrations of nano-TiO_2 (anatase; 10–40 mg/mL) on germination parameters of parsley in a tissue culture and observed a considerable enhancement in the percentage of germination, germination rate index, root and shoot length, FW, vigour index and Chl content of plants with increasing nano-anatase concentration, dose 30 mg/mL being the best. The soaking of flax (*Linum usitatissimum*) seeds in the suspensions of 25 and 32 nm anatase showed that the dose of 100 mg/L TiO_2 NPs had positive effects on seed germination, which can be connected with antimicrobial properties of the anatase crystalline structure of TiO_2 that improved the plant resistance to stress (Clement et al. 2013). Based on the fact that priming enhances germination, Hosseini et al. (2013) investigated the effects of soaking strawberry seeds in different concentrations (3–9%) of nano-atanase on germination parameters and found that the most effective concentration was 7.5%. No adverse effects of TiO_2 NPs on seed germination of oilseed rape, lettuce and kidney bean were detected, and treatments with 100, 500, 1000 and 2500 mg/L TiO_2 NPs significantly increased root growth of *L. sativa*. On the other hand, TiO_2 NPs did not affect *Brassica campestris* and *Phaseolus vulgaris* plants, and even though the plants absorbed TiO_2 NPs, activities of enzymes and Chl content did not change (Song et al. 2013). Larue et al. (2012a) studied the effects of TiO_2 NPs on seed germination and the first days of root elongation, which could be considered as the most sensitive stages of plant development, and found that the germination rate of wheat seeds was not affected. However the application of 50 and 100 mg/L of nanosized anatase (14 nm) or rutile (22 nm) caused an increase in the root length representing more than 50% compared to plantlets germinated and cultivated in pure water. Similar results concerning the fact that exposure to TiO_2 NPs (14 nm or 25 nm anatase) was connected with increased root elongation and did not affect germination, evapotranspiration and plant biomass of hydroponically cultivated *T. aestivum* and *B. napus* plants were published by Larue et al. (2012b).

Application of TiO_2 NPs (0–4000 mg/L) to 7-day-old hydroponically grown cucumber plants significantly increased root length (average >300%) (Servin et al. 2012). According to Larue et al. (2012a), locally generated oxidative stress and enlargement of cell wall pores caused by TiO_2 NPs increased water flow into roots as well as turgor in roots of *T. aestivum*, which resulted in the stimulation of root elongation.

Compared to the control, application of 10 ppm TiO_2 NPs shortened the germination time of wheat seeds from 1.4 to 0.9 days, and length of shoots and seedlings treated with 2 and 10 ppm TiO_2 NPs was enhanced in comparison with the untreated control or plants treated with the same concentrations of bulk TiO_2 (Feizi et al. 2012). Treatment with 40 ppm TiO_2 NPs was found to improve the mean germination time of fennel by 32% compared to the control, and exposure to 60 ppm TiO_2 NPs significantly improved germination percentage and shoot DW (Feizi et al. 2013). Also both TiO_2 treatments, i.e. application in bulk form or as TiO_2 NPs at concentration 60 mg/L, improved germination percentage and decreased the mean germination time of medicinal plant *Salvia officinalis* (Hassan et al. 2013). Wu et al. (2014) achieved better germination of *L. sativa* seeds by the bombardment of seeds by TiO_2 NPs via the electrospray for 2–4 min. The application of a single-capillary electrospray system with a particle deposition stage (where seeds are placed) using TiO_2 NPs enabled weakening of the structure of the seed coat and finally resulted in breaking of the coat-imposed seed dormancy. Application of this method using CuO NPs and AuNPs also resulted in the improved germination percentage of lettuce.

The application of 2000 mg/L TiO_2 NPs of mean particle size 20 nm promoted both seed germination and seedling vigour of canola (Mahmoodzadeh et al. 2013). The morphological (root length, plant height, fresh biomass) and physiological (photosynthetic gas exchange, Chl content, nitrate reductase activity) parameters of oilseed rape (*B. napus*) and its antioxidant system (SOD, guaiacol peroxidase, catalase) were improved due to treatment with low (up to 4000 mg/L) TiO_2 NPs concentrations (Li et al. 2015b). According to researchers, enhancement of morphological characteristics suggested that TiO_2 NPs probably induced the absorption of water and fertilizer and catalysed the photosynthetic process, whereby stimulation of plant growth can be connected also with increased inorganic nitrogen conversion into organic nitrogen in proteins and Chl.

Treatment with TiO_2 NPs (anatase) was found to enhance the activities of some enzymes (nitrate reductase, glutamate dehydrogenase, glutamine synthase and glutamic-pyruvic transaminase) and also promoted the absorption of nitrate by spinach plants, accelerated transformation of inorganic nitrogen (such as nitrogen in NO_3^- and NH_4^+) into organic nitrogen (such as protein and Chl) and enhanced the FW and DW of spinach (Yang et al. 2006). TiO_2 NPs applied as foliar spray at concentration 10 mg/L on the leaves of *V. radiata* had beneficial effect on plant production and biochemical characteristics resulting in enhanced shoot and root length (17% and 50%, respectively), root area (43%) and root nodule (67%) as well as in raising of Chl and total soluble leaf protein content (46% and 94%, respectively); it also contributed to the increase of microbial population in the rhizosphere (Raliya et al. 2015). Gao et al. (2008) sprayed spinach (*Spinacia oleracea*) plants in the stage of two leaves with the suspension containing 0.03% of nano-anatase once a week until the plants had eight leaves and found that this treatment led to increased activity of Rubisco (ribulose-1,5-bisphosphate carboxylase/oxygenase) activase, which significantly stimulated carboxylation of Rubisco and consequently the rate of photosynthesis and was reflected in improved growth of spinach (*S. oleracea*).

Morteza et al. (2013) studied the effects of TiO_2 (0.01% or 0.03%) applied in the form of aqueous solutions containing bulk TiO_2 or TiO_2 NPs on *Z. mays* plants at different stages of plant growth (vegetative stage, appearance of male flowers and female flowers) and found that Chl content (Chl*a* and Chl*b*), total Chl (Chl*a* + Chl*b*), Chl*a*/Chl*b*, carotenoids and anthocyanins were significantly affected by TiO_2, and the application of spray with TiO_2 NPs at the reproductive stage (appearance of male and female flowers) resulted in higher amount of pigment in comparison with the control. This indicates beneficial effects of TiO_2 NPs from the aspect of increased crop yield, especially maize yield. TiO_2 NPs also promoted photosynthesis in *S. lycopersicum* leaves under mild heat stress causing an increase in the net photosynthetic rate, conductance to H_2O and transpiration rate of plant leaves, and they also increased regulated and decreased nonregulated PSII energy dissipation (Qi et al. 2013). Colloidal solution of laser-synthesized anatase TiO_2 NPs applied in concentration up to 1.5 mmol/L stimulated germination of *B. oleracea* var. *capitata* seeds and caused an increase of root length in the germinated seeds, number of leaves, leaf area, plant height and amounts of assimilation pigments (Chl*a*, Chl*b* and carotenoids) levels, while at higher TiO_2 NPs concentrations (>2.0 mmol/L), phytotoxic effects were manifested (Singh et al. 2012). Ze et al. (2011) observed that TiO_2 NPs could induce considerable increase of light-harvesting complex II (LHCII) b gene expression and LHCII content in the thylakoid membrane in *A. thaliana* exceeding that obtained at treatment with bulk TiO_2, suggesting that TiO_2 NPs could promote the light absorption of chloroplasts, by increasing LHCII-regulated distribution of light energy from PSI to PSII and accelerate transformation from light energy to electronic energy, resulting in more effective water photolysis and O_2 evolution. The application of 0.25–4% TiO_2 NPs (rutile) on aged spinach seeds resulted in an increase of the germination rate and the germination and vigour indexes, while during the growth stage, the plant DW was increased as a result of increased Chl formation, the Rubisco activity and the photosynthetic rate, 2.5% TiO_2 NPs treatment being the most effective. The bulk TiO_2 did not exhibit significant effects (Zheng et al. 2005). Anatase TiO_2 NPs significantly improved photosynthetic electron transport (PET) through the whole chain, photoreduction activity of PSII, oxygen evolution and photophosphorylation activity of spinach chloroplasts also under ultraviolet light (Zheng et al. 2007). Application of 0.25% TiO_2 NPs (rutile) accelerated the rate of oxygen evolution rate (OER) in spinach chloroplasts, caused higher non-cyclic photophosphorylation activity of chloroplasts compared to cyclic photophosphorylation activity and activated activities of Mg^{2+}-ATPase and chloroplast coupling factor I (CF1)-ATPase in thylakoid membranes suggesting a beneficial effect of TiO_2 NPs on the activation of the photochemical reaction of spinach chloroplasts (Hong et al. 2005). Similarly, also Lei et al. (2007) observed that nano-anatase exhibited positive effects on photosynthesis, which was reflected in a considerable improvement of PET, photoreduction activity of PSII, OER and photophosphorylation activity of spinach chloroplasts, and these effects were estimated not only under visible light but also under ultraviolet light.

In addition, TiO_2 NPs also dramatically increased callugenesis and the size of calli of barley (*H. vulgare*) without causing negative change in the quality of the callus (Mandeh et al. 2012).

9.5.2 Silver Nanoparticles

Adverse effects of silver on biological processes in microorganisms connected with the alteration of cell membrane structure and functions (Jampílek and Kráľová 2017a, d; Konotop et al. 2014; Franci et al. 2015) or inhibition of the expression of proteins associated with ATP production (Yamanaka et al. 2005) and with interaction of Ag^+ ions with thiol (sulfhydryl) groups (Russell and Hugo 1994) are widely utilized in different fields of human activities, including agriculture (e.g. Jampílek and Kráľová 2015, 2017a, d). Moreover, $AgNO_3$ has been reported to be a potential regulator of ethylene activity and plant growth modulator, and it seems that Ag^+ ion-mediated responses are involved in polyamines and ethylene- and Ca-mediated pathways and have a decisive role in the regulation of some physiological process (Kumar et al. 2009). The Ag^+ ions also enhance auxin efflux independently of effects on ethylene response (Strader et al. 2009).

At present there is growing interest to utilize antifungal properties of AgNPs for plant disease management (e.g. Karimi et al. 2012; Aziz et al. 2016), because well-dispersed and stabilized AgNPs solution can act as excellent fungicide due to good adhesion on bacterial and fungal cell surface (Kim et al. 2008) and so contribute to healthy plant growth. The toxic impact of AgNPs on plants, similarly to the effects of other metal NPs, is connected with their chemical composition enabling the release of toxic Ag^+ ions as well as with stress caused by some specific properties of these NPs, such as surface, size and shape (e.g. Masarovičová and Kráľová 2013; Masarovičová et al. 2014). Syu et al. (2014) investigated the effects of the size and shape of AgNPs on growth and gene expression of *Arabidopsis* plant. Decahedral AgNPs (45 nm) exhibited the highest stimulation of root growth, while spherical ones (8 nm) did not stimulate root growth but induced uppermost levels of anthocyanin accumulation in plants. On the other hand, the highest antimicrobial activity was observed with triangular (47 nm) and spherical AgNPs. The highest levels of Cu/Zn SOD were obtained after treatment with spherical AgNPs, while the lowest Cu/Zn SOD levels were observed for decahedral NPs. Moreover, AgNPs were found to activate *Arabidopsis* gene expression involved in cell proliferation, metabolism and hormone signalling pathways.

Rani et al. (2016) observed differences in the effects of synthetic and biosynthesized AgNPs on growth, physiology and oxidative stress of water hyacinth (*Eichhornia crassipes*) after treatment with 1, 10 and 100 mg/L. In the fifth day of treatment with 100 mg/L biosynthesized AgNPs, plant growth stimulation was observed; AgNPs also increased the levels of carbohydrate content at 1 and 10 mg/L, and enhanced protein content at application of 100 mg/L was estimated, while the contents of phenol and Chl were reduced. On the other hand, chemically synthesized AgNPs caused plant growth reduction and increased content of carbohydrates at 10 mg/L without affecting protein content; however at all studied concentrations, they decreased the amounts of total phenol and Chl contents. The researchers suggested that with respect to biosynthesized AgNPs, the toxicity of synthetic AgNPs to plants could be caused mainly by Ag^+ occurring in the suspensions. Biosynthesized AgNPs applied at lower concentrations (100 mg/L) exhibited growth-stimulating activity also on plants of *B. rapa* ssp. *pekinensis*, while concentrations of 250 or 500 mg/L reduced both root and shoot growth as well as fresh biomass (Baskar et al. 2015). Exposure to 1.0 mg/L biosynthesized AgNPs was found to significantly increase plant fresh biomass of turnip (*B. rapa* ssp. *rapa*), while the application of higher concentrations was toxic (Thiruvengadam et al. 2015). The treatment of seeds of *Pennisetum glaucum* with biosynthesized AgNPs caused improved germination (Parveen and Rao 2015). Biologically synthesized AgNPs applied in the concentration range 10–30 μg/mL improved not only seed germination but also plant growth of *Boswellia ovalifoliolata* and treated seeds sprouted within 10 days compared to control, indicating that AgNPs contributed to facilitated penetration of water and nutrients through seed coat, which was reflected in accelerated seed germination and plant growth (Savithramma et al. 2012).

Application of AgNPs to *S. lycopersicum* plants being under NaCl stress improved percentage of germination, germination rate, root length as well as FW and DW of plants, and based on the gene expression patterns associated with AgNPs exposure, it can be suggested that AgNPs may be involved in the response to stress and might be useful to enhance plant tolerance against salinity (Almutairi 2016a). Early-stage enhanced growth of *G. max* plants that were treated with AgNPs (15 nm) at 2 ppm and exposed to flooding was estimated. Based on proteomic study, it was concluded that 107 root proteins showing different changes were mainly associated with stress, signalling and cell metabolism and AgNPs treatment was able to downregulate the genes of the alcohol dehydrogenase 1 and pyruvate decarboxylase 2 genes that were upregulated under flooding. In AgNPs-treated plants, the estimated amount of cytotoxic by-products of glycolysis was lower compared to flooded plants. Based on these observations, it could be supposed that AgNPs lowered the oxygen-deprivation stress in *G. max* plants (Mustafa et al. 2015). Under waterlogging conditions, foliar application of AgNPs (50 or 100 ppm) caused increased height and corm number of saffron, indicating that some effects caused by flooding stress on saffron growth may be modulated by AgNPs application (Sorooshzade et al. 2012). The effects of stable AgNPs with mean diameter of 17 nm prepared using γ-radiation with gum acacia as the stabilizing and protecting agent and $AgNO_3$ on vegetative growth and the yield of two cultivars (Bronco and Nebraska) of *P. vulgaris* after foliar application ranging from 5 to 60 ppm were manifested in considerably enhanced plant height, root length, number of leaves/plant, the leaves' area, total FWs and DWs/plant and yield (i.e. number, FWs and DWs of pods/plant and 100-seed weight). Moreover, altered protein patterns were estimated in two studied cultivars, and increased levels of growth-stimulating compounds in the Nebraska variety were reflected finally in the improvement of both growth parameters and yield of AgNPs-treated plants (El-Batal et al. 2016). An increase in AgNPs concentration from 20 to 60 ppm applied in the form of spray resulted in the improved seed yield of basil plants, and the increased level of AgNPs resulted in the reduction of polyphenol compound content (Nejatzadeh-Barandozi et al. 2014).

From two plant species, *C. sativus* and *T. aestivum*, that were exposed to AgNPs and Ag^+ (applied as $AgNO_3$) at the germination and vegetative growth stages, stimulation of root elongation in cucumber plants was observed after the application of 200 mg/L AgNPs and 5 mg/L Ag^+. Due to the exposure of two investigated plant species to higher AgNPs concentrations, the plants were more susceptible to the toxicity of AgNPs at the vegetative growth stage than at the germination stage (Cui et al. 2014). The application of AgNPs (25–400 ppm) positively affected the root and shoot length, FW and vigour index of 7-day-old *B. juncea* seedlings, which was reflected in the increased root length (326%) and vigour index (133%) of the treated plants as well as in the enhanced photosynthetic quantum efficiency and Chl contents in leaves of *B. juncea* plants compared to the control plants and in the reduced ROS levels and the decreased content of proline (Sharma et al. 2012). Increasing concentration of AgNPs from 20 to 60 ppm increased shoot and root lengths, leaf surface area, Chl, carbohydrate and protein contents of *P. vulgaris* and *Z. mays* plants (Salama 2012). Razzaq et al. (2016) reported that AgNPs of 10–20 nm applied at concentration 25 ppm significantly increased the number of seminal roots, leaf area, root biomass, FW and DW of wheat plants, number of grains per spike and 100-grain weight as compared to the control. Treatment with AgNPs up to 30 μg/mL accelerated root growth of *O. sativa* and markedly affected root branching and DW, while at 60 μg/mL, restriction of root's ability to grow was estimated (Mirzajani et al. 2013). However, additionally it was observed that due to treatment with AgNPs, the rhizosphere bacteria were differently affected: *Bacillus thuringiensis* SBURR1 was totally eliminated, and *Bacillus amyloliquefa*ciens SBURR5 became the most populated one, and the damage of bacterium cell wall was indicated by estimated leakage of reducing sugars and proteins through the bacterial membrane.

Treatment of hydroponically cultivated poplars (*Populus deltoides* x nigra) and *A. thaliana* with sublethal AgNPs concentrations resulted in stimulation of root elongation, FW and evapotranspiration, and at subinhibitory concentrations of AgNPs, accumulation of Ag in poplar tissues was found to increase with increasing exposure concentration and with decreasing AgNP size (Wang et al. 2013).

Treatment with AgNPs applied as foliar spray at concentrations 20–60 mg/mL was found to be effective for preventing dark storage-induced petal abscission of pelargonium in plants that were stored for 5 days in a growth chamber in darkness at temperature 20 ± 2 °C and relative humidity 65%. In AgNPs-treated plants, higher contents of leaf Chls and carotenoids as well as higher activity of ascorbate peroxidase and guaiacol peroxidase were estimated compared to the untreated control, indicating a possible beneficial effect of AgNPs on alleviation of dark storage-induced oxidative stress (Hatami and Ghorbanpour 2013). Seifsahandi and Sorooshzadeh (2013) studied the effects of foliar AgNPs (0.2, 0.4 and 0.6 mmol/L) and $AgNO_3$ (0, 10, 20 and 30 mmol/L) application on the vegetative and phytochemical properties of borage, where spraying was applied at the onset of the flowering stage (65 days after cultivation) and was maintained until flowering (98 days after cultivation). Such treatments significantly enhanced the vegetative (leaf number, greenness of leaves, plant DW, inflorescence DW and petal abscission) and phytochemical (phenol, tannin and alkaloid content, mucilage percentage and swelling index) properties of borage, and the best results have been achieved at treatment with 0.6 mmol/L AgNPs. Shams et al. (2013) observed that growth indexes (except fruit pH) and Ag concentration increased significantly in cucumber plants sprayed with suspension of 50 nm AgNPs in distilled water.

Yin et al. (2012) examined the effects of 20 nm polyvinylpyrrolidone (PVP)-coated and 6 nm gum arabic-coated AgNPs on the germination and early growth of 11 wetland plants and compared them with the effects of $AgNO_3$. They found that for some species, the effects of Ag on plants in soil were either reduced with respect to pure culture or in some cases reversed, and consequently, growth stimulation instead of inhibition was observed. This could be connected with the fact that some ligands occurring commonly in soil solution (e.g. -SH, S^{2-}, Cl^- and PO_4^{3-}) have negative effect on the bioavailability and attenuate the toxicity of Ag at lower concentrations. The exposure to 5 mg/L PVP-coated AgNPs (20 nm) for 10 days led to the upregulation of 286 genes and downregulation of 81 genes in *A. thaliana*, while at treatment with the same Ag^+ dose, only 84 genes were upregulated and 53 genes downregulated as compared to nonexposed plants. A considerable overlapping between genes differentially expressed in response to AgNPs and Ag^+ (13% and 21% of total up- and downregulated genes, respectively) indicated that AgNP-induced stress is connected with both Ag^+ toxicity and NPs-specific effects. However, three genes that were greatly upregulated at AgNPs treatment, but not in the presence of Ag^+, belong to the thalianol biosynthetic pathway, which is expected to be involved in the plant defence system (Kaveh et al. 2013). In greenhouse experiments, Gusev et al. (2016) observed that AgNPs of average diameter 10 nm stabilized with cationic polymer polyhexamethylene biguanide hydrochloride stimulated the growth of fodder beet (*Beta vulgaris*), which could be connected with the changed activity of oxidases and, consequently, with the changed amount of auxins in plant tissues.

Ag-SiO_2 core-shell NPs with the average size of 101.8 ± 8.9 nm applied as elicitor to the hairy root cultures of *Artemisia annua* stimulated artemisinin production in 20-day-old hairy root cultures up to 13.3 mg/L, which represents a 3.9-fold increase over the control (Zhang et al. 2013).

The dry matter yield of maize plants that were exposed to magnetic field and AgNPs was significantly higher than that of plants treated with Kemira, Librel and Humax fertilizers. The beneficial effect of treatment with magnetic field and AgNPs was reflected also in the increased maize fresh yield (by 35% compared to the control) and in higher percentages for ears (Berahmand et al. 2012).

9.5.3 Gold-Based Nanomaterials

Binder et al. (2007) found that Au^+ ions support ethylene binding, however, unlike Ag^+, do not block ethylene action on plants. They affect seedlings independently of ethylene signalling, and it could be supposed that ethylene signalling is not blocked by Au^+ ions due to their size, which is smaller than that of Ag^+ ions. However, when gold was applied in nanoform, it was found to exhibit beneficial effects on plants (Arora et al. 2012; Wan et al. 2014; Kumar et al. 2013). Different concentrations of AuNPs applied through foliar spray on *B. juncea* plants increased the number of leaves per plant without affecting the mean leaf area. With application of 10 ppm AuNPs, the seed yield showed the best increase, while concentration up to 25 ppm AuNPs resulted in increased sugar content, and improvement of the redox status of the treated plants was observed (Arora et al. 2012). Similarly, gold nanorods significantly promoted the root elongation of watermelon, although high concentrations were found to be phytotoxic and the enzymes activities of plants indicated that oxidative stress happened in the treated plants (Wan et al. 2014). The treatment of *A. thaliana* with 10 μg/mL AuNPs of size 24 nm enhanced the total seed yield threefold over the control, and application of these AuNPs at 10 and 80 μg/mL resulted in considerably better germination rate, vegetative growth and free radical scavenging activity (Kumar et al. 2013). Moreover, the researchers reported that *A. thaliana* plants treated with AuNPs showed significant correlation between expression of key plant regulatory molecules, microRNAs (miRs), seed germination, growth and plant antioxidant potential.

9.5.4 CeO_2 Nanoparticles

The rare earth metals, cerium and lanthanum, applied at very low concentrations were reported to be beneficial for plants. For example, increasing of La and Ce concentrations from ca. 0.007 to ca. 0.6 mmol/L in the continuously flowing nutrient solutions significantly improved DW of corn roots, while the DW of shoots and the DW of whole plants were not significantly affected (Diatloff et al. 1995a, b). Beneficial effects on plant growth that were also observed at application of low concentrations of bulk CeO_2 were attributed to the higher relative Chl content and enhanced photosynthesis as well as its antioxidative effects. An increase in the total antioxidant capacity of plants due to treatment with low CeO_2 NPs concentrations was described as well, and it was postulated that the nanosized CeO_2 probably worked as a radical scavenger (Corral-Diaz et al. 2014).

CeO_2 NPs exhibit both pro-oxidant and antioxidant effects on different cell systems or organisms, and the majority of studies focused on nano-CeO_2 effects on plants are devoted to its negative impact on production of some plants and biochemical characteristics (e.g. Zhao et al. 2013, 2015; Majumdar et al. 2014; Nhan et al. 2015), while some papers report also about its beneficial effects.

According to Wang et al. (2012b), CeO_2 NPs at concentrations 0.1–10 mg/L applied on tomato had either a negligible or a small beneficial effect on plant growth and *S. lycopersicum* production; however, CeO_2 NPs taken up by plant roots were found to be translocated to shoots and edible tissues, and significantly higher Ce concentration was estimated in the fruits of tomato treated with 10 mg/L than in the control. The germination of radish (*Raphanus sativus*) seeds treated with concentrations 50, 100 and 200 mg/L of pristine- and citric acid-coated CeO_2 NPs applied in the form of suspensions was not influenced; however, the treatment with 200 mg/L of citric acid-coated CeO_2 NPs (molar ratio CeO_2/citric acid = 1:7) resulted in significant enhancement of biomass, increased water content as well as reduced Ce uptake (by 94%) compared to bare NPs, indicating that this coating reduced phytotoxic effects of CeO_2 (Trujillo-Reyes et al. 2013). Nevertheless, Corral-Diaz et al. (2014) reported that treatment with CeO_2 NPs (62–500 mg/kg) did not affect growth, gas exchange, photosynthesis, flavonoids, phenols and nutrients' accumulation in tubers and leaves of radish plants, while tubers' antioxidant capacity increased, indicating the improvement of the radical scavenging potency of radish by CeO_2 NPs.

Rico et al. (2015) studied production and biochemical characteristics of soil-grown barley (*H. vulgare*) treated with CeO_2 NPs (0, 125, 250 and 500 mg/kg). The researchers observed that application of 500 mg/kg CeO_2 NPs resulted in significant increase (331%) of shoot biomass compared to control plants; however, treated plants did not form grains. The application of 250 mg/kg CeO_2 NPs led to higher Ce accumulation in grains (up to 294%) with a simultaneous significant increase in some primary and secondary macro- (P, K, Ca, Mg, S) and micronutrients (Fe, Zn, Cu) as well as to the enhanced contents of methionine, aspartic acid, tyrosine, threonine, arginine and linolenic acid in grains. Similar results were obtained in a study focused on CeO_2 NPs impact on yield and nutritional parameters in wheat. The application of 500 mg/kg CeO_2 NPs to soil stimulated plant growth (+9%), shoot biomass (+13%) and grain yield (+37%) compared to control, while application of 500 mg/kg CeO_2 NPs caused modification of the amino acid composition and increased linolenic acid content. Also modified sulphur and manganese storage in grains was estimated due to CeO_2 NPs treatment (Rico et al. 2014).

Gomez-Garay et al. (2014) investigated the impact of CeO_2 NPs on the growth of in vitro cultivated plantlets of *Medicago arborea*. Addition of CeO_2 NPs to the culture medium did not affect germination rate and shoot DW; however treatment with low CeO_2 NPs concentrations resulted in the increased number of trifoliate leaves and the root length, while the root DW was found to decrease. The root growth was significantly stimulated also by application of 500 mg/L CeO_2 NPs in *C. sativus* and *Z. mays* plants, but it was reduced in *Medicago sativa* and *Lycopersicon esculentum* plants (Lopez-Moreno et al. 2010a). Application of CeO_2 NPs (500–4000 mg/L) on *G. max* plants significantly increased root growth, and due to treatment with 4000 mg/L, root length increased even by 75% (Lopez-Moreno et al. 2010b). In soil amended with 125 mg/kg CeO_2 NPs, stimulation of root growth of cilantro (*Coriandrum sativum* L.) was estimated (Morales et al. 2013). The exposure to CeO_2 NPs at 250 ppm significantly increased the total plant biomass of *A. thaliana*, while the root length was unaffected. On the other hand, the application of 500 ppm of CeO_2 NPs did not significantly affect fresh biomass but reduced the average root length by about 60%, indicating greater sensitivity of plant roots to CeO_2 NPs toxicity (Ma et al. 2013b).

9.5.5 Al-/Al_2O_3-Based Nanomaterials

In general, aluminium is considered to be phytotoxic, and according to Silva (2012), its toxicity is manifested mainly as root growth inhibition, ROS production, modifications on root cell wall and plasma membrane, nutrient unbalances, accumulation of callose and disturbance of cytoplasmic Ca^{2+} homeostasis. Information on Al toxicity with an emphasis on plant response to Al stress was summarized, for example, by Mossor-Pietraszewska (2001) and Silva (2012). On the other hand, it was observed that Al applied at low concentrations can stimulate plant growth (e.g. Foy 1984; Kinraide 1993). According to Kinraide (1993), because growth stimulation by Al^{3+} occurred only at pH that reduced root growth, the nature of these beneficial effects is explained by the alleviation of H^+ toxicity by Al^{3+}, and the Al^{3+} ions were found to increase cell membrane electrical polarity and stimulated H^+ extrusion (required for root growth at low pH) (Yan et al. 1992). Application of 2 mg/L and 5 mg/L Al was found to enhance the growth of two cultivars of *Betula pendula* originating from soils containing only low Al levels. Kidd and Proctor (2000) and Ghanati et al. (2005) reported about the positive impact of Al on the growth enhancement of tea plants. A possible reason of this stimulation was an enhancement of activities of antioxidant enzymes caused by Al, which finally increased membrane integrity and delayed lignification and ageing.

The results of the study of the effects of nanoscale alumina (Al_2O_3 NPs) on the growth and the enzymatic antioxidant system of wheat seedlings showed that root elongation was significantly improved at application of 50 and 1000 mg/L Al_2O_3 NPs, while at treatment with 200 and 500 mg/L, the lengths of roots were comparable with those of the control plants (Riahi-Madvar et al. 2012). Nano-Al_2O_3 stimulated the root growth of radish and rape and practically did not influence growth of cucumber roots while significantly retarded root growth of ryegrass and lettuce (Lin and Xing 2007). Significant stimulation of root length, enhancement in the number of fronds per colony as well as biomass accumulation accompanied with increased photosynthetic efficiency in *L. minor* plants due to treatment with 20 nm Al_2O_3 NPs were observed by Juhel et al. (2011). Based on stimulation of root growth of *A. thaliana* after treatment with Al_2O_3 NPs with 400 mg/L, 2000 mg/L and 4000 mg/L, Lee et al. (2010b) assumed that inert nano-Al_2O_3 could act similarly as nanosized perlite, which enhances gas transfer, prevents water loss and hinders soil compaction. An increase of the mean root length and mean biomass of tobacco (*Nicotiana tabacum*) plants exposed to 0.1%, 0.5% and 1% Al_2O_3 NPs was estimated as well; however, simultaneously it caused significant reduction of the leaf count of the seedlings. A high increase in expression of microRNAs (miR395, miR397, miR398 and miR399) during exposure to 1% Al_2O_3 NPs compared to the control suggested that these miRNAs could strengthen the capability of tobacco plants to withstand stress caused by Al_2O_3 NPs (Burklew et al. 2012).

9.5.6 Silicon-Based Nanomaterials

Silicon has been recognized as a beneficial nutrient for plant growth and development (Liang et al. 2007) due to its ability to activate some defence mechanisms and regulate some physiological processes related to defence mechanisms in plants (Ma 2004). Moreover, Si deposited on the walls of epidermis and vascular tissues plays an important role as a physicomechanical barrier (Ma and Yamaji 2008) and has a significant function in plant protection. It can reduce the transpiration rate of plants resulting in better tolerance to water stress during periods of low soil moisture (Marafon and Endres 2013). Beneficial effects of Si on grasses were manifested in enhanced yield and fitness, because it can considerably contribute to pathogen, drought and pest resistance, and these advantageous effects on plants are observed also when it is applied in a nanoform. As the predominant form of Si that is uptaken and transported within the plants, silicic acid was reported (Schaller et al. 2013). Thus, Si attenuates stresses connected with salt excess, drought and insufficient supply of nutrients or associated with climatic conditions, reduces toxicity of metals and metalloids and may also delay plant senescence processes. According to Savvas and Ntatsi (2015), attenuation of abiotic stresses in vascular plants by Si is connected with SiO_2 deposition inside the plant tissues. This not only contributes to mechanical strength and erectness to leaves but also affects the mobility of water/nutrient in plants and contributes to their enhanced antioxidant activity. Moreover, Si could form complexes or co-precipitate toxic metals not only in soil but also in plant tissues and can also modulate gene expression and signalling through phytohormones. In addition, Cui et al. (2015) prepared low surface area SiO_2 NPs with particle size ca. 480 nm and a specific surface area of 48 m^2/g using wheat straw ash with activated carbon and NPK compound fertilizer as by-products.

Haghighi and Pessarakli (2013) studied the beneficial effects of different concentrations of Si and nano-Si on growth and selected gas exchange characteristics of cherry tomatoes under salt stress condition and found that Si application increased the FWs and DWs, root volume and concentration of Chl and attenuated the negative effect of salinity on the plant FW, photosynthetic rate, concentration of Chl and leaf water content, and improvement in the photosynthesis rate, plant water use efficiency and mesophyll conductance under saline stress were observed as well. However, no significant difference was found between applications of nano-Si and Si. Soaking of sunflower (*H. annuus*) seeds in low concentration (0.2 and 0.4 mmol/L) SiO_2 NPs solutions significantly reduced mean germination time and improved root length, mean daily germination, seedling vigour index and final germination percentage (Janmohammadi and Sabaghnia 2015). After application of nano-Si (20 nm) under NaCl stress on tomato (*S. lycopersicum*) seeds, Almutairi (2016b) found that tomato seeds exhibited better germination percentage, and germination, the root length and FW of plants were increased. Moreover, the gene expression patterns associated with exposure to nano-Si indicated that these NPs could be potentially involved in the plant's response to stress and can positively affect plants' tolerance to salinity. Positive effects of nano-Si on germination, growth and yield of *Vicia faba* plants cultivated under salt stress conditions were also reported by Qados and Moftah (2015).

The foliar application of 2.5 mmol/L SiNPs alleviated Cd-induced stress in seedlings of *O. sativa* cv. Youyou 128 cultivated in solution supplemented with or without 20 μmol/L $CdCl_2$. SiNPs promoted growth and nutrition with essential metals (Fe, Zn, Mg) and enhanced Chl*a* content in rice plants under Cd stress while decreasing Cd accumulation and translocation of Cd from root to shoot and lowering malondialdehyde (MDA) levels, but higher glutathione content and different antioxidant enzyme activities were determined (Wang et al. 2015). SiNPs alleviated also Cr^{6+} phytotoxicity in pea plants and promoted plant growth, and the protection of pea plants against toxic Cr^{6+} ions by application of SiNPs was connected with reduced Cr accumulation resulting in reduced phytotoxicity and reduced oxidative stress, which was reflected in enhanced levels of antioxidants and nutrients (Tripathi et al. 2015). In grasses, the availability of Si had a considerable effect on Ca and micronutrient content and stoichiometry, what could significantly influence the nutrient cycling in grass-dominated ecosystems (Brackhage et al. 2013).

SiO_2 NPs enhanced silica bioavailability in maize seeds under hydroponic conditions, whereby an increase in germination percentage (95%), DW (6%) and silica accumulation (18%) in SiO_2 NPs-treated seeds compared to seeds treated with bulk SiO_2 was estimated (Suriyaprabha et al. 2012). Azimi et al. (2014) studied interaction of SiO_2 NPs with seed prechilling on germination and early seedling growth of tall wheatgrass (*Agropyron elongatum*) and found that treatment with SiO_2 NPs significantly increased seed germination (about 86% at 40 and 60 mg/L) and the application of lower SiO_2 NP concentrations up to 40 mg/L promoted plant weight by approximately 50% compared to untreated plants; however, concentrations of 60 mg/L or 80 mg/L already exhibited negative effects. Moreover, seed prechilling with simultaneous application of SiO_2 NPs mostly broke the seed dormancy of tall wheatgrass. Soaking of roots of 1-year-old Changbai larch (*Larix olgensis*) seedlings for 6 h at 500 μL/L of nanostructured SiO_2 resulted in considerable stimulation of seedling growth (mean height by 43%, root collar diameter by 31%, main root length by 14% and number of lateral roots by 32%) and improved seedling quality compared to the control plants (Lin et al. 2004). Beneficial effects of silica NPs on basil (*Ocimum basilicum*) cultivated under salinity stress were reflected in significantly increased leaf DW and FW and Chl content of SiO_2 NPs-treated plants (Kalteh et al. 2014). SiO_2 NPs (10–20 nm) applied at concentrations 10–40 mg/L were found to enhance germination, shoot and root growth as well as shoot and root DW of rice plants, while further increasing of SiO_2 NPs concentration led to a reduction of these positive effects (Adhikari et al. 2013). Sabaghnia and Janmohammadi (2014) investigated the effect of SiO_2 NPs on salinity tolerance in early growth of some lentil genotypes and found that treatment with SiO_2 NPs increased germination percentage, shoot and root length as well as FW and DW of seedlings that were under salinity stress.

SiO_2 NPs (20–40 nm)-treated soil reveals enhanced silica uptake and elongated roots, which make the plant to resist in stress conditions like drought (Suriyaprabha et al. 2013). The investigation of maize phytochemical responses to foliar application of high surface (360 m^2/g) SiO_2 NPs (20–40 nm) applied at 15 g/L showed improved expression of organic compounds and SiO_2 contents compared to the bulk and the control, and soil amendment was found to be better than foliar application (Suriyaprabha et al. 2014a). Moreover, the results suggest that plants treated with SiO_2 NPs showed a higher expression of phenolic compounds and a lower expression of stress-responsive enzymes against fungi (*Aspergillus* spp. and *Fusarium* spp.), where maize plants expressed enhanced resistance to *Aspergillus* spp. as compared to *Fusarium* spp. This indicates that silica NPs could be applied against phytopathogens as an effective antifungal agent (Suriyaprabha et al. 2014b). SiO_2 NPs synthesized using rice husk (50 nm) stimulated the percentage of seed germination (100%) in *Z. mays* more effectively than conventional Si sources and showed advantageous effect on beneficial bacterial population and increased the nutrient value of soil (Karunakaran et al. 2013). SiO_2 NPs (12 nm) had beneficial effects on the seed germination of tomato plants, and treatment with 8 g/L SiO_2 NPs significantly improved seed germination percentage, average germination time, seed germination index, seed vigour index and plant FW and DW (Siddiqui and Al-Whaibi 2014). Nanoscale SiO_2 application improved defence mechanisms of squash (*Cucurbita pepo* cv. white bush marrow) plants against toxicity induced by salt stress by augmenting the net photosynthetic rate, stomatal conductance, transpiration rate and water use efficiency, and total Chl and proline levels and carbonic anhydrase activity in the leaves of plants were enhanced as well. Reduced MDA and H_2O_2 levels and lower electrolyte leakage due to treatment with nanoscale SiO_2 are reflected in better seed germination and improved growth characteristics of plants, which indicates that nanosilica reduces oxidative damage by the expression of some antioxidant enzymes (e.g. catalase, SOD, peroxidase, glutathione reductase and ascorbate peroxidase) (Siddiqui et al. 2014).

Monodispersed mesoporous SiO_2 NPs of the size of 20 nm containing interconnected pores with a diameter of ca. 2.5 nm did not affect adversely seed germination. They were found to penetrate into the roots using symplastic and apoplastic pathways and then via xylem conducting tissues to the above-ground parts (stems, leaves) of plants, indicating that these NPs can serve as a new delivery tool for the transport of biomolecules of different sizes into plants (Sun et al. 2014). Yi et al. (2015) prepared functionalized mesoporous SiO_2 NPs with redox-responsive short-chain gatekeepers for agrochemical delivery. The researchers used novel decanethiol ($C_{10}H_{22}S$) gatekeeper system grafted onto mesoporous SiO_2 NPs for the study of controlled release of salicylic acid, in which $C_{10}H_{22}S$ was conjugated only to the external surfaces of the SiO_2 NPs through glutathione-cleavable disulphide linkages. The release of salicylic acid from such nanoformulation was connected with the induction of sustained expression of the plant defence gene PR-1 ongoing up to 7 days after introduction, while at application of free salicylic acid, the PR-1 expression steadily decreased after 3 days. Mesoporous SiO_2 NPs functionalized with amine cross-linked fluorescein isothiocyanate with particle size of 20 nm that contained a network of interconnected pores with a diameter of ca. 2 nm had no effect on seed germination in lupin and were not phytotoxic. In the *T. aestivum* and lupin plants cultivated hydroponically in nutrient solution containing these SiO_2 NPs, the

SiO_2 NPs were detected not only within cells and cell walls of the roots but also in vascular transport elements, xylem and other associated cells (Hussain et al. 2013).

Although the roots of *A. thaliana* were able to uptake relatively high amount of SiO_2 NPs with particle sizes of 14 nm, 50 nm and 200 nm, SiO_2 NPs were found to be not phytotoxic up to concentration 1000 ppm (Slomberg and Schoenfisch 2012). Lower concentrations (such as 50 and 100 mg/L) of SiO_2 NPs were found to have some positive effects on *T. aestivum* seedlings, while treatment with concentrations exceeding 200 mg/L was phytotoxic, which was reflected in the reduction of FW and DW of roots and shoots, in declined levels of Chls and carotenoids in leaves as well as in higher proline content and increased lipid peroxidation and catalase activity in leaves (Karimi and Mohsenzadeh 2016).

9.5.7 Selenium Nanoparticles

Selenium (Se) is a metalloid that did not belong to essential nutrients of higher plants, although it can exhibit some beneficial effects on plants. At low concentrations, plant growth-regulating, antioxidant or anti-senescent activity of Se was observed, and it was also found that Se can participate in modulating of abiotic stresses as well as in the defence against pathogens, while higher Se concentrations exhibit phytotoxic effects (Kaur et al. 2014). Treatment with low Se concentrations increased not only the production of *B. rapa* seeds by 43% but also total respiratory activity in leaves and flowers compared to control (Lyons et al. 2009). At low Se concentrations, its antioxidant activity was manifested in the inhibition of lipid peroxidation in ryegrass, while higher accumulation of lipid peroxidation products observed at treatment with higher Se concentrations was connected with its pro-oxidant activity (Hartikainen et al. 2000). The comprehensive review of El-Ramady et al. (2014) is focused on the use and biological effects of Se and nano-Se.

Beneficial effects of nano-Se on plants were investigated by Domokos-Szabolcsy et al. (2012) and Haghighi et al. (2014). In tobacco callus cultures, the application of nano-Se in the concentration range 265–530 μmol/L resulted in significant stimulation (about 40%) of the organogenesis and the growth of root system. On the other hand, stimulation was not observed with selenate treatment, and the application of higher selenate concentrations (265–530 μmol/L) led to total inhibition of both callus growth and root regeneration (Domokos-Szabolcsy et al. 2012). It is known that antioxidative mechanism participates in attenuation of abiotic stresses by Se. Haghighi et al. (2014) exposed some tomato (*Lycopersicum esculentum* Mill. cv. 'Halil') plants in hydroponic solution to Se and nano-Se for 3 days and then exposed the plants to high (40 °C) and/or to low temperature (10 °C) for 24 h, and afterwards, the cultivation continued (for 10 days) at optimal temperature (25/17 ± 2 °C day/night). Subsequently some growth characteristics of tomato plants were investigated. They found that the application of 1 μmol/L nano-Se increased the Chl content by 27% in plants exposed to low temperature stress and nano-Se reduced the detrimental effect caused by low or high temperature more effectively than Se. Also at treatments with 4, 8 and 12 μmol/L nano-Se, the Chl content was not changed markedly compared to this estimated in plants cultivated at optimal temperature. On the other hand, treatment with 2.5 μmol/L nano-Se significantly improved shoot and root FW and DW as well as root volume. Chitooligosaccharide SeNPs of average size about 100 nm were found to be suitable for application as plant nutrient regulator in rice planting, because their use in the booting stage increased rice production by 8–15% and total antioxidant capacity by 30–60% as well as Se content (three- to tenfold) (Tong et al. 2008). The use of SeNPs as nutritional agent suitable for cultivation of Se-rich Chinese cabbage was proposed by Bi et al. (2010), and Cheng and Cheng (2010) patented fertilizers containing nanosized Se suitable to promote the longevity of house flowering plants.

9.6 Conclusion

Nanotechnology and nanoscale science afford unambiguously a great potential in innovative and improved solutions as well as in troubleshooting. Nanosized materials change their physical and chemical properties in comparison with bulk materials and have helped to improve and innovate a variety of agricultural preparations used for improvement of growth and productivity of crops. As discussed above, applications of small quantities of various nanoscale materials have benefits for plant growth, while at utilization of higher quantities, adverse effects are manifested, which can result in a significant yield decrease. Thus, attention must be devoted also to the impact of risk factors associated with applications of nanomaterials on the environment and possible adverse/hazardous effects. It can be concluded that also in case of fertilizers, nutrients and plant growth-promoting compounds, the statement credited to Paracelsus: "The dose makes the poison" holds true, because their application at higher concentrations is usually phytotoxic.

References

Abdel-Aziz HMM, Hasaneen MNA, Omer AM (2016) Nano chitosan-NPK fertilizer enhances the growth and productivity of wheat plants grown in sandy soil. Span J Agric Res 14:e0902

Adhikari T, Kundu S, Biswas AK, Tarafdar JC, Rao AS (2012) Effect of copper oxide nano particle on seed germination of selected crops. J Agric Sci Technol A 2:815–823

Adhikari T, Kundu S, Rao AS (2013) Impact of SiO_2 and Mo nano particles on seed germination of rice (*Oryza sativa* L.). Int J Agric Food Sci Technol 4:809–816

Adhikari T, Kundu S, Rao AS (2016a) Zinc delivery to plants through seed coating with nano-zinc oxide particles. J Plant Nutr 39:139–149

Adhikari T, Sarkar D, Mashayekhi H, Xing BS (2016b) Growth and enzymatic activity of maize (*Zea mays* L.) plant: solution culture test for copper dioxide nano particles. J Plant Nutr 39:102–118

Alidoust D, Isoda A (2013) Effect of gamma Fe_2O_3 nanoparticles on photosynthetic characteristic of soybean (*Glycine max* (L.) Merr.): foliar spray versus soil amendment. Acta Physiol Plant 35:3365–3375

Alipour ZT (2016) The effect of phosphorus and sulfur nanofertilizers on the growth and nutrition of *Ocimum basilicum* in response to salt stress. J Chem Health Risks 6:125–131

Almeelbi T, Bezbaruah A (2014) Nanoparticle-sorbed phosphate: iron and phosphate bioavailability studies with *Spinacia oleracea* and *Selenastrum capricornutum*. ACS Sustain Chem Eng 2:1625–1632

Almutairi ZM (2016a) Influence of silver nano-particles on the salt resistance of tomato (*Solanum lycopersicum*) during germination. Int J Agric Biol 18:449–457

Almutairi ZM (2016b) Effect of nano-silicon application on the expression of salt tolerance genes in germinating tomato (*Solanum lycopersicum* L.) seedlings under salt stress. Plant Omics J 9:106–114

Antisari LV, Carbone S, Gatti A, Vianello G, Nannipieri P (2015) Uptake and translocation of metals and nutrients in tomato grown in soil polluted with metal oxide (CeO_2, Fe_3O_4, SnO_2, TiO_2) or metallic (Ag, Co, Ni) engineered nanoparticles. Environ Sci Pollut Res 22:1841–1853

Arora S, Sharma P, Kumar S, Nayan R, Khanna PK, Zaidi MGH (2012) Gold-nanoparticle induced enhancement in growth and seed yield of *Brassica juncea*. Plant Growth Regul 66:303–310

Auffan M, Rose J, Bottero JY, Lowry GV, Jolivet JP, Wiesner MR (2009) Towards a definition of inorganic nanoparticles from an environmental, health and safety perspective. Nat Nanotechnol 4:634–641

Aydinalp C, Marinova S (2009) The effects of heavy metals on seed germination and plant growth on alfalfa plant (*Medicago sativa*). Bulg J Agric Sci 15:347–350

Azimi R, Borzelabad MJ, Feizi H, Azimi A (2014) Interaction of SiO_2 nanoparticles with seed prechilling on germination and early seedling growth of tall wheatgrass (*Agropyron elongatum* L.). Pol J Chem Technol 16:25–29

Aziz N, Faraz M, Pandey R, Sakir M, Fatma T, Varma A, Barman I, Prasad R (2015) Facile algae-derived route to biogenic silver nanoparticles: synthesis, antibacterial and photocatalytic properties. Langmuir 31:11605–11612

Aziz N, Pandey R, Barman I, Prasad R (2016) Leveraging the attributes of *Mucor hiemalis*-derived silver nanoparticles for a synergistic broad-spectrum antimicrobial platform. Front Microbiol 7:1984. doi:10.3389/fmicb.2016.01984

Basak BB, Pal S, Datta SC (2012) Use of modified clays for retention and supply of water and nutrients. Curr Sci 102:1272–1278

Bashir K, Takahashi R, Nakanishi H, Nishizawa NK (2013) The road to micronutrient biofortification of rice: progress and prospects. Front Plant Sci 4:15–17

Bashmakov DI, Lukatkin AS, Revin VV, Duchovskis P, Brazaitytė A, Baranauskis K (2005) Growth of maize seedlings affected by different concentrations of heavy metals. Ekologija 3:22–27

Baskar V, Venkatesh J, Park SW (2015) Impact of biologically synthesized silver nanoparticles on the growth and physiological responses in *Brassica rapa ssp pekinensis*. Environ Sci Pollut Res 22:17672–17682

Becana M, Moran JF, Iturbe-Ormaetxe I (1998) Iron dependent oxygen free radical generation in plants subjected to environmental stresses: toxicity and antioxidants. Plant Soil 201:137–147

Begum P, Fugetsu B (2012) Phytotoxicity of multi-walled carbon nanotubes on red spinach (*Amaranthus tricolor* L) and the role of ascorbic acid as an antioxidant. J Hazard Mater 243:212–222

Begum P, Ikhtiari R, Fugetsu B, Matsuoka M, Akasaka T, Watari F (2012) Phytotoxicity of multi-walled carbon nanotubes assessed by selected plant species in the seedling stage. Appl Surf Sci 262:120–124

Berahmand AA, Panahi AG, Sahabi H, Feizi H, Moghaddam PR, Shahtahmassebi N, Fotovat A, Karimpour H, Gallehgir O (2012) Effects silver nanoparticles and magnetic field on growth of fodder maize (*Zea mays* L.) Biol Trace Elem Res 149:419–424

Berber MR, Hafez IH, Minagawa K, Mori T (2014) A sustained controlled release formulation of soil nitrogen based on nitrate-layered double hydroxide nanoparticle material. J Soils Sediments 14:60–66

Berkner S, Schwirn K, Voelker D (2016) Nanopharmaceuticals: tiny challenges for the environmental risk assessment of pharmaceuticals. Environ Toxicol Chem 35:780–787

Bhardwaj D, Sharma P, Sharma M, Tomar R (2014) Removal and slow release studies of phosphate on surfactant loaded hydrothermally synthesized silicate nanoparticles. J Taiwan Inst Chem E 4:2649–2658

Bi D, Liu S, Liu Y, Yin X (2010) Preparation of selenium-rich Chinese cabbage using selenium nanoparticle containing nutrient. CN Patent 101734971. 16 June 2010

Biederman LA, Harpole WS (2013) Biochar and its effects on plant productivity and nutrient cycling: a meta-analysis. GCB Bioenergy 5:202–214

Binder BM, Rodriguez FI, Bleecker AB, Patterson SE (2007) The effects of group 11 transition metals, including gold, on ethylene binding to the ETR1 receptor and growth of *Arabidopsis thaliana*. FEBS Lett 581:5105–5109

Bindraban PS, Dimkpa C, Nagarajan L, Roy A, Rudy Rabbinge R (2015) Revisiting fertilisers and fertilisation strategies for improved nutrient uptake by plants. Biol Fertil Soils 51:897–911

Borges R, Brunatto SF, Leitao AA, de Carvalho GSG, Wypych F (2015) Solid-state mechanochemical activation of clay minerals and soluble phosphate mixtures to obtain slow-release fertilizers. Clay Miner 50:153–162

Borm PJA, Robbins D, Haubold S, Kuhlbusch T, Fissan H, Donaldson K, Schins R, Stone V, Kreyling W, Lademann J, Krutmann J, Warheit D, Oberdoster E (2006) The potential risks of nanomaterials: a review carried out for ECETOC. Part. Fibre Toxicol 3:11

Bortolin A, Aouada FA, Mattoso LHC, Ribeiro C (2013) Nanocomposite PAAm/methyl cellulose/montmorillonite hydrogel: evidence of synergistic effects for the slow release of fertilizers. J Agric Food Chem 61:7431–7439

Brackhage C, Schaller J, Bäucker E, Dudel EG (2013) Silicon availability affects the stoichiometry and content of calcium and micro nutrients in the leaves of common reed. Silicon 5:199–204

Brayner R, Fiévet F, Coradin T (2013) Nanomaterials: a danger or a promise? A chemical and biological perspective. Springer, London

Burke DJ, Zhu S, Pablico-Lansigan MP, Hewins CR, Samia ACS (2014) Titanium oxide nanoparticle effects on composition of soil microbial communities and plant performance. Biol Fertil Soils 50:1169–1173

Burklew CE, Ashlock J, Winfrey WB, Zhang B (2012) Effects of aluminum oxide nanoparticles on the growth, development, and microRNA expression of tobacco (*Nicotiana tabacum*). PLoS One 7:e34783

Burman U, Saini M, Kumar P (2013) Effect of zinc oxide nanoparticles on growth and antioxidant system of chickpea seedlings. Toxicol Environ Chem 95:605–612

Buzea C, Pacheco I, Robbie K (2007) Nanomaterials and nanoparticles: sources and toxicity. Biointerphases 2:MR17–MR71

Canas JE, Long MQ, Nations S, Vadan R, Dai L, Luo MX, Ambikapathi R, Lee EH, Olszyk D (2008) Effects of functionalized and nonfunctionalized single-walled carbon nanotubes on root elongation of select crop species. Environ Toxicol Chem 27:1922–1931

Carvajal MF, Martínez-Sánchez F, Alcaraz CF (1994) Effect of Ti on some physiological activity indicators of *Capsicum annuum* L. plants. J Hortic Sci 69:427–432

Černíková A, Opatřilová R, Jampílek J (2014) Rapid informative screening of nano-alaptide as potential transdermal permeation enhancer of acetylsalicylic acid and paracetamol. Mil Med Sci Lett 83:34–39

Černíková A, Dohnal J, Jampílek J (2015) Permeation of indomethacin through skin using nanonized alaptide. ADMET DMPK 3:344–350

Chakravarty D, Erande MB, Late DJ (2015) Graphene quantum dots as enhanced plant growth regulators: effects on coriander and garlic plant. J Sci Food Agric 95:2772–2778

Charitidis CA, Georgiou P, Koklioti MA, Trompeta AF, Markakis V (2014) Manufacturing nanomaterials: from research to industry. Manuf Rev 1:11

Chaudhry Q, Castle L (2011) Food applications of nanotechnologies: an overview of opportunities and challenges for developing countries. Trends Food Sci Technol 22:595–603

Chellaram C, Murugaboopathi G, John AA, Sivakumar R, Ganesan S, Krithika S, Priya G (2014) Significance of nanotechnology in food industry. APCBEE Procedia 8:109–113

Chen H, Seiber JN, Hotze M (2014) ACS select on nanotechnology in food and agriculture: a perspective on implications and applications. J Agric Food Chem 62:1209–1212

Chen GS, Qiu JL, Liu Y, Jiang RF, Cai SY, Liu Y, Zhu F, Zeng F, Luan TG, Ouyang GF (2015) Carbon nanotubes act as contaminant carriers and translocate within plants. Sci Rep 5:15682

Cheng G, Cheng J (2010) Nanosize selenium-rich compound fertilizer for promoting longevity of house flowering plants. CN Patent 101851136. 6 Oct 2010

Clement L, Hurel C, Marmier N (2013) Toxicity of TiO_2 nanoparticles to cladocerans, algae, rotifers and plants – effects of size and crystalline structure. Chemosphere 90:1083–1090

Coles D, Frewer LJ (2013) Nanotechnology applied to European food production: a review of ethical and regulatory issues. Trends Food Sci Technol 34:32–43

Corradini E, de Moura MR, Mattoso LHC (2010) A preliminary study of the incorporation of NPK fertilizer into chitosan nanoparticles. Express Polym Lett 4:509–515

Corral-Diaz B, Peralta-Videa JR, Alvarez-Parrilla E, Rodrigo-García J, Morales MI, Osuna-Avila P, Niu G, Hernandez-Viezcas JA, Gardea-Torresdey JL (2014) Cerium oxide nanoparticles alter the antioxidant capacity but do not impact tuber ionome in *Raphanus sativus* (L). Plant Physiol Biochem 84:277–285

Cui D, Zhang P, Ma YH, He X, Li YY, Zhao YC, Zhang ZY (2014) Phytotoxicity of silver nanoparticles to cucumber (*Cucumis sativus*) and wheat (*Triticum aestivum*). J Zhejiang Univ Sci A 15:662–670

Cui JL, Cui LY, Cheng FP, Liu LJ, Sun HL, Li S, Wen ZS, Sun JC (2015) A green route for preparation of low surface area SiO_2 microspheres from wheat straw ash with activated carbon and NPK compound fertilizer as by-products. RSC Adv 5:80238–80244

Czerpak R, Bajguz A, Chodkowski K, Popow H (1994) Influence of nickel and cobalt on the growth and biochemical changes of *Chlorella pyrenoidosa* (Chlorophyceae). Pol Arch Hydrobiol 41:161–169

De Jong WH, Borm PJ (2008) Drug delivery and nanoparticles: applications and hazards. Int J Nanomedicine 3:133–149

Dehkourdi EH, Mosavi M (2013) Effect of anatase nanoparticles (TiO_2) on parsley seed germination (*Petroselinum crispum*) *in vitro*. Biol Trace Elem Res 155:283–286

Delfani M, Firouzabadi MB, Farrokhi N, Makarian H (2014) Some physiological responses of black-eyed pea to iron and magnesium nanofertilizers. Commun Soil Sci Plant Anal 45:530–540

Dhoke SK, Mahajan P, Kamble R, Khanna A (2013) Effect of nanoparticles suspension on the growth of mung (*Vigna radiata*) seedlings by foliar spray method. Nanotechnol Dev 3:e1

Diatloff E, Smith FW, Asher CJ (1995a) Rare-earth elements and plant growth. 2. Responses of corn and mungbean to low concentrations of lanthanum in dilute, continuously flowing nutrient solutions. J Plant Nutr 18:1977–1989

Diatloff E, Smith FW, Asher CJ (1995b) Rare-earth elements and plant growth. 3. Responses of corn and mungbean to low concentrations of cerium in dilute, continuously flowing nutrient solutions. J Plant Nutr 18:1991–2003

Dolez PI (2015) Nanoengineering: global approaches to health and safety issues. Elsevier, Amsterdam

Domokos-Szabolcsy E, Marton L, Sztrik A, Babka B, Prokisch J, Fari M (2012) Accumulation of red elemental selenium nanoparticles and their biological effects in *Nicotinia tabacum*. Plant Growth Regul 68:525–531

Ekinci M, Dursun A, Yildirim E, Parlakova F (2014) Effects of nanotechnology liquid fertilizers on the plant growth and yield of cucumber (*Cucumis sativus* L.) Acta Sci Pol Hortorum Cultus 13:135–141

El-Batal AI, Gharib FA, Ghazi SM, Hegazi AZ, AbdEl Hafz AGM (2016) Physiological responses of two varieties of common bean (*Phaseolus vulgaris* L.) to foliar application of silver nanoparticles. Nanomater Nanotechnol 6:13

Elfeky SA, Mohammed MA, Khater MS, Osman YAH, Elsherbini E (2013) Effect of magnetite nano-fertilizer on growth and yield of *Ocimum basilicum* L. Int J Indig Med Plants 46:1286–1293

El-Kereti MA, El-feky SA, Khater MS, Osman YA, El-Sherbini EA (2013) ZnO nanofertilizer and He Ne laser irradiation for promoting growth and yield of sweet basil plant. Recent Pat Food Nutr Agric 5:169–181

El-Ramady H, Domokos-Szabolcsy E, Abdalla NA, Alshaal TA, Shalaby TA, Sztrik A, Prokisch J, Fari M (2014) Selenium and nano-selenium in agroecosystems. Environ Chem Lett 12:495–510

El-Sheekh MM, El-Naggar AH, Osman MEH, El-Mazaly E (2003) Effects of Co^{2+} in the algae *Monoraphidium minutum* and *Nitzschia perminuta*. Braz J Plant Physiol 15:159–166

European Commission (2008) Definition of a nanomaterial. http://ec.europa.eu/environment/chemicals/nanotech/faq/definition_en.htm. Accessed 30 May 2016

Fan LL, Wang YH, Shao XW, Geng YQ, Wang ZC, Ma Y, Liu J (2012) Effects of combined nitrogen fertilizer and nano-carbon application on yield and nitrogen use of rice grown on saline-alkali soil. J Food Agric Environ 10:558–562

Feizi H, Moghaddam PR, Shahtahmassebi N, Fotovat A (2012) Impact of bulk and nanosized titanium dioxide (TiO_2) on wheat seed germination and seedling growth. Biol Trace Elem Res 146:101–106

Feizi H, Kamali M, Jafari L, Rezvani Moghaddam P (2013) Phytotoxicity and stimulatory impacts of nanosized and bulk titanium dioxide on fennel (*Foeniculum vulgare* Mill). Chemosphere 91:506–511

Foy CD (1984) Physiological effects of hydrogen, aluminum, and manganese toxicities in acid soil. In: Adams F (ed) Soil acidity and liming, 2nd edn. American Society of Agronomy, Madison, pp 57–97

Franci GL, Falanga AR, Galdiero S, Palomba L, Rai M, Morelli G, Galdiero M (2015) Silver nanoparticles as potential antibacterial agents. Molecules 20:8856–8874

Fröhlich E (2013) Cellular targets and mechanisms in the cytotoxic action of non-biodegradable engineered nanoparticles. Curr Drug Metab 14:976–988

Gao FQ, Liu C, Qu CX, Zheng L, Yang F, Su MG, Hong FH (2008) Was improvement of spinach growth by nano-TiO_2 treatment related to the changes of Rubisco activase? Biometals 21:211–221

Garcia M, Forbe T, Gonzalez E (2010) Potential applications of nanotechnology in the agro-food sector. Food Sci Technol 30:573–581

Ghafariyan MH, Malakouti MJ, Dadpour MR, Stroeve P, Mahmoudi M (2013) Effects of magnetite nanoparticles on soybean chlorophyll. Environ Sci Technol 47:10645–10652

Ghanati F, Morita A, Yokota H (2005) Effects of aluminum on the growth of tea plant and activation of antioxidant system. Plant Soil 276:133–141

Ghodake G, Seo YD, Park DH, Lee DS (2010) J Nanoelectron Optoelectron 5:157–160

Ghorbanpour M, Hadian J (2015) Multi-walled carbon nanotubes stimulate callus induction, secondary metabolites biosynthesis and antioxidant capacity in medicinal plant *Satureja khuzestanica* grown *in vitro*. Carbon 94:749–759

Ghormade V, Deshpande MV, Paknikar KM (2011) Perspectives for nano-biotechnology enabled protection and nutrition of plants. Biotechnol Adv 29:792–803

Giroto AS, Fidélis SC, Ribeiro C (2015) Controlled release from hydroxyapatite nanoparticles incorporated into biodegradable, soluble host matrixes. RSC Adv 5:104179–104186

Gomez-Garay A, Pintos B, Antonio MJ, Lobo C, Villalobos N, Martin L (2014) Uptake of CeO_2 nanoparticles and its effect on growth of *Medicago arborea in vitro* plantlets. Biol Trace Elem Res 161:143–150

Guan J, Liu GL, Cai K, Gao CZ, Liu RT (2015) Probing the interactions between carboxylated multi-walled carbon nanotubes and copper-zinc superoxide dismutase at a molecular level. Luminiscence 30:693–698

Guerinot ML, Yi Y (1994) Iron: nutritious, noxious, and not readily available. Plant Physiol 104:815–820

Gui X, Deng Y, Rui Y, Gao B, Luo W, Chen S, Nhan le V, Li X, Liu S, Han Y, Liu L, Xing B (2015) Response difference of transgenic and conventional rice (*Oryza sativa*) to nanoparticles (gamma Fe_2O_3). Environ Sci Pollut Res 22:17716–17723

Gusev AA, Kudrinsky AA, Zakharova OV, Klimov AI, Zherebin PM, Lisichkin GV, Vasyukova IA, Denisov AN, Krutyakov YA (2016) Versatile synthesis of PHMB-stabilized silver nanoparticles and their significant stimulating effect on fodder beet (*Beta vulgaris* L.) Mater Sci Eng C Mater Biol Appl 62:152–159

Hafeez A, Razzaq A, Mahmood T, Jhanzab HM (2015) Potential of copper nanoparticles to increase growth and yield of wheat. J Nanosci Adv Technol 1:6–11

Haghighi M, Pessarakli M (2013) Influence of silicon and nano-silicon on salinity tolerance of cherry tomatoes (*Solanum lycopersicum* L.) at early growth stage. Sci Hortic 161:111–117

Haghighi M, Heidarian S, da Silva JAT (2012) The effect of titanium amendment in *N*-withholding nutrient solution on physiological and photosynthesis attributes and micronutrient uptake of tomato. Biol Trace Elem Res 150:381–390

Haghighi M, Abolghasemi R, da Silva JAT (2014) Low and high temperature stress affect the growth characteristics of tomato in hydroponic culture with Se and nano-Se amendment. Sci Hortic 178:231–240

Hanif HU, Arshad M, Ali MA, Ahmed N, Qazi IA (2015) Phyto-availability of phosphorus to *Lactuca sativa* in response to soil applied TiO_2 nanoparticles. Pak J Agric Sci 52:177–182

Hartikainen H, Xue T, Piironen V (2000) Selenium as an antioxidant and pro-oxidant in ryegrass. Plant Soil 225:193–200

Hassan F, Shahram A, Farzin A, Saeed JP (2013) Comparative effects of nanosized and bulk titanium dioxide concentrations on medicinal plant *Salvia officinalis* L. Ann Rev Res Biol 3:814–824

Hatami M, Ghorbanpour M (2013) Effect of nanosilver on physiological performance of pelargonium plants exposed to dark storage. J Hortic Res 21:15–20

Hazeem LJ, Waheed FA, Rashdan S, Bououdina M, Brunet L, Slomianny C, Boukherroub R, Elmeselmani WA (2015) Effect of magnetic iron oxide (Fe_3O_4) nanoparticles on the growth and photosynthetic pigment content of *Picochlorum sp.* Environ Sci Pollut Res 22:11728–11739

Hendry GAF, Brocklebank KJ (1985) Iron-induced oxygen radical metabolism in waterlogged plants. New Phytol 101:199–206

Hong FH, Zhou J, Liu C, Yang F, Wu C, Zheng L, Yang P (2005) Effect of nano-TiO_2 on photochemical reaction of chloroplasts of spinach. Biol Trace Elem Res 105:269–279

Hosseini HR, Chehrazi M, Dehkourdi EH, Hosseini M (2013) Application of anatase nanoparticles (TiO_2) on strawberry seed germination (*Fragaria ananassa* L.). Adv Hortic Sci 27:143–146

Hrubý M, Cígler P, Kuzel S (2002) Titanium in plant nutrition: the contribution to understanding the mechanism of titanium action in plants. J Plant Nutr 25:577–598

Husen A, Siddiqi KS (2014) Carbon and fullerene nanomaterials in plant system. J Nanobiotechnol 12:16

Hussain HI, Yi ZF, Rookes JE, Kong LXX, Cahill DM (2013) Mesoporous silica nanoparticles as a biomolecule delivery vehicle in plants. J Nanopart Res 15:1676

Hussien MM, El-Ashry SM, Haggag WM, Mubarak DM (2015) Response of mineral status to nano-fertilizer and moisture stress during different growth stages of cotton plants. Int J ChemTech Res 8:643–650

Iijima S, Yudasaka M, Yamada R, Bandow S, Suenaga K, Kokai F, Takahashi K (1999) Nano-aggregates of single-walled graphitic carbon nano-horns. Chem Phys Lett 309:165–170

Jain V, Nainawatee HS (2000) Cobalt reduces nitrate inhibition of nodulation in mungbean (*Vigna radiata*). Biol Fertil Soils 31:522–524

Jampílek J, Kráľová K (2015) Application of nanotechnology in agriculture and food industry, its prospects and risks. Ecol Chem Eng 22:321–361

Jampílek J, Kráľová K (2017a) Nano-antimicrobials: activity, benefits and weaknesses. In: Grumezescu AM (ed) Nanostructures in therapeutic medicine, Nanostructures for antimicrobial therapy, vol 2. Elsevier, London. Chapter 2, in press

Jampílek J, Kráľová K (2017b) Application of nanobioformulations for controlled release and targeted biodistribution of drugs. In: Keservani RK (ed) Recent advances and applications of nanobiomaterials. CRC Press, Boca Raton. Chapter 5, in press

Jampílek J, Kráľová K (2017c) Impact of nanoparticles on living organisms and human health. In: Nalwa HS (ed) Encyclopedia of nanoscience and nanotechnology. American Scientific Publishers, Valencia. in press

Jampílek J, Kráľová K (2017d) Nanopesticides: preparation, targeting and controlled release. In: Grumezescu AM (ed) Nanotechnology in food industry, New Pesticides and soil sensors, vol 10. Elsevier, London. Chapter 4, pp 81–127

Jampílek J, Opatřilová R, Coufalová L, Černíková A, Dohnal J (2013) Utilization of alaptide as transdermal penetration modifier in pharmaceutical compositions for human and veterinary applications containing anti-inflammatory drugs and/or antimicrobial chemotherapeutics. WO2013020527 A1. 14 Feb 2013

Jampílek J, Opatřilová R, Řezáčová A, Oktábec Z, Dohnal J (2014) Alaptide: methods of effecting its solubility, membrane permeation and pharmaceutical compositions for human and/or veterinary applications. WO2014019556 A1. 6 Feb 2014

Jampílek J, Záruba K, Oravec M, Kuneš M, Babula P, Ulbrich P, Brezaniová I, Opatřilová R, Tříska J, Suchý P (2015) Preparation of silica nanoparticles loaded with nootropics and their *in vivo* permeation through blood–brain barrier. Biomed Res Int 2015:812673

Janmohammadi M, Sabaghnia N (2015) Effect of pre-sowing seed treatments with silicon nanoparticles on germinability of sunflower (*Helianthus annuus*). Bot Lith 21:13–21

Janrao K, Gadhave MV, Banerjee SK, Gaikwad DD (2014) Nanoparticle induced nanotoxicity: an overview. Asian J Biomed Pharm Sci 4:1–7

Jeyasubramanian K, Thoppey UUG, Hikku GS, Selvakumar N, Subramania A, Krishnamoorthy K (2016) Enhancement in growth rate and productivity of spinach grown in hydroponics with iron oxide nanoparticles. RSC Adv 6:15451–15459

Jiang W, Liu D, Liu X (2001) Effects of copper on root growth, cell division, and nucleolus of *Zea mays*. Biol Plant 44:105–109

Jiao J, Li CZ, Huang GB (2006a) Protective effects and their mechanisms of cobalt on soybean seedling's leaf under drought stress. Chin J Appl Ecol 17:796–800

Jiao J, Li CZ, Huang GB (2006b) The protective effects of ethylene production inhibitors on *Vicia faba* seedling leaves under heat stress. J Plant Ecol 30:465–471

Joseph S, Anawar HM, Storer P, Blackwell P, Chia C, Lin Y, Munroe P, Donne S, Horvat J, Wang JL, Solaiman ZM (2015) Effects of enriched biochars containing magnetic iron nanoparticles on mycorrhizal colonisation, plant growth, nutrient uptake and soil quality improvement. Pedosphere 25:749–760

Juhel G, Batisse E, Hugues Q, Daly D, van Pelt FN, O'Halloran J, Jansen MA (2011) Alumina nanoparticles enhance growth of *Lemna minor*. Aquat Toxicol 105:328–336

Kalteh M, Alipour ZT, Ashraf S, Aliabadi MM, Nosratabadi AF (2014) Effect of silica nanoparticles on basil (*Ocimum basilicum)* under salinity stress. J Chem Health Risks 4:49–55

Kalyanaraman SB, Sivagurunathan P (1993) Effect of cadmium, copper, and zinc on the growth of blackgram. J Plant Nutr 16:2029–2042

Kampeerapappun P, Phanomkate N (2013) Slow release fertilizer from core-shell electrospun fibers. Chiang Mai J Sci 40:775–782

Karimi J, Mohsenzadeh S (2016) Effects of silicon oxide nanoparticles on growth and physiology of wheat seedlings. Russ J Plant Physiol 63:119–123

Karimi N, Minaei S, Almassi M, Shahverdi AR (2012) Application of silver nano-particles for protection of seeds in different soils. Afr J Agric Res 7:1863–1869

Karunakaran G, Suriyaprabha R, Manivasakan P, Yuvakkumar R, Rajendran V, Prabu P, Kannan N (2013) Effect of nanosilica and silicon sources on plant growth promoting rhizobacteria, soil nutrients and maize seed germination. IET Nanobiotechnol 7:70–77

Kasuya D, Yudasaka M, Takahashi K, Kokai F, Iijima S (2002) Selective production of single-wall carbon nanohorn aggregates and their formation mechanism. J Phys Chem B 106:4947–4951

Kaur N, Sharma S, Kaur S, Nayyar H (2014) Selenium in agriculture: a nutrient or contaminant for crops? Arch Agron Soil Sci 60:1593–1624

Kaveh R, Li YS, Ranjbar S, Tehrani R, Brueck CL, Van Aken B (2013) Changes in *Arabidopsis thaliana* gene expression in response to silver nanoparticles and silver ions. Environ Sci Technol 47:10637–10644

Khodakovskaya MV, de Silva K, Biris AS, Dervishi E, Villagarcia H (2012) Carbon nanotubes induce growth enhancement of tobacco cells. ACS Nano 6:2128–2135

Khodakovskaya MV, Kim BS, Kim JN, Alimohammadi M, Dervishi E, Mustafa T, Cernigla CE (2013) Carbon nanotubes as plant growth regulators: effects on tomato growth, reproductive system, and soil microbial community. Small 9:115–123

Kidd PS, Proctor J (2000) Effects of aluminium on the growth and mineral composition of *Betula pendula* Roth. J Exp Bot 51:1057–1066

Kim HS, Kang HS, Chu GJ, Byun HS (2008) Antifungal effectiveness of nanosilver colloid against rose powdery mildew in greenhouses. Solid State Phenom 135:15–18

Kim JH, Oh Y, Yoon H, Hwang I, Chang YS (2015) Iron nanoparticle-induced activation of plasma membrane H^+-ATPase promotes stomatal opening in *Arabidopsis thaliana*. Environ Sci Technol 49:1113–1119

Kinraide TB (1993) Aluminium enhancement of plant growth in acid rooting media. A case of reciprocal alleviation of toxicity by two toxic cations. Physiol Plant 88:619–625

Kisan B, Shruthi NK, Sharanagoud AH, Revanappa SB, Ramachandra R, Hosmani AC, Bheemann AN, Pramod (2015) Effect of nano zinc oxide on the leaf physical and nutritional quality of spinach. Bioscan Int Quart J Life Sci 10:1437–1439

Kole C, Kole P, Randunu KM, Choudhary P, Podila R, Ke PC, Rao AM, Marcus RK (2013) Nanobiotechnology can boost crop production and quality: first evidence from increased plant biomass, fruit yield and phytomedicine content in bitter melon (*Momordica charantia*). BMC Biotechnol 13:37

Konotop YO, Kovalenko MS, Ulynets VZ, Meleshko AO, Batsmanova LM, Taran NY (2014) Phytotoxicity of colloidal solutions of metal-containing nanoparticles. Cytol Genet 48:99–102

Kottegoda N, Munaweera I, Madusanka N, Karunaratne V (2011) A green slow-release fertilizer composition based on urea-modified hydroxyapatite nanoparticles encapsulated wood. Curr Sci 101:73–78

Koval'skii VV, Grinkevich NI, Gribovskaya IF, Dinevich LS, Shandova AN (1971) Cobalt in medicinal plants and its effect on the accumulation of biologically active compounds. Rastit Resour 7:503–510

Kráľová K, Šeršeň F, Blahová M (1994) Effects of Cu(II) complexes on photosynthesis in spinach chloroplasts: aqua(aryloxyacetato)copper(II) complexes. Gen Physiol Biophys 13:483–491

Kráľová K, Masarovičová E, Šeršeň F, Ondrejkovičová I (2008) Effect of different Fe(III) compounds on photosynthetic electron transport in spinach chloroplasts and on iron accumulation in maize plants. Chem Pap 62:358–363

Kumar V, Parvatam G, Ravishankar GA (2009) $AgNO_3$- a potential regulator of ethylene activity and plant growth modulator. Electron J Biotechnol 12:1

Kumar V, Kumari A, Guleria P, Yadav SK (2012) Evaluating the toxicity of selected types of nanochemicals. Rev Environ Contam Toxicol 215:39–121

Kumar V, Guleria P, Kumar V, Yadav SK (2013) Gold nanoparticle exposure induces growth and yield enhancement in *Arabidopsis thaliana*. Sci Total Environ 461:462–468

Küpper H, Küpper F, Spiller M (1996) Environmental relevance of heavy metal-substituted chlorophylls using the example of water plants. J Exp Bot 295:259–266

Kuzma J, VerHage P (2006) Nanotechnology in agriculture and food production: anticipated applications. Washington, DC. http://www.nanotechproject.org/process/assets/files/2706/94_pen4_agfood.pdf. Accessed 14 June 2016

Lahiani MH, Dervishi E, Chen JH, Nima Z, Gaume A, Biris AS, Khodakovskaya MV (2013) Impact of carbon nanotube exposure to seeds of valuable crops. ACS Appl Mater Interfaces 5:7965–7973

Lahiani MH, Chen JH, Irin F, Puretzky AA, Green MJ, Khodakovskaya MV (2015) Interaction of carbon nanohorns with plants: uptake and biological effects. Carbon 81:607–619

Larue C, Laurette J, Herlin-Boime N, Khodja H, Fayard B, Flank AM, Brisset F, Carriere M (2012a) Accumulation, translocation and impact of TiO_2 nanoparticles in wheat (*Triticum aestivum spp.*): influence of diameter and crystal phase. Sci Total Environ 431:197–208
Larue C, Veronesi G, Flank AM, Surble S, Herlin-Boime N, Carriere M (2012b) Comparative uptake and impact of TiO_2 nanoparticles in wheat and rapeseed. J Toxicol Environ Health A 75:722–734
Lee J, Mahendra S, Alvarez PJJ (2010a) Nanomaterials in the construction industry: a review of their applications and environmental health and safety considerations. ACS Nano 4:3580–3590
Lee CW, Mahendra S, Zodrow K, Li D, Tsai YC, Braam J, Alvarez PJ (2010b) Developmental phytotoxicity of metal oxide nanoparticles to *Arabidopsis thaliana*. Environ Toxicol Chem 29:669–675
Lehmann J, Rillig MC, Thies J, Masiello CA, Hockaday WC, Crowley D (2011) Biochar effects on soil biota – a review. Soil Biol Biochem 43:1812–1836
Lei Z, Su MY, Chao L, Liang C, Hao H, Xiao W, Liu XQ, Fan Y, Gao FQ, Hong FS (2007) Effects of nanoanatase TiO_2 on photosynthesis of spinach chloroplasts under different light illumination. Biol Trace Elem Res 119:68–76
Li JL, Chang PR, Huang J, Wang YQ, Yuan H, Ren HX (2013a) Physiological effects of magnetic iron oxide nanoparticles towards watermelon. J Nanosci Nanotechnol 13:5561–5567
Li L, Sillanpaa M, Tuominen M, Lounatmaa K, Schultz E (2013b) Behavior of titanium dioxide nanoparticles in *Lemna minor* growth test conditions. Ecotoxicol Environ Saf 88:89–94
Li X, Yang YC, Gao B, Zhang M (2015a) Nanoparticles at low concentrations. PLoS One 10:e0122884
Li J, Naeem MS, Wang X, Liu L, Chen C, Ma N, Zhang C (2015b) Nano-TiO_2 is not phytotoxic as revealed by the oilseed rape growth and photosynthetic apparatus ultra-structural response. PLoS One 10:e0143885
Liang Y, Sun W, Zhu YG, Christie P (2007) Mechanisms of silicon-mediated alleviation of abiotic stresses in higher plants: a review. Environ Pollut 147:422–428
Libralato G, Devoti AC, Zanella M, Sabbioni E, Micetic I, Manodori L, Pigozzo A, Manenti S, Groppi F, Ghirardini AV (2016) Phytotoxicity of ionic, micro- and nano-sized iron in three plant species. Ecotoxicol Environ Saf 123:81–88
Lin DH, Xing BS (2007) Phytotoxicity of nanoparticles: inhibition of seed germination and root growth. Environ Pollut 150:243–250
Lin J, Jiang W, Liu D (2003) Accumulation of copper by roots, hypocotyls, cotyledons and leaves of sunflower (*Helianthus annuus* L). Bioresour Technol 86:151–155
Lin BS, Diao SQ, Li CH, Fang LJ, Qiao SC, Yu M (2004) Effect of TMS (nanostructured silicon dioxide) on growth of Changbai larch seedlings. J For Res 15:138–140
Liu R, Lal R (2014) Synthetic apatite nanoparticles as a phosphorus fertilizer for soybean (*Glycine max*). Sci Rep 4:5686
Liu RQ, Lal R (2015) Potentials of engineered nanoparticles as fertilizers for increasing agronomic productions. Sci Total Environ 514:131–139
Liu RQ, Zhang HY, Lal R (2016) Effects of stabilized nanoparticles of copper, Zinc, manganese, and iron oxides in low concentrations on lettuce (*Lactuca sativa*) seed germination: nanotoxicants or nanonutrients? Water Air Soil Pollut 227:42
Loeffler J, Sutter U, Hedderich R, Fiedeler U, Malsch I, Túquerres G, Koskinen J, Linder M, Lojkowski W, Moritz T, Zins M, Bernabeu E, Larena A (2005) Overview on promising nanomaterials for industrial applications. Steinbeis-Europa-Zentrum, Karlsruhe
Lopez-Moreno ML, de la Rosa G, Hernandez-Viezcas JA, Peralta-Videa JR, Gardea-Torresdey JL (2010a) X-ray absorption spectroscopy (XAS) corroboration of the uptake and storage of CeO_2 nanoparticles and assessment of their differential toxicity in four edible plant species. J Agric Food Chem 58:3689–3693
Lopez-Moreno ML, De La Rosa G, Hernandez-Viezcas JA, Castillo-Michel H, Botez CE, Peralta-Videa JR, Gardea-Torresdey JL (2010b) Evidence of the differential biotransformation and

genotoxicity of ZnO and CeO_2 nanoparticles on soybean (*Glycine max*) plants. Environ Sci Technol 44:7315–7320

Luther W, Nass R, Campbell R, Dellwo U, Schuster F, Tenegal F, Kallio M, Lintunen P, Oleg Remškar SM, Zumer M, Hoet P, Brüske-Hohlfeld I, Lipscomb S, Malanowski N, Zweck A (2004) Industrial application of nanomaterials – chances and risks. Future Technologies Division, VDI Technologiezentrum GmbH, Düsseldorf

Lyons GH, Genc Y, Soole K, Stangoulis JCR, Liu F, Graham RD (2009) Selenium increases seed production in *Brassica*. Plant Soil 318:73–80

Ma JF (2004) Role of silicon in enhancing the resistance of plants to biotic and abiotic stresses. Soil Sci Plant Nutr 50:11–18

Ma JF, Yamaji N (2008) Functions and transport of silicon in plants. Cell Mol Life Sci 65:3049–3057

Ma XM, Gurung A, Deng Y (2013a) Phytotoxicity and uptake of nanoscale zero-valent iron (nZVI) by two plant species. Sci Total Environ 443:844–849

Ma C, Chhikara S, Xing B, Musante C, White JC, Dhankher OP (2013b) Physiological and molecular response of *Arabidopsis thaliana* (L.) to nanoparticle cerium and indium oxide exposure. ACS Sustain Chem Eng 1:768–778

Mahmoodzadeh H, Nabavi M, Kashefi H (2013) Effect of nanoscale titanium dioxide particles on the germination and growth of canola (*Brassica napus*). J Ornamental Hortic Plants 3:25–32

Majumdar S, Peralta-Videa JR, Bandyopadhyay S, Castillo-Michel H, Hernandez-Viezcas JA, Sahi S, Gardea-Torresdey JL (2014) Exposure of cerium oxide nanoparticles to kidney bean shows disturbance in the plant defense mechanisms. J Hazard Mater 278:279–287

Maksymiec W (1997) Effect of copper on cellular processes in higher plants. Photosynthetica 34:321–342

Malekian R, Abedi-Koupai J, Eslamian SS, Mousavi SF, Abbaspour KC, Afyuni M (2011) Ion-exchange process for ammonium removal and release using natural Iranian zeolite. Appl Clay Sci 51:323–329

Mandeh M, Omidi M, Rahaie M (2012) *In vitro* influences of TiO_2 nanoparticles on barley (*Hordeum vulgare* L.) tissue culture. Biol Trace Elem Res 150:376–380

Manikandan A, Subramanian KS (2016) Evaluation of zeolite based nitrogen nano-fertilizers on maize growth, yield and quality on inceptisols and alfisols. Int J Plant Soil Sci 9:IJPSS.22103

Marafon AC, Endres L (2013) Silicon: fertilization and nutrition in higher plants. Rev Cienc Agrar 56:380–388

Marschner H (1995) Mineral nutrition of higher plants, 2nd edn. Academic, London

Martínez-Fernández D, Vítková M, Pilar Bernal M, Komárek M (2015) Effects of nano-maghemite on trace element accumulation and drought response of *Helianthus annuus* L. in a contaminated mine soil. Water Air Soil Pollut 226:101

Masarovičová E, Kráľová K (2013) Metal nanoparticles and plants. Ecol Chem Eng S 20:9–22

Masarovičová E, Kráľová K, Šeršeň F (2011) Plant responses to toxic metal stress. In: Pessarakli M (ed) Handbook of plant and crop stress, 3rd edn. CRC Press, Boca Raton, pp 595–634

Masarovičová E, Kráľová K, Zinjarde SS (2014) Metal nanoparticles in plants. Formation and action. In: Pessarakli M (ed) Handbook of plant and crop physiology, 3rd edn. CRC Press, Boca Raton, pp 683–731

Miralles P, Johnson E, Church TL, Harris AT (2012) Multiwalled carbon nanotubes in alfalfa and wheat: toxicology and uptake. J R Soc Interf 9:3514–3527

Mirzajani F, Askari H, Hamzelou S, Farzaneh M, Ghassempour A (2013) Effect of silver nanoparticles on *Oryza sativa* L. and its rhizosphere bacteria. Ecotoxicol Environ Saf 88:48–54

Mondal A, Basu R, Das S, Nandy P (2011) Beneficial role of carbon nanotubes on mustard plant growth: an agricultural prospect. J Nanopart Res 13:4519–4528

Morales MI, Rico CM, Hernandez-Viezcas JA, Nunez JE, Barrios AC, Tafoya A, Flores-Marges JP, Peralta-Videa JR, Gardea-Torresdey JL (2013) Toxicity assessment of cerium oxide

nanoparticles in cilantro (*Coriandrum sativum* L.) plants grown in organic soil. J Agric Food Chem 61:6224–6230
Morla S, Ramachandra Rao CSV, Chakrapani R (2011) Factors affecting seed germination and seedling growth of tomato plants cultured *in vitro* conditions. J Chem Biol Phys Sci B 1:328–334
Morteza E, Moaveni P, Farahani HA, Kiyani M (2013) Study of photosynthetic pigments changes of maize (*Zea mays* L.) under nano TiO_2 spraying at various growth stages. Springer Plus 2:247
Mossor-Pietraszewska T (2001) Effect of aluminium on plant growth and metabolism. Acta Biochim Pol 48:673–686
Mukherjee A, Peralta-Videa JR, Bandyopadhyay S, Rico CM, Zhao LJ, Gardea-Torresdey JL (2014) Physiological effects of nanoparticulate ZnO in green peas (*Pisum sativum* L.) cultivated in soil. Metallomics 6:132–138
Mukherjee A, Majumdar S, Servin AD, Pagano L, Dhankher OP, White JC (2016) Carbon nanomaterials in agriculture: a critical review. Front Plant Sci 7:172
Mustafa G, Sakata K, Hossain Z, Komatsu S (2015) Proteomic study on the effects of silver nanoparticles on soybean under flooding stress. J Proteomics 122:100–118
Nair R, Mohamed MS, Gao W, Maekawa T, Yoshida Y, Ajayan PM, Kumar DS (2012) J Nanosci Nanotechnol 12:2212–2220
National Nanotechnology Initiative (2008) Nanotechnology: big things from a tiny world. Arlington. https://www.nano.gov/sites/default/files/pub_resource/nanotechnology_bigthingsfromatinyworld-print.pdf. Accessed 30 May 2016
Nejatzadeh-Barandozi F, Darvishzadeh F, Aminkhani A (2014) Effect of nano silver and silver nitrate on seed yield of (*Ocimum basilicum* L.). Org Med Chem Lett 4:11
Nhan LV, Ma CX, Rui YK, Liu ST, Li XG, Xing BS, Liu LM (2015) Phytotoxic mechanism of nanoparticles: destruction of chloroplasts and vascular bundles and alteration of nutrient absorption. Sci Rep 5:11618
Oloumi H, Mousavi EA, Mohammadinejad R (2014) Multi-walled carbon nanotubes enhance Cd^{2+} and Pb^{2+} uptake by canola seedlings. Agrochimica 58:91–102
Onishchenko DV, Reva VP, Kuryavyi VG, Voronov BA (2015) Sprouting of pepper and tomato seeds with the use of multiwalled carbon nanotubes. Russ Agric Sci 41:146–149
Opatřilová R, Černíková A, Coufalová L, Dohnal J, Jampílek J (2013) *In vitro* permeation of micronized and nanonized alaptide from semi-solid formulations. Sci World J 2013:787283
Pais I (1983) The biological importance of titanium. J Plant Nutr 6:3–131
Palit S, Sharma A, Talukder G (1994) Effects of cobalt on plants. Bot Rev 60:149–181
Palmqvist NGM, Bejai S, Meijer J, Seisenbaeva GA, Kessler VG (2015) Nano titania aided clustering and adhesion of beneficial bacteria to plant roots to enhance crop growth and stress management. Sci Rep 5:10146
Pandey AC, Sanjay SS, Yadav RS (2010) Application of ZnO nanoparticles in influencing the growth rate of *Cicer arietinum*. J Exp Nanosci 5:488–497
Parisi C, Vigani M, Rodríguez-Cerezo E (2015) Agricultural nanotechnologies: what are the current possibilities? NanoToday 10:124–127
Parveen A, Rao S (2015) Effect of nanosilver on seed germination and seedling growth in *Pennisetum glaucum*. J Clust Sci 26:693–701
Peralta JR, Gardea-Torresdey JL, Tiemann KJ, Gomez E, Arteaga S, Rascon E, Parsons JG (2001) Uptake and effects of five heavy metals on seed germination and plant growth in alfalfa (*Medicago sativa* L.). Bull Environ Contam Toxicol 66:727–734
Pereira EI, Minussi FB, da Cruz CCT, Bernardi ACC, Ribeiro C (2012) Urea-montmorillonite-extruded nanocomposites: a novel slow-release material. J Agric Food Chem 60:5267–5272
Pereira EI, da Cruz CCT, Solomon A, Le A, Cavigelli MA, Ribeiro C (2015) Novel slow-release nanocomposite nitrogen fertilizers: the impact of polymers on nanocomposite properties and function. Ind Eng Chem Res 54:3717–3725
Pérez-de-Luque A, Hermosín MC (2013) Nanotechnology and its use in agriculture. In: Bagchi D, Bagchi M, Moriyama H, Shahidi F (eds) Bio-nanotechnology: a revolution in food, biomedical and health sciences. Wiley-Blackwell, West Sussex, pp 299–405

Peško M, Kráľová K (2013) Physiological response of *Brassica napus* L. plants to Cu(II) treatment. Proc ECOpole 7:155–161

Pourali S, Roozbahani A (2015) Evaluation of quantitative and qualitative traits of potato (*Solanum tuberosum* L.) under amino acids and iron application. Crop Res (Hisar) 50:101–106

Pourkhaloee A, Haghighi M, Saharkhiz MJ, Jouzi H, Doroodmand MM (2011) Carbon nanotubes can promote seed germination via seed coat penetration. Seed Technol 33:155–169

Pradhan S, Patra P, Mitra S, Dey KK, Jain S, Sarkar S, Roy S, Palit P, Goswami A (2014) Manganese nanoparticles: impact on non-nodulated plant as a potent enhancer in nitrogen metabolism and toxicity study both *in vivo* and *in vitro*. J Agric Food Chem 62:8777–8785

Pradhan N, Singh S, Ojha N, Shrivastava A, Barla A, Rai V, Bose S (2015) Facets of nanotechnology as seen in food processing, packaging, and preservation industry. Biomed Res Int 2015:365672

Prasad R (2014) Synthesis of silver nanoparticles in photosynthetic plants. J Nanoparticle. Article ID 963961. http://dx.doi.org/10.1155/2014/963961

Prasad TNVKV, Sudhakar P, Sreenivasulu Y, Latha P, Munaswamy V, Reddy KR, Sreeprasad TS, Sajanlal PR, Pradeep T (2012) Effect of nanoscale zinc oxide particles on the germination, growth and yield of peanut. J Plant Nutr 35:905–927

Prasad R, Kumar V, Prasad KS (2014) Nanotechnology in sustainable agriculture: present concerns and future aspects. Afr J Biotechnol 13:705–713

Prasad R, Pandey R, Barman I (2016) Engineering tailored nanoparticles with microbes: quo vadis? WIREs Nanomed Nanobiotechnol 8:316–330

Qados AMSA, Moftah AE (2015) Influence of silicon and nano-silicon on germination, growth and yield of faba bean (*Vicia faba* L.) under salt stress conditions. Am J Exp Agric 5:509–524

Qi MF, Liu YF, Li TL (2013) Nano-TiO_2 improve the photosynthesis of tomato leaves under mild heat stress. Biol Trace Elem Res 156:323–328

Raliya R, Tarafdar JC, Gulecha K, Choudhary K, Ram R, Mal P, Saran RP (2013) Scope of nanoscience and nanotechnology in agriculture. J Appl Biol Biotechnol 1:41–44

Raliya R, Biswas P, Tarafdar JC (2015) TiO_2 nanoparticle biosynthesis and its physiological effect on mung bean (*Vigna radiata* L.). Biotechnol Rep 5:22–26

Rani PU, Yasur J, Loke KS, Dutta D (2016) Effect of synthetic and biosynthesized silver nanoparticles on growth, physiology and oxidative stress of water hyacinth: *Eichhornia crassipes* (Mart) Solms. Acta Physiol Plant 38:UNSP 58

Rashidi L, Khosravi-Darani K (2011) The applications of nanotechnology in food industry. Crit Rev Food Sci Nutr 51:723–730

Rashidzadeh A, Olad A, Salari D, Reyhanitabar A (2014) On the preparation and swelling properties of hydrogel nanocomposite based on sodium alginate-g-poly(acrylic acid-co-acrylamide)/clinoptilolite and its application as slow release fertilizer. J Polym Res 21:1–15

Raskar SV, Laware SL (2014) Effect of zinc oxide nanoparticles on cytology and seed germination in onion. Int J Curr Microbiol Appl Sci 3:467–473

Razzaq A, Ammara R, Jhanzab HM, Mahmood T, Hafeez A, Hussain S (2016) A novel nanomaterial to enhance growth and yield of wheat. J Nanosci Technol 2:55–58

Riahi-Madvar A, Rezae F, Jalali V (2012) Effects of alumina nanoparticles on morphological properties and antioxidant system of *Triticum aestivum*. Iran J Plant Physiol 3:595–603

Rico CM, Lee SC, Rubenecia R, Mukherjee A, Hong J, Peralta-Videa JR, Gardea-Torresdey JL (2014) Cerium oxide nanoparticles impact yield and modify nutritional parameters in wheat (*Triticum aestivum* L.) J Agric Food Chem 62:9669–9675

Rico CM, Barrios AC, Tan WJ, Rubenecia R, Lee SC, Varela-Ramirez A, Peralta-Videa JR, Gardea-Torresdey JL (2015) Physiological and biochemical response of soil-grown barley (*Hordeum vulgare* L.) to cerium oxide nanoparticles. Environ Sci Pollut Res 22:10551–10558

Rizwan M, Ali S, Qayyum MF, Ibrahim M, Zia-ur-Rehman M, Abbas T, Ok YS (2016) Mechanisms of biochar-mediated alleviation of toxicity of trace elements in plants: a critical review. Environ Sci Pollut Res 23:2230–2248

Rout GR, Sahoo S (2015) Role of iron in plant growth and metabolism. Rev Agric Sci 3:1–24
Russell AD, Hugo WB (1994) Antibacterial activity and action of silver. Prog Med Chem 31:351–370
Sabaghnia N, Janmohammadi M (2014) Effect of nano-silicon particles application on salinity tolerance in early growth of some lentil genotypes. Ann Univ Mariae Curie-Sklodowska C 69:39–55
Salama HMH (2012) Effects of silver nanoparticles in some crop plants, common bean (*Phaseolus vulgaris* L.) and corn (*Zea mays* L.) Int Res J Biotechnol 3:190–197
Sarkar S, Datta SC, Biswas DR (2015) Effect of fertilizer loaded nanoclay/superabsorbent polymer composites on nitrogen and phosphorus release in soil. Proc Indian Nat Sci Acad B Biol Sci 85:415–421
Savithramma N, Ankanna S, Bhumi G (2012) Effect of nanoparticles on seed germination and seedling growth of *Boswellia ovalifoliolata* an endemic and endangered medicinal tree taxon. Nano Vision 2:61–68
Savvas D, Ntatsi G (2015) Biostimulant activity of silicon in horticulture. Sci Hortic 196:66–81
Saxena M, Maity S, Sarkar S (2014) Carbon nanoparticles in 'biochar' boost wheat (*Triticum aestivum*) plant growth. RSC Adv 4:39948–39954
Schaller J, Brackhage C, Paasch S, Brunner E, Bäucker E, Dudel EG (2013) Silica uptake from nanoparticles and silica condensation state in different tissues of *Phragmites australis*. Sci Total Environ 442:6–9
Scrimshaw NS (1991) Iron deficiency. Sci Am 265:46–52
Seifsahandi M, Sorooshzadeh A (2013) Comparison between the influences of silver nanoparticles and silver nitrate on the growth and phytochemical properties of borage (*Borago officinalis* L.) Curr Nanosci 9:241–247
Sekhon BS (2014) Nanotechnology in agri-food production: an overview. Nanotechnol Sci Appl 7:31–53
Selva Preetha P, Subramanian KS, Sharmila Rahale C (2014) Characterization of slow release of sulphur nutrient – a zeolite based nano-fertilizer. Int J Dev Res 4:229–233
Šeršeň F, Kráľová K, Bumbálová A, Švajlenová O (1997) The effect of Cu(II) ions bound with tridentate Schiff base ligands upon the photosynthetic apparatus. J Plant Physiol 15:299–305
Servin AD, Castillo-Michel H, Hernandez-Viezcas JA, Diaz BC, Peralta-Videa JR, Gardea-Torresdey JL (2012) Synchrotron micro-XRE and micro-XANES confirmation of the uptake and translocation of TiO_2 nanoparticles in cucumber (*Cucumis sativus*) plants. Environ Sci Technol 46:7637–7643
Shahrekizad M, Gholamalizadeh Ahangar A, Mir N (2015) Improving agronomic traits of sunflower (*Helianthus annuus*). J Nanostruct 5:117–127
Shams G, Ranjbar M, Amiri A (2013) Effect of silver nanoparticles on concentration of silver heavy element and growth indexes in cucumber (*Cucumis sativus* L. negeen). J Nanopart Res 15:1630
Sharma P, Bhatt D, Zaidi MGH, Saradhi PP, Khanna PK, Arora S (2012) Silver nanoparticle-mediated enhancement in growth and antioxidant status of *Brassica juncea*. Appl Biochem Biotechnol 167:2225–2233
Sheykhbaglou R, Sedghi M, Tajbakhsh Shishevan M, Sharifi SR (2010) Effects of nano-iron oxide particles on agronomic traits of soybean. Notulae Sci Biol 2:112–113
Siddiqui MH, Al-Whaibi MH (2014) Role of nano-SiO_2 in germination of tomato (*Lycopersicum esculentum* seeds Mill). Saudi J Biol Sci 21:13–17
Siddiqui MH, Al-Whaibi MH, Faisal M, Al Sahli AA (2014) Nano-silicon dioxide mitigates the adverse effects of salt stress on *Cucurbita pepo* L. Environ Toxicol Chem 33:2429–2437
Sigua GC, Stone KC, Hunt PG, Cantrell KB, Novak JM (2015) Increasing biomass of winter wheat using sorghum biochars. Agron Sustain Dev 35:739–748
Silva S (2012) Aluminium toxicity targets in plants. J Bot 2012:219462
Singh D, Kumar S, Singh SC, Lal B, Singh NB (2012) Applications of liquid assisted pulsed laser ablation synthesized TiO_2 nanoparticles on germination, growth and biochemical parameters of *Brassica oleracea var. capitata*. Sci Adv Mater 4:522–531

Singh NB, Amist N, Yadav K, Singh D, Pandey JK, Singh SC (2013) Zinc oxide nanoparticles as fertilizer for the germination, growth and metabolism of vegetable crops. J Nanoeng Nanomanuf 3:353–364

Slomberg DL, Schoenfisch MH (2012) Silica nanoparticle phytotoxicity to *Arabidopsis thaliana*. Environ Sci Technol 46:10247–10254

Smirnova EA, Gusev AA, Zaitseva ON, Lazareva EM, Onishchenko GE, Kuznetsova EV, Tkachev AG, Feofanov AV, Kirpichnikov MP (2011) Multi-walled carbon nanotubes penetrate into plant cells and affect the growth of *Onobrychis arenaria* seedlings. Acta Nat 3:99–106

Song GL, Gao Y, Wu H, Hou WH, Zhang CY, Ma HQ (2012) Physiological effect of anatase TiO_2 nanoparticles on *Lemna minor*. Environ Toxicol Chem 31:2147–2152

Song U, Shin M, Lee G, Roh J, Kim Y, Lee EJ (2013) Functional analysis of TiO_2 nanoparticle toxicity in three plant species. Biol Trace Elem Res 155:93–103

Sonkar SK, Roy M, Babar DG, Sarkar S (2012) Water soluble carbon nano-onions from wood wool as growth promoters for gram plants. Nanoscale 4:7670–7675

Sonkaria S, Ahn SH, Khare V (2012) Nanotechnology and its impact on food and nutrition: a review. Recent Pat Food Nutr Agric 4:8–18

Sorooshzadeh A, Hazrati S, Oraki H, Govahi M, Ramazani A (2012) Foliar application of nano-silver influence growth of saffron under flooding stress. In: Proceeding of the 4th international conference "Nanocon-2012". Oct 2012, Brno, pp 510–512 (PB-19). http://nanocon2012.tanger.cz/files/proceedings/04/reports/1254.pdf

Sperotto RA, Ricachenevsky FK, Waldow VA, Fett JP (2012) Iron biofortification in rice: it's a long way to the top. Plant Sci 190:24–39

Stoltzfus RJ, Dreyfuss ML (1998) Guidelines for the use of iron supplements to prevent and treat iron deficiency anemia. ILSI Press, Washington, DC

Strader LC, Beisner ER, Bartel B (2009) Silver ions increase auxin efflux independently of effects on ethylene response. Plant Cell 21:3585–3590

Sun ZM, Zhang K, Liu JT, Si HS, Wang YQ (2012) Effects of nitrogen regulators on fertilizer nitrogen transformation in meadow cinnamon soil and on pakchoi growth. Yingyong Shengtai Xuebao 23:2497–2503

Sun DQ, Hussain HI, Yi ZF, Siegele R, Cresswell T, Kong LX, Cahill DM (2014) Uptake and cellular distribution, in four plant species, of fluorescently labeled mesoporous silica nanoparticles. Plant Cell Rep 33:1389–1402

Suriyaprabha R, Karunakaran G, Yuvakkumar R, Rajendran V, Kannan N (2012) Silica nanoparticles for increased silica availability in maize (*Zea mays* L.) seeds under hydroponic conditions. Curr Nanosci 8:902–908

Suriyaprabha R, Karunakaran G, Yuvakkumar R, Prabu P, Rajendran V, Kannan N (2013) Application of silica nanoparticles for increased silica availability in maize. In: Chauhan AK, Murli C, Gadkari SC (eds) Proceedings of the 57th Dae solid state physics symposium, Mumbai, Dec 2012. AIP Conf Proc 1512, pp 424–425

Suriyaprabha R, Karunakaran G, Yuvakkumar R, Rajendran V, Kannan N (2014a) Foliar application of silica nanoparticles on the phytochemical responses of maize (*Zea mays* L.) and its toxicological behavior. Synth React Inorg M 44:1128–1131

Suriyaprabha R, Karunakaran G, Kavitha K, Yuvakkumar R, Rajendran V, Kannan N (2014b) Application of silica nanoparticles in maize to enhance fungal resistance. IET Nanobiotechnol 8:133–137

Syu YY, Hung JH, Chen JC, Chuang HW (2014) Impacts of size and shape of silver nanoparticles on *Arabidopsis* plant growth and gene expression. Plant Physiol Biochem 83:57–64

Tarafdar JC, Raliya R, Mahawar H, Rathore I (2014) Development of zinc nanofertilizer to enhance crop production in pearl millet (*Pennisetum americanum*). Agric Res 3:257–262

Thiruvengadam M, Gurunathan S, Chung IM (2015) Physiological, metabolic, and transcriptional effects of biologically-synthesized silver nanoparticles in turnip (*Brassica rapa ssp rapa* L.). Protoplasma 252:1031–1046

Tiwari S, Mohanty P (1993) Cobalt chloride induced stimulation of photosystem II electron transport in *Synechocystis* PCC 6803 cells. Photosynth Res 38:463–469

Tong CY, Xiao SY, Wang B, Xiao XJ, Zheng YQ, Xue CG, Wang QM, Liu XM (2008) Preparation and application of chitooligosaccharides-selenium nanoparticles for plant nutrition regulator. J Hunan Univ (Nat Sci) 35:60–64

Torabian S, Zahedi M, Khoshgoftar AH (2016) Effects of foliar spray of two kinds of zinc oxide on the growth and ion concentration of sunflower cultivars under salt stress. J Plant Nutr 39:172–180

Tripathi S, Sarkar S (2015) Influence of water soluble carbon dots on the growth of wheat plant. Appl Nanosci 5:609–616

Tripathi S, Sonkar SK, Sarkar S (2011) Growth stimulation of gram (*Cicer arietinum*) plant by water soluble carbon nanotubes. Nanoscale 3:1176–1181

Tripathi DK, Singh VP, Prasad SM, Chauhan DK, Dubey NK (2015) Silicon nanoparticles (SiNp) alleviate chromium (VI) phytotoxicity in *Pisum sativum* (L.) seedlings. Plant Physiol Biochem 96:189–198

Tripathi S, Kapri S, Datta A, Bhattacharyya S (2016) Influence of the morphology of carbon nanostructures on the stimulated growth of gram plant. RSC Adv 6:43864–43873

Trujillo-Reyes J, Vilchis-Nestor AR, Majumdar S, Peralta-Videa JR, Gardea-Torresdey JL (2013) Citric acid modifies surface properties of commercial CeO_2 nanoparticles reducing their toxicity and cerium uptake in radish (*Raphanus sativus*) seedlings. J Hazard Mater 263:677–684

Tuhy L, Samoraj M, Baśladyńska S, Chojnacka K (2015) New micronutrient fertilizer biocomponents based on seaweed biomass. Pol J Environ Stud 24:2213–2221

Unsworth JB, Corsi C, Van Emon JM, Farenhorst A, Hamilton DJ, Howard CJ, Hunter R, Jenkins JJ, Kleter GA, Kookana RS, Lalah JO, Leggett M, Miglioranza KS, Miyagawa H, Peranginangin N, Rubin B, Saha B, Shakil NA (2016) Developing global leaders for research, regulation, and stewardship of crop protection chemistry in the 21st century. J Agric Food Chem 64:52–60

Vaculíková E, Grünwaldová V, Král V, Dohnal J, Jampílek J (2012a) Primary investigation of the preparation of nanoparticles by precipitation. Molecules 17:11067–11078

Vaculíková E, Grünwaldová V, Král V, Dohnal J, Jampílek J (2012b) Preparation of candesartan and atorvastatin nanoparticles by solvent evaporation. Molecules 17:13221–13234

Vaculíková E, Placha D, Pisárčik M, Peikertová P, Dedková K, Devínsky F, Jampílek J (2014) Preparation of risedronate nanoparticles by solvent evaporation technique. Molecules 19:17848–17861

Vaculíková E, Černíková A, Placha D, Pisárčik M, Dedková K, Peikertová P, Devínsky F, Jampílek J (2016) Cimetidine nanoparticles for permeability enhancement. J Nanosci Nanotechnol 16:7840–7843

Valadkhan M, Mohammadi K, Nezhad MTK (2015) Effect of priming and foliar application of nanoparticles on agronomic traits of chickpea. Biol Forum Int J 7:599–602

Vallee BL, Auld DS (1990) Zinc coordination, function, and structure of zinc enzymes and other proteins. Biochemistry 29:5647–5659

Ventola CL (2012a) The nanomedicine revolution: Part 1: Emerging concepts. Pharm Ther 37:512–525

Ventola CL (2012b) The nanomedicine revolution: Part 2: Current and future clinical applications. Pharm Ther 37:582–591

Ventola CL (2012c) The nanomedicine revolution: Part 3: Regulatory and safety challenges. Pharm Ther 37:631–639

Vestel J, Caldwell DJ, Constantine L, D'Aco VJ, Davidson T, Dolan DG, Millard SP, Murray-Smith R, Parke NJ, Ryan JJ, Straub JO, Wilson P (2016) Use of acute and chronic ecotoxicity data in environmental risk assessment of pharmaceuticals. Environ Toxicol Chem 35:1201–1212

Vijayarengan P (2012) Growth and biochemical variations in radish under zinc applications. Intern J Res Plant Sci 2:43–49

Vijayarengan P (2013) Changes in growth, biochemical constituents and antioxidant potentials in cluster bean *Cyamopsis tetragonoloba* L. Taub under zinc stress. Int J Curr Sci 5:37–49

Vijayarengan P, Mahalakshmi G (2013) Zinc toxicity in tomato plants. World Appl Sci J 24:649–653

Wan YJ, Li JL, Ren HX, Huang J, Yuan H (2014) Physiological investigation of gold nanorods toward watermelon. J Nanosci Nanotechnol 14:6089–6094

Wang XP, Han HY, Liu XQ, Gu XX, Chen K, Lu DL (2012a) Multi-walled carbon nanotubes can enhance root elongation of wheat (*Triticum aestivum*) plants. J Nanopart Res 14:841

Wang Q, Ma X, Zhang W, Pei HC, Chen YS (2012b) The impact of cerium oxide nanoparticles on tomato (*Solanum lycopersicum* L.) and its implications for food safety. Metallomics 4:1105–1112

Wang J, Koo Y, Alexander A, Yang Y, Westerhof S, Zhang QB, Schnoor JL, Colvin VL, Braam J, Alvarez PJJ (2013) Phytostimulation of poplars and *Arabidopsis* exposed to silver nanoparticles and Ag^+ at sublethal concentrations. Environ Sci Technol 47:5442–5449

Wang XG, Lu SY, Gao CM, Xu XB, Zhang XJ, Bai X, Liu MZ, Wu L (2014) Highly efficient adsorption of ammonium onto palygorskite nanocomposite and evaluation of its recovery as a multifunctional slow-release fertilizer. Chem Eng J 252:404–414

Wang S, Wang FY, Gao SC (2015) Foliar application with nano-silicon alleviates Cd toxicity in rice seedlings. Environ Sci Pollut Res 22:2837–2845

Watanabe Y, Yamada H, Ikoma T, Tanaka J, Stevens GW, Komatsu Y (2014) Preparation of a zeolite NaP1/hydroxyapatite nanocomposite and study of its behavior as inorganic fertilizer. J Chem Technol Biotechnol 89:963–968

Weissig V, Guzman-Villanueva D (2015) Nanopharmaceuticals (part 2): products in the pipeline. Int J Nanomedicine 10:1245–1257

Weissig V, Pettinger TK, Murdock N (2014) Nanopharmaceuticals (part 1): products on the market. Int J Nanomedicine 9:4357–4373

Wu S, Huang L, Head J, Chen D, Kong I, Tang Y (2012) Phytotoxicity of metal oxide nanoparticles is related to both dissolved metals ions and adsorption of particles on seed surfaces. J Pet Environ Biotechnol 3:126

Wu SG, Huang L, Head J, Ball M, Tang YJ, Chen DR (2014) Electrospray facilitates the germination of plant seeds. Aerosol Air Qual Res 14:632–641

Xiang L, Zhao HM, Li YW, Huang XP, Wu XL, Zhai T, Yuan Y, Cai QY, Mo CH (2015) Effects of the size and morphology of zinc oxide nanoparticles on the germination of Chinese cabbage seeds. Environ Sci Pollut Res 22:10452–10462

Yamanaka M, Hara K, Kudo J (2005) Bactericidal actions of a silver ion solution on *Escherichia coli*, studied by energy-filtering transmission electron microscopy and proteomic analysis. Appl Environ Microbiol 71:7589–7593

Yan F, Schubert S, Mengel K (1992) Effect of low root medium pH on net proton release, root respiration and root growth of corn (*Zea mays* L.) and broad bean (*Vicia faba* L.) Plant Physiol 99:415–421

Yan SH, Zhao L, Li H, Zhang Q, Tan JJ, Huang M, He SB, Li LJ (2013) Single-walled carbon nanotubes selectively influence maize root tissue development accompanied by the change in the related gene expression. J Hazard Mater 246:110–118

Yang F, Hong FS, You WJ, Liu C, Gao FQ, Wu C, Yang P (2006) Influences of nano-anatase TiO_2 on the nitrogen metabolism of growing spinach. Biol Trace Elem Res 110:179–190

Yao Y, Gao B, Inyang M, Zimmerman AR, Cao X, Pullammanappallil P, Yang L (2011) Removal of phosphate from aqueous solution by biochar derived from anaerobically digested sugar beet tailings. J Hazard Mater 190:501–507

Yi ZF, Hussain HI, Feng CF, Sun DQ, She FH, Rookes JE, Cahill DM, Kong LG (2015) Functionalized mesoporous silica nanoparticles with redox-responsive short-chain gatekeepers for agrochemical delivery. ACS Appl Mater Interfaces 7:9937–9946

Yin LY, Colman BP, McGill BM, Wright JP, Bernhardt ES (2012) Effects of silver nanoparticle exposure on germination and early growth of eleven wetland plants. PLoS One 7:e47674

Yurela I (2005) Copper in plants. Braz J Plant Physiol 17:145–156

Yuvaraj M, Subramanian KS (2014) Fabrication of zinc nano fertilizer on growth parameter of rice. Trends Biosci 7:2564–2565

Yuvaraj M, Subramanian KS (2015) Controlled-release fertilizer of zinc encapsulated by a manganese hollow core shell. Soil Sci Plant Nutr 64:319–326

Zahra Z, Arshad M, Rafique R, Mahmood A, Habib A, Qazi IA, Khan SA (2015) Metallic nanoparticle (TiO_2 and Fe_3O_4) application modifies rhizosphere phosphorus availability and uptake by *Lactuca sativa*. J Agric Food Chem 63:6876–6882

Ze YG, Liu C, Wang L, Hong MM, Hong FS (2011) The regulation of TiO_2 nanoparticles on the expression of light-harvesting complex II and photosynthesis of chloroplasts of *Arabidopsis thaliana*. Biol Trace Elem Res 143:1131–1141

Zhai GS, Gutowski SM, Walters KS, Yan B, Schnoor JL (2015) Charge, size, and cellular selectivity for multiwall carbon nanotubes by maize and soybean. Environ Sci Technol 49:7380–7390

Zhang B, Zheng LP, Li WY, Wang JW (2013) Stimulation of artemisinin production in *Artemisia annua* hairy roots by Ag-SiO_2 core-shell nanoparticles. Curr Nanosci 9:363–370

Zhang Y, Liang XY, Yang XG, Liu HY, Yao JM (2014) An eco-friendly slow-release urea fertilizer based on waste mulberry branches for potential agriculture and horticulture applications. ACS Sustain Chem Eng 2:1871–1878

Zhang M, Gao B, Chen JJ, Li YC (2015) Effects of graphene on seed germination and seedling growth. J Nanopart Res 17:78

Zhao LJ, Sun YP, Hernandez-Viezcas JA, Servin AD, Hong J, Niu GH, Peralta-Videa JR, Duarte-Gardea M, Gardea-Torresdey JL (2013) Influence of CeO_2 and ZnO nanoparticles on cucumber physiological markers and bioaccumulation of Ce and Zn: a life cycle study. J Agric Food Chem 61:11945–11951

Zhao LJ, Sun YP, Hernandez-Viezcas JA, Hong J, Majumdar S, Niu GH, Duarte-Gardea M, Peralta-Videa JR, Gardea-Torresdey JL (2015) Monitoring the environmental effects of CeO_2 and ZnO nanoparticles through the life cycle of corn (*Zea mays*) plants and in situ μ-XRF mapping of nutrients in kernels. Environ Sci Technol 49:2921–2928

Zheng L, Hong FS, Lu SP, Liu C (2005) Effect of nano-TiO_2 on strength of naturally aged seeds and growth of spinach. Biol Trace Elem Res 104:83–91

Zheng L, Su MY, Liu C, Chen L, Huang H, Wu X, Liu XQ, Yang F, Gao FQ, Hong FS (2007) Effects of nanoanatase TiO_2 on photosynthesis of spinach chloroplasts under different light illumination. Biol Trace Elem Res 119:68–76

Zhou LL, Cai DQ, He L, Zhong NQ, Yu M, Zhang X, Wu ZY (2015) Fabrication of a high-performance fertilizer to control the loss of water and nutrient using micro/nano networks. ACS Sustain Chem Eng 3:645–653

Zhu ZQ, Sun HX, Li GX, Liang WD, Bao XM, An J, La PQ, Dai JF, Li A (2013) Preparation of polyacrylamide/graphite oxide superabsorbent nanocomposites with salt tolerance and slow release properties. J Appl Polym Sci 129:2328–2334

Zou HT, Ling Y, Dang XL, Yu N, Zhang YL, Zhang YL, Dong JH (2015) Solubility characteristics and slow-release mechanism of nitrogen from organic-inorganic compound coated urea. Int J Photoenergy 2015:705471

Synthesis, Characterization, and Application of Chitosan Nanomaterials Loaded with Zinc and Copper for Plant Growth and Protection

10

Ram Chandra Choudhary, R.V. Kumaraswamy, Sarita Kumari, Ajay Pal, Ramesh Raliya, Pratim Biswas, and Vinod Saharan

Abstract

In recent years, chitosan-based nanomaterials have been the most researched biomaterials in the field of medical, pharmaceutical, and agriculture. Metal-based chitosan nanomaterials have attracted much attention due to its dual activity as plant growth promoter and plant protection agent. In addition, chitosan encapsulated metals are less toxic due to slow release phenomenon and showed long-lasting effect in plants. Blending of Zn and Cu with nanochitosan has additional advantages of providing nutrition to plants and help in vigor growth of plant for further protection from abiotic and biotic stress. In addition, Cu/Zn chitosan nanoparticles have been successfully tested against many plant pathogenic bacteria and fungi. Moreover, Cu/Zn chitosan nanoparticles are involved in inducing amylase and protease enzymes related to mobilization of food for seed germination. A recent study revealed that Zn/Cu chitosan nanomaterials are enhancing defense enzymes of plant which

R.C. Choudhary • R.V. Kumaraswamy • S. Kumari • V. Saharan (✉)
Department of Molecular Biology and Biotechnology, Maharana Pratap University of Agriculture and Technology, Udaipur, Rajasthan 313001, India
e-mail: ramchandra026@gmail.com; kumarabiotech@gmail.com; saritajakhar5@gmail.com; vinodsaharan@gmail.com

A. Pal
Department of Chemistry and Biochemistry, College of Basic Sciences and Humanities, Chaudhary Charan Singh Haryana Agricultural University, Hisar, Haryana 125004, India
e-mail: ajaydrdo@rediffmail.com

R. Raliya • P. Biswas
Department of Energy, Environmental and Chemical Engineering, Washington University in St.Louis, St. Louis, MO 63130, USA
e-mail: rameshraliya@wustl.edu; pbiswas@wustl.edu

R. Prasad et al. (eds.), *Nanotechnology*, DOI 10.1007/978-981-10-4573-8_10

protect them from diseases. In this chapter, we have explained thoroughly the synthesis of Cu and Zn chitosan NPs, their characterization and applications in plant growth, and protection.

Keywords
Chitosan • Micronutrient • Plant growth • Copper • Zinc

10.1 Introduction

Nanotechnology plays an important role in modern agriculture to address global challenges such as climate change, severity of plant diseases, and the limited availability of important plant nutrients (Parisi et al. 2015).To gain higher crop production, continuous use of agrochemicals increases environmental contamination, cost of food production and undesirable side effects (Carvalho 2006). In the recent years, numerous biopolymers such as starch, cellulose, alginate, chitin, and chitosan have been used for the development of new materials with environmental sustainability and desirable functionality (Babu et al. 2013). Among such polymers, chitosan is a deacetylated form of chitin, which is linear copolymer of 2-acetamido-2-deoxy-β--D-glucopyranose and 2-amino-2-deoxy-β-D-glucopyranose. It is the second most abundant polymer in nature after cellulose and found as exoskeleton of crustaceans, cuticles of insects, and cell walls of fungi (Piras et al. 2014). Due to its unique properties such as abundance, biocompatibility, biodegradability, hydrophilicity, safe, and nontoxic, chitosan nanoparticles (NPs) are used in several application including antifungal activity (Saharan et al. 2013), antibacterial activity (Qi et al. 2004; Du et al. 2009; Ali et al. 2011), plant growth-promoting activity (Van et al. 2013; Saharan et al. 2015, 2016), and nano-fertilizers (Corradini et al. 2010).

Preparation of chitosan NPs is generally achieved through ionic gelation method which involves inter- and intramolecular cross-linking between amino group of chitosan and phosphate group of TPP (T). This method is fast, convenient, and controllable as compared to others (Agnihotri et al. 2004; Ing et al. 2012). The incorporation of cross-linking agent such as TPP results in alteration of physicochemical parameters, namely, particle size, zeta potential, and surface morphology that will finally affect its functional properties (Rodrigues et al. 2012).

Transition metals such as copper, zinc, and iron have numerous roles in crop plants (Vasconcelos 2014). In cellular system, Cu plays an exclusive role as cofactor of several enzymes through electron transport chain and redox reaction (Badawy and Rabea 2011; Rajasekaran and Santra 2015). Likewise, Zn is a major component of various enzymes such as oxidoreductases, transferases, and hydrolases. It participates in the formation of carbohydrate and chlorophyll. Similarly, it is also a component of zinc finger proteins known to bind with DNA and RNA for gene regulation. Excess Zn causes generation of ROS, membrane disruption, chromosome alteration, etc. Furthermore, Cu and Zn have been traditionally used as major components of many agrochemicals in agriculture as crop improvement or protection tools. Metal-based nanoparticles including zinc oxide, silver, copper oxide, aluminum oxide, silicon dioxide, etc. have been observed more or less toxic to plants including rice and maize (Yang et al. 2015).

It has been shown that chitosan reported as excellent metal chelating or encapsulating agent for micro- or nanosize particles for Cu or Zn individually (Saharan et al. 2013; Rajasekaran and Santra 2015). So, Cu- and Zn-based chitosan NPs can exhibit dual role in plant as antimicrobial agents and plant growth enhancers (Du et al. 2009; Brunel et al. 2013; Saharan et al. 2013, 2015).

10.2 Synthesis of Cu and Zn Chitosan NPs

Chitosan-based nanomaterials can be prepared through a series of methods classified as follows:

- Chemical methods: in situ polymerization, interfacial polymerization, and interfacial polycondensation
- Physicochemical methods: ionic gelation, complex coacervation, and spray-cooling
- Physical methods: spray-drying, centrifugal extrusion–spheronization, fluid bed coating, etc.

Among them, ionic gelation is the most common method for preparation of Cu- and Zn-based chitosan NPs (Du et al. 2009; Saharan et al. 2013).

10.2.1 Ionic Gelation Method

Different concentrations of chitosan have been tried by various groups (Brunel et al. 2013; Saharan et al. 2015, 2016) and further dissolved in 1% (v/v) acetic acid at room temperature and stirred until the solution becomes transparent. Different concentration of TPP solution in deionized water mix drop wise by syringe/nozzle to chitosan solution under magnetic stirring. The solution is kept overnight for proper cross-linking at room temperature. After completion of reaction, chitosan solution is transformed into spherical gel particles (colloidal form) as shown in Fig. 10.1.

Generally, the pH of this solutions remained in the range of 3.5–4 and depending on the concentration of chitosan and TPP. Separation of nanoparticles can be achieved by alternative centrifugation and ultra-sonication followed by two to three time washing with Di water to bring a pH at around 4.5–5. Finally, the material can be freeze-dried and stored for further use. Copper- and zinc-loaded chitosan NPs can be obtained by addition of copper sulfate and zinc sulfate solution during chitosan NP synthesis process.

In this method, chitosan NPs are formed mainly through mechanism of intra- and intermolecular electrostatic interaction between the positively charged amino group of chitosan and negatively charged phosphate group of TPP.

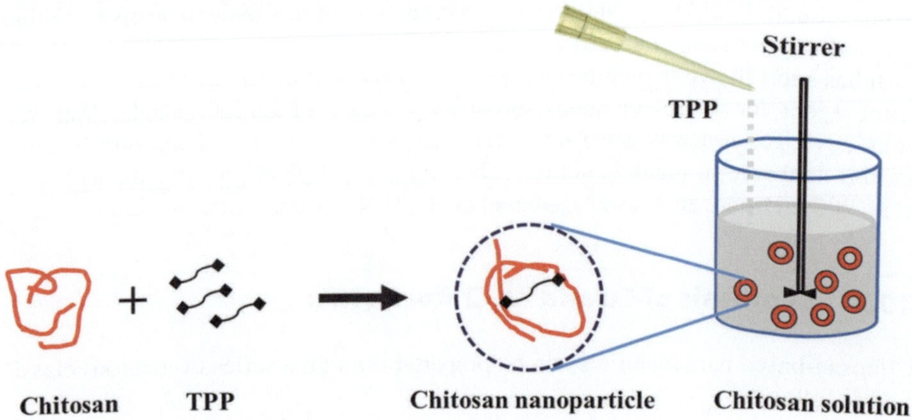

Fig. 10.1 Preparation of chitosan NPs by ion gelation method

10.3 Characterization of Cu and Zn Chitosan NPs

The developed nanomaterial can be characterized for various physicochemical characters such as mean size, polydispersity index (PDI), ζ-potential, functional group analysis, internal and surface morphology by dynamic light scattering (DLS), Fourier transform infrared (FTIR), scanning electron microscopy (SEM), and transmission electron microscopy (TEM) (Saharan et al. 2013). For elemental analysis of nanoparticles, energy dispersive spectroscopy (SEM-EDS) is the preferred method which provides information about the elemental composition of nanoparticles and their distribution. Atomic absorption spectroscopy (AAS) can be used for release profile or encapsulation efficacy of metals in Cu and Zn chitosan NPs (Jaiswal et al. 2012).

10.3.1 Dynamic Light Scattering (DLS)

The average particle size, particle size distribution, polydispersity index (PDI), and ζ-potential of the chitosan NPs can be measured by DLS. The main principle of DLS or photon correlation spectroscopy (PCS) is based on Brownian motion of particles, micelles, or molecules in suspension, which results from random zigzag motion by collision with solvent molecules that are also randomly moving. The laser beam is used for illumination of sample, and the fluctuations of the scattered light from random motion of particles are detected by a photon detector at a particular angle θ (scattered angles) as shown in Fig. 10.2.

All samples were analyzed in suspension immediately after sonication in order to avoid particle growth due to aggregation. Mean values of each parameter were obtained from the analysis of three different batches at scattering angle of 90° at 25 °C. Hydrodynamic diameter of Cu- and Zn-based chitosan NPs was reported in

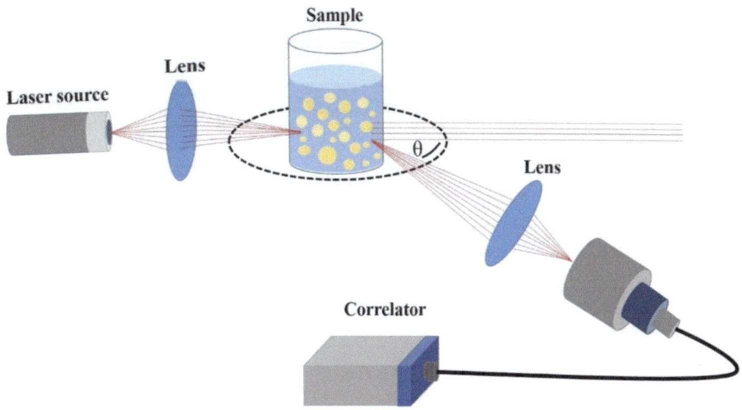

Fig. 10.2 Illustration of the dynamic light scattering principle

the range of 200–600 nm (Saharan et al. 2013; Dhillon et al. 2014). PDI is an important characteristic of NPs, which explains the monodispersity/polydispersity of particles in aqueous medium. Higher value of PDI (>0.5) normally showed the polydispersity and lower value (<0.5) represents the monodispersity of particles. In chitosan NPs, monodisperse particles tend to show PDI value from 0.2 to 0.4. Zeta potential is another feature of NPs which decides the stability of nanomaterials. NPs will remain stable at higher ζ-potential due to higher electrostatic repulsion between particles. Cu-/Zn-based chitosan NPs normally express ζ-potential from +20 to 60 mV. In our study, the average particle size, PDI, and ζ-potential of Cu–chitosan NPs were 338.2 nm, 0.258, and 34.4 mV, whereas in Zn–chitosan NPs, the average particle size, PDI, and ζ-potential were found to be 407.2 nm, 0.136, and 35.2 mV, respectively, as shown in Fig. 10.3.

10.3.2 Fourier Transform Infrared (FTIR) Spectroscopy

FTIR studies can be performed for chemical structure of bulk chitosan and chitosan NPs and interaction between different components of nanoparticulate systems (Saharan et al. 2015). The functional groups of metal-based chitosan NPs could be illustrated as shifting of vibration from higher to lower wave number. In FTIR, samples are gently triturated with KBr and compressed into disks and scanned against a blank KBr pellet background. For each spectrum, a 32-scan interferogram was collected in transmittance/absorbance mode at 25 °C in the 4000–400 cm^{-1} region.

In bulk chitosan, a band at 3424 cm^{-1} indicates stretching vibration of combined peaks of -NH_2 and -OH groups, while some specific peaks at 1647 and 896 cm^{-1} indicate amide (-$CONH_2$) and anhydro-glucosidic ring, respectively (Fig. 10.4a).

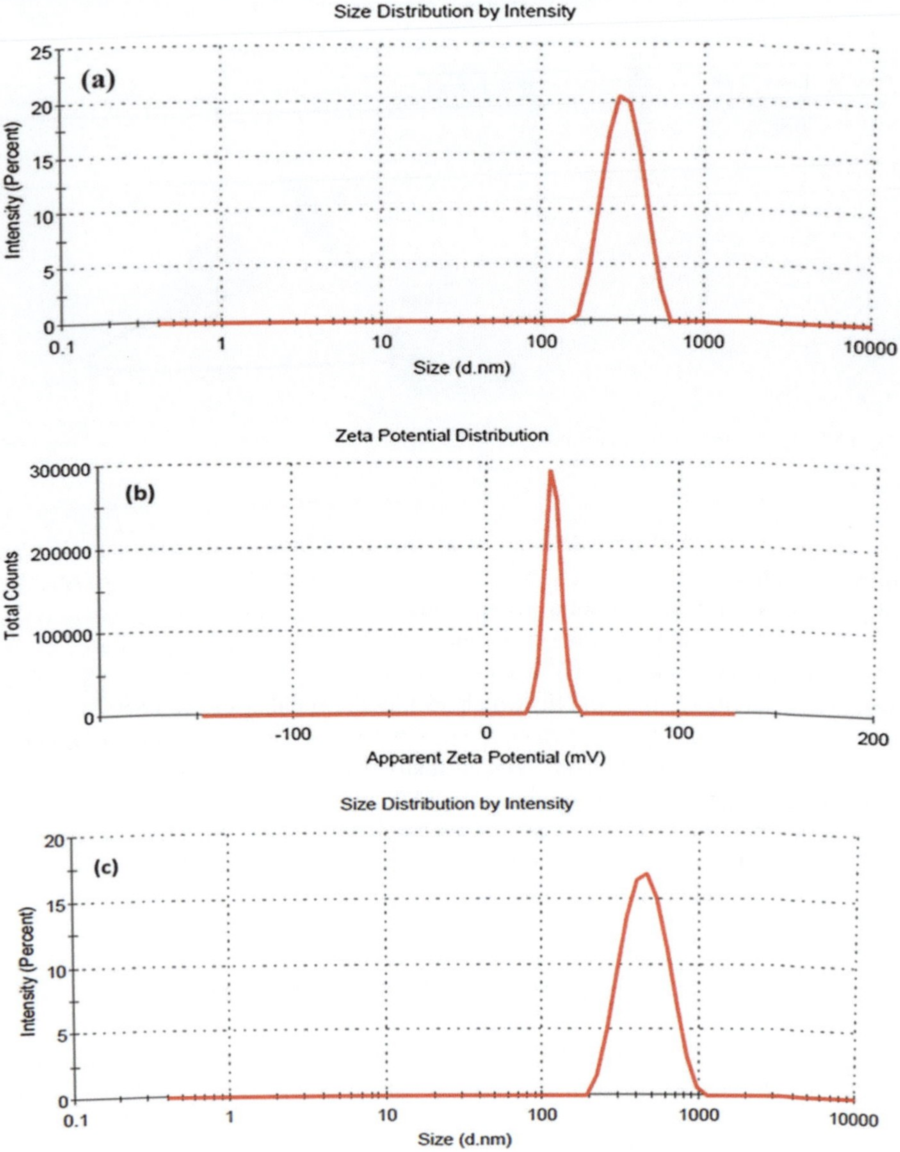

Fig. 10.3 Average particle size and zeta potential of Cu–chitosan (**a**, **b**) and Zn–chitosan NPs (**c**, **d**)

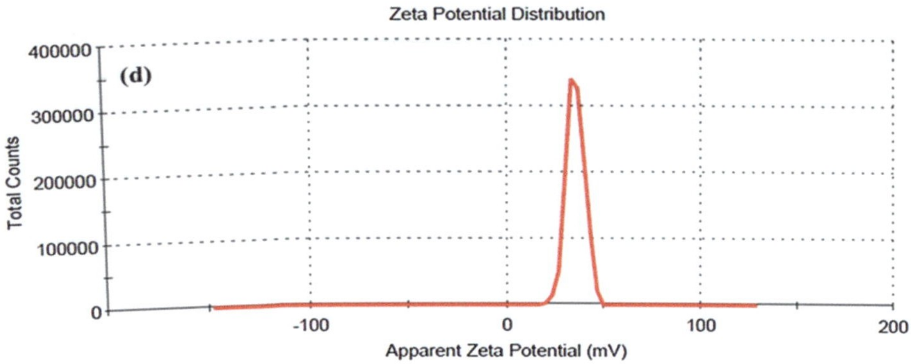

Fig. 10.3 (continued)

In case of Cu–chitosan NPs, peaks at 1639 (-CONH$_2$) and 1540 cm^{-1} (-NH$_2$) cm^{-1} are sharper and may be due to interaction of copper with chitosan (Fig. 10.4b). While, in Zn–chitosan NPs, peaks at 1640, 1541, and 893 showed sharper and shifting of bands at 1639, 1540, and 894 cm^{-1} (as in Fig. 10.4b) for -CONH$_2$, -NH$_2$, and anhydro-glucosidic ring, respectively, resulting from interaction of Zn with chitosan (Fig. 10.4c). The broadening of the -OH peak is the result of intense hydrogen bonding in bulk chitosan, whereas reduced amount of hydrogen bonding results from cross-linking of chitosan with TPP in chitosan NPs.

10.3.3 Scanning Electron Microscope (SEM)

In SEM, high-energy electron beam is used to analyze the surface of NPs, as chitosan-based NPs are nonconductive in nature; therefore, the material is subjected to sputter coating. It involves applying of ultrathin coating (2–20 nm) of electrically conductive metals like Au, Pt, Ag, and Cr on non-/poorly conducting specimen. Jaiswal et al. (2012) reported a porous surface of Cu–chitosan composite by SEM analysis.

In a recent study by Saharan et al., the group elucidated nano- and microsize pores on the Cu–chitosan NP surface at different magnification (Fig. 10.5). Saharan et al. (2015) described the Cu sorption mechanism by a porous surface of chitosan nanomaterials based on SEM study. Which was further justify through EDS spectra of porous chitosan nanomaterials where higher Cu deposition was found in porous area of NPs and confirm the presence of C, P, O, and N in Cu in chitosan NPs (Fig. 10.6a, b).

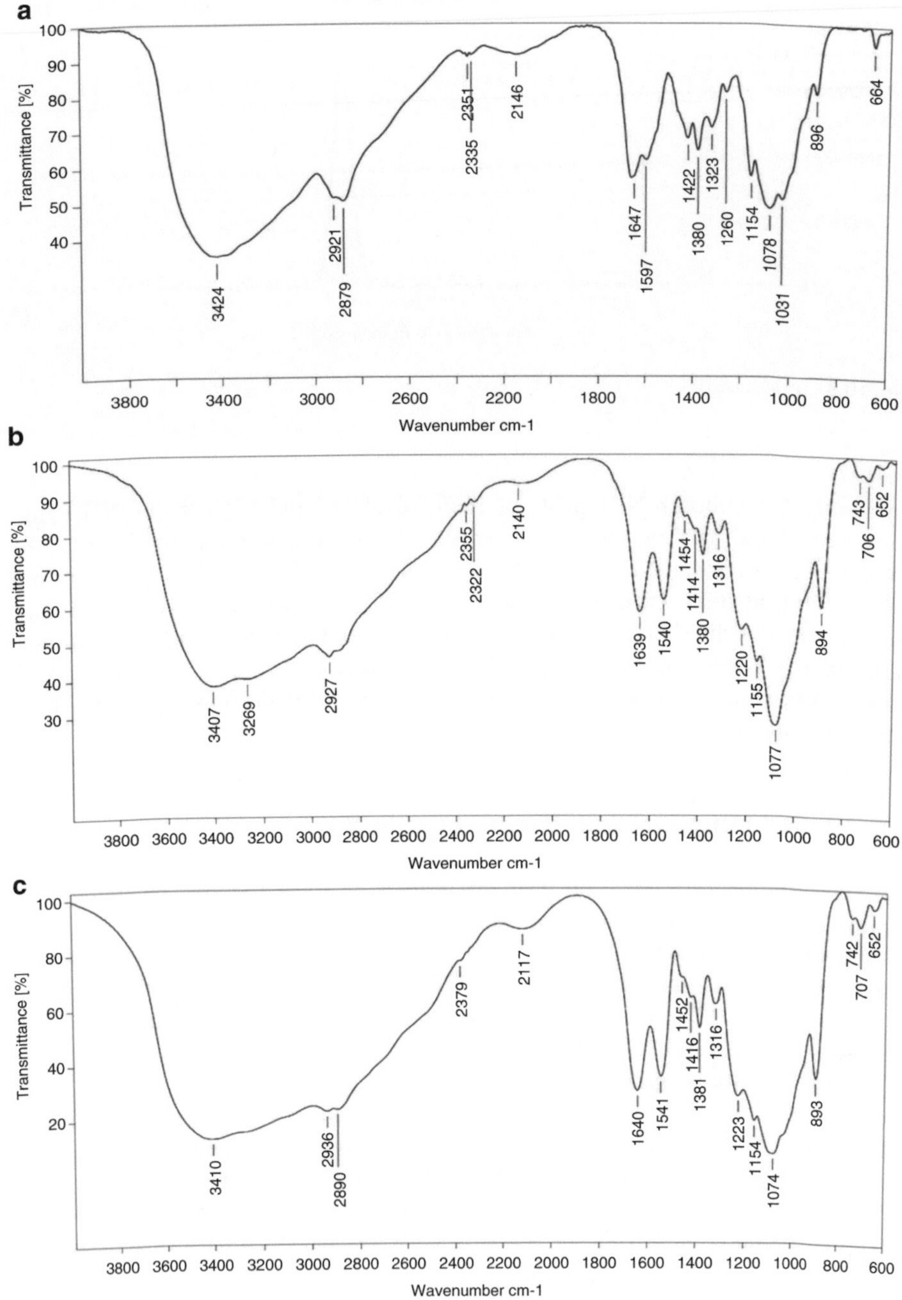

Fig. 10.4 FTIR spectra: Chitosan (**a**) Cu–chitosan (**b**) and Zn–chitosan NPs (**c**)

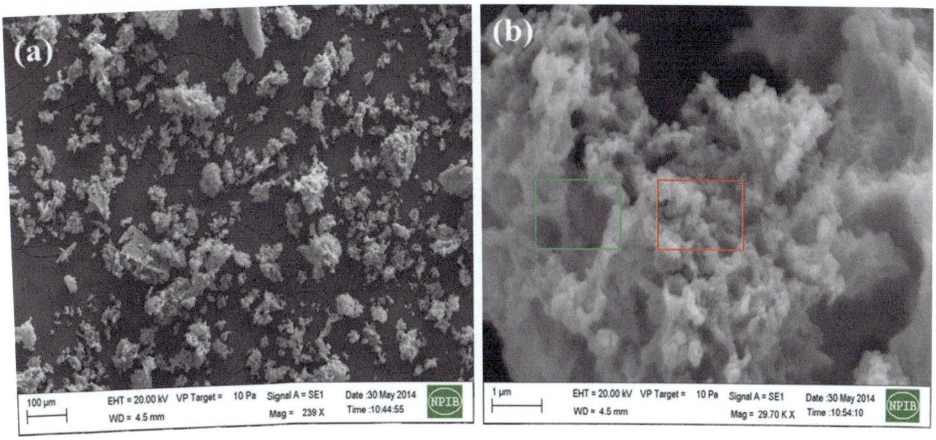

Fig. 10.5 SEM micrographs: (**a**) Cu–chitosan NPs at 230X (**b**) Porous Cu–chitosan at 29.70KX (microsize in *green* rectangular and nanosize pores in *red* rectangular (Adopted from Saharan et al. 2015)

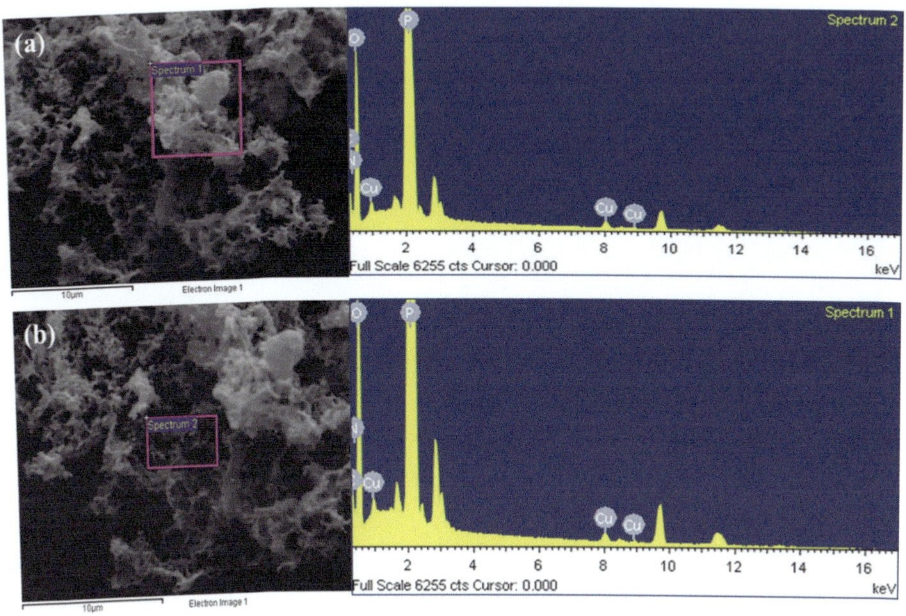

Fig. 10.6 SEM-EDS elemental analysis of Cu–chitosan NPs (**a**) Spectra of nonporous surface and spectra of porous surface (**b**) (Adopted from Saharan et al. 2015)

10.3.4 Transmission Electron Microscopy (TEM)

TEM is said to be an excellent tool for detection of internal structure as well as the size of Cu- and Zn-based chitosan NPs because of its high resolution. It can also be used for determination of structure, encapsulation of a particular metal ion in chitosan NPs, and interaction and localization of nanoparticle in the living system. In TEM, the beam of electron is used and interacts with the sample to form an image on a photographic plate. Typically, TEM sample preparation consists of droplet of sample suspension on a carbon film-covered copper grid followed by drying a particle suspension before observation by TEM (Fan et al. 2012). Generally, chitosan NPs show spherical shape under TEM. Liu and Gao (2009) have showed the surface appearance of chitosan NPs as spherical shape and smooth surface with size a range from 150 to 350 nm. Saharan et al. (2013) also reported the spherical structure of Cu–chitosan NPs. Similarly, Brunel et al. (2013) reported that Cu-loaded chitosan NPs have spherical structure with size ranging from 100 to 500 nm, and darker background of Cu–chitosan NPs might be due to the presence of copper in NPs. Thus, these NPs appeared white in darker background (negative staining). In another study, Manikandan and Sathiyabama (2015) showed spherical shape of Cu–chitosan NPs using high-resolution electron microscope (HRTEM). The difference between the size measured by DLS and TEM might be due to DLS measurement used in the aqueous solution, whereas in TEM the chance of aggregation increased with water vaporization in preparation of the sample.

Low-magnification TEM micrograph shows aggregated particles and at higher magnification shows porous network structure (Fig. 10.7a, b).

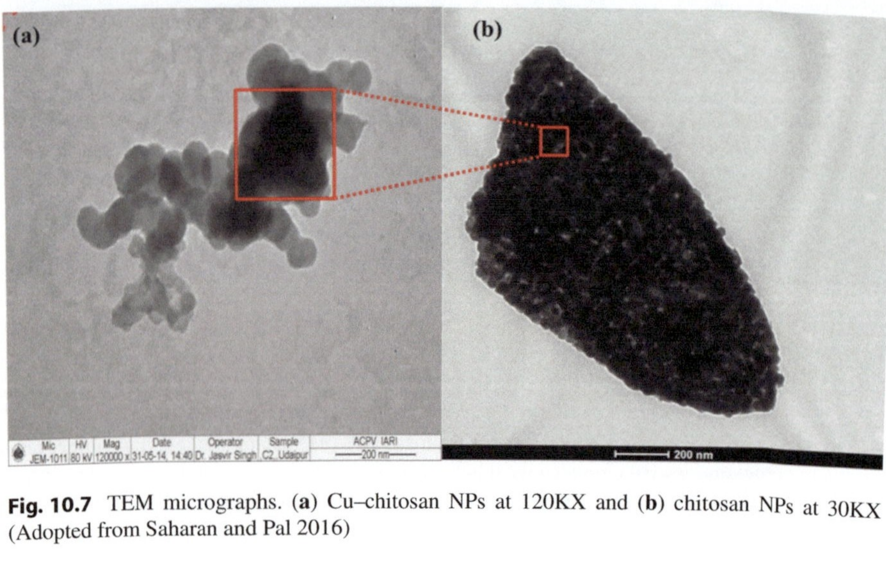

Fig. 10.7 TEM micrographs. (**a**) Cu–chitosan NPs at 120KX and (**b**) chitosan NPs at 30KX (Adopted from Saharan and Pal 2016)

10.3.5 Atomic Absorption Spectroscopy (AAS)

Atomic absorption spectroscopy is one of the most widely used methods for quantitative analysis of various elements. It has been used for measure release profile or encapsulation efficacy of metals in chitosan NPs. This method is principally based on excitation of samples by radiation and reading of the spectra produced by it. The atoms of metals absorb a particular wavelength of light and make transitions to excited states, and this amount of energy is measured in the form of photons of light in the samples, since each atom or element has different levels of energy and gives rise to different spectra (Jaiswal et al. 2012). Cu encapsulation was reported from 70 to 80% in various studies as confirm by AAS (Saharan et al. 2013). This might be due to porous structure of chitosan NPs. The effect of various factors like pH, agitation rate time, and temperature effect on release of metal ions from nanoformulations can be studied by AAS (Saharan et al. 2013).

10.4 Application of Cu and Zn Chitosan NPs

Chitosan biopolymer has been reported as excellent nanodelivery system for delivery of micronutrients and agrochemicals for crop improvement. This has been achieved by converting chitosan biopolymer into nanosized materials to encapsulate biological active components. In this regard, chitosan nanomaterials have been synthesized by incorporating micronutrients like Cu and Zn (Table 10.1).

Copper and zinc are active components of many agrochemicals like fertilizer and pesticides and serve as cofactors of several enzymes in plant. Therefore, Cu and Zn ultimately lead higher plant growth and participated in plant disease control. Heavy metal toxicity is a major constraint for limitation for crop production. At excess concentration, it reported phytotoxic at biochemical, physiological, and morphological levels (Adrees et al. 2015). Copper toxicity induces alteration in the leaf area, stem size, and leaf length in maize (Barbosa et al. 2013). Feigl et al. (2013) reported reduced stem size in Indian mustard and rapeseed under copper stress. Additionally, fresh weight of roots, shoots, and leaves of wheat (Azooz et al. 2012) and maize seedling decreased with excess Cu (Dresler et al. 2014). Zinc at higher level negatively affects growth parameter and structure of plant parts. In tomato, elevated level of Zn decreased the length of roots and shoots and area of leaves (Vijayarengan and Mahalakshmi 2013). Likewise, Todeschini et al. (2011) reported Zn toxicity in poplar which leads to changed leaf morphology, ultrastructure, and formation of calcium oxalate crystals. Therefore, encapsulation of these micronutrients into nanosize, slow-releasing structure would be use to overcome the higher dose in fixed time line.

Table 10.1 Application of chitosan NPs and Cu- and Zn-based chitosan NPs

Purpose	Material	Finding	References
Antibacterial activity	Chitosan and Cu-loaded chitosan NPs	Inhibited growth *of E. coli*, *Salmonella choleraesuis*, *Salmonella typhimurium*, and *Staphylococcus aureus*	Qi et al. (2004)
Antibacterial activity	Ag+-, Cu2+-, Zn2+-, Mn2+-, and Fe2+-loaded chitosan NPs	Antibacterial activity was significantly enhanced by the all metal ions loaded, except Fe2+	Du et al. (2009)
Growth and biophysical characteristics	Chitosan NPs	Enhanced chlorophyll content, photosynthesis intensity, and uptake of nutrients in coffee	Van et al. (2013)
Antimicrobial activity	Cu–chitosan NPs	Effective against *Fusarium graminearum*	Brunel et al. (2013)
Antifungal activity	Chitosan-based NPs	Effective against *Alternaria alternata*, *Macrophomina phaseolina*, and *Rhizoctonia solani*	Saharan et al. (2013)
Plant growth and antifungal activity	Cu–chitosan NPs	Increased seed germination, seedling length, fresh and dry weight, and antifungal activity against *Alternaria solani* and *Fusarium oxysporum*	Saharan et al. (2015)
Plant immune response	Chitosan NPs	Induction of defense enzyme and defense-related genes in *Camellia sinensis*	Chandra et al. (2015)
Plant growth-promoting activity	Cu–chitosan NPs	Increased seedling growth by up regulation of amylase and protease enzyme in maize seed germination	Saharan et al. (2016)

10.4.1 Antimicrobial Activity

Over the last decades, considerable interests have been focused on various metal-based chitosan nanomaterials as potential antimicrobial agent. Nanochitosan acquired remarkable advantage over bulk chitosan due to large surface area and small size. Being nanosize of chitosan, it can easily interact with plant as well as microbial system that will lead to enhanced plant immune system, growth promotion, and antimicrobial activity (Van et al. 2013; Saharan and Pal 2016). More surface area and positively charged nanomaterials of Cu and Zn chitosan could interact easily to negatively charged cellular components of bacteria and fungi. This interaction provides an excellent antimicrobial activity. The antibacterial activity of Cu–chitosan NPs was tested against various bacteria, viz., *E. coli*, *Salmonella choleraesuis*, *Salmonella typhimurium*, and *Staphylococcus aureus*, and found that these nanoparticles could significantly inhibit their growth (Qi et al. 2004). The mycelial growth of several fungi including *A. alternate*, *M. phaseolina*, *R. solani*, and *F. oxysporum* was comprehensively controlled by various treatments of

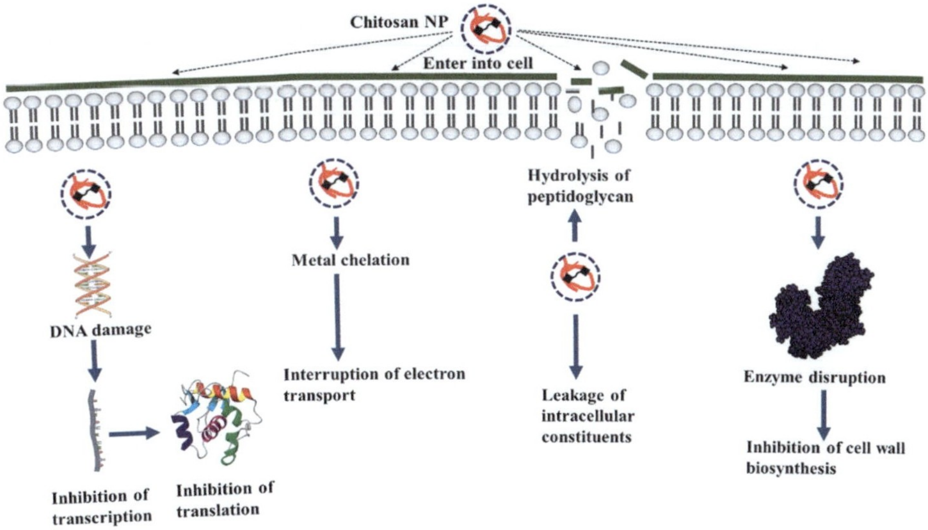

Fig. 10.8 A proposed mechanism for action of chitosan NPs for phytopathogenic bacteria and fungi

Cu–chitosan NPs and exhibited higher antifungal activity compared to bulk chitosan, saponin, and $CuSO_4$ (Saharan et al. 2013). In another study, Cu–chitosan NPs are found to be effective against *Alternaria solani* and *Fusarium oxysporum* than the bulk chitosan and $CuSO_4$ (Saharan et al. 2015).

Similarly, Zn-loaded chitosan NPs showed a wide spectrum of effective antimicrobial activities against various bacterial species including *E. coli*, *S. choleraesuis*, and *S. aureus*, which showed higher antibacterial activity than bulk chitosan, chitosan NPs, and Zn ions (Du et al. 2009). The cationic nature of chitosan and anionic component of microorganism's surface as cell membrane (LPS) and cell surface protein leads to electrostatic interaction which plays a crucial role in antimicrobial activity (Kong et al. 2008). Similarly, antimicrobial activity also depends on physicochemical properties of chitosan as well as the nature of microorganism. Cell surface hydrophobicity makes another mechanism of interaction between chitosan surface and bacteria (Saharan and Pal 2016). The increased interaction leads to membrane destabilization and subsequently to leakage of intracellular substances leading to the death of bacteria (Kong et al. 2008). Similarly, for antifungal and antiviral activity, the negatively charged components of fungal and viral surface like protein and glycoprotein also interact with chitosan (Sudarshan et al. 1992). In general, gram-negative bacteria were found to be more sensitive than gram-positive bacteria. A hypothetical model for mechanism of action of chitosan NPs against plant pathogen is shown in Fig. 10.8.

10.4.2 Seed Germination and Seedling Growth

Bulk chitosan induced several physiological and biochemical responses in plants. Seed germination and seedling growth were reported to enhance by chitosan materials in plants. In soybean, chitosan could enhance growth and yield (Dzung and Thang 2002); in rice, chitosan stimulated the growth and yield along with reinforcing the defense response (Boonlertnirun et al. 2008), while in case of rapeseed, seed soaked with chitosan could increase germination rate and length and weight of hypocotyls and radicles (Sui et al. 2002). Further, seed priming with chitosan increases seed germination and seedling vigor in pearl millet (Manjunatha et al. 2008) and germination rate in cucumber, chili, pumpkin, and cabbage (Chandrkrachang 2002).

Wheat growth was promoted in terms of germination capacity, root length, and seedling height by the action of oligochitosan (Ma et al. 2014).To determine the growth and toxicity of any type of nanoparticles, germination assay is the fundamental procedure in plants (Feizi et al. 2012). Positive response of bulk chitosan on seed germination and seedling growth leads to development of nanoformulation of chitosan. For enhancing the crop yield, early growth of seedling, viz., percent germination, shoot length, root length, fresh weight, dry weight, and seed vigor index, is important; therefore, the use of nanoformulation of chitosan could play a decisive role. Few reports indicated that nanoformulation of chitosan has more pronounced effect on seed germination and seedling growth as compared to bulk chitosan (Saharan et al. 2013, 2015, 2016).

The effect of chitosan NPs on seed germination and seedling vigor has been studied in various crops for promotion of plant growth by increasing and uptake of nutrients and water through adjusting cell osmotic pressure (Guan et al. 2009; Katiyar et al. 2015). Few studies have been carried out on the use of Cu/Zn chitosan NPs in seed germination and seedling growth. A recent study revealed that Cu–chitosan NPs enhance the maize seedling significantly after 4 h of seed treatment. Cu–chitosan NPs at concentration of 0.01, 0.04, 0.08, 0.12, and 0.16% comprehensively induced various characters of seedling like seed germination, root length, shoot length, root number, seedling length, fresh and dry weight, and SVI. At concentration 0.04, 0.08, and 0.12%, higher value of shoot length, root length, root number, seedling length, fresh weight, and SVI has been recorded. 0.01% $CuSO_4$ and 0.16% of Cu–chitosan NPs showed toxic effect on seedling growth. However, percent germination and dry weight were not significantly varied but comparatively higher in chitosan NP treatments. The growth promotory effect of bulk chitosan has been recorded significantly lower as compared to chitosan NPs. Similarly, as compared to control and $CuSO_4$, bulk chitosan has been reported to have higher value for all parameters except for percent germination and SVI (Saharan et al. 2016). In another study, Cu–chitosan NPs, at 0.08, 0.10, and 0.12% treatments, showed significantly growth promotory effect on seed germination, seedling length, and fresh and dry weight in tomato plants (Fig. 10.9).

Zn–chitosan formulation has been evaluated for growth and development in dry bean (Ibrahima and Ramadan 2015). In Robusta coffee, nanochitosan significantly

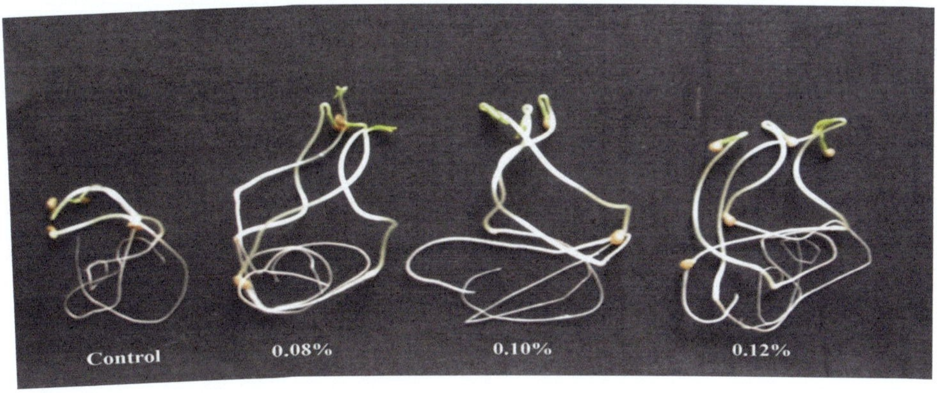

Fig. 10.9 Effect of Cu–chitosan NPs on growth of tomato seedling (Source: Saharan et al. 2015)

increased chlorophyll content, photosynthetic intensity, nutrient uptake, and seedling growth (Van et al. 2013).

10.4.3 Biochemical Response

Chitosan is actively involved in biochemical and molecular events in plant cells. These events lead to synthesis of various biomolecules which contributed to plant growth and protection from biotic and abiotic stress. Application of chitosan showed several effects, which mainly include accumulation of phytoalexins, PR protein synthesis, hypersensitive response, and synthesis of jasmonic acid (JA) and abscisic acid (ABA) in plants. A further study revealed that chitosan actually induces the expression of various genes involved in plant growth and protection including defense response genes encoding PAL (phenylalanine ammonia-lyase) enzyme and protease inhibitors (Hadwiger et al. 2002; Katiyar et al. 2015). Other biochemical and molecular events involved in chitosan application observed in plants are increases in cytosolic Ca^{2+} ion concentration, plasma membrane H^+-ATPase inhibition, transduction, activation of MAP kinases, and chromatin alterations. In maize, Cu–chitosan NPs have increased soluble sugar content by the action of α-amylase activity and free amino acid content by protease action (Saharan et al. 2016).

10.4.3.1 Storage Food Mobilizing Enzyme

For successful seed germination, the first requirement is degradation of reserve food through storage food mobilizing enzyme or de novo production of these enzymes for nourishment of the embryo as energy source. The chitosan component of nanomaterial is also found to enhance seed germination and subsequent growth by enhancing the activity of starch, protein, and lipid-degrading enzyme. The activity of hydrolytic enzymes such as α-amylase and protease is also found to increase the soluble sugar and free amino acids, respectively (Fig. 10.10). In wheat and maize, chitosan has been reported enhancing the activities of hydrolytic and antioxidant

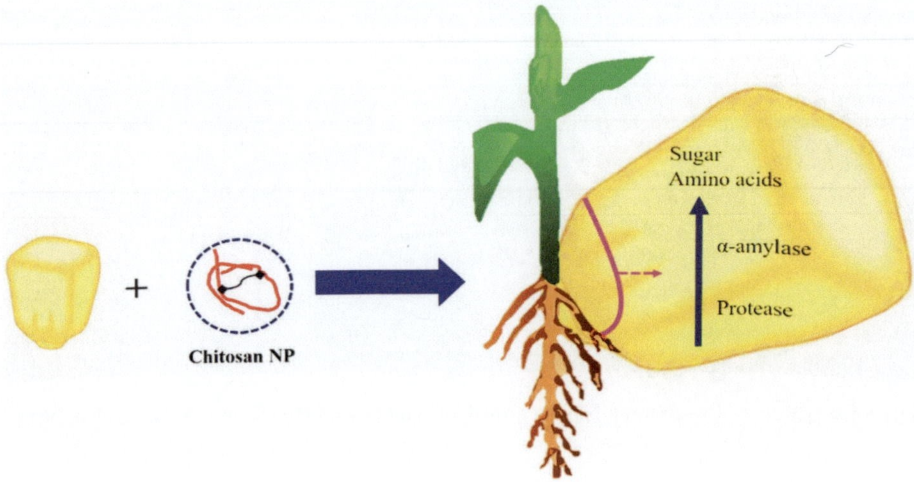

Fig. 10.10 Schematic representation of activity of chitosan NPs on seed germination and seedling development in maize

enzymes for release of assimilates for the growing embryo in seeds (Guan et al. 2009; Hameed et al. 2013). In another study, α-amylase and protease activity were recorded at 0, 1, 3, 5, 7, and 9 days of seed germination in maize.

The α-amylase activity was increased from the first to fifth day, and the maximum activity was measured at the fifth day in Cu–chitosan NP treatments (0.01, 0.04, 0.08, 0.12, and 0.16%) as compared to other treatments (control, bulk chitosan, and $CuSO_4$). While, at 0 day, negligible activity was measured in all treatment including chitosan NPs. After the fifth day, the activity was declined due to less availability of starch. At 0.16%, Cu–chitosan NP sand $CuSO_4$, lower activity of α-amylase activity was observed among treatments. Bulk chitosan induces higher α-amylase activity as compared to control, $CuSO_4$, and 0.16% Cu–chitosan NPs. Similarly, protease activity was also influenced by Cu–chitosan NPs and found maximum at the third day of germination in all treatment of Cu–chitosan NPs except at 0.16%. On the third day, the minimum protease activity was observed in 0.16% Cu–chitosan NPs followed by $CuSO_4$. In all treatments, protease activity declined after the third day and reached minimum at 9 days of germination. These results suggest that the increased activity of α-amylase and protease activity that may be induced by the action of Cu–chitosan NPs leads to and finally induces mobilization of starch and protein (Saharan et al. 2016). Zn–chitosan NPs showed higher seedling growth in maize as compared to bulk Zn. Nanoformulation of Zn–chitosan may be involved by enhancing the activity of storage food mobilizing enzyme (unpublished data). On the basis of these facts, it can be assumed that chitosan may be involved in upregulation of genes related to carbohydrate metabolism in plants. Likewise, Cu and Zn as structural and catalytic component of many regulatory proteins may accelerate metabolic process in germinating seeds through component of chitosan NPs.

10.4.3.2 Plant Defense System

Chitosan acts as plant defense booster through enhancing defense responses and plant immunity, which results from several defense-related enzymes under biotic and abiotic stresses. In several plant species, chitosan induces defense mechanism including cucumber, tomato (Ben-Shalom et al. 2003), strawberry fruits (El Ghaouth et al. 1992), chili (Photchanachai et al. 2006), and rose shrub (Wojdyla 2004). In rice, it stimulates hydrogen peroxide (H_2O_2) production (Lin et al. 2005), nitric acid (NO) pathway in tobacco (Zhang et al. 2011), and drought resistance in coffee (Dzung et al. 2011). Furthermore, synthesis of phytoalexin (Kim et al. 2005), jasmonic acid–ethylene signaling in rapeseed (Yin et al. 2013), activation of mitogen-activated protein kinases (MAPKs) (Yin et al. 2010). It may be involved in the signaling pathway of biosynthesis of phenolic compounds and can induce plant pathogenesis-related proteins (chitinase and chitosanase) that can degrade the cell wall of phytopathogens (Dixon et al. 1994). Moreover, it also participates in broad-based resistance against different phytopathogens in systemic acquired resistance (SAR) which is long-lasting and developed in uninfected plant parts and becomes more resistant to secondary infection (Rakwal et al. 2002). Chitosan also triggers lignification and suberization processes which provide mechanical support and act as barrier for invading pathogen.

In *C. sinensis*, chitosan NPs are found to induce activation of defense-related enzymes and antioxidant enzymes, namely, peroxidase (POX), polyphenol oxidase (PPO), phenylalanine ammonia-lyase (PAL), β-1,3-glucanase and superoxide dismutase (SOD), and catalase (CAT), respectively. These enzymes are responsible for scavenging of reactive oxygen species (ROS) by the oxidation of phenolic compounds for resistance to pathogens in the host (Chandra et al. 2015). In tobacco and *Brassica napus*, oligochitosan is found to enhance the activities of PAL, PPO, POX, CAT, and SOD (Yin et al. 2008). In finger millet, chitosan NPs are found to enhance POX activity by the induction of reactive oxygen species, and later it may be reason for delayed symptom of blast disease caused by *Pyricularia grisea* (Cke.) Sacc (Manikandan and Sathiyabama 2016). In our research, Cu–chitosan and Zn–chitosan NPs enhance various defense enzymes significantly in maize plant as compared to bulk chitosan and bulk Cu/Zn. Therefore, Cu/Zn chitosan NPs are currently under research for protection from fungal diseases in maize crop (unpublished data). Altogether, chitosan has profound role in plant defense through multiples line of action (Fig. 10.11).

10.5 Future Prospects

To feed the world's growing population, agriculture production must be doubled up to 2,050. Introduction of nanotechnology in agriculture could bring revolution in production sustainability and fulfill the future demand. Nanotechnology has been exploited immensely in other fields but its use in agriculture is in budding stage. The development of "on-command" or "on-demand" delivery system is a prime requirement for agriculture for sustainable growth (Prasad et al. 2017). Indiscriminate

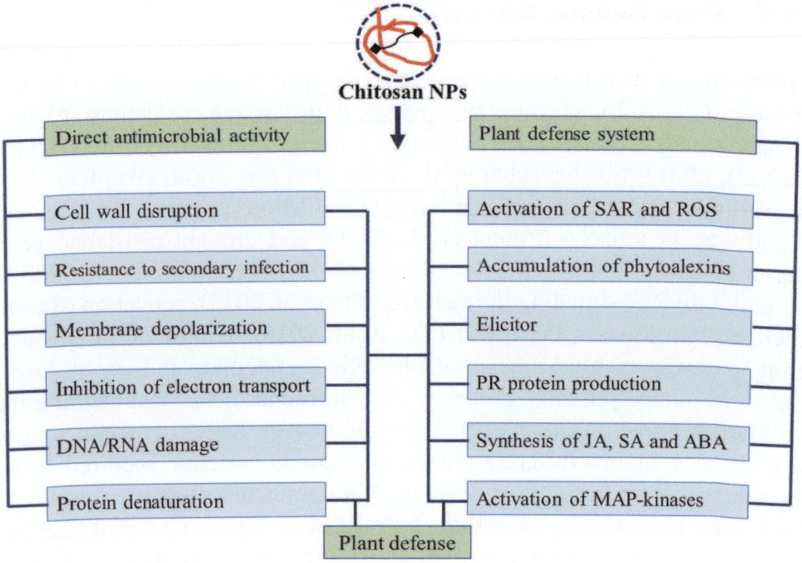

Fig. 10.11 Roles of chitosan in antimicrobial action and plant defense system

application of agrochemicals in agriculture has done unrepairable damaged to the ecosystem which needs to be ameliorated. The controlled use of agrochemicals could be possible by the development of smart delivery system using biomaterials. Encapsulation of various biological active compounds including agrochemicals in chitosan is an option for controlled and sustained delivery. In connection to this, Cu- and Zn-based chitosan NPs possessing broad-spectrum antimicrobial activity act as plant defense booster and plant growth with regard to controlling plant diseases. Cu and Zn have traditionally been used as a component of agrochemicals. Cu and Zn, as components of nanoparticulate system, may reduce the risk of hazardous agrochemicals for crop improvement and protection. Due to high metal chelation ability of chitosan, metallic-based chitosan nanomaterials can be developed to fulfill the deficiency of micronutrient, antimicrobial action, and plant growth.

Appropriate modification with chitosan nanomaterials by functionalizing it with metals and other biological active compound could significantly enhance its bioactivity toward plants through improvement of physical and chemical properties.

References

Adrees M, Ali S, Rizwan M, Ibrahim M, Abbas F, Farid M, Zia-ur-Rehman M, Irshad MK, Bharwana SA (2015) The effect of excess copper on growth and physiology of important food crops: a review. Environ Sci Pollut Res 22:8148–8162

Agnihotri SA, Nadagouda NM, Tejraj M (2004) Recent advances on chitosan-based micro- and nanoparticles in drug delivery. J Control Release 100:5–28

Ali SW, Rajendran S, Joshi M (2011) Synthesis and characterization of chitosan and silver loaded chitosan nanoparticles for bioactive polyester. Carbohydr Polym 83:438–446

Azooz MM, Abou-Elhamd MF, Al-Fredan MA (2012) Biphasic effect of copper on growth, proline, lipid peroxidation and antioxidant enzyme activities of wheat (*Triticum aestivum* cv. Hasaawi) at early growing stage. Aust J Crop Sci 6:688–694

Babu RB, O'Connor K, Seeram R (2013) Current progress on bio-based polymers and their future trends. Prog Biomater 2:8

Badawy MEI, Rabea EI (2011) A biopolymer chitosan and its derivatives as promising antimicrobial agents against plant pathogens and their applications in crop protection. Int J Carbohydr Chem 1:1–29

Barbosa RH, Tabaldi LA, Miyazaki FR, Pilecco M, Kassab SO, Bigaton D (2013) Foliar copper uptake by maize plants: effects on growth and yield. Cienc Rural 43:1561–1568

Ben-Shalom N, Ardi R, Pinto R, Aki C, Fallik E (2003) Controlling gray mould caused by *Botrytis cinerea* in cucumber plants by means of chitosan. Crop Prot 22:285–290

Boonlertnirun S, Boonraung C, Suvanasara R (2008) Application of chitosan in rice production. J Met Mater Miner 18:47–52

Brunel F, Gueddari NEE, Moerschbacher BM (2013) Complexation of copper (II) with chitosan nanogels: toward control of microbial growth. Carbohydr Polym 92:1348–1356

Carvalho FP (2006) Agriculture, pesticides, food security and food safety. Environ Sci Pol 9:685–692

Chandra S, Chakarborty N, Dasgupta A, Sarkar J, Panda K, Acharya K (2015) Chitosan nanoparticle: a positive modulator of innate immune responses in plants. Sci Report 5:1–13

Chandrkrachang S (2002) The applications of chitin in agriculture in Thailand. Adv Chitin Sci 5:458–462

Corradini E, de Moura MR, Mattoso LHC (2010) A preliminary study of the incorporation of NPK fertilizer into chitosan nanoparticles. Express Polym Lett 4(8):509–515

Dhillon GS, Kaur S, Brar SK (2014) Facile fabrication and characterization of chitosan-based zinc oxide nanoparticles and evaluation of their antimicrobial and antibiofilm activity. Int Nano Lett 4:107

Dixon RA, Harrison MJ, Lamb CJ (1994) Early events in the activation of plant defenses. Annu Rev Phytopathol 32:479–510

Dresler S, Hanaka A, Bednarek W, Maksymiec W (2014) Accumulation of low-molecular-weight organic acids in roots and leaf segments of *Zea mays* plants treated with cadmium and copper. Acta Physiol Plant 36:1565–1575

Du WL, Niu SS, Xu YL, Xu ZR, Fan CL (2009) Antibacterial activity of chitosan tripolyphosphate nanoparticles loaded with various metal ions. Carbohydr Polym 75:385–389

Dzung NA, Thang NT (2002) In: Suchiva VK, Chandrkrachang S, Methacanon P, Peter MG (eds) Advances in chitin science, Bangkok, pp 463–467

Dzung NA, Phuong Khanh VT, Dzung TT (2011) Research on impact of chitosan oligomers on biophysical characteristics, growth, development and drought resistance of coffee. Carbohydr Polym 84:751–755

El Ghaouth A, Arul J, Grenier J, Asselin A (1992) Antifungal activity of chitosan on two postharvest pathogen of strawberry fruits. Phytopathology 82:398–402

Fan W, Yan W, Xu Z, Ni H (2012) Formation mechanism of monodisperse, low molecular weight chitosan nanoparticles by ionic gelation technique. Colloids Surf B: Biointerfaces 90:21–27

Feigl G, Kumar D, Lehotai N, Kolbert Z (2013) Physiological and morphological responses of the root system of Indian mustard (*Brassica juncea* L. Czern.) and rapeseed (*Brassica napus* L.) to copper stress. Ecotoxicol Environ Saf 94:179–189

Feizi H, Moghaddam PR, Shahtahmassebi N, Fotovat A (2012) Impact of bulk and nanosized titanium dioxide (TiO_2) on wheat seed germination and seedling growth. Biol Trace Elem Res 146:101–106

Guan YJ, Hu J, Wang XJ, Shao CX (2009) Seed priming with chitosan improves maize germination and seedling growth in relation to physiological changes under low temperature stress. J Zhejiang Univ Sci B 10:427–433

Hadwiger LA, Klosterman SJ, Choi JJ (2002) The mode of action of chitosan and its oligomers in inducing plant promoters and developing disease resistance in plants. In: Suchiva K, Chandrkrachang S, Methacanon P, Peter MG (eds) Advances in *chitin science*. MTEC, Bangkok, pp 452–457

Hameed A, Sheikh MA, Hameed A, Farooq T, Basra SMA, Jamil A (2013) Chitosan priming enhances the seed germination, antioxidants, hydrolytic enzymes, soluble proteins and sugars in wheat seeds. Agrochimica 57:32–46

Ibrahima EA, Ramadan WA (2015) Effect of zinc foliar spray alone and combined with humic acid or/and chitosan on growth, nutrient elements content and yield of dry bean (*Phaseolus vulgaris* L.) plants sown at different dates. Sci Hortic 184:101–105

Ing LY, Zin NM, Sarwar A, Katas H (2012) Antifungal activity of chitosan nanoparticles and correlation with their physical properties. Int J Biomater 2012:1–9

Jaiswal M, Chauhan D, Sankararamakrishnan N (2012) Copper chitosan nanocomposite: synthesis, characterization, and application in removal of organophosphorous pesticide from agricultural runoff. Environ Sci Pollut Res 19:2055–2062

Katiyar D, Hemantarajan A, Sing B (2015) Chitosan as a promising natural compound to enhance potential physiological responses in plant: a review. Indian J Plant Physiol 20(1):1–9

Kim HJ, Chen F, Wang X, Rajapakse NC (2005) Effect of chitosan on the biological properties of sweet basil (*Ocimum basilicum* L.) J Agric Food Chem 53:3696–3701

Kong M, Chen XG, Liu CS, Liu CG, Meng XH, Yu LJ (2008) Antibacterial mechanism of chitosan microspheres in a solid dispersing system against *E. coli*. Colloids Surf B: Biointerfaces 65:197–202

Lin W, Hu X, Zhang W, Rogers WJ, Cai W (2005) Hydrogen peroxide mediates defence responses induced by chitosans of different molecular weights in rice. J Plant Physiol 162:937–944

Liu H, Gao C (2009) Preparation and properties of ionically cross-linked chitosan nanoparticles. Polym Adv Technol 20:613–619

Ma L, Li J, Yy YCM, Wang Y, Xm L, Li N (2014) Germination and physiological response of wheat (*Triticum aestivum*) to pre-soaking with oligochitosan. Int J Agric Biol 16:766–770

Manikandan A, Sathiyabama S (2015) Green synthesis of copper-chitosan nanoparticles and study of its antibacterial activity. J Nanomed Nanotechnol 6:1

Manikandan A, Sathiyabama M (2016) Preparation of Chitosan nanoparticles and its effect on detached rice leaves infected with *Pyricularia grisea*. Int J Biol Macromol 84:58–61

Manjunatha G, Roopa KS, Prashanth GN, Shekar Shetty H (2008) Chitosan enhances disease resistance in pearl millet against downy mildew caused by *Sclerospora graminicola* and defence-related enzyme activation. Pest Manag Sci 64:1250–1257

Parisi C, Vigani M, Cerezo ER (2015) Agricultural nanotechnologies: what are the current possibilities? Nano Today 10:124–127

Photchanachai S, Singkaew J, Thamthong J (2006) Effects of chitosan seed treatment on *Colletotrichum sp*. and seedling growth of chili cv. 'Jinda. Acta Hortic 712:585–590

Piras AM, Maisettab G, Sandreschia S, Esinb S, Gazzarria M, Batonib G, Chiellini F (2014) Preparation, physical-chemical and biological characterization of chitosan nanoparticles loaded with lysozyme. Int J Biol Macromol 67:124–131

Prasad R, Bhattacharyya A, Nguyen QD (2017) Nanotechnology in sustainable agriculture: recent developments, challenges, and perspectives. Front Microbiol 8:1014. doi:10.3389/fmicb.2017.01014

Qi L, Xu Z, Jiang X, Hu C, Zou X (2004) Preparation and antibacterial activity of chitosan nanoparticles. Carbohydr Res 339:2693–2700

Rajasekaran P, Santra S (2015) Hydrothermally treated chitosan hydrogel loaded with copper and zinc particles as a potential micronutrient-based antimicrobial feed additive. Front Vet Sci 2:621

Rakwal R, Tamogami S, Agrawal GK, Iwahashi H (2002) Octadecanoid signaling component "burst" in rice (*Oryza sativa* L.) seedling leaves upon wounding by cut and treatment with fungal elicitor chitosan. Biochem Biophys Res Commun 295:1041–1045

Rodrigues S, Rosa da Costa AM, Grenha A (2012) Chitosan/carrageenan nanoparticles: effect of cross-linking with tripolyphosphate and charge ratios. Carbohydr Polym 89:282–289

Saharan V, Pal A (2016) Chitosan based nanomaterials in plant growth and protection. Springer Brief Plant Sci. doi:10.1007/978-81-322-3601-6_3

Saharan V, Mehrotra A, Khatik R, Rawal P, Sharma SS, Pal A (2013) Synthesis of chitosan based nanoparticles and their *in vitro* evaluation against phytopathogenic fungi. Int J Biol Macromol 62:677–683

Saharan V, Sharma G, Yadav M, Choudhary MK, Sharma SS, Pal A, Raliya R, Biswas P (2015) Synthesis and in vitro antifungal efficacy of Cu-chitosan nanoparticles against pathogenic fungi of tomato. Int J Biol Macromol 75:346–353

Saharan V, Kumaraswamy RV, Choudhary RC, Kumari S, Pal A, Raliya P, Biswas P (2016) Cu-chitosan nanoparticle mediated sustainable approach to enhance seedling growth in maize by mobilizing reserved food. J Agric Food Chem 64(31):6148–6155

Sudarshan NR, Hoover DG, Knorr D (1992) Antibacterial action of chitosan. Food Biotechnol 6:257–272

Sui XY, Zhang WQ, Xia W, Wang Q (2002) Effect of chitosan as seed coating on seed germination and seedling growth and several physiological and biochemical indexes in rapeseed. Plant Physiol Commun 38:225–227

Todeschini V, Lingua G, Agostino GD, Carniato F, Roccotiello E, Berta G (2011) Effects of high zinc concentration on poplar leaves: a morphological and biochemical study. Environ Exp Bot 71(1):50–56

Van SN, Minh HD, Anh DN (2013) Study on chitosan nanoparticles on biophysical characteristics and growth of Robusta coffee in green house. Biocatal Agric Biotechnol 2(4):289–294

Vasconcelos MW (2014) Chitosan and chitooligosaccharide utilization in phytoremediation and biofortification programs: current knowledge and future perspectives. Front Plant Sci 5:616

Vijayarengan P, Mahalakshmi G (2013) Zinc toxicity in tomato plants. World Appl Sci J 24(5):649–653

Wojdyla AT (2004) Chitosan (biochikol 020 PC) in the control of some ornamental foliage diseases. Commun Agric Appl Biol Sci 69:705–715

Yang Z, Chen J, Dou R, Gao X, Mao C, Wang L (2015) Assessment of the phytotoxicity of metal oxide nanoparticles on two crop plants, maize (*Zea mays* L.) and rice (*Oryza sativa* L.) Int J Environ Res Public Health 12:15100–15109

Yin H, Bai XF, Du YG (2008) The primary study of oligochitosan inducing resistance to *Sclerotinia scleraotiorum* on *B. napus*. J Biotechnol 136:600–601

Yin H, Zhao X, Bai X, Du Y (2010) Molecular cloning and characterization of a *Brassica napus* L. map kinase involved in oligochitosan-induced defense signaling. Plant Mol Biol Report 28:292–301

Yin H, Li Y, Zhang HY, Wang WX, Lu H, Grevsen K, Zhao X, Du Y (2013) Chitosan oligosaccharides-triggered innate immunity contributes to oilseed rape resistance against *Sclerotinia Sclerotiorum*. Int J Plant Sci 174:722–732

Zhang H, Zhao X, Yang J, Yin H, Wang W, Lu H, Du Y (2011) Nitric oxide production and its functional link with OIPK in tobacco defense response elicited by chitooligosaccharide. Plant Cell Rep 30(6):1153–1162

Nanotechnology for Enhancing Crop Productivity

11

Suresh Kaushik and Setyowati Retno Djiwanti

Abstract

Agriculture is currently facing a number of challenges like low nutrient use efficiency, stagnation in crop yields, multi-nutrient deficiencies, climate change, and water availability. One of the frontier technologies like nanotechnology can be explored to detect precisely and supply the accurate quantity of plant nutrients and pesticides to enhance crop productivity in agriculture. Nanotechnology involves the designing, production, characterization and application of devices, structures, and systems by controlling the size and shape at nanometer scale. Nanotechnology using nanodevices and nanomaterials provides new avenues for potential novel applications in agriculture such as efficient delivery of pesticide and fertilizer using nanomaterial-based formulations such as nano-fertilizers, nano-pesticides, and nano-herbicides. New innovative smart delivery systems and sensitive nano-biosensor-based technology have great potential to solve the problems faced in crop production. This chapter summarizes some new developments in smart delivery systems and nano biosensor-based technology for enhancing crop productivity.

Keywords

Nanotechnology • Nanomaterials • Nano-pesticide • Nano-fertilizer • Crop productivity

S. Kaushik (✉)
Division of Soil Science and Agricultural Chemistry, Indian Agricultural Research Institute, New Delhi 110012, India
e-mail: drsckaushik@gmail.com

S.R. Djiwanti
Plant Protection Division, Indonesian Spice and Medicinal Crop Research Institute (ISMECRI), Tentara Pelajar Street No. 3, Bogor 16111, West Java, Indonesia
e-mail: retnomuslim@yahoo.com

R. Prasad et al. (eds.), *Nanotechnology*, DOI 10.1007/978-981-10-4573-8_11

11.1 Introduction

Agriculture presently is facing a number of challenges like stagnation in crop yields, low nutrient use efficiency, multi-nutrient deficiencies, the onset of pest and disease, water availability, shrinkable arable land, climate change, and shortage of labors. Moreover, demand for more food is increasing day by day due to the ever-increasing global population. This rapidly increasing global population needs sufficient food supply to feed millions of new mouths. For global food security, crop production must be increased severalfold sustainably with reduced inputs. Hence, there is an urgent need for proper management to utilize optimum plant nutrients and pesticides for enhancing crop production so as to cope with this alarming situation. For this, one of the frontier technologies like nanotechnology can be explored to detect precisely and supply the accurate quantity of plant nutrients and pesticides to increase crop production and conservation of inputs. Nanotechnology is the science in which materials are manipulated at nanoscale displaying different properties from bulk materials due to their size and surface area. This is the fast-growing field of science and technology. Nanotechnology and its derived applications are very important in the area of agriculture to address a number of issues pertaining to sustainable agricultural inputs aiming for enhancing crop productivity. Nanotechnology has several applications in various stages of crop production, processing, storing, packaging, and transporting of agricultural products. This chapter briefly reviews and describes some of the applications of nanotechnology in crop production practices to enhance crop productivity.

11.2 Nanotechnology

The term "nanotechnology" was first defined in 1974 by Norio Taniguchi of the Tokyo Science University as the study of manipulating matter on an atomic and molecular scale (Agrawal and Rathore 2014). The definition of nanotechnology is based on the prefix "nano" which is from the Greek word meaning "dwarf." A nanometer is one billionth of a meter. The word nanotechnology is generally used when referring to materials with the size of 1–100 nanometers (nm) and deals with the alteration of materials at nanoscale and creation of nano-sized materials generally less than 100 nm at least in one dimension. These materials display different properties from bulk materials due to their size. These differences include physical strength, chemical reactivity, electrical conductance, magnetism, and optical effects. Nanotechnology involves the manipulation or self-assembly of individual atom or molecule or molecular cluster into structures to create materials and devices with new or vastly different properties.

Nanoparticles are synthesized from organic or inorganic nanomaterials (NMs) using a number of chemical and physical methods (Table 11.1). The techniques for synthesis of nanoparticles are usually involved either a bottom-up or a top-down methods (Fig.11.1). In top-down approach, the size of particles is reduced by different chemical and physical procedures like milling, homogenization using high

Table 11.1 Some examples of nanomaterials and their applications

Nanomaterials	Applications
Inorganic	
Metal nanoparticles	
AgO, TiO_2, ZnO, CeO_2; Fe_2O_3	Delivery of biomolecules (proteins, peptides, nucleic acids), biosensors, diagnostic techniques, pesticide degradation
FePd, Fe-Ni; Silica; CdTe, CdSe	
Clay	
Montmorillonite layered double hydroxides	Delivery of pesticides, fertilizers, plant growth-promoting factors
Organic	
Carbon nanotubes	
Nanofibers	Biocatalysts, sensing
Lipids	Delivery of DNA, and pesticides, essential oils
Liposomes	
Lippopolyplexes	
Solid lipid nanoparticles	
Polymeric	Biocompatible, biodegradable, non-toxic for delivery of DNA/RNA
Natural	
Cellulose, starch, gelatin, albumin	
Chitin, chitosan	
Synthetic	Delivery of pesticides and DNA/RNA
Dendrimers	
Polyethylene oxide	
Polyethylene glycol	
Polylactides	

Source: Ghormade et al. (2011)

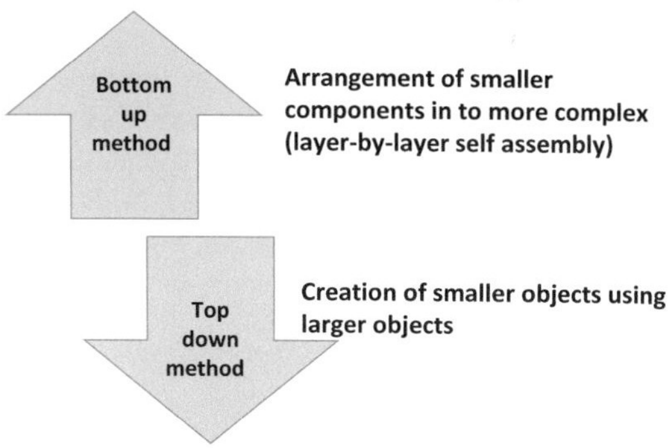

Fig. 11.1 Two approaches of synthesis of nanoparticles

pressure, and sonication, while in bottom-up method, the nanostructure building blocks of the nanoparticles are formed first and, subsequently, assembled to produce the final nanoparticles. Metal oxide nanoparticles such as ZnO, TiO_2, MgO, and AgO are some examples of inorganic materials. Metallic nanomaterials are very interesting materials with unique electronic and electro-catalytic properties depending on their size and morphology. They can be utilized for nanostructured materials with specific forms like quantum dots (QDs). Some other examples of inorganic materials are clay nanoparticles having a structure of stacked platelets. Organic materials like carbon nanotubes (CNTs), polymers, and lipids are also materials with various applications in nanotechnology.

Quantum dots are inorganic nanocrystals used for development of optical biosensors which are widely used to detect organic compounds and biomolecules such as proteins, enzymes, and amino acids. Similarly, surface plasmon resonance (SPR) is a robust tool that can measure the binding kinetics of two molecules without the help of any fluorescent tag. Thus, this technique can be adopted for biomolecular interaction analysis taking less time to detect binding events. Dendrimers are synthetic, organic macromolecules with highly branched and defined three-dimensional structures which provide a high degree of surface functionality and versatility. Similarly, aptamers are single-stranded nucleic acid which fit for the target in all the way forming three-dimensional structures with strict bonding of target specific binding with high affinity. This kind of nanosensors gives more specific and effective detection of plant diseases.

Carbon nanotubes are hollow cylindrical tubes with one (single-walled CNTs), two (double-walled CNTs), or several (multi-walled CNTs) concentric graphite layers capped by fullerenic hemisphere. Easy electron transfer reactions and high surface area by virtue of unique metallic, structural, and electronic characteristics of CNTs make them an important category of nanomaterials. Lipid nanoparticles used as delivery systems are composed of aqueous dispersion of dry powder as well as solid lipids like triglycerides, waxes, steroids, fatty acids, and emulsifiers and are prepared by homogenization process using high pressure. Polymeric nanoparticles prepared from natural and synthetic polymers by wet synthetic methods are commonly used because surface modification can be carried out easily in subsequent steps for stability. Nanoparticles prepared from natural sources and biopolymers have some merits of biocompatibility and biodegradability.

A biosensor is a device that integrates a biological recognition element with a physical or chemical transducer to detect a biological product. A biosensor comprises broadly three components: the biological recognition element, the transducer, and the signal processing electronics (Fig. 11.2). Different categories of nanomaterials such as nanotubes, quantum dots, or other biological nanomaterials have been used in biosensor technology to produce nano-biosensors (Prasad et al. 2016). These nanomaterials can contribute to either the bio-recognition element or the transducer or both of a biosensor. Nanoparticle-based biosensors are particularly attractive because they can

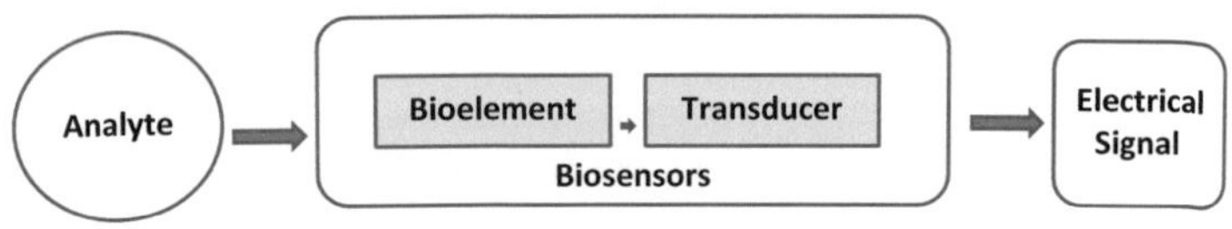

Fig. 11.2 A typical representation of a biosensor

be easily synthesized in bulk using standard chemical methods. Biosensors may be categorized according to the mechanism of biological selectivity (bioreceptor) or mode of physiochemical signal transduction (transducer). Bioreceptor is a molecular species which exploits a biochemical mechanism of recognition and is accountable for binding the concerned analyte to the sensor for measurement. Bioreceptor can broadly be categorized into various distinct classes such as antibody-antigen bioreceptor, enzymatic bioreceptor, nucleic acid bioreceptor, cellular bioreceptor, biometric bioreceptor, and bacteriophage bioreceptor. The transducer plays a crucial part in the detection as well as identification process of a biosensor. The transduction methods such as optical, electrochemical, and mass based are the most favored and universal methods. Several types of biosensors are being developed for different applications in enhancing agricultural crop productivity.

11.3 Potential Applications of Nanotechnology in Crop Productivity

As traditional approach for crop productivity is reaching their limits, agriculture has to adopt novel approaches based on nanotechnology to meet the demand of an ever-growing world population. Precision farming has been a long-derived aim to maximize output while minimize inputs through monitoring environmental variables and applying targeted action. One application of agricultural nanotechnology addresses low use efficiency of inputs. For example, controlled release mechanisms through nanoscale carriers avoid temporal overdose as well as reduce the amount of agricultural chemicals used and consequently minimize inputs and waste. The potential applications of nanotechnology in agricultural research include slow and control-released chemical fertilizers, micronutrients and biofertilizers for high efficiency and efficacy, delivery of chemical pesticides encapsulated in nanomaterials for slow and controlled release, stabilization of biopesticides with nanomaterials, agricultural diagnostics for pest and disease detection, water retention, and nano-genetic manipulation of agricultural crops (Prasad et al. 2014, 2017; Bhattacharyya et al. 2016) (Fig. 11.3).

11.3.1 Nano-fertilizers

Much of the fertilizers are not available to plants because of lost as runoff and leaching causing pollution. Nanotechnology has provided the feasibility of exploiting nanomaterials as fertilizer carriers for controlled release of nutrients. These smart fertilizers increase nutrient use efficiency and reduce costs of environmental protection. Nanomaterials having potential contribution in slow release of fertilizers can be used for encapsulation of fertilizers which is done by various ways such as encapsulation of nutrient inside nano-porous materials and coating of fertilizer with thin polymer film. Fertilizers can be delivered as emulsion or particle of nanoscale dimension. Fertilizers with sulfur nano-coatings are used for slow release of nutrients. Kaolin and polymeric biocompatible nanoparticles are some nanomaterials

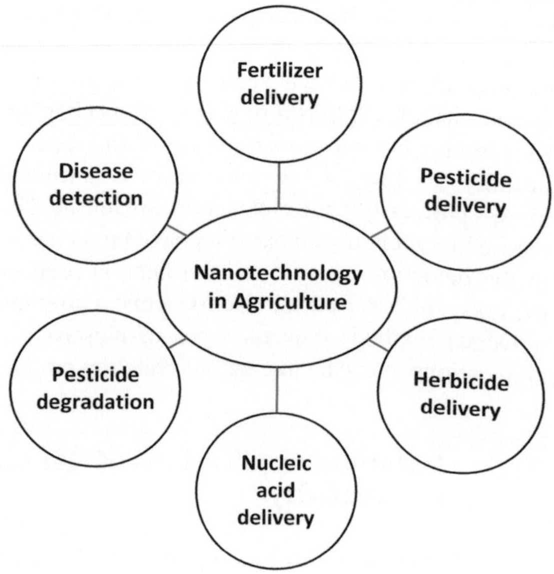

Fig. 11.3 Application of nanotechnology in enhancing crop productivity

with potential applications in slow and controlled release of fertilizers (Wilson et al. 2008). Biodegradable, polymeric chitosan nanomaterials are used for the preparation of controlled release of the NPK fertilizer (Corradini et al. 2010).

Smart nano-fertilizers can be converted into intelligent smart nano-fertilizers by combining nanodevices so as to synchronize the release of fertilizers with their uptake by crops. Dr. Carlos Montreal of Agriculture and Agri-Food Canada is one of the several research scientists developing such intelligent smart fertilizers that release nitrogen only when the plant needs it and in the amount the plant requires (Montreal et al. 2016). The plants communicate their surrounding environment by producing various kinds of chemical signals. Many chemical compounds that are associated with nitrogen uptake have been identified. These compounds can be used to synchronize the release of fertilizer with nitrogen uptake by the crop. Nano-biosensor binding to these compounds can be developed so as to control the release of fertilizers. Nano-sized biosensors made up of specific chemical composition with polymer coating have been developed by this team. This system allows the fertilizers to be released into the soil when the plant requires it. The research team is trying to make intelligent fertilizers with the biodegradable three-dimensional polymer coating less than 100 nm thick.

Biofertilizers are the formulations with beneficial microorganism like plant growth-promoting rhizobacteria (PGPR), *Rhizobium*, *Azotobacter*, *Azospirillum*, and blue-green algae, fungal mycorrhizae, and phosphate-solubilizing bacteria (PSB), e.g., *Pseudomonas* sp. and *Bacillus* sp. (Wu et al. 2005; Prasad et al. 2015). There are some constraints in their usage due to lack of suitable carrier materials, short shelf life, and problems in transportation and storage. Formulations resistant to desiccation can be developed by using polymeric nanoparticle coatings in biofertilizer preparation. Similarly, micronutrients can be trapped in nanomaterials for their slow release to promote optimum plant growth and development (Peteu et al. 2010).

11.3.2 Nano-pesticides

Nanomaterial based formulations have potential applications for efficient delivery of pesticides. The process of absorption, attachment, encapsulation, or entrapment of the active ingredient into the nano-matrix occurs in nano-pesticide formulations (Fig. 11.4). These formulations release the active ingredient slowly and in a controlled way. The advantages of nanomaterial-based formulations are improvement in efficiency and efficacy as well as lower toxicity. Nano-pesticides are prepared either by very small particles of pesticidal active ingredients or some other nano-structured molecules with pesticidal properties. Nano-pesticide-based formulations have better dispersion and wettability quality as compared to conventional formulations (Bergeson 2010a). Nanomaterials used for formulating nano-pesticides have desired properties such as biodegradability, solubility, permeability, and thermal stability (Bouwmeester et al. 2009). Nano-pesticides have large specific surface area enhancing affinity to the target. Nano-emulsions, nano-encapsulates, nano-containers, and nano-cages are some of the nano-pesticide delivery methods which have been widely discussed and reviewed recently (Bergeson 2010b). The ability to delay or control delivery of pesticides to the target organisms is achieved by these nano-pesticide delivery methods (Singh et al. 2015). Spraying of pesticides involves large droplets associated with splash loss, so droplet size is diminished by using nano-pesticides which contribute for effective spraying and, thus, decrease of splash losses. In order to keep the residue level below the critical limits as permissible by the regulatory criteria in foodstuffs, the nano-formulation should be degraded fast in the soil but slowly in plants. The sodium dodecyl sulfate (SDS) can play a role to enhance the photodegradation of the NPs in soil. Pesticide (imidacloprid) nano-formulations were prepared using SDS-modified silver/titanium dioxide (Ag/TiO_2) along with chitosan and alginate by microencapsulation technique to increase the photodegradation by Guan et al. (2010). Keeping in view for the formulation stability which is an important aspect at the nano-level, polymer stabilizers like polyvinylpyrrolidone, polyvinyl alcohol, and poly(acrylic acid)-b-poly(butyl acrylate) were used for formulating a stable nano-pesticide by Liu et al. (2008a).

Plants also provide a source of biomolecules which are not persistent in water and soil but biologically efficient (Isman 2000). These phytochemicals like essential oils and secondary metabolites face problems of cost-effectiveness and stability. Incorporation of essential oil obtained from *Artemisia arborescens* L. into solid lipid

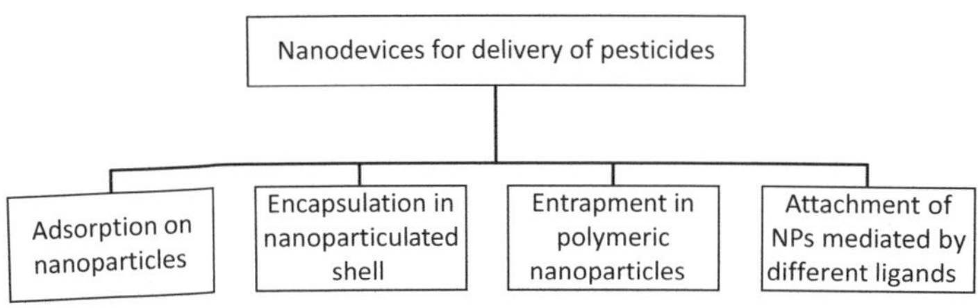

Fig. 11.4 Different nanodevices for delivery of pesticides

nanoparticles diminished the fast evaporation of essential oil (Lai et al. 2006). Similarly, amorphous nano-silica obtained from various sources such as phytoplankton, volcanic soil is used as a biopesticide useful against stored grain, fungal organism, and worms (Liu et al. 2008). There are some biological control agents such as fungi, bacteria, and viruses which are used against insect pests. Bacterial and viral formulations prepared from these organisms are usually susceptible to desiccation, ultraviolet light inactivation, or even heat. Nano-formulations may provide new ways to enhance the stability of these biological agents. Mycopesticides (fungal biocontrol agents) are promising biological pesticides as there is no need of ingestion; instead, they act by contact, are very specific, and can be easily mass-produced (Baric et al. 2008). Microbial products such as enzymes, antibiotics, inhibitors, and toxins are promising as biopesticides against plant pathogens and pests (Bhattacharyya et al. 2016). Microbial-based formulations need stabilization and directed delivery mechanism toward identified targets. Chitosan and clay being biocompatible nanomaterials can be used as stabilizing and delivery agents and, hence, have potential applications in the development of biopesticide formulations.

11.3.3 Degradation of Pesticides

Pesticide-contaminated soil and water are generally treated using phytoremediation, photochemical processes, and oxidation processes. Biorecalcitrant pollutants can be degraded using nanoparticles (Hee and Cheng 2006). It has been showed that pesticides like chlorpyrifos and atrazine can be degraded using nano-sized zerovalent iron. Similarly, some pesticide residues can also be photo-catalytically decomposed using TiO_2 doped with Fe_2O_3 either incorporated into the pesticide formulations or sprayed directly on crops (Sasson et al. 2007).

11.3.4 Detection of Pesticide Residues

Pesticide residues are characterized by a high persistence in the environment and toxic to both wildlife and human. Pesticide residues in soil affect the soil microbial biodiversity. Some pesticides particularly organochlorine suppress symbiotic nitrogen fixation resulting in the reduced crop production. Nitrification bacteria are very sensitive to pesticides and herbicides, and sulfonylurea herbicide has been found to inhibit this process (Gigliotti and Allievi 2001). Some fungicides like chlorothalonil and dinitrophenyl affect nitrification and denitrification microbial processes (Kinney et al. 2005; Lang and Cai 2009). Some pesticides like benomyl and dimethoate affect negatively symbiotic mycorrhizal fungi (Menendez et al. 1999; Chiocchio et al. 2000).

The use of fertilizers, pesticides, herbicides, and other chemicals contribute soil and water pollution. Pesticides can get into water by runoff from treated area, leaching through the soil and through drift during pesticide spraying. More recent studies also reported the presence of pesticides in surface water and groundwater close to agriculture lands all over the world (Cerejeira et al. 2003; Konstantinou et al. 2006; Gilliom 2007; Woudneh et al. 2009; Añasco et al. 2010). Soil organisms such as nematodes, mites, earthworms, spiders, and insects enhance soil aggregation and porosity and, hence, increasing infiltration and reducing runoff. It was found that the mixture of insecticides and fungicides at different concentrations caused a neurotoxic effect in earthworms and was physiologically damaging due to their high toxicity (Schreck et al. 2008).

For organic farming certification, inspection of pesticide residue in soil and water is also necessary. Pesticide residue detection using nanosensors have higher sensitivity, low detection limits, fast response, and super-selectivity (Liu et al. 2008b). Nanomaterial-based nanosensors can be used to detect such pesticide residues in soil and water instead of traditional gas or liquid chromatography-mass spectroscopy techniques. Although traditional techniques are accurate and reliable, these techniques involve time-consuming steps. Enzyme-based biosensors for organochlorine, carbamate, and organophosphate residue detection have been reviewed in detail by Dyk and Pletschke (2011).

11.3.5 Nano-herbicides

Weeds are the unwanted plants along with the desired plant crops and herbicides are used to kill these weeds. Conventional herbicides when sprayed have a chance of killing food crops too, and there might be a huge loss in the crop yield. Nano-herbicide destroys the entire weeds from their roots but not affecting the desired food crops. As nanoparticles are target specific, they can be used to kill the weeds and destroy it to get better yield. Herbicides like atrazine and triazine could be encapsulated to get efficient release to the plants in the same way as nano-fertilizers and nano-pesticides.

11.3.6 Detection of Plant Pathogens

Nanoparticles can be used as an identification tool for detection of bacterial, viral, and fungal plant pathogens in agriculture (Boonham et al. 2008; Prasad 2014). Nano gold-based immuno sensors were used to detect Karnal bunt disease in wheat using surface plasmon resonance by Singh et al. (2010). Plants respond to different stress conditions through physiological changes. Induction of systemic defense is one such response regulated by plant hormones like salicylic acid, jasmonic acid, and methyl jasmonate. This indirect stimulus was used to develop a sensitive electrochemical sensor, using modified gold electrode with copper nanoparticles, to monitor the level of salicylic acid in oil seeds to detect the fungi (*Sclerotinia sclerotiorum*) by Wang et al. (2010). They successfully measured the salicylic acid accurately using this type of sensor. Hence, research could be aimed for detecting pathogens and their by-products using nano biosensor technology.

11.3.7 Water Retention

The crop production needs large amount of freshwater, so research should focus to improve water usage conservation. Some precise water delivery systems are water absorption efficiency of plant, water distribution near roots, in situ water holding capacity, encapsulated water released on demand, and interaction with field intelligence through distributed nano-sensor systems (Savage et al. 2009). Nano-porous materials are capable of storing water and releasing it slowly, depending on drought level controlled by wireless nanosensors. This permits less water use and decrease losses into the environment. Nano-zeolites are crystalline aluminum silicates that allow the exchange of ions and reversible dehydration. They can improve the water retention of sandy soils and improve the porosity in clay soils. Hence, zeolites could be used for nano-organozeolite fertilizers and their use could reduce surface runoff (Prasad et al. 2014).

11.3.8 Enhancing Seed Germination

Seed is the most important input determining crop productivity. There are many studies on the effects of nanomaterials on the seed germination and growth to promote its use for crop production. Zheng et al. (2012) reported the effect of nano-TiO_2 on the growth of spinach seeds. It was reported that nano-TiO_2-treated seeds produced plant that has 73 % more dry weight compared to the control over germination period of 30 days. Khodakovskya et al. (2009) reported the use of CNTs for improving the germination of tomato seeds through better permeation of moisture. The carbon nanotubes serve as new pores for water permeation by penetration of seed coat, and these CNTs act as a passage to channelize the water into the seeds. Shah and Belozerova (2009) studied the effect of different metal nanoparticles like Si, Pd, Au, and Cu on lettuce seed germination indicating positive influence at different concentration range. This facilitated germination could be exploited in rainfed agricultural system.

11.3.9 Delivery of Genetic Material

Nanotechnology is likely to play an important role in the development of genetically modified crops. The development of biotic and abiotic variety of crop plant involves the delivery of genetic material either DNA or RNA resulting in the alteration of gene expression (Palerice and Gatehouse 2008). There are many obstacles for gene delivery to transform plant (Ghormade et al. 2011). Microinjection, microprojectile bombardment, viral gene delivery, and *Agrobacterium*-mediated transformation are some techniques applied to transform plant. These methods are either applied only for dicotyledonous plant transformation or have low efficiency. For efficient gene transformation, nanoparticles such as gold nanoparticle-embedded carbon matrices as a carrier were used successfully for the delivery of genetic material (Vijay Kumar et al. 2010).

11.4 Ethical and Safety Issues

The application of nanotechnology in agriculture is crucial because it directly affects human health. For the delivery of pesticides and fertilizers to the plants, nanomaterials are used to decrease the dosage for controlled slow delivery. But the risk assessment for using nanomaterials is still not studied and defined properly. Hence, some concerns related to the use of nanomaterials in agriculture include phytotoxicity and potential residue carry-over in food stuff which have been reviewed in detail by Bouwmeester et al. (2009). There is a need to evaluate the toxico-kinetics and toxico-dynamics of nanomaterials used in crop production. Nanomaterials might have toxic effects in agricultural crop production (Nel et al. 2006). Nanomaterials with good wettability and dispersion and with biodegradable, biocompatible, nontoxic, stable, and well-understood toxico-kinetics and toxico-dynamics would be desirable and ideal for their efficient and effective applications in agriculture (Khot et al. 2012).

11.5 Future Research

Nanotechnology could be developed for real-time monitoring of the crop growth and field conditions including temperature, moisture level, soil fertility, crop nutrient status, insects, plant diseases, and weeds. Smart precision farming will make use of global satellite positioning system and remote sensing devices and computers to measure environmental conditions enabling us to know whether crops are growing at maximum efficiency. Networks of wireless nanosensors positioned across crop fields will provide necessary data. This data will facilitate for best agronomic intelligence processes with the goal to minimize resource inputs and maximize outputs. It is also well known that humidity, light, temperature, soil conditions, fertilization, insects, and plant diseases all affect the release of volatile organic compounds which could be detected by electronic nose. In the future, electronic noses will detect crop diseases, identify insect infestation, and monitor food quality and could also be used in food industry to assess the freshness spoilage of fruits and vegetables during the processing and packaging process. Similarly, smart dust technology will be used for monitoring various parameters such as temperature, humidity, insect, and disease infestation in the field. Smart nano-biosensors will be scattered like dust across the farms, working like the eyes, ears, and noses of the farming world. These tiny wireless smart nano-biosensors are capable to communicate the information they sense and will be programmed to respond various parameters such as nutrient deficiency, insect attack and disease infestation, and variation in humidity and temperature for real-time monitoring of the crop growth and field conditions (Ingale and Chaudhari 2013).

11.6 Conclusions

Nanotechnology is becoming a promising technology with the ability to create massive changes in agricultural crop productivity such as pesticide, biopesticide, and fertilizer delivery systems. Nanoscale devices can be used to make agricultural system "smart." Smart sensors and smart delivery systems will help in enhancing crop productivity in agriculture by providing accurate information. This can help in maintaining farms and fields with precise control and report timely needs of crops and thus helping farmer to make better decisions. Smart nanosensors and real-time monitoring system will have a large impact on smart precision methodologies. Networks of wireless nanosensors positioned across cultivated fields will provide data leading to best agronomic intelligence processes and practices with the goal to reduce resource inputs and enhance crop productivity.

References

Agrawal S, Rathore P (2014) Nanotechnology pros and cons to agriculture: a review. Int J Curr Microbiol App Sci 3(3):43–55

Añasco N, Uno S, Koyama J, Matsuoka T, Kuwahara N (2010) Assessment of pesticide residues in freshwater areas affected by rice paddy effluents in Southern Japan. Environ Monit Assess 160(1):371–383

Baric TK, Sahu B, Swain V (2008) Nanosilica-from medicine to pest control. Parasitol Res 103:253–258

Bergeson LL (2010a) Nanosilver: US EPA's pesticide office considers how best to proceed. Envir Qual Manage 19(3):79–85

Bergeson LL (2010b) Nanosilver pesticide products: what does the future hold? Qual Manage 19(4):73–82

Bhattacharyya A, Duraisamy P, Govindarajan M, Buhroo AA, Prasad R (2016) Nano-biofungicides: emerging trend in insect pest control. In: Prasad R (ed) Advances and applications through fungal nano-biotechnology, Springer, Cham, pp 307–319

Boonham N, Glover R, Tomlinson J, Mumford R (2008) Exploiting generic platform technologies for the detection and identification of plant pathogens. Eur J Plant Pathol 121:355–363

Bouwmeester H, Deekkers S, Noordam MY, Hagens WI, Bulder AS, de Heer C, ten Voorde SECGS, Wijnhoven WP, Marvin HJP, Sips AJAM (2009) Review of health safety aspects of nanotechnogies in food production. Regul Toxicol Pharacol 53:52–62

Cerejeira MJ, Viana P, Batista S, Pereira T, Silva E, Valério MJ, Silva A, Ferreira M, Silva-Fernande AM (2003) Pesticides in Portuguese surface and ground waters. Water Res 37(5):1055–1063

Chiocchio V, Venedikian N, Martinez AE, Menendez A, Ocampo JA, Godeas A (2000) Effect of the fungicide benomyl on spore germination and hyphal length of the Arbuscular mycorrhizal fungus *Glomus mosseae*. Int Microbiol 3:173–175

Corradini E, Moura MR, Mattoso LHC (2010) A preliminary study of the incorporation of NPK fertilizer into chitosan nanoparticles. Express Polym Lett 4:509–515

Dyk V, Pletschkeb JS (2011) Review on the use of enzymes for the detection of organochlorine, organophosphate and carbamate pesticides in the environment. Chemosphere, Elmsford 82:291–307

Ghormade V, Deshpande MV, Paknikar KM (2011) Perspectives for nano-biotechnology enabled protection and nutrition of plants. Biotechnol Adv 29:792–803

Gigliotti C, Alliev L (2001) Differential effects of the herbicides Bensulphuron and Cinosulphuron on soil microorganisms. J Environ Sci Health B 36(6):775–782

Gilliom RJ (2007) Pesticides in U.S. streams and groundwater. Environ Sci Technol 41(10):3408–3414

Guan H, Chi D, Yu J, Li H (2010) Dynamics of residues from a novel nano-imidacloprid formulation in soybean fields. Crop Prot 29:942–946

Hee Joo S, Cheng IF (2006) Nanotechnology for environmental remediation. Springer, New York

Ingale AG, Chaudhari A (2013) Biogenic synthesis of nanoparticles and potential application: an eco-friendly approach. J Nanomedi Nanotech 4(2):165. http://dx.doi.org/10.4172/2157-7439.1000165

Isman MB (2000) Plant essential oil for pest and disease management. Crop Prot 19:603–608

Khodakovskaya M, Dervishi E, Mahmood M, Xu Y, Li Z, Watanabe F, Biris AS (2009) Carbon nanotubes are able to penetrate plant seed coat and dramatically affect seed germination and plant growth. ACS Nano 3:3221–3227

Khot LR, Shankaran S, Maja JM, Ehsani R, Schuster EW (2012) Applications of nanomaterials in agricultural production and crop protection: a review. Crop Prot 35:64–70

Kinney CA, Mandernack KW, Mosier AR (2005) Laboratory investigations into the effects of the pesticides mancozeb, chlorothalonil, and prosulfuron on nitrous oxide and nitric oxide production in fertilized soil. Soil Biol Biochem 37(5):837–850

Konstantinou IK, Hela DG, Albanis TA (2006) The status of pesticide pollution in surface waters (rivers and lakes) of Greece. Part I. Review on occurrence and levels. Environ Pollut 141(3):555–570

Lai F, Wissing SA, Muller RH, Fadda AM (2006) *Artemisia arborescens* L. essential oil-loaded solid lipid nanoparticles for potential agricultural application: preparation and characterization. AAPS Pharm SciTech 7:1–9

Lang M, Cai Z (2009) Effects of chlorothalonil and carbendazim on nitrification and denitrification in soils. J Environ Sci 21(4):458–467

Liu Y, Tong Z, Prud'homme RK (2008a) Stabilized polymeric nanoparticles for controlled and efficient release of bifenthrin. Pest Manag Sci 64:808–812

Liu S, Yuan L, Yue X, Zheng Z, Tang Z (2008b) Recent advances in nanosensors for organophosphate pesticide detection. Adv Power Techno 119:419–441

Menendez A, Martínez A, Chiocchio V, Venedikian N, Ocampo JA, Godeas A (1999) Influence of the insecticide dimethoate on Arbuscular mycorrhizal colonization and growth in soybean plants. Int Microbiol 2:43–45

Montreal CM, DeRosa M, Mallubhotla SC, Bindraban PS, Dimkpa C (2016) Nanotechnologies for increasing the crop use efficiency of fertilizer-micronutrients. Biol Fertil Soil 52(3):423–437

Nel A, Xia T, Madler I, Li N (2006) Toxic potential of materials at the nano level. Science 311:622–627

Palerice DRG, Gatehouse JA (2008) RNAi-mediated crop protection against insects. Trends Biotechnol 26:393–309

Peteu SF, Oancea F, Sicuia OA, Constantinescu F, Dinu S (2010) Responsive polymers for crop protection. Polymers 2:229–251

Prasad R (2014) Synthesis of silver nanoparticles in photosynthetic plants. J Nanoparticles, Article ID 963961, http://dx.doi.org/10.1155/2014/963961

Prasad R, Kumar V, Prasad KS (2014) Nanotechnology in sustainable agriculture: present concerns and future aspects. Afr J Biotechnol 13(6):705–713

Prasad R, Kumar M, Varma A (2015) Role of PGPR in soil fertility and plant health. In: Egamberdieva D, Shrivastava S, Varma A (eds) Plant growth-promoting rhizobacteria and medicinal plants. Springer, Cham, pp 247–260

Prasad R, Pandey R, Barman I (2016) Engineering tailored nanoparticles with microbes: quo vadis. WIREs Nanomed Nanobiotechnol 8:316–330. doi:10.1002/wnan.1363

Prasad R, Bhattacharyya A, Nguyen QD (2017) Nanotechnology in sustainable agriculture: recent developments, challenges, and perspectives. Front Microbiol 8:1014. doi:10.3389/fmicb.2017.01014

Sasson Y, Levy-Ruso G, Toledano O, Ishaaya I (2007) Nanosuspension: emerging novel agro-chemical formulations. In: Isaaya I, Nauen R, Horowitz AR (eds) Insecticides design using advanced technologies. Springer, Dordrecht, pp 1–32

Savage N, Diallo M, Dencan J, Street A, Sustich R (2009) Nanotechnology applications for clean water. William Andrew, Norwich, NY

Schreck E, Geret F, Gontier TM (2008) Neurotoxic effect and metabolic responses induced by a mixture of six pesticides on the earthworm *Aporrectodea caliginosa nocturna*. Chemosphere 71(10):1832–1839

Shah V, Belozerova I (2009) Influence of metal nanoparticles on the soil microbial community and germination of lettuce seeds. Water Air Soil Pollut 197:143–148

Singh S, Singh M, Agarwal VV, Kumar A (2010) An attempt to develop surface plasmon resonance based immune-sensor for Karnal bunt (*Tilletia indica*) diagnosis based on the experience of nano-gold based lateral flow immune-dipstic test. Thin Solid Films 519:1156–1159

Singh S, Singh BK, Yadav SM, Gupta AK (2015) Application of nanotechnology and their role in disease management. Res J Nanosci Nanotech 5(1):1–5

Taniguchi N (1974) On the basic concept of nano-technology. In: Proceedings of the international conference on production engineering, Tokyo, Part II, Japan Society of Precision Engineering

Vijayumar PS, Abhilash OU, Khan BM, Prasad BLV (2010) Nanogold-loaded sharp-edged carbon bullets as plant-gene carriers. Adv Funct Mater 20:2416–2423

Wang Z, Wei F, Liu SY, Xu, Q, Huang, J-Y, Dong, XY, Hua, JH, Yang, Q, Zhao, YD, Chen, H (2010) Electrocatalytic oxidation of phytohormone salicylic acid at copper nanoparticles-modified gold electrode and its detection in oilseed rape infected with fungal pathogen *Sclerotinia sclerotiorum*. Talanta 80:1277–1280

Wilson MA, Tran NH, Milev AS, Kannagara GSK, Volk H, Lu GHM (2008) Nanomaterials in soils. Geoderma 146:291–301

Woudneh MB, Ou Z, Sekela M, Tuominen T, Gledhill M (2009) Pesticide multi residues in waters of the lower Fraser Valley, British Columbia, Canada. Part I. Surface water. J Environ Qual 38(3):940–947

Wu SC, Cao ZH, Li ZG, Cheung KC, Wong MH (2005) Effect of biofertilizer containing N-fixer, P and K solubilizers and AM fungi on maize growth: a greenhouse trial. Geoderma 125:155–166

Zheng L, Hong F, Lu S, Liu C (2012) Effect of nano-TiO_2 on strength of naturally aged seeds and growth of spinach. Biol Trace Elem Res 104:83–91

Nanomaterial-Based Biosensors in Agriculture Application and Accessibility in Rural Smallholding Farms: Food Security

12

M.S. Mufamadi and P.R. Sekhejane

Abstract

In the absence of inexpensive screening tools, food contamination poses immense threat to food safety and security and ultimately inclines burden on the public health, particularly for the populace in low- and middle-income countries, e.g. sub-Sahara Africa (SSA) countries. Current traditional methods for detection of contaminants in food and to ensure food quality and safety are associated with time-consuming procedures that are expensive and not accessible to those in rural areas. This chapter reviews the latest development and highlights the impact of various nanomaterials used during constructing biological sensors for screening each of these above food contaminants, in detail. The presence of nanomaterials is promising to offer device that is affordable, highly sensitive, specific and user-friendly. This chapter also highlights the accessibility of this technology, particularly to those in the rural and smallholder farmers. Furthermore, also try to address the potential contributions that nanotechnology can have in food safety and security.

Keywords

Biosensor • Nanotechnology • Nanomaterials • Agriculture • Food safety • Security

M.S. Mufamadi (✉)
Nanotechnology and Biotechnology, NABIO Consulting (Pty) Ltd, Pretoria East 0010, South Africa
e-mail: steven@nabioconsulting.co.za; Steven.Mufamadi@gmail.com

P.R. Sekhejane
Africa Institute of South Africa, Human Sciences Research Council, Pretoria 0001, South Africa

R. Prasad et al. (eds.), *Nanotechnology*, DOI 10.1007/978-981-10-4573-8_12

12.1 Introduction

Investment in science and technology towards armed conflicts far exceeds investment towards food safety and security. That is the reality of 63% of Sub-Sahara Africa (SSA) population that relies heavily on agriculture as a source of income and employment, and therefore this has direct translation to survival and food security. The devastating change in climate conditions, urbanisation, use of pesticides and dwindling economy has also played a significant role in impacting negatively towards food security in SSA. Another element that is depressing the production of adequate food is postharvest infections by pathogenic microorganisms and processing methods (socio-economic), environments (environmental), etc.; these factors amount to over 50% of food wastage. Food losses are mainly due to poor or lack of infrastructure and logistics, technology, skills, knowledge and management capacity of supply chain actors (Kiaya 2014).

At the same time, there is introduction and evolution of scientific methods that are being developed to reduce loss of food such as nanotechnology. Nanotechnology has been researched and well documented for over two decades, with applications being more prominent in the biological sciences and extending to plant sciences. The technology can be used for reducing the plant diseases as drug/treatment delivery modalities, molecular marker conjugates or diagnostic tool for disease (Madhuri et al. 2010; Prasad et al. 2014). Diagnosis of a disease in its very early stage can play an important role in improving prognosis, thus providing a fair chance of preserving the crops. Though the toxicity of nanotechnology products is still highly debateable, however, alternatively making use of nanotechnology as nanosensors offers a better alternative compared to application as crop disease drugs. Loss of life through diarrhoeal diseases is a classic example in low- and middle-income countries in SSA, and this is commonly associated with consuming food that was cleansed with or grown in bacteria contaminated water. The consequences that come with failure to address the food safety and security hamper on improving life quality, which is one of the most important objectives of global research and goals. Careful attention to food chain process is an unavoidable necessity "from field to fork" as quality of life is closely tied to food quality and safety. Food pollutants such as mycotoxins and pesticides are most prominent hazards and of serious concern in the food industry for consumer health (Cozzini et al. 2008).

Nanotechnology is one of those divergent technologies, i.e. it can be merged with other technologies, for instance, nano-biotechnology and nanomedicine. Therefore, it does not necessarily have to replace existing traditional methods but it can improve them. However, adoption of technology-based innovation remains a challenge for several and simple reasons such as accessibility, capacity of skills and readiness to adopt. The population of SSA is growing rapidly and as such; that means there is an unmet growing need or demand for food and consumption. Due to soaring percentage of food losses, advances in agriculture sector require to develop tools that can analyse molecular and cellular biology of crops; importantly these tools should afford to alert the users of early disease detection, monitor and where possible guide the diagnosis path.

This chapter will review the latest developments, application or use of various nanomaterial-based technologies that are used in agriculture, particularly for smallholder farmers. For the purpose of this section, the biosensors that will be highlighted are those that are commonly used for mycotoxin detection, because mycotoxins are common toxins that impact on food safety and security. Secondly the analysis will be based on whether these technologies are beneficial, accessible (affordability and ease of use) and safe. We mention these aspects from the perspective of sustainable development in the context of Africa's development.

12.2 Nano-based Biosensor in Agriculture

Biosensors are analytical device which uses a living organism or biological molecules to convert a biological response into an electrical signal for quantification through a processor. Therefore, the key functional components are the interactive sensor (bio-recognition), the transducer and the signal readability. There are various biosensors; for example, immunosensors are a subset of biosensors that utilise antibody-antigen (Ab-Ag) as a binary recognising model. Analytical immunosensors which utilise either antigen or antibody as the bio-specific sensing element, when an antibody (Ab) is used as a biomolecular recognition element for a specific analyte (antigen-Ag) to form a stable complex, the device is then referred to as an immunosensor. The use of novel biosensor could have a broad appeal to the food industry research and development, food safety inspection agencies, food producers and policy makers overseeing food safety and security (Prasad et al. 2017).

However, the use of current traditional methods for detection of food safety is associated with the following challenges: expensive, time-consuming procedure which requires multiple sample preparation steps prior to detection and requires trained personnel and complicated instrument which are not accessible particularly to those leaving in rural areas (Koedrith et al. 2014).

Nanotechnology is amongst the most profound alternative solutions towards game changing in the early twenty-first century as it has the potential to make an impact in many fields such as health, water, energy, food and agriculture.

Nanotechnology is the manipulation of materials, device and systems at the level of individual atoms and molecules, nanoscale (1–100 nm) in order to produce novel characteristics (Otles and Yalcin 2012; Prasad et al. 2016). The uses of nanomaterials during construction of biological sensors promise to bypass many barriers associated with traditional methods. The following are some of the advantages of nano-based biosensors over traditional methods: offer high sensitivity and specificity, rapid and accurate detection time and real-time signal and they are user-friendly. In addition, the technology is capable to detect a single or multiple analytes (e.g. biological or chemical materials) that could be potentially harmful to plants, animals and human at a very low concentration, with small sample preparation and small portable instrumentation (Vo-Dinh 2005). Furthermore, the presence of nanomaterials allow the introduction of many new signal transduction technologies in biosensors (Paddle 1996; Khanna 2008; Li et al. 2010; Otles and Yalcin 2012). In agriculture, nano-based biosensors offer

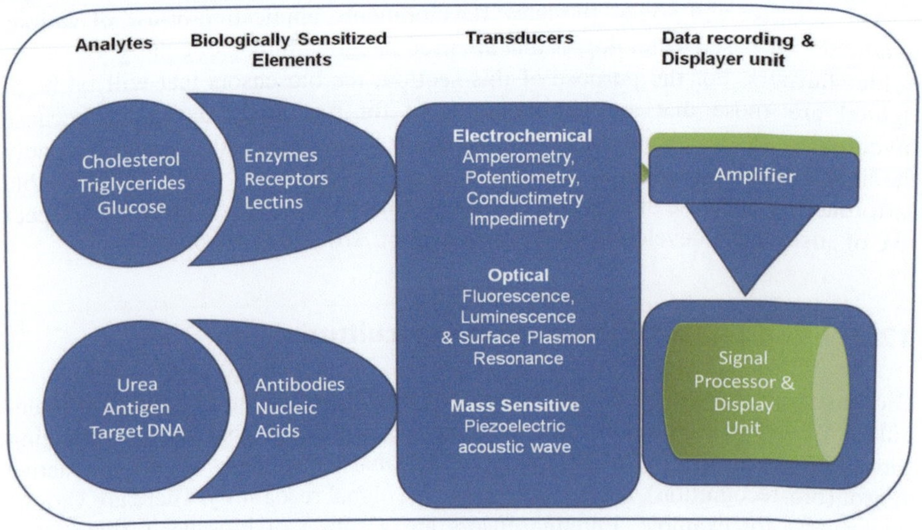

Fig. 12.1 Schematic diagram demonstrating principles of nanobiosensors, from biological sensitised probe, transducers and data displayer unit (consists of amplifier, data processor and display unit)

opportunities for direct and indirect detection of the food-borne pathogenic microorganisms, pesticide, drug residues, toxic contaminants and heavy metal ions in foods, in a short period of time. In addition, nano-based biosensors can also be used to monitor antibiotic resistance, soil condition, crop stress, plant growth and nutrient content or food quality (Teodoro et al. 2010; Tarafdar et al. 2013; Prasad 2014; Prasad et al. 2014, 2016, 2017).

A typical nano-based biosensor consists of three components: (1) biological sensitive probe, (2) transducer and (3) data recording and displayer unit. The biological sensitive probes are sensing elements or bio-mimic components that facilitate biological component interacting with the analytes and determines the signal proportional to the analyte concentration (e.g. sensitivity, specificity and surface area during molecular interaction). The following are biological components/probe examples: (1) antibody-antigen interactions, (2) nucleic acid interactions (3) enzymatic interactions, (4) cellular interactions (i.e. microorganisms, proteins) and (5) synthetic bioreceptors. A transducer is a physical component responsible for converting the recognition signal events into a digital signal such as electrochemical (amperometry, potentiometry, conductometry/impedimetry), optical (fluorescence, luminescence and surface plasmon resonance) and mass sensitive (piezoelectric/acoustic wave). Data recording and displayer unit consists of amplifier, signal processor and display that are responsible for data transferred and results display (Fig. 12.1) (Wang 2005; Rai et al. 2012).

The following are the examples of a nano-based biosensor used in agriculture: magnetic, gold nanoparticles, Quantum Dots, DNA-aptamer and carbon nanotubes. The presence of these nanoparticles can result to the construction of nanobiosensor

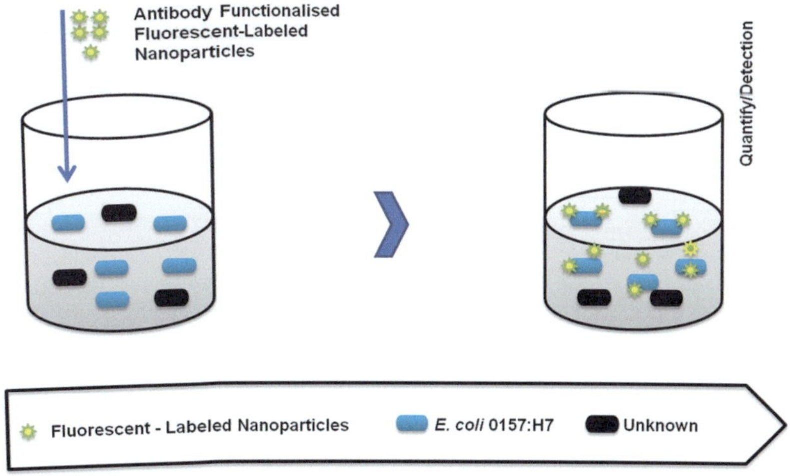

Fig. 12.2 Scheme illustrating optical biosensor composed of antibody functionalised fluorescent-labelled nanoparticles for detection in food-borne pathogens in a contaminated food

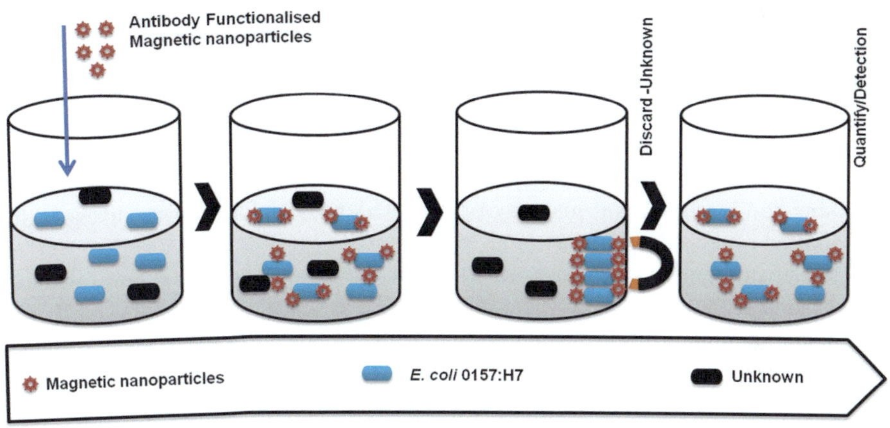

Fig. 12.3 Scheme illustrating biosensor composed of functionalised magnetic nanoparticles for target detection of food-borne pathogens in a contaminated food

devices that are stable with high rapid detection, sensitivity and selectivity activities (Figs. 12.2 and 12.3).

12.3 Application of Nano-based Biosensor in Agriculture

12.3.1 Nano-based Biosensors for the Detection of Bacteria

Food-borne microorganism pathogens are posing a big threat to the public health, particularly in the SSA countries where they are no appropriate analytical tools to

Table 12.1 Nano-based biosensors for sensing bacteria pathogens and their toxins in contaminated food

Nanomaterials in biosensors	Bacteria pathogens	Detection limit and linearity range	Sample	References
Liposomes, carbon nanotube	*Vibrio cholera*	Detection limit of 10^{-16} gram (g) and linear range of 10^{-14}–10^{-7} g/mL	Cholera toxin	Viswanathan et al. (2006)
Quantum dot nanoparticles	*Salmonella*	Detection limit of 1.4×10^3 CFU/mL in phosphate buffered saline (PBS) containing 1% bovine serum albumin (BSA) and 4×10^3 CFU/mL in food extracts	PBS-BSA containing *Salmonella typhimurium* and food extracts	Kim et al. (2013)
Chitosan-gold nanoparticles	*Salmonella*	Detection limit of 143 cells/mL and linear range from 10^3 to 10^6 cells/mL	Skimmed milk containing *Salmonella*	Afonso et al. (2013)
Magnetic nanoparticle (MNPs), Gold nanoparticles (AuNPs)	*Escherichia coli (E.coli) O157:H7*	Sensitivity of 10^1 colony CFU/mL and linear range between 10^1 and 10^6 CFU/mL	Broth containing *E. coli O157:H7*	Wang and Alocijia (2015)

analysing food sample, and majority of the incidences of food-borne diseases are not reported or recorded (Kumar et al. 2012).

Bacteria are the leading causes of food poisoning, gastroenteritis and other illnesses, such as *Salmonella*, *Escherichia coli*, *Campylobacter* and *Vibrio cholerae*. In human, salmonellosis is the primary cause of food-borne bacteria worldwide. The consumption of contaminated meat and poultry is the primary source of transmission of the salmonellosis from carrier animals to humans (Hogue et al. 1997; Mead 2004; Nowak et al. 2007).

Despite recent advances of technology in food safety, traditional methods are very expensive and time-consuming. The use of nano-based biosensor promising to offer rapid detection and specificity to food-borne bacteria pathogens during food chain, including harvesting, processing, distributing and storing. Different nano-based biosensors have recently been developed for sensing bacterial pathogens and their toxins in contaminated food (Table 12.1).

In a study by Viswanathan et al. (2006), they have developed an electrochemical immunosensor composed of liposomes and poly(3,4-ethylenedioxythiophene)-coated carbon nanotubes (CTs) for cholera toxin detection. The device is constructed as follows: potassium ferrocyanide-encapsulated and ganglioside (GM1)-functionalised liposomes and CTs. The CTs were detected via sandwich-type assay, whereby toxin detection first bound introduced by the anti-CT antibody and thereafter by GM1-functionalised liposome. The detection limit of this immunosensor was as follows: 10^{-16} g of cholera toxin.

Kim et al. (2013) has developed a quantum dot nanoparticle-based biosensor for rapid detection of pathogenic *salmonella* species in food products. This device was based on superparamagnetic particles for separate and concentrate cell and anti-*Salmonella* polyclonal antibodies immobilised by streptavidin-biotin to the quantum dot surface for selectivity activities. The research show high fluorescence activities which is equivalent to bacteria concentration, and the *Salmonella* detection limits were 1.4×10^3 CFU/mL in PBS-BSA and 4×10^3 CFU/mL in food extract. Afonso et al. (2013) have developed an electrochemical detection of *Salmonella* using gold nanoparticles in skimmed milk. The results showed the *Salmonella* detection limit of 143 cells mL^{-1} and a linear range from 10^3 to 10^6 cells mL^{-1}.

In a recent study by Wang and Alocijia (2015), they have developed a novel method for rapid detection of *Escherichia coli* O157:H7 in broth employing of nanoparticle-labelled biosensor. The device was constructed employing a functionalised magnetic nanoparticle (MNPs) and gold nanoparticles (AuNPs) conjugated with monoclonal antibodies (Abs) on their surface structure for detection of the targeted pathogen in the sample. The results of this sensor showed a rapid detection of *E. coli*, with a sensitivity of 10^1 colony CFU/mL and linear range between 10^1 and 10^6 CFU/mL.

12.3.2 Nano-based Biosensors for the Detection of Mycotoxins

Mycotoxins are toxic chemical products produced by organisms of the fungus kingdom known as moulds. They are natural contaminates to food and animal feed products, particularly for those that are stored under conditions that are highly favourable to mould growth (Akbas and Ozdemir 2006).

The following are some of the mycotoxins that are well-known to be highly toxic to animals and humans: aflatoxins, ochratoxin and zearalenone (Zimmerli and Dick 1996; Hussein and Brasel 2001; Ventura et al. 2004). Ingestion of such mycotoxin-contaminated food and feed products poses high threat to human and animal health, and because of that, mycotoxins are regulated in food in many countries. Majority of them are labelled as carcinogenic, mutagenic, hepatotoxic and nephrotoxin (Hussein and Brasel 2001). Because of such concerns, there is a great need to develop of novel analytic instruments that can improve their detection in contaminated foods and animal feeds.

For the detection or monitoring of mycotoxins in contaminated food, several nano-based biosensors have been proposed: nanoparticle, immunosensor, optical, magnetic, quantum dots and carbon nanotubes (Zhilong and Zhujum 1997; Maragos and Thompson 1999; Xu and Han 2004; Sapsfor et al. 2006; Adanyi et al. 2007; Owino et al. 2008; Radoi et al. 2008; Viswanathan and Radecki 2008; Kaushnik et al. 2008).

In the case of aflatoxin detection, Parker and Tothill (2009) have developed an electrochemical immunosensor to test the presence of aflatoxin M_1 contaminates in the milk and also in matrix interference. The sensor was constructed using antibody-modified screen-printed carbon-working electrode with carbon counter and silver-silver chloride pseudo-reference electrode. The results showed that this technique is sensitivity when compared to a commercial ELISA, with a detection limit of 39–1000 ng l^{-1} and interference free.

In another study, Xu et al. (2013) have developed an optical biosensor constructed with gold nanorod (GNR)-conjugated antibody for aflatoxin B1 detection in peanut (Xu et al. 2013). This technique showed great advantage in terms of sensitivity and selectivity against aflatoxin B1, with a detection limit of 0.16 ng/mL. The results also showed high recovery in real peanut sample at range from 94.2 to 117.3%. Eldin et al. (2014) have developed AuNPs conjugated with anti-aflatoxin B1 polyclonal antibody for testing the sensor sensitivity in rice, peanuts and maize (Eldin et al. 2014). In this study, sensor showed to detect anti-aflatoxin B1 with high sensitivity of 5 ng/mL in assay duration of 2 min.

In the case of ochratoxin detection, electrochemical immunosensors constructed with nanostructure-conjugated polyclonal antibodies and gold nanoparticles were developed by Bone et al. (2010). The outcomes of this study showed a different limit of detection and linear working range. The polyclonal antibody-based immunosensor has a detection limit of 0.86 ng/mL and linear working range between 0.9 and 9.0 ng/mL, while gold nanoparticles have a detection limit of 0.20 ng/mL of mycotoxin ochratoxin and linear working range between 0.3 and 8.5 ng/mL. Turan and Şahin (2016) have recently developed another technique for detect ochratoxin A using molecularly imprinted biocompatible magnetic nanoparticles. The technique employed U-vis spectrophotometer. The results for this study showed that the presence of magnetic nanoparticles were able to enhance adsorption time and capacity and also specify recognition to ochratoxin A.

12.3.3 Nano-based Biosensors for Detection of Pesticides and Other Contaminants

Pesticides are the variety of toxic substances used to control agricultural pests such as weeds, fungus and insects. The advantages using pesticides in agriculture is to protect plants by killing agricultural pests, herbicides for weeds, fungicides for fungus and insecticides. However, pesticide residues in agriculture products pose a human health risk; therefore, they are in need of pesticide residue detection or monitoring in the prior causing hazard to human. Current traditional methods for pesticide residue detection are hampered by many disadvantages: expensive, time-consuming and many more.

Several nano-based biosensors have been developed for detecting and monitoring pesticides such as herbicides, fungicides, insecticides and other contaminate detection in food (McGrath et al. 2012; Zhao et al. 2015; Vimala et al. 2016).

A novel enzyme biosensor based on multiwalled carbon nanotubes (CTs) for detection of pesticides residue in food was developed by Zhang et al. (2008). This study was focusing on acetylcholinesterase (AChE)-direct immobilisation through modified glassy carbon electrode and for sensitivity towards carbaryl pesticide residue. The result proved that nanobiosensor is suitable for pesticides detection, with immobilised enzyme activity at 0.01 U and detection limit of 10–12 g/L.

Norouzi et al. (2010) have developed a novel flow injection electrochemical biosensor for the detection of monocrotophos and organophosphate pesticide. The structure was based on chitosan-gold nanoparticle film combine with Fast Fourier Transform Continuous Cyclic Voltammetry (FFTRCCV) technique. The utilising of chitosan-gold nanoparticles was able to increase the immobilisation level. In addition results also showed increase in sensitivity with detection limit of 10 nm, with less than 70 s responding time. Furthermore, the results showed also structure stability with more 50-day storage stability.

In case of detection of paraoxon and parathion pesticides in foods and water, employing an optical nanobiosensor was recently developed by Zheng et al. (2011). A simple biosensor was constructed using nanostructured films of the enzyme AChE and CdTe quantum dots. The technique showed high sensitivity and the detection limits at present in picomolar concentrations: 5×10^{-11} M for paraoxon and 4.47×10^{-12} M for parathion. However, the technique was shown to be associated with poor selectivity when characterisation of organophosphate was done in mixture of different types of organophosphates (OPs).

In another study Guan et al. (2012) have developed acetylcholinesterase biosensors for detection of dichlorvos pesticides residues. The biosensor based five bilayer of chitosan and acetylcholinesterase liposome bioreactor. The presence of liposome-based bioreactors in structure were for encapsulating the enzyme AChE. This study showed that the use of the acetylcholinesterase biosensors as a great tool for analysis, of enzyme inhibitor, in this case the inhibition of the pesticide was found be at the concentration range between 0.25 and 1.50 and from 1.75 to 10.00 μM, with a detection limit at estimated range between 0.86 ± 0.098 μg/L. In addition, the enzymatic reaction was also estimated employing Michaelis-Menten constant, K_m, and found to be 0.36 mM.

In a recent study, Song et al. (2015) have developed a simple electrochemical biosensor for quantitative determination of carbamate pesticide in the real samples. The biosensor was contracted based on citrate-capped gold nanoparticles (AuNPs)/(3-mercaptopropyl)-trimethoxysilane (MPS)/gold electrode (Au) sensing layer. The result showed that the device was able to sense the quantity of carbamate pesticides in a real sample. In addition, the results also showed that the technique allowed to quantitatively determine pesticides at a very low potential, with the linear range between 0.003 and 2.00 μM. Detection limits or sensitivity plays a crucial role as it may be detrimental to human life if toxins are not detected at levels that may potentially induce diseases.

12.3.4 Nano-based Biosensor for Herbicides Detections

Songa et al. (2009a, b, c) have developed a novel biosensor for the detection of glyphosate, glufosinate herbicide and its metabolite, e.g. aminomethylphosphonic acid (AMPA). The biosensors were constracted utilising the electrochemical deposition in polymer and electrostatic attachment of horseradish peroxidase (HRP) and nanocomposites to measure the activity of the enzyme onto. The first study was to detect glyphosate and glufosinate herbicide on corn samples (Songa et al. 2009a). The results showed a sensitive biosensor for herbicides detection and the limit of detection of 0.1 Î¼g L-1 (10–10 M) for both glyphosate and glufosinate. The second study was to detect glyphosate herbicide and its metabolite AMPA (Songa et al. 2009b). The detection limits of the biosensor in this study were as follows: 0.16 μg/L of glyphosate and 1.0 μg/Lof AMPA. The third study was focused on glyphosate herbicide detection using HRP immobilised on sulfonated polymer matrix. The presence of glyphosate herbicide inhibited the activity of HRP resulting into the decrease in its responses to water. The technique exhibited determination of glyphosate herbicide at a range between of 0.25 and 14.0 μg/L(−1), with a detection limit of 1.70 μg/L(−1).

In another study Haddaoui and Raouafi (2015) have designed a disposable enzymatic nanobiosensor for the detection of herbicide chlortoluron. Herbicide chlortoluron is widely used in agriculture to prevent weed growth in cereal fields. This study demonstrated an electrochemically nanostructured method utilising modified screen-printed carbon electrodes (SPCEs) and ZnO nanoparticles that allow the analysis of an enzymatic activity inhibition in tyrosinase and for the detection of part per billion (ppb) levels of herbicide chlortoluron. The results from this study showed the designed nano-based biosensor with a detection limit of 0.47 nM. The presence of chlortoluron herbicide showed to induce inhibition of tyrosinase with a level ranging from 1 to 100 nM.

12.3.5 Nano-based Biosensors for Detection of Veterinary Drug Resistance and Residues

Veterinary drugs or antibiotics used in farming animal have the big role to play; these include prevention and treatment of animal diseases. In addition, they can also be used to enhance animal growth rates and feed efficiency and animal health (McEwena and Fedorka-Cray 2002).

However, veterinary antibiotic use in animals may pose serious risks to animal and human health, particularly if they developed antibiotic resistance against microbial pathogens. In animals, antibiotic resistance could result into outbreak or disease spread amongst animals. In human, antibiotic resistance may pose high risks to human health, particularly if the drug residues or their metabolites can be transferred into food (Levy and Marshall 2004; Courvalin 2008). Furthermore, consumption of meat contaminated with drug residues might cause increased risk of resistance of normal flora in human or may induce infection resistance in animals and humans (Landers et al. 2012; Aziz et al. 2016).

Emerging concern regarding antibacterial resistance increases; therefore, there is a great need to develop a rapid screening instrument to monitor antibiotic resistance and residues in food animals prior human consumption (Cháfer-Pericás et al. 2010). In most SRA countries, regulation programme and information on the effect in veterinary drug residues in food animals are little or not available. Farmers and entrepreneurs depend on the veterinary drugs available in the black market for their success in farming animals, particularly in poultry industry. Drugs are used without consultation with a veterinary or misused/abused by not adhering to the regulator guidelines of the veterinary drug recommendation (Idowu et al. 2010).

Nano-based biosensors offer an opportunity for the development of smart analytical tools to determine the level or monitor the presence of the veterinary drug residues in food in a real-time (Wu et al. 2014, 2015). In addition, it is also promising to offer cheaper devices that are highly sensitive, specific and accurate (Huet et al. 2010). There are several nano-based biosensors that have been developed specifically to determine level/concentration of the different types of veterinary antibiotic and drug residues in meat, poultry products, milk and many more. The following are some of the novel biosensors: electrochemical, optical and aptamer biosensors. Furthermore, nano-based biosensors have offered an opportunity to detect veterinary drug residue/s in food either using single or multi-analyte detection approaches (Mungroo and Neethirajan 2014).

Aptamers are preferred in antibiotic detection due to their binding affinity, easy labelling (e.g. fluorescence or colorimetry) and stability (Hou et al. 2013). Song et al. (2012) demonstrated an aptamer biosensor with the capability to validate the presence of ampicillin in milk using its single-stranded DNA (ssDNA) aptamer. This study showed high sensitivity and specificity to ampicillin and achieved employing of gold nanoparticle based on dual fluorescence-colorimetric method. The results show the lower limits of detection for ampicillin in the milk: 2 ng/mL and 10 ng/mL by fluorescence and colorimetry, respectively.

12.4 Nanotechnology in Food Safety and Security: Policy Rationale

Nanotechnology is recognised and supported by the African Union (AU) through the Science Technology and Innovation Strategy for Africa (STISA 2024). The first priority research and innovation areas in the STISA 2024 objectives are eradication of hunger and food and nutrition security. The framework of AU's STISA 2024 also fits into the global context with regard to the 2015 sustainable development goals (SDGs). Of the 17 goals that were drawn, 5 goals cross-cut through poverty and health, i.e. no poverty (goal 1), zero hunger (goal 2), good health and wellbeing (goal 3), decent work and economic growth (goal 8), and responsible consumption and production. The current challenges such as climate change, food security and sustainability require engagement of scientific interventions such as nanotechnology; however, it is important that the technological interventions advocated for do not undermine other SDG pillars, overall human health and development.

The aspect of food safety is critical to realise food security; therefore, as much as research and development drive the innovative methods such as nanotechnology, safety and security should not be compromised. According to food and agriculture organisation, food security is the state achieved when food systems function in such a manner that all people, at all times, have physical and economic access to sufficient, safe and nutritious food to meet their dietary needs for an active and healthy life (FAO 1996, 2002). The canon of food systems consists of three elements: (1) food availability (production, distribution and exchange), (2) food accessibility (affordability, allocation and preference) and (3) food utilisation (nutritional value, social value and food safety) (Gregory et al. 2005; Sastry et al. 2011). The three aforementioned elements can be impacted by emerging technologies such as nanotechnology, i.e. if the nanotechnology used is expensive; it thus impacts on the accessibility and utilisation. As mentioned earlier in this chapter that nanotechnology is a convergent technology, development of biosensors coupled with nanotechnology emerged, which is referred to nanobiosensors. In case of food safety the used of nano-based biosensors promising to offers high sensitive, specificity, rapid and accurate detection time of the mycotoxins when compared and security, food storage is highly threatened by natural contaminants particularly mycotoxins, but the use of nano-based biosensors promising a rapid detection of the mycotoxins in small quantities when compared to traditional methods.

The contamination by mycotoxins significantly influences the production of crops and animal feed and the quality of agricultural products and compromises the safety standards of foods, thus leading to overwhelming economic losses. Beyond the economic losses, they pose a great threat to human health. So, the problem is manifolded as neither food losses nor loss of life though contaminated food is favourable. It is therefore imperative for the timely, rapid and accurate detection of the mycotoxin contamination in crops and animal feed. Chromatographic techniques and chromatography mass spectrometry-linked techniques are the conventional analytical methods for detection of mycotoxins. However, these methods are based on the physical characteristics of toxins and have proven to be successful, but they require highly specialised (expensive) instrumentation and special skills. For instance, these techniques require long and complicated sample pretreatment procedures and high determination cost, which are not suitable for the high-throughput detection of large samples (Lin and Guo 2016). In recent study conducted by Masikini et al. (2015) showed better parameter in detecting fumonisins employing nano-based biosensor. Masikini et al. (2015) devised a polyaniline-carbon nanotube immunosensor which was doped with palladium telluride quantum dots (PdTe-QDs). The findings obtained from the experiments showed that the detection of FB1 antigen were a positive feedback that can be put to use for trace detection of fumonisins. The nano-based PdTe-DQ immunosensor had demonstrated the lowest limit of detection of 0.46 pg L^{-1} as well as sensitivity of 0.0162 kΩ L ng^{-1} for FB1, which was a good sensitivity.

The developments highlighted throughout this chapter provide us with numerous examples of potential advantages of nanotechnology, and the growing mechanisms of applying it for the benefit of agriculture have not yet made it to the market. According to Parisi et al. (2015), some studies show that public opinion is not negative towards nanotechnology and the introduction of nanotechnology in the market will most likely receive consumer attention and acceptance if benefits are clearly demonstrated to the consumer, e.g. smallholding farmers. The rapid progress of nanotechnology in other key industries may over time be transferred to agricultural applications as well and facilitate their development. Inbaraj and Chen (2016) also report on the use of nano-based sensors for bacterial pathogen detection and recommending them for various advantages such as rapid, sensitive and user-friendly detection, thus enabling portability for infield application. But this is arguable as there is a high risk of misanalysis as there may be numerous interferences when it comes to real-sample analysis, reproducibility and toxicity. Parisi et al. (2015), while they also commend the possibilities that the nanotechnology, also acknowledge that they may be expensive and complicated for smallholding farmers to make use of. Therefore, there is a great necessity for long-term studies to be conducted in testing the feasibility, practicality and affordability of the nanobiosensors as perhaps handheld devices for use by farmers in SSA. We note that this may not be a priority for SSA population to make use of this technology, as addressing poverty and making income are rather more crucial for the survival. Secondly, there are existing mechanisms that some farmers make use that are relatively affordable, but not efficient with regard to crop yield. Parisi et al. (2015) demonstrated multiple ways in which nano-based biosensors may be used holistically in for food safety and security, from crop production, fertilisers and water purification to diagnostics. So, it is critical that when advocating and placing nanotechnology in the farming or agriculture sector, it does not only become narrow in terms of benefiting the technology producers but also end users.

12.5 Conclusion

The use of nano-based biosensors in food safety enabled to improve detection capacity of food-borne pathogenic microorganisms, pesticide, drug residues and toxin contaminants in agriculture products that pose a threat to human health. The presence of nanomaterials in the construction of biosensor technology resulted into a better device, which is easy to handle and highly sensitive with improved detection speed. In addition, provide biosensors with capability of sensing single analyte of biological or chemical toxic contaminants in food sample. Furthermore, the presence of biological molecules for bio-recognition during antigen-antibody interactions offers the opportunity to enhance specificity or selectivity detection of the analytes or food-borne pathogens.

In case of accessibility in rural smallholding farm, majority of nano-based biosensors are still at the development stage, but the presence of nanomaterials is promising to offer biosensors that are affordable, high sensitive, specific, user-friendly and do not require a professional to operate them. Because of these advantages, nano-based biosensor technology is promising to offer novel devices that have a potential to benefit significantly small farms in remote or rural areas. In addition, this technology will give local farms a chance to produce quality food that adhere with local or international regulatory or safety standard guidelines as well as maximised crop production to meet the rising global demands of food.

References

Adanyi N, Levkovets IA, Rodriguez-Gil S, Ronald A, Váradi M, Szendro I (2007) Development of immunosensor based on OWLS technique for determining aflatoxin B1 and ochratoxin A. Biosens Bioelectron 22(6):797–802

Afonso AS, Perez-Lopez FRC, Mattoso LHC, Hernandez M (2013) Electrochemical detection of *Salmonella* using gold nanoparticles. Biosens Bioelectron 40(1):121–126

Akbas M, Ozdemir M (2006) Effect of different ozone treatments on aflatoxin degradation and physicochemical properties of pistachios. J Sci Food Agric 86(13):2099–2104

Aziz N, Pandey R, Barman I, Prasad R (2016) Leveraging the attributes of *Mucor hiemalis*-derived silver nanoparticles for a synergistic broad-spectrum antimicrobial platform. Front Microbiol (in press)

Bonel L, Vidal J, Duato P, Castillo J (2010) Ochratoxin A nanostructured electrochemical immunosensors based on polyclonal antibodies and gold nanoparticles coupled to the antigen. Anal Methods 2:335–341

Courvalin P (2008) Predictable and unpredictable evolution of antibiotic resistance. J Intern Med 264:4–16

Cozzini P, Ingletto G, Singh R, Asta CD (2008) Mycotoxin detection plays "cops and robbers": Cyclodextrin chemosensors as specialized police? Int J Mol Sci 9(12):2474–2494

Eldin TAS, Elshoky HA, Ali MA (2014) Nanobiosensor based on gold nanoparticles probe for aflatoxin B1 detection in food. Int J Curr Microbiol App Sci 3(8):219–230

FAO (1996) Rome Declaration on World Food Security and World Food Summit Plan of Action. World Food Summit 13–17 November 1996

FAO (2002) The State of Food Insecurity in the World 2001. Rome

Gregory PJ, Ingram JSI, Brklacich (2005) Climate change and food security. Philos Trans R Soc B 360(1463):2139–2148

Guan H, Zhang F, Yu J, Chi D (2012) The novel acetylcholinesterase biosensors based on liposome bioreactors-chitosan nanocomposites film for detection of organophosphates pesticides. Food Res Int 49(1):15–21

Haddaoui M, Raouafi N (2015) Chlortoluron-induced enzymatic activity inhibition in tyrosinase/ZnO NPs/SPCE biosensor for the detection of ppb levels of herbicide. Sensors Actuators B Chem 219:171–178

Hogue A, White P, Petter JG, Schlosser N, Gast R, Ebel E et al (1997) Epidemiology and control of egg-associated with *Salmonella* enteritidis in the United States of America. Rev Sci Tech 16:542–553

Hou H, Bai X, Xing C, Gu N, Zhang B, Tang J, Bai X, Xing C, Gu N, Zhang B et al (2013) Aptamer-based cantilever array sensors for oxytetracycline detection. Anal Chem 85:2010–2014

Huet A, Fodey T, Haughey SA, Weigel S, Elliott C, Delahaut P (2010) Advances in biosensor-based analysis for antimicrobial residues in food. Trends Anal Chem 29(11):1281–1294

Hussein HS, Brasel JM (2001) Toxicity, metabolism, and impact of mycotoxins on humans and animals. Toxicology 167(2):101–134

Idowu F, Junaid K, Paul A, Gabriel O, Paul A, Sati N, Maryam M, Jarlath U (2010) Antimicrobial screening of commercial eggs and determination of tetracycline residue using two microbiological methods. Int J Poult Sci 9(10):959–962

Inbaraj BS, Chen BH (2016) Nanomaterial-based sensors for detection of foodborne bacterial pathogens and toxins as well as pork adulteration in meat products. J Food Drug Anal 24(1):15–28

Kaushnik A, Solanski P, Ansari A, Ahmad S, Malhorta B (2008) Chitosan-iron oxide nanobiocomposite based immunosensor for ochratoxin-a. Electrochem Commun 10(9):1364–1368

Khanna VK (2008) New-generation nano-engineered bio-sensors, enabling nanotechnologies and nanomaterials. Sens Rev 28(1):39–45

Kiaya V (2014) Post-harvest losses and strategies to reduce them. Technical Paper on Postharvest Losses, Action Contre la Faim (ACF).

Kim G, Park SB, Moon J, Lee S (2013) Detection of pathogenic *Salmonella* with nanobiosensors. Anal Methods 5:5717–5723

Koedrith P, Thasiphu T, Tuitemwong K, Boonprasert R, Tuitemwong P (2014) Recent advances in potential nanoparticles and nanotechnology for sensing foodborne pathogens and their toxin in food and crops: current technologies and limitations. Sensor and Materials 26(10):711–736

Kumar S, Dilbaghi N, Barnela M, Bhanjana G, Kumar R (2012) Biosensors as novel platforms for detection of food pathogens and allergens. BioNanoSci 2(4):196–217

Landers TF, Cohen B, Wittum TE, Larson EL (2012) A review of antibiotic use in food animals: perspective, policy and potential. Public Health Reports January–February 127:1–22

Levy SB, Marshall B (2004) Antibacterial resistance worldwide: causes, challenges and responses. Nat Med 10:122–129

Li Y, Schluesener HJ, Xu S (2010) Gold nanoparticle-based biosensors. Gold Bull 43(1):29–41

Lin X, Guo X (2016) Advances in biosensors, chemosensors and assays for the determination of *Fusarium* mycotoxins. Toxins 8(6):161

Madhuri S, Ajoy KC, Kumar R (2010) Nanotechnology in agricultural diseases and food safety. J Phytology 2(4):83–92

Maragos CM, Thompson VS (1999) Fiber-optic immunosensor for mycotoxins. Nat Toxins 7(6):371–376

Masikini M, Mailu SN, Tsegaye A et al (2015) A fumonisins immunosensor based on polyanilino-carbon nanotubes doped with palladium telluride quantum dots. Sensors 15:529–546

McEwen SA, Fedorka-Cray PJ (2002) Antimicrobial use and resistance in animals. Clin Infect Dis 34(3):93–106

McGrath TF, Elliott CT, Fodey TL (2012) Biosensors for the analysis of microbiological and chemical contaminants in food. Anal Bioanal Chem 403:75–92

Mead GC (2004) Microbiological quality of poultry meat: a review. Braz J Poult Sci 6(3):135–142

Mungroo NA, Neethirajan S (2014) Biosensors for the detection of antibiotics in poultry industry-a review. Biosensors 4:472–493

Norouzi P, Pirali-Hamedani M, Ganjal MR, Faridbod F (2010) A novel acetylcholinesterase biosensor for determination of monocrotophos using FFT continuous cyclic voltammetry. Int J Electrochem Sci 5:1434–1446

Nowak B, Müffling T, Chaunchom S, Hartung J (2007) *Salmonella* contamination in pigs at slaughter and on the farm: a field study using an antibody ELISA test and a PCR technique. Int J Food Microbiol 115(3):259–267

Otles S, Yalcin B (2012) Review on the application of nanobiosensors in food analysis. Acta Sci Pol Technol Aliment 11(1):7–18

Owino J, Arotiba O, Hendricks N, Songa E, Jahed N, Waryo TT, Ngece R, Baker P, Iwuoha E (2008) Electrochemical immunosensor based on polythionine/gold nanoparticles for the determination of aflatoxin B1. Sensors 8(12):8262–8274

Paddle B (1996) Biosensors for chemical and biological agents of defence interest. Biosens Bioelectron 11(11):1079–1113

Parisi C, Vigani M, Rodriguez-Cerezo (2015) Agricultural nanotechnologies: what are the current possibilities? NanoToday 10(2):124–127

Parker CO, Tothill IE (2009) Development of an electrochemical immunosensor for aflatoxin M (1) in milk with focus on matrix interference. Biosens Bioelectron 24(8):2452–2457

Prasad R (2014) Synthesis of silver nanoparticles in photosynthetic plants. J Nanopart Article ID 963961. http://dx.doi.org/10.1155/2014/963961

Prasad R, Kumar V, Prasad KS (2014) Nanotechnology in sustainable agriculture: present concerns and future aspects. Afr J Biotechnol 13(6):705–713

Prasad R, Pandey R, Barman I (2016) Engineering tailored nanoparticles with microbes: quo vadis. WIREs Nanomed Nanobiotechnol 8:316–330. doi:10.1002/wnan.1363

Prasad R, Bhattacharyya A, Nguyen QD (2017) Nanotechnology in sustainable agriculture: recent developments, challenges, and perspectives. Front Microbiol 8:1014. doi:10.3389/fmicb.2017.01014

Radoi A, Targa M, Prieto-Simon B, Marty JL (2008) Enzyme-linked immunosorbent assay (ELISA) based on superparamagnetic nanoparticles for aflatoxin M1 detection. Talanta 77(1):138–143

Rai V, Acharya S, Dey N (2012) Implications of nanobiosensors in agriculture. J Biomaterials and Nanobiotechnology 3:315–324

Sastry RK, Rashmi HB, Rao NH (2011) Nanotechnology for enhancing food security in India. Food Policy 36:391–400

Song KM, Jeong E, Jeon W, Cho M, Ban C (2012) Aptasensor for ampicillin using gold nanoparticle based dual fluorescence-colorimetric methods. Anal Bioanal Chem 402(6):2153–2161

Song Y, Chen J, Wang LA (2015) Simple electrochemical biosensor based on AuNPs/MPS/Au electrode sensing layer for monitoring carbamate pesticides in real samples. J hazardous 304:103–109

Songa EA, Somerset S, Waryo T, Baker PG, Iwuoha EI (2009a) Amperometric nanobiosensor for quantitative determination of glyphosate and glufosinate residues in corn samples. Pure Appl Chem 81(1):123

Songa EA, Waryo T, Jahed N, Baker PGL, Kgarebe B, Iwuoha EI (2009b) Electrochemical nanobiosensor for glyphosate herbicide and its metabolite. Electroanalysis 21(3–5):671–674

Songa EA, Arotiba OA, Owino JH, Jahed N, Baker PG, Iwuoha EI (2009c) Electrochemical detection of glyphosate herbicide using horseradish peroxidase immobilized on sulfonated polymer matrix. Bioelectrochemistry 75(2):117–123

Tarafdar JC, Sharma S, Raliya R (2013) Nanotechnology: interdisciplinary science of applications. Afr J Biotechnol 12(3):219–226

Teodoro S, Micaela B, David KW (2010) Novel use of nano-structured alumina as an insecticide. Pest Manag Sci 66(6):577–579

Turan E, Sahin F (2016) Molecularly imprinted biocompatible magnetic nanoparticles for specific recognition of Ochratoxin A. Sensors Actuators B Chem 227:668–676

Ventura M, Gomez A, Anaya I, Diaz J, Broto F, Agut M, Comellas L (2004) Determination of aflatoxins B1, G1, B2 and G2 in medicinal herbs by liquid chromatography-tandem mass spectrometry. J Chromatography A 1048(1):25–29

Vimala V, Clarke SK, Urvinder Kaur S (2016) Pesticides detection using acetylcholinesterase nanobiosensor. Biosens J 5:1–4

Viswanathan S, Radecki J (2008) Nanomaterials in electrochemical biosensors for food analysis. Pol J Food Nutrition Sci 58(2):157–164

Viswanathan S, Wu L, Huang M, Ho J (2006) Electrochemical immunosensor for cholera toxin using liposomes and poly(3,4-ethylenedioxythiophene)-coated carbon nanotubes. Anal Chem 78(4):1115–1121

Vo-Dinh T (2005) Optical nanosensors for detecting proteins and biomarkers in individual living cells. Methods Mol Biol 300:383–402

Wang J (2005) Nanomaterial-based amplified transduction of biomolecular interactions. Small 1(11):1036–1043

Wang Y, Alocijia EC (2015) Gold nanoparticle-labeled biosensor for rapid and sensitive detection of bacterial pathogens. J Biol Eng 9:16

Wu Y, Tang L, Huang L, Han Z, Wang J, Pan H (2014) A low detection limit penicillin biosensor based on single graphene nanosheets preadsorbed with hematein-ionic liquids-penicillinase. Mater Sci Eng C Mater Biol Appl 1(39):92–99

Wu S, Zhang H, Duan S, Fang CC, Dai WZ (2015) Aptamer-based fluorescence biosensor for chloramphenicol determination using upconversion nanoparticles. Food Control 50:597–604

Xu S, Han X (2004) A novel method to construct a third-generation biosensor: self-assembling gold nanoparticles on thiol-functionalized poly (styrene-coacrylic acid) nanospheres. Biosens Bioelectron 19(9):1117–1120

Xu X, Liu X, Li Y, Ying Y (2013) A simple and rapid optical biosensor for detection of aflatoxin B1 based on competitive dispersion of gold nanorods. Biosens Bioelectron 47:361–367

Zhang S, Shan L, Tian Z, Zheng Y, Shi L et al (2008) Study of enzyme biosensor based on carbon nanotubes modified electrode for detection of pesticides residue. Chin Chem Lett 19:592–594

Zhao G, Wang H, Liu G (2015) Advances in biosensor-based instruments for pesticide residues rapid detection. Int J Electrochem Sci 10:9790–9807

Zheng Z, Zhoub Y, Li X, Liua S, Tangb Z (2011) Highly-sensitive organophosphorous pesticide biosensors based on nanostructured films of acetylcholinesterase and CdTe quantum dots. Biosens Bioelectron 26:3081–3085

Zhilong G, Zhujun Z (1997) Cyclodextrin-based optosensor for determination of tryptophan. Microchim Acta 126(3):325–328

Zimmerli B, Dick R (1996) Ochratoxin A in table wine and grape juice: occurrence and risk assessment. Food Addit Contam 13(6):655–668

Nanosensors: Frontiers in Precision Agriculture

13

Manoj Kaushal and Suhas P. Wani

Abstract

In the last decennium, nanotechnology has earned strength in and become the influential gizmo in current agriculture. Nanotechnology can boost agricultural production by improving nutrient use efficiency with nanoformulations of fertilizers; agrochemicals for crop enhancement, detection and treatment of diseases, host-parasite interactions at the molecular level using nanosensors, plant disease diagnostics, contaminants removal from soil and water, postharvest management of vegetables and flowers, and reclamation of salt-affected soils; etc. Nanobiosensors can be also employed for sensing a wide variety of pathogens, fertilizers, moisture and soil pH aiming to remove plant protection product applications, reduce loss of nutrients, and enhance crop yields through good nutrient management. Here we review nanotechnology applications for agriculture production, metal oxide-based nanosensors for protection of crops from diseases caused by bacteria and counter microbial attacks.

Keywords

Nanosensors • Precision farming • Nanotechnocrates • Plant growth hormones • Pathogen detection • Soil conditions

13.1 Introduction

Nanotechnology is a novel, innovative, interdisciplinary scientific approach which can be described as the design, formulation and appliance of instruments, devices, setups and systems by the size and shape of the material at the nanometre (10^{-9} of a

M. Kaushal (✉) • S.P. Wani
Research Program-Asia, International Crops Research Institute for the Semi-Arid Tropics, Patancheru 502324, Telangana, India
e-mail: kaushal.mbg@gmail.com

R. Prasad et al. (eds.), *Nanotechnology*, DOI 10.1007/978-981-10-4573-8_13

metre) scale where unique phenomena enable novel applications (Prasad et al. 2014, 2016). The word 'nano' is derived from the Greek word meaning 'dwarf', which means one billionth segment of a metre. Nanoparticles (NPs) are defined as the atomic or molecular aggregates with at least one dimension between 1 and 100 nm (Thakkar et al. 2010; Prasad 2014; Prasad et al. 2016). Engineered nanomaterials (ENMs) and engineered nanoparticles (ENPs) are categorized into four types: (1) carbon-based materials, such as carbon nanotubes (CNTs); (2) semiconductor, metal oxide-based materials such as nanogold, nanozinc and nanoscale metal oxides such as TiO_2, Al_2O_3 and ZnO and quantum dots (QDs); (3) polymer NPs, e.g. dendrimers, to perform specific chemical functions; and (4) composites combined NMs with other NMs or NPs or with bulk-type materials (Lin and Xing 2008; Nair et al. 2010). CNTs consist of hollow cylinders formed of carbon atoms. These appear like coiled tubes graphite ducts, and their walls appear like hexadic carbon rings which are commonly formed in huge bundles and can be categorized as single-walled carbon nanotube (SWCNT) and multi-walled carbon nanotubes (MWCNT). SWCNTs comprise of sole, cylindrical graphene coat, whereas MWCNTs constitute multiple graphene coats. CNTs possess high tensile strength, hence are considered 100 times tougher than steel and are good conductors of electricity. CNT-based nanodevices have been extensively employed as they possess good electrical conductivity, elevated electrochemical catalytic efficiency, biosuitability and non-toxicity. Nanotechnology utilizes NMs or NPs which exhibit physiochemical properties such as electrochemical (Rathee et al. 2016), chemical luminescence (Roda et al. 2016) and optical (Tereshchenko et al. 2016), which are entirely different than the actual material (Krejcova et al. 2015). These NPs possess unique properties such as magnetism, enlarged specific surface area, grater surface energy and quantum confinement. Nanotechnology thus has two major aspects: (1) synthesis of ENPs and (2) application of ENPs to achieve desired goals. Nanotechnology is major key enabling technologies that exhibit potential applications in multiple sectors like biotechnology, electronics, medical science and energy technologies by converging science and engineering (Suman et al. 2010; Aziz et al. 2015; Prasad et al. 2016). In addition it has the high ability to positively affect the agricultural sector with aim of improving crop production and food security while promoting social and economic equity (Prasad et al. 2014; Parisi et al. 2015).

13.2 Nanotechnology in Agricultural Sector

Technical innovations in agriculture are prime necessity to combat worldwide challenges of population boost, climate change and restricted food availability. Sustainable agricultural intensification is a concept aimed at increasing food production from the same existing farmland without adverse environmental impacts (The Royal Society 2009). The application of nanotechnology approach to agricultural crop handouts is one of the proposed apparatus for sustainable agricultural intensification. Nanotechnology potential is explored in agriculture sector to chop applications of plant protection aggregates, lowers nutrient losses, promotes rapid pest detection and ameliorates yields through improved nutrient management (Ghormade et al. 2011). The broad range of nanotechnology applications in agriculture include (1) nanoclays and nanozeolites to enhance the water-holding ability of soil (Sekhon 2014), which acts as a slack release source of water that reduces the hydric scantiness interval during crop season and herbicide delivery; (2) nanosensors for soil analysis, water management and transmission, and pesticide and nutrient drop; (3) nanomagnets for deportation of soil contaminants; (4) NPs for modern insecticides, pesticides and insect repellents; and (5) nanomaterials enabled discharge of genetic material for crop enhancement (Fig. 13.1). The utilization of target-specific NPs can shorten the loss to nontarget plant tissues. Nanoscale delivery systems such as encapsulation of agrochemicals (fertilizers and pesticides) by utilizing surface ionic attachments, dendrimers and other mechanisms display calm and targeted liberation of agrochemicals, which results in passive uptake of active ingredients and in turn reduces the amount of agrochemical application by minimizing the input and waste (Chen and Yada 2011).

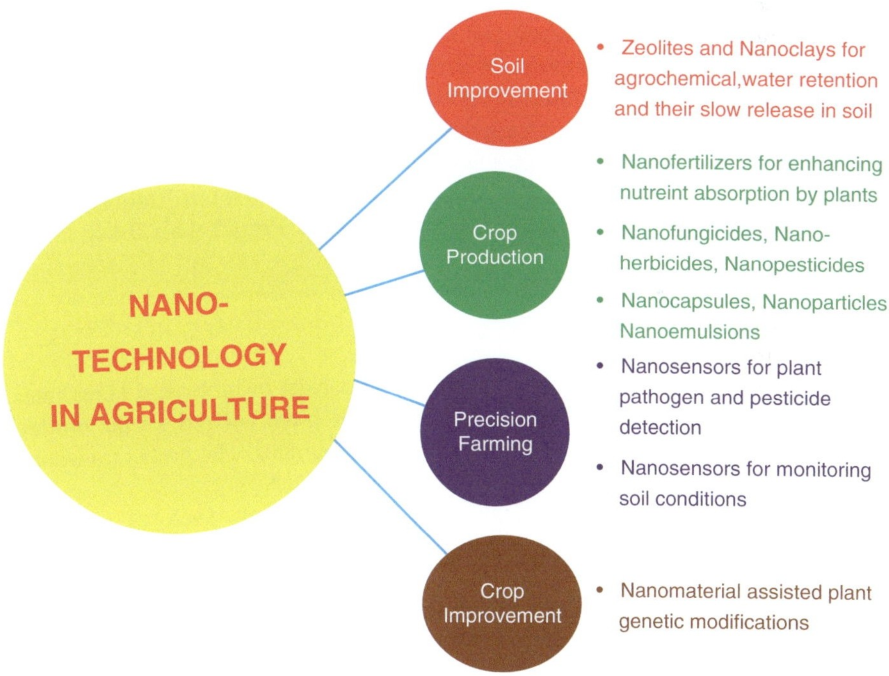

Fig. 13.1 Applications of nanotechnology in agriculture

13.3 Precision Agriculture

Precision agriculture or precision farming comprises of infotech utilized to the administration of commercial agriculture with a high-aspired aim to elevate long-desired goal to maximize output (crop yields) while reducing input (i.e. fertilizers, pesticides, herbicides, etc.) by monitoring environmental variables and employing spotted activity. Precision agriculture enabled technologies including (a) remote sensing, (b) geosynchronous positioning system (GPS) and (c) geographical information system (GIS). GIS techniques measure localized environmental conditions to identify the kind and area of problems in crop growth that makes pathogen control and fertilizer practice site-specific, accurate and efficient which in turn is helpful for yield analysis, pests monitoring, drought and forecasting (Sekhon 2014). The major components of precision farming include variable-rate technology (VRT) and yield monitors, which depend with the combination of GIS and GPS machinery, including the application of VRT field instruments. VRT is a device, which manipulates the variability notification and utilizes the inputs depending on the crop requisite that utilizes centralized results to analyse soil status and plant progression, fertilizer, chemical and water utilization, which in turn gives increased yields and reduces input expenses, thus lowering waste and labour expenditure. Nanotech-enabled systems help in increasing the adoption of autonomous sensors that are linked into GPS systems to render efficient monitoring services directed towards crop growth and soil conditions. Precision farming utilizes nanosensors and nano-based smart delivery systems which assist farmers to utilize agricultural resources such as water, nutrients and agrochemicals to increase crop productivity with improved fertilization supervision and minimizing inputs (Rai et al. 2012; Jones 2014; Prasad et al. 2014).

13.4 Nanosensors

Biosensors are compact analytical devices which consist of biological constitutes, viz. proteins, oligonucleotides, cells and tissue, which are utilized for the production of an output or signal which can be utilized by humans. The probe, transducer and detector are major components of a biosensor (Fig. 13.2).

The probe consists of biologically sensitized elements such as receptors, enzymes, antibodies, nucleic acids, microorganisms and organelles either a biologically developed element or bio-mime constituent. It collects signals arising out from analyte then forwards them to the transducer. The latter measures the changes occurring as a result of response on the biological component and transforms the energy into calculable electrical output. Transducers are categorized as electrochemical (electrodes), mass (piezoelectric crystals), optical (optrodes) and thermistors. The signals coming out from transducer are entrapped in the detector and then transfers to a microprocessor. These signals after amplification are analysed, and the data collected is shifted to output where it is displayed or saved (Velasco-Garcia 2014). The world's first biosensor was developed by the Research Center of Advanced Bionics of the National Institute of Advanced Industrial Science and Technology, Japan, in

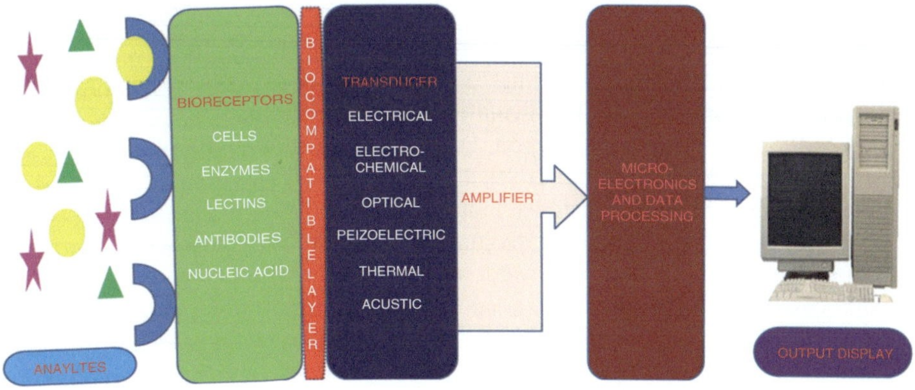

Fig. 13.2 Diagrammatic representation of major components of biosensor

2004 which can diagnose the possible occurrence of soil diseases. Soil diagnosis with this biosensor was based on the principle of computing the relative activity of favourable and unfavourable soil microbes decided on the grounds of differential oxygen utilization during respiration. Biosensors incorporating nanoparticles (NPs) are defined as nanobiosensors/nanosensors. The presence of NPs boosts the overall efficiency of biosensors probably due to the increased surface for reaction. Nanotubes, nanowires, nanoparticles or nanocrystals are mostly utilized to determine the signal transduction as they possess unique thermal, electrical and optical features, beneficial to increase sensitivity, decrease response period and amend the detection range (Yao et al. 2014). Nanosensors are replacing traditional sensors as they are portable, sensitive and specific. Nanosensors utilizing electrochemically functional SWCNTs utilizing either NPs or NTs to detect gases, viz. ammonia, nitrogen oxides, sulphur dioxide and volatile organics, have tremendous implementation in checking agricultural pollutants (Wanekaya et al. 2006). Nano-smart dust (utilizing small wireless sensors and transponders) and gas sensors can be utilized to determine the levels of environmental pollution (Mousavi and Rezaei 2010). Gas sensor arrays possessing NPs were developed through inkjet printing of metal ion-chelated DNA/SWCNTs on microfabricated electrodes, chased by the reduction of metal ions to metal. The consequences on the sensitivity and distinctivity of the gas sensors towards different gases, viz. H_2, H_2S, NH_3 and NO_2, showed the increase in both parameters towards some samples by functionalizing along with variable metal NPs. The united responses provide a distinct pattern for every sample that helps the system in identification and quantification of a particular gas (Su et al. 2013).

13.4.1 Nanosensors in Agriculture

Among the major tasks of nanotech devices includes elevation in the application of autonomous nanosensors positioned throughout cultivated fields combined with global positioning system for absolute time and comprehensive surveying of crop progress which provides high quality and essential data, such as optimum period of planting and harvesting of crops, leading to better agronomic practices (El Beyrouthya and El Azzi 2014). Intel company has developed chips possessing nanoscale characteristics and has installed big wireless sensor nodes (denoted as 'motes') all over a vineyard in Oregon, USA. The sensors record temperature once each minute and are the primary moves directed to full automation of the vineyard. Crossbow Technologies has designed motes which can be employed on agricultural field for irrigation estimation, frost diagnosing and alert pesticide employment, harvest period, bioremediation and water quality computation. Nanobarcodes and nanoprocessing could be utilized to keep a check on agricultural manufacturing (Li et al. 2005). The concept of grocery barcoding was employed for efficiently analysing and identification of crop diseases. Nanobarcodes were constructed which were able to tag variable pathogens in an agricultural field, which can be monitored by employing tools dependent on fluorescence.

13.4.1.1 Nanosensors to Monitor Soil Conditions and Plant Growth Hormone

Real-time monitoring of crop progression and farm status such as moisture content, soil fertility, temperature, crop nutrient capacity, pathogens, plant diseases, etc. can be done through the promotion of nanotech. Utilization of nanosensors to accurately measure the soil criterion (pH, nutrients, residual pesticides and soil moisture), checking of pathogens and estimation of nitrogen uptake helps farmers to utilize inputs more efficiently, thus fostering sustainable agriculture (Bellingham 2011; Prasad et al. 2017). These nanosensors can be used to check the period and volume of water application and agrochemical treatments based on the temporal and spatial MN needs of crops which is necessary to achieve the mission of precision agriculture. Modern research efforts are conducted to design precise water delivery systems for irrigation in future. The elements essential for this advancement constitute water storage, in situ water-holding ability, water delivery near the roots, water consumption capability of plants and enclosed water liberated according to crop requirement. Monreal et al. (2015) proposed that there are cross-talks between roots and microorganisms in wheat or other crops' rhizosphere which are an important part of chemical signalling networks. This causes liberation of microbial or plant metabolites into the soil. Figure 13.3 indicates that this concept of chemical connection is utilized to develop brilliant nanofertilizer distribution platforms for MNs, such as Zn and Fe. Brilliant nanofertilizer has been advised to possess nanobiosensors suspended in a biopolymer that coats a coating around fertilizer bits. The basic procedure of nutrient liberation depends on the identification and joining of a particular plant SOS by a nanobiosensor suspended in a biopolymer film that covers Zn-fertilizer NPs. After attachment, the fertilizer-ZnO-NPs (or other MN-NPs) get liberated in a reaction to root signal. After the development of such MN nanofertilizer delivery systems, their evaluation needs to be done in various crops, climate and farm conditions to amend micronutrient use efficiency, hence reducing fertilizer wastage to the environment. Nanotechnocrates possess the ability to investigate the plant's regulation of phytohormones that are responsible for root growth and

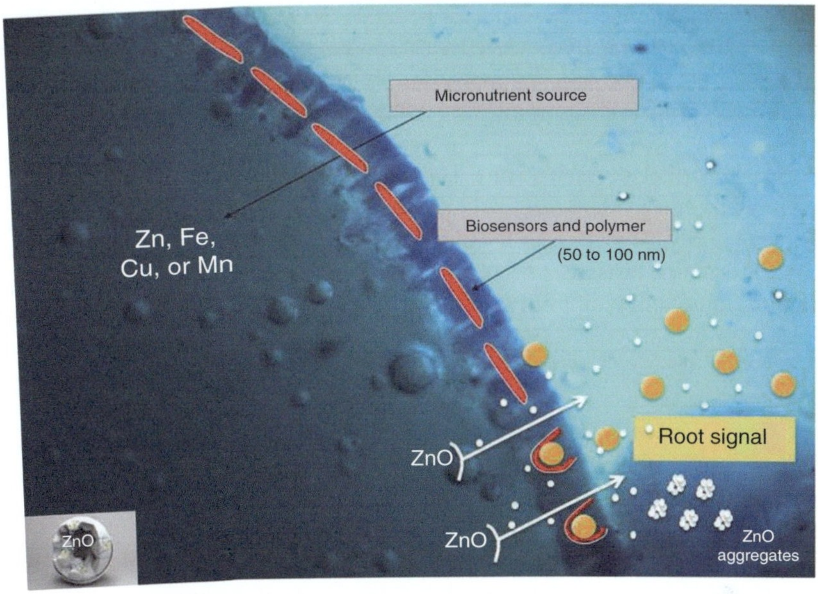

Fig. 13.3 A conceptual model for the synchronized release of ZnO nanoparticles according to crop demand. The binding of specific root chemical signals (*yellow*) with a nanobiosensor (*red*) housed in a thin polymer film (*blue*) coating ZnO-fertilizer nanoparticles (*dark grey*). The selective signal-biosensor binding process results in the release, dissolution, plant uptake and aggregation of ZnO NPs (*white spheres*) in the soil solution of crop rhizospheres (Adopted from Monreal et al. 2015)

seedling organization reacting with auxin has designed which will help researchers to understand the mechanism of plant roots adaption to their environment (McLamore et al. 2010). Nanotechnology-based microelectromechanical system (MEMS) sensors possess the ability to detect and also to react towards environmental changes by utilizing microelectronic circuits. Researchers have successfully developed wireless nanosensors consisting of micromachined MEMS cantilever beams coated with water-sensitive nanopolymer for moisture optimization.

13.4.1.2 Nanobiosensors for Plant Pathogen Detection

Nano-chips are kinds of microarrays which consist of fluorescent oligo capture probes that are used to identify hybridization. These nano-chips are highly specific and sensitive to detect the single nucleotide changes occurring in bacteria and viruses (Lopez et al. 2009). An electrochemical immunosensor has been designed that uses screen-printed electrode (SPE) which has a coating of agarose/nano-Au membrane and horseradish peroxidase (HRP)-marked VP antibody (HR-anti-VP) to detect the presence of *Vibrio parahaemolyticus* (Zhao et al. 2007). Seo et al. (2008) designed a biochip sensor setup, comprising two Ti contact pads and Ti nanowell device on $LiNbO_3$ substrate. When the bacteria were immune from the phages (uninfected bacteria), tiny voltage changes were noticed in the nanowell showing a power spectral density. Karnal bunt disease in wheat crop was identified by nanogold-dependent immunosensor utilizing surface plasmon resonance (SPR) (Singh et al. 2010). Wang et al. (2010) designed an electrochemical sensor utilizing modified gold electrode with copper NPs for the detection of pathogenic fungus *Sclerotonia sclerotiorum* by monitoring salicylic acid level in oil seeds.

13.4.1.3 Nanobiosensors for Pesticide Residue Detection

Nanosensors to detect pesticide residue have several advantages such as compact designs, sensitivity, low detection range, super selectivity and fast responses (Liu et al. 2008). Enzyme biosensors to check the presence of pesticide is established on the computation of enzyme inhibition included in the enzymatic reaction. The optical biosensors use several sensor techniques such as resonant mirrors, SPR and waveguides for the examination of biomolecular interactions. In SPR, a surface plasmon wave arises at the surface of two media having dielectric constants with contrary signs. Several biosensors to detect pesticide mixtures have been designed, viz. methyl parathion (Kang et al. 2010; Parham and Rahbar 2010), parathion (Li et al. 2006a; Wang and Li 2008), fenitrothion (Kumaravel and Chandrasekaran 2011), pirimicarb (Sun and Fung 2006) and dichlorvos and paraoxon (Vamvakaki and Chaniotakis 2007). Some biosensors are developed on the basis of the acetylcholinesterase (AChE) inhibition and chemometric result calculations utilizing artificial neural networks. A biosensor for organophosphorus pesticides (0.5 nM) was designed based on AChE binding on MWCNT-modified thick film strip electrode which showed the presence of paraoxon at parts per billion (0.145) levels (Joshi et al. 2005). An electrochemical sensor to detect the presence of pesticides methyl parathion and chlorpyrifos was designed, which is developed on AChE anchored on polyaniline deposited on vertically arranged CNTs wrapped with ssDNA. The pH around the electrode surface gets altered due to the inhibition of enzyme reaction which helps to analyse pesticide. The assay range of the biosensor for the presence of these pesticides was discovered as 1.10–12 mol/dm^3 (Viswanathan et al. 2009). A highly sensitive AChE biosensor modulated with hollow Au nanospheres, possessing diagnosis range of 0.06 μg/dm^3 for chlorpyrifos and 0.08 μg/dm^3 for carbofuran, has been designed (Sun et al. 2014). Yu et al. (2009) designed a biosensor to check the presence of organophosphorus pesticide in vegetables utilizing surface-functionalized CNTs to manage the effective anchorage of AChE on the surface of glassy carbon electrode. Liposomes are microscopic, fluid-filled pouches of phospholipid layers with endless walls. Electroactive marker enclosed immunoliposomes are utilize as signal enhancer for electrochemical immunoassays. MWCNTs were assembled onto liposome bioreactors to develop an AChE biosensor to detect the presence of organophosphates. An amperometric biosensor developed on coat-by-coat arrangement of SWCNT-poly (diallyl dimethyl ammonium chloride) and AChE was prepared to check the detection of carbaryl having a diagnosis range of 4.9×10^{-15} M. Zn-Se quantum dot immobilized AChE was utilized to detect organophosphate pesticides using graphene-chitosan nanocomposite-modified electrode. The detection limit of this biosensor employing methyl parathion as a model enzyme inhibitor was 0.2 nmol/dm^3.

QDs are inorganic semiconductor NPs possessing a shape of sphere having a diameter of 2–8 nm that fluoresce on being stimulated by an excitation light source. Its structure comprises of an inorganic core that decides the colour to be emitted and an inorganic shell. In addition an aqueous organic film is also present where biomolecules get joined to target the various biomarkers having size-dependent fluorescent properties, and hence emission of desired wavelengths (between 400 and 2,000 nm) occurred with varying configuration and amplitude (Patolsky et al. 2006). Single wavelength has the ability to excite QDs of different sizes, and emissions governed at varying dots are utilized as high-resolution biological fluorescent probes because of their inherent optical properties compared with common organic dye wavelengths. Their huge quantum output and restricted emission bands form sharper colours and cause to increased sensitivity and the feasibility of multiplexing of examination. An enzyme biosensor was designed to monitor trichlorfon amperometrically utilizing poly (N-vinyl-2-pyrrolidone) (PVP)-capped CdS QDs (Li et al. 2006b). An electrochemical biosensor dependent on colloidal AuNP that altered solgel interface was designed to detect the presence of monocrotophos, carbaryl and methyl parathion (Du et al. 2008a). The aggregated AuNPs on a solgel-derived silicate arrangement gave easy passage to electron deportation and caused the hydrolysis reaction, enhancing the sensitivity of the amperometric response. An efficient biosensor was designed to detect the presence of monocrotophos by joining distinct features of AuNPS and QDs. Newly developed electrochemical sensor working on CdTe QDs (cadmium telluride quantum dots)-AuNP electrode showed more sensitivity in comparison to QDs or AuNPs alone (Du et al. 2008b). Vinayaka et al. (2009) used CdTe QDs in a fluoroimmunoanalysis for sensing 2,4-dichlorophenoxyacetic acid (2,4-D) up to 250 pg l^{-1}.

An electrochemical immunosensor was designed to detect diuron, which is a substitutive phenyl urea herbicide (Sharma et al. 2011). Cheap ablated electrodes fabricated on polystyrene substrate were modulated which contained Prussian blue (PB)-AuNP coat. The electrically deposited PB-AuNP coat increased electron transport in the surrounding of the gold electrode enhancing the sensitivity. Kaushik et al. (2009) designed a sensor possessing nucleic acid on a chitosan nanobiocomposite coat possessing iron oxide NPs, accumulated on the glass surface, having a coat of indium tin oxide. It was used to detect pyrethroid, cypermethrin and permethrin. This disposable nanobiocomposite bioelectrode detected cypermethrin and permethrin fastly (N50 s) at 0.0025 ppm. Organophosphate hydrolase (OPH) was joined to a pH-perceptive fluorophore enveloped in silica NPs (100–500 nm) to develop an enzyme biosensor for detecting organophosphates. OPH showed stability up to 60 days and possessed a detection range (34 μM) of ppm for paraoxon (Ramanathan et al. 2009). A nanobiosensor working on the principle of atomic force microscopy tip operative with the acetolactate synthase enzyme was designed to detect the presence of metsulfuron-methyl by acquiring force curves (da Silva et al. 2013).

In 1990, Ellington's group (Otles and Yalcin 2010) described a system named SELEX (selection evolution of ligands by exponential enrichment) for the development of aptamers (Wanekaya et al. 2006; Farrell et al. 2013; Su et al. 2013). Nanoaptamers are sensors which are synthetic single-stranded nucleic acids, which have the ability to turn into unique and well-defined three-dimensional structures to form binding pockets and clefts working on doctrine of particular goal, joining with strong attraction (Chiu and Huang 2009). To detect the presence of herbicide and pesticide, respectively, an effectual aptamer sensor was designed to analyse the toxicity status in food (McKeague et al. 2011). An aptamer-controlled reversible prohibition of Au nanozyme action to detect acetamiprid was reported. The peroxidase-like nanozyme action of AuNPs and strong attraction and the specificity of acetamiprid-marked S-18 aptamer were conjugated for the analysis of 0.1 ppm acetamiprid between the analysis period of 10 min. AuNPs decorated MCNT-reduced graphene oxide nanoribbon composites were utilized which formed aptamer immobilization enabled to design a hypersensitive label-clear electrochemical impedimetric aptasensor to sense acetamiprid, where the alteration of electron transport resistance was correlated to the synthesis of acetamiprid-aptamer complex on the surface of reformed electrode. The aptasensor showed a linear reaction for acetamiprid within the limits of $5 \cdot 10^{-14}$–$1 \cdot 10^{-5}$ mol/dm^3 with an immensely meagre analysis limit of $1.7 \cdot 10^{-14}$ mol/dm^3 (Weerathunge et al. 2014).

13.5 Conclusions

Nanotechnology is a novel discovery which has the ability to reform the agricultural arena using modern gadgets. The broad range of nanotechnology applications in agriculture are pesticide delivery encased in nanomaterials for confined release, biopesticides balance nanomaterials, delayed discharge of nanomaterial aided fertilizers, and nanomaterials aided transmission of genetic material for crop amendment. Nanosensors can be utilized to determine nutrient, moisture and physiological status of plants which assist in taking up appropriate and timely corrective measures. Intelligent nanosensors can be used to distribute nutrients according to the crop necessity which are necessary to achieve the mission of precision agriculture. Thus, nanosensors that can be distributed in the agricultural fields have the ability to precisely detect the presence of crop pathogens and monitor the status of soil nutrients. In addition, nanosensors can be used in agriculture for interpretations of the interactions among root and soil organisms, nutrient operations, disease protection and the sustainance of crop stature. We should try to improve speed, accuracy and range of nanosensors. Applicability of nanotechnology in consumer products has evoked some ethical and societal concerns including health and environmental concerns and intellectual property rights. Thus despite of the potential scope of nanotechnology in agriculture, there is a need to check its applicability to humans as well as to the environment. Risk management strategies and apprehensions for the use of NPs should be put in place in parallel to the technological developments so as to transform nanotechnology to be employed in agricultural arena for continuous, elevated crop outputs, utility, social integrity, bio-protection and environmental sustainability.

References

Aziz N, Faraz M, Pandey R, Sakir M, Fatma T, Varma A, Barman I, Prasad R (2015) Facile algae-derived route to biogenic silver nanoparticles: synthesis, antibacterial and photocatalytic properties. Langmuir 31:11605–11612

Bellingham BK (2011) Proximal soil sensing. Vadose Zone J 10:1342–1342. doi:10.2136/vzj2011.0105br

Chen H, Yada R (2011) Nanotechnologies in agriculture: new tools for sustainable development. Trends Food Sci Technol 22:585–594

Chiu T-C, Huang C-C (2009) Aptamer-functionalized nano-biosensors. Sensors 9:10356–10388

da Silva AC, Deda DK, da Roz AL et al (2013) Nanobiosensors based on chemically modified AFM probes: a useful tool for metsulfuron-methyl detection. Sensors (Basel) 13(2):1477–1489

Du D, Chen S, Cai J, Zhang A (2008a) Electrochemical pesticide sensitivity test using acetylcholinesterase biosensor based on colloidal gold nanoparticle modified sol-gel interface. Talanta 74(4):766–772. doi:10.1016/j.talanta.2007.07.014

Du D, Chen S, Song D, Li H, Chen X (2008b) Development of acetylcholinesterase biosensor based on CdTe quantum dots/gold nanoparticles modified chitosan microspheres interface. Biosens Bioelectron 24(3):475–479. doi:10.1016/j.bios.2008.05.005

El Beyrouthya M, El Azzi D (2014) Nanotechnologies: novel solutions for sustainable agriculture. Adv Crop Sci Technol 2:e118. doi:10.4172/2329-8863.1000e118

Farrell D, Hoover M, Chen H, Friedersdorf L (2013) Overview of resources and support for nanotechnology for sensors and sensors for nanotechnology: improving and protecting health, safety, and the environment. US National Nanotechnology Initiative, Arlington. http://nano.gov/sites/default/files/pub_resource/nsi_nanosensors_resources_for_web.pdf

Ghormade V, Deshpande MV, Paknikar KM (2011) Perspectives for nano-biotechnology enabled protection and nutrition of plants. Biotechnol Adv 29:792–803

Jones PBC (2014) A nanotech revolution in agriculture and the food industry. Information Systems for Biotechnology, Blacksburg. http://www.isb.vt.edu/articles/jun0605.htm

Joshi KA, Tang J, Haddon R, Wang J, Chen W, Mulchandani A (2005) A disposable biosensor for organophosphorus nerve agents based on carbon nanotubes modified thick film strip electrode. Electroanalysis 17:54–58

Kang TF, Wang F, Lu LP, Zhang Y, Liu TS (2010) Methyl parathion sensors based on gold nanoparticles and Nafion film modified glassy carbon electrodes. Sensor Actuat B-Chem 145:104e109

Kaushik A, Solanki PR, Ansarib AA, Malhotra BD, Ahmad S (2009) Iron oxide-chitosan hybrid nanobiocomposite based nucleic acid sensor for pyrethroid detection. Biochem Eng J 46:132–140

Krejcova L, Michalek P, Rodrigo MM, Heger Z, Krizkova S, Vaculovicova M, Hynek D, Adam V, Kizek R (2015) Nanoscale virus biosensors: state of the art. Nanobiosensors in Disease Diagnosis 4:47–66. doi:10.2147/NDD.S56771

Kumaravel A, Chandrasekaran M (2011) A biocompatible nano TiO_2/nafion composite modified glassy carbon electrode for the detection of fenitrothion. J Electroanal Chem 650:163e170

Li Y, Cu YT, Luo D (2005) Multiplexed detection of pathogen DNA with DNA-based fluorescence nanobarcodes. Nat Biotechnol 23(7):885–889

Li C, Wang C, Hua S (2006a) Development of a parathion sensor based on molecularly imprinted nano-TiO_2 self-assembled film electrode. Sensor Actuat B-Chem 117:166e171

Li XH, Xie Z, Min H, Li C, Liu M, Xian YJ (2006b) Development of quantum dots modified acetylcholinesterase biosensor for the detection of Trichlorfon. Electroanalysis 18(22):2163–2167. doi:10.1002/elan.200603615

Lin D, Xing B (2008) Root uptake and phytotoxicity of ZnO nanoparticles. Environ Sci Technol 42:5580–5585

Liu Y, Tong Z, Prud'homme RK (2008) Stabilized polymeric nanoparticles for controlled and efficient release of bifenthrin. Pest Manag Sci 64:808–812

Lopez MM, Llop P, Olmos A, Marco-Noales E, Cambra M, Bertolini E (2009) Are molecular tools solving the challenges posed by detection of plant pathogenic bacteria and viruses? Curr Issues Mol Biol 11:13e46

McKeague Maureen, Giamberardino Amanda, De Rosa C Maria (2011) Advances in aptamer-based biosensors for food safety. In: Vernon Somerset (ed) Environ Biosensors. INTECH, Janeza Trdine 9, 51000 Rijeka, Croatia – EUROPEAN UNION. ISBN: 978–953–307-4863

McLamore ES, Diggs A, Calvo Marzal P et al (2010) Non-invasive quantification of endogenous root auxin transport using an integrated flux microsensor technique. Plant J 63(6):1004–1016

Monreal CM, DeRosa M, Mallubhotla SC, Bindraban PS, Dimkpa C (2015) The application of nanotechnology for micronutrients in soil-plant systems, VFRC Report 2015/3. Virtual Fertilizer Research Center, Washington, DC, p 44

Mousavi SR, Rezaei M (2010) Nanotechnology in agriculture and food production. J Appl Environ Biol Sci 1(10):414–419

Nair R, Varghese SH, Nair BG, Maekawa T, Yoshida Y, Kumar DS (2010) Nanoparticulate material delivery to plants. Plant Sci 179:154–163. doi:10.1016/j.plantsci.2010.04.012

Otles S, Yalcin B (2010) Nano-biosensors as new tool for detection of food quality and safety. Log Forum 6(4):67–70

Parham H, Rahbar N (2010) Square wave voltammetric determination of methyl parathion using ZrO2-nanoparticles modified carbon paste electrode. J Hazard Mater 177:1077e1084

Parisi C, Vigani M, Rodriguez-Cerezo E (2015) Agricultural nanotechnologies: what are the current possibilities? NanoToday 10:124–127. doi:10.1016/j.nantod.2014.09.009

Patolsky F, Zheng G, Lieber C (2006) Nanowire-based biosensors. Anal Chem 78:4260–4269

Prasad R (2014) Synthesis of silver nanoparticles in photosynthetic plants. J Nanoparticles 2014:Article ID 963961. http://dx.doi.org/10.1155/2014/963961

Prasad R, Kumar V, Prasad KS (2014) Nanotechnology in sustainable agriculture: present concerns and future aspects. Afr J Biotechnol 13(6):705–713

Prasad R, Pandey R, Barman I (2016) Engineering tailored nanoparticles with microbes: quo vadis. WIREs Nanomed Nanobiotechnol 8:316–330. doi:10.1002/wnan.1363

Prasad R, Bhattacharyya A, Nguyen QD (2017) Nanotechnology in sustainable agriculture: recent developments, challenges, and perspectives. Front Microbiol 8:1014. doi:10.3389/fmicb.2017.01014

Rai V, Acharya S, Dey N (2012) Implications of nanobiosensors in agriculture. J Biomater Nanobiotchnol 3:315–324

Ramanathan M, Luckarift HR, Sarsenova A, Wild JR, Ramanculov ER, Olsen EV et al (2009) Lysozyme-mediated formation of protein–silica nano-composites for biosensing applications. Colloids Surf B: Biointerfaces 73:58–64

Rathee K, Dhull V, Dhull R, Singh S (2016) Biosensors based on electrochemical lactate detection: a comprehensive review. Biochem Biophys Rep 5:35–54. doi:10.1016/j.bbrep.2015.11.010

Roda A, Mirasoli M, Michelini E, Massimo DF, Zangheri M, Cevenini L, Barbara PS (2016) Progress in chemical luminescence-based biosensors: a critical review. Biosens Bioelectron 76:164–179. doi:10.1016/j.bios.2015.06.017

Sekhon BS (2014) Nanotechnology in agri-food production: an overview. Nanotechnol Sci Appl 7:31–53. doi:10.2147/NSA.S39406

Seo S, Dobozi-King M, Young RF, Kish LB, Cheng M (2008) Patterning a nanowell sensor biochip for specific and rapid detection of bacteria. Microelectron Eng 85(7):1484–1489. doi:10.1016/j.mee.2007.12.046

Sharma P, Sablok K, Bhalla V, Suri CR (2011) A novel disposable electrochemical immunosensor for phenyl urea herbicide Diuron. Biosens Bioelectron 26(10):4209–4212. doi:10.1016/j.bios.2011.03.019

Singh D, Singh SC, Kumar S, Lal B, Singh NB (2010) Effect of titanium dioxide nanoparticles on the growth and biochemical parameters of *Brassica oleracea*. In: Riberio C, de-Assis OBG,

Mattoso LHC, Mascarenas S (eds), Symposium of International conference on Food and Agricultural Applications of Nanotechnologies. Sao Pedro, SP, Brazil

Su HC, Zhang M, Bosze W, Lim JH, Myung NV (2013) Metal nanoparticles and DNA co-functionalized single-walled carbon nanotube gas sensors. Nanotechnology 24(50):502–505

Suman PR, Jain VK, Varma A (2010) Role of nanomaterials in symbiotic fungus growth enhancement. Curr Sci 99:1189–1191

Sun H, Fung Y (2006) Piezoelectric quartz crystal sensor for rapid analysis of pirimicarb residues using molecularly imprinted polymers as recognition elements. Anal Chim Acta 576:67e76

Sun D, Hussain H, Yi Z, Siegele R, Cresswell T, Kong L, Cahill D (2014) Uptake and cellular distribution, in four plant species, of fluorescently labelled mesoporous silica nanoparticles. Plant Cell Rep 33:1389–1402

Tereshchenko A, Bechelany M, Roman V, Volodymyr K, Valentyn NS, Yakimova R (2016) Optical biosensors based on ZnO nanostructures: advantages and perspectives. A review. Sensors Actuators B Chem 229:664–671. doi:10.1016/j.snb.2016.01.099

Thakkar MN, Mhatre S, Parikh RY (2010) Biological synthesis of metallic nanoparticles. Nanotechol Biol Med 6:257–262

The Royal Society (2009) Reaping the benefits: science and the sustainable intensification of global agriculture. The Royal Society, London

Vamvakaki V, Chaniotakis NA (2007) Pesticide detection with a liposome-based nano-biosensor. Biosens Bioelectron 22:2848e2853

Velasco-Garcia MN (2014) Optical biosensors for probing at the cellular level: a review of recent progress and future prospects. Semin Cell Dev Biol 20(1):27–33

Vinayaka AC, Basheer S, Thakur MS (2009) Bioconjugation of CdTe quantum dot for the detection of 2,4-dichlorophenoxyacetic acid by competitive fluoroimmunoassay based biosensor. Biosens Bioelectron 24:1615–1620

Viswanathan S, Radecka H, Radecki J (2009) Electrochemical biosensor for pesticides based on acetylcholinesterase immobilized on polyaniline deposited on vertically assembled carbon nanotubes wrapped with ssDNA. Biosens Bioelectron 24:2772e2777

Wanekaya AK, Chen W, Myung NV, Mulchandani A (2006) Nanowire-based electrochemical biosensors. Electroanalysis 18(6):533–550

Wang M, Li Z (2008) Nano-composite ZrO_2/Au film electrode for voltammetric detection of parathion. Sensor Actuat B-Chem 133:607e612

Wang Z, Wei F, Liu SY, Xu Q, Huang JY, Dong XY, Yua JH, Yang Q, Zhao YD, Chen H (2010) Electrocatalytic oxidation of phytohormone salicylic acid at copper nanoparticles-modified gold electrode and its detection in oilseed rape infected with fungal pathogen *Sclerotinia sclerotiorum*. Talanta 80:1277e1281

Weerathunge P, Ramanathan R, Shukla R, Sharma TK, Bansal V (2014) Aptamer-controlled reversible inhibition of gold nanozyme activity for pesticide sensing. Anal Chem 86(24):11937–11941. doi:10.1021/ac5028726

Yao J, Yang M, Duan YX (2014) Chemistry, biology, and medicine of fluorescent nanomaterials and related systems: new insights into biosensing, bioimaging, genomics, diagnostics, and therapy. Chem Rev 114:6130–6178. doi:10.1021/cr200359p

Yu Y, Anthony SM, Bae SC, Luijten E, Granick S (2009) Biomolecular science of liposome nanoparticle constructs. Mol Cryst Liq Cryst 507:18–25

Zhao S, Jhang H, Wang W, Mao B (2007) Cloning and developmental expression of the Xenopus Nkx6 genes. Dev Genes E 6:217

Application of Nanomaterials Toward Development of Nanobiosensors and Their Utility in Agriculture

14

Ravindra Pratap Singh

Abstract

Nanobiosensor is a modified biosensor, a compact analytical device incorporating a biologically sensitized element onto a physicochemical transducer with miniature structure when compared to conventional biosensors. This can be used for sensing the variety of fertilizers, herbicide, pesticide, insecticide, pathogens, etc. and support sustainable agriculture for enhancing crop productivity. Thus nanomaterial utilities have emerged as a boon to our society. Day to day total land and water resources for agriculture are declining and reported huge loss in agricultural product due to the ever use of increasing level of herbicides, pesticides, and heavy metals. Recently nanomaterial-based technology emerged as an evolving field to revolutionize agricultural systems (sustainable) which enhance the quality of the agricultural products. This book chapter focuses on the utility of nanobiosensors in agriculture to monitoring of soil quality, pH, humidity, microbial load, etc. to enhance productivity. Nanomaterials have also a profound impact on energy, the economy, and the environment. New prospects for integrating nanotechnologies into nanobiosensors must be explored pertaining to any potential disadvantages to the environment or to human health. With targeted efforts by governments and academics in developing such agroproducts, thus nanotechnology will be transformative in the field of agriculture by focused R&D toward the goals for attaining sustainable agriculture.

Recently, nanomaterial-based biosensors exhibit fascinating prospects over traditional biosensors. Nanobiosensors have potential advantages such as enhanced detection sensitivity/specificity in its applications in different fields including environmental and bioprocess control, quality control of food, agriculture, biodefence, and, particularly, medical fields. But here we are concerned

R.P. Singh (✉)
Department of Biotechnology, Indira Gandhi National Tribal University (Central University), Amarkantak, Anuppur, Madhya Pradesh 484887, India
e-mail: rpsnpl69@gmail.com; ravindra69_99@yahoo.com

R. Prasad et al. (eds.), *Nanotechnology*, DOI 10.1007/978-981-10-4573-8_14

with the role of nanomaterials and nanobiosensor in agriculture and agroproducts. Some of their potential applications are discussed in this book chapter.

Keywords
Nanotechnology • Nanomaterials • Biosensor • Nanobiosensor • Agriculture • Health • Environment etc

14.1 Introduction

Nanotechnology (NT) using nanomaterials have emerged as a boon to our society with immense potential in agriculture. Development of nanobiosensors can be utilized for sensing a wide variety of fertilizers, herbicide, pesticide, insecticide, pathogens, etc. toward support of sustainable agriculture for increasing crop productivity. As we know that agricultural net land and water resources are rapidly declining, due to this huge loss, increasing use of fertilizers, herbicides, pesticides, rodenticides, and heavy metals has been demanded with new advanced technologies (Prasad et al. 2014). Nanomaterials exhibit a profound impact on medicine, energy, space, environment, and the economy. Nanobiosensors are advanced biosensors, which are used nanomaterials by using several techniques with rapid responses and very high sensitivities which are helpful for food quality assurance. In another way, nanobiosensors are the product of nanomaterials that will be focused R&D toward attaining the goal for the sustainable agriculture. The aim of this chapter is to present the directions of the development of nanobiosensors and their utility to detect a range of biological and chemical compounds in the agricultural food industry (Singh et al. 2011a, b, c, 2012a, b; Singh 2011; Sagadevan and Periasamy 2014; Prasad et al. 2017). Fig. 14.1 showed few examples of nanomaterials to be used in the fabrication of nanobiosensor for application in agriculture, health and environment.

In addition, NT is a rapidly evolving field which revolutionizes agricultural food systems to enhance the quality of their products and natural resource. As a result, it may boost rural economy locally as well as globally. Nanobiosensors are monitoring soil quality (pH, humidity, microbial load, etc.) to enhance crop productivity. Biosensors are analytical devices with less sensitivity, specificity, selectivity, and reproducibility as shown in Fig. 14.2 whereas nanobiosensor is an advanced version of a biosensor with ultrahigh sensitivity, specificity, selectivity, and reproducibility (accuracy and precision). In an immobilized bioreceptor probes onto solid matrix are selective for target analyte molecules at the nanoscale or atomic scale which open up new tools for real bioanalytical applications for the detection of analytes like urea, glucose, pesticides, metabolites and microorganisms/pathogens, etc. (Singh 2012, 2016; Singh et al. 2008, 2009, 2011a, b, c).

The nanobiosensors show unique properties like being highly specific, stable under normal storage conditions, specific interaction between analytes, independent of stirring, pH, and temperature; reaction time is very less or minimal with accurate, precise, reproducible, linear range and without electrical noise. These are tiny, biocompatible, nontoxic, nonantigenic, cheap, portable, and easily operated.

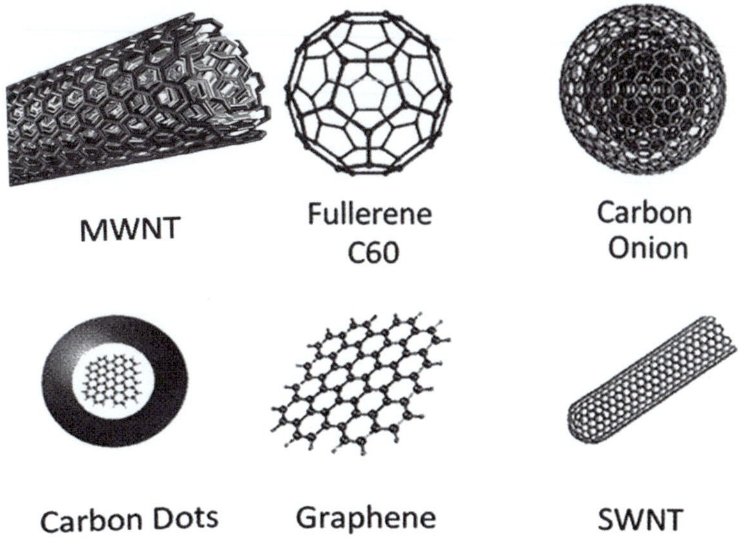

Fig. 14.1 Few examples of nanomaterials to be used in nanobiosensor fabrication

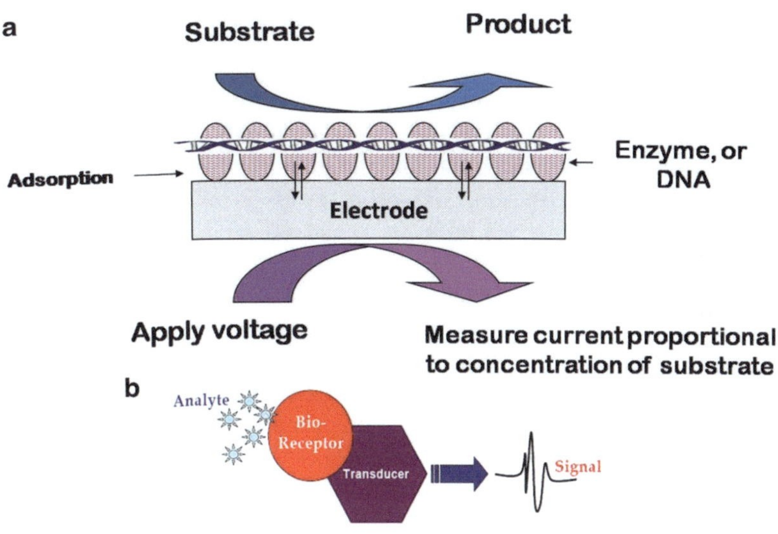

Fig. 14.2 Schematic presentation of biosensor

Nanobiosensor comprises of biologically sensitized elements (probe), transducer, and detector as shown in Fig. 14.3. The biological elements including receptors, antibodies, enzymes, nucleic acids, organelles, cell, tissue, and microorganisms, which are receiving signals from the sample of interest and transmits it to desired trasducer (Singh et al. 2012a, b, 2014).

The transducer acts as an interface depending upon mode of action, measuring the physical change that occurs with the reaction at the bioreceptor/sensitive

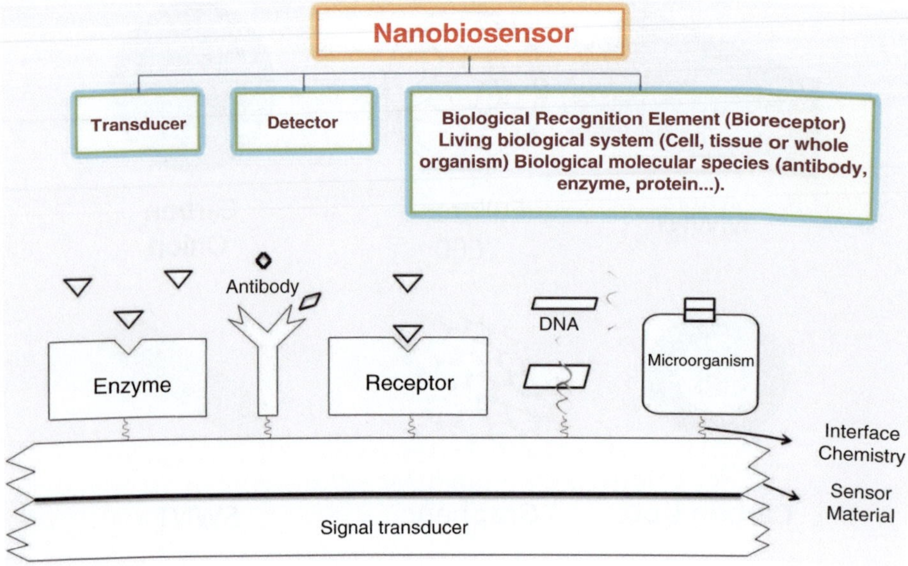

Fig. 14.3 Schematic presentation of nanobiosensor

biological probe and then transforming that energy into measurable electrical output. The detector traps the signal events from the transducer, which are then passed to a microprocessor where they are amplified and analyzed. Then the data is transferred to output/displayed/stored. Nanosensors are ultrasensitive which can detect even virus particles or very low concentrations of an analyte of interest at an atomic scale with the highest efficiency due to increased surface to volume ratio. But nanobiosensors are very sensitive and error prone and still in infancy stage (Singh and Pandey 2011; Singh et al. 2011a, b, c).

The role of nanobiosensor or nanomaterial-based biosensor in agriculture has marked advantages such as enhanced detection sensitivity/specificity and which reflects a wide potential for its applications in different fields including environmental, quality control of food, biodefense, and medical applications. But in this book chapter, the author is concerned with the role of nanobiosensor in agriculture and agriproducts. Few of the applications of nanobiosensors are as diagnostic tools for soil quality and plant disease control which may also be used to diagnose soil disease caused by viruses, bacteria, and fungi (Singh et al. 2011a, b, c; Singh and Choi 2010).

14.2 Nanofertilizers

Nanofertilizer is a product that gives nutrients to crops which is encapsulated within nanomaterials such as nanoparticles, nanotubes, or nanoporous materials or coated with a thin protective polymer film or given as particles or emulsions of nanoscale dimensions. Nanofertilizers reduce nitrogen loss due to soil leaching/erosion.

Recently carbon nanotubes were reported to penetrate tomato seeds (Khodakovskaya et al. 2009), and zinc oxide nanoparticles were reported to enter the root tissue of rye grass (Lin and Xing 2008). However, the nanofertilizers showed sustained release of nutrients on demand.

Nanofertilizer allows selective nitrogen release as nutrient to the plants. Slow controlled release of fertilizers improves healthy soil by reducing toxic effect concern with overuse of fertilizer. Zeolites are natural, aluminum silicates that act better plant growth to not only improve the efficiency, value of fertilizer, water infiltration, and retention but also improves yield, retain nutrients with long-term soil quality, and reduce loss of nutrients in soil. In addition, when Zeolites linked to a nanobiosensor which can modernize agriculture to sense the deficiency in either plant or soil and control, the release of water/nutrients retained in the zeolite. Pesticides inside nanoparticles have been developed and used as a smart delivery system for the greater production of crops and less injury to agricultural workers (Lauterwasser 2005).

Nanodevice detects contaminants, biomolecules, pests, nutrient, and plant stress due to drought, temperature, and pressure. They may be potentially helpful for farmers to enhance competence. Organophosphorus pesticides have been monitored by liposome-based biosensors (Vamvakaki and Chaniotakis 2007). Zhang et al. have developed a method for the detection of *Escherichia coli* (*E. coli*) using bismuth nanofilm-modified GCE (Zhang et al. 2007). Seo et al. have constructed a biochip sensor system for specific and rapid detection of bacteria (Seo et al. 2008). Giardi and Piletska have reported PS-II (photosystem II)-based biosensor for the detection of not only several groups of herbicides but also those which have a potential to monitor polluting chemicals concerned with industrial and urban effluents, sewage sludge, landfill leak water, groundwater, and irrigation water (Giardi and Piletska 2006). Although, the detection of DNA and protein has been reported for (nanobiosensors) ssDNA-CNTs probes/optical biosensors to detect specific kinds of DNA oligonucleotides (Cao et al. 2008) and MWNT/ZnO/CHIT-composite-film-modified GCE for immobilization ssDNA probes to effectively discriminate different DNA sequences (Zhang et al. 2008), a nanobiosensor with bionanocomposite layer of MWNT in chitosan deposited on a SPCE for the detection of DNA damage (Galandova et al. 2008), a nanobiosensor with GNPs functionalized with alkanethiol-capped LNA/DNA chimeras in a tail-to-tail hybridization mode for single-stranded DNA (McKenzie et al. 2007). Ma et al. have reported nano-SiO_2/p-aminothiophenol (PATP) film for the detection of the PAT gene sequences by a label-free EIS method (Ma et al. 2008). Maki et al. have reported for the first time nanowire-based field-effect transistor for the ultra-sensitive detection of special protein molecules which might play a vital role in detection of plant pathogens, certain plant diseases linked to mineral deficiency and biomarkers (Maki et al. 2008).

14.3 Food Products and Packaging

Food analysis is increasing in the food industry. The SPR biosensor monitors interactions of a specific binding protein with the vitamin immobilized on a sensor chip. Recently, biosensors analyze the presence of antibiotics in a very short time. Immunobiosensors based on the surface immobilization of monoclone antibodies onto indium tin oxide (ITO) electrodes detect *Escherichia coli* O157:H7. Carbohydrate, proteins, lipids, nucleic acid, antibodies, and receptors in the biological system at nanoscale when utilized engineered nanomaterials in agriculture raised risks to environment or human health. An insufficient knowledge like toxicity, biocompatibility, bioaccumulation, and exposure risk of nanomaterials are not well documented. The nanoparticles and the nanoscale materials used in the construction of nanobiosensors have to be properly characterized and tested in biological environments, and the probable toxicity has to be examined. The possible hazard of nanoparticles to biological organisms has significantly drawn interest from academics, industry, governmental ethical committees, and nongovernmental organizations worldwide. Besides the commercialization of nanobiosensors, they may act as substitutes for agricultural commodities which might be "disastrous" on the economy of developing countries and check the nanoproduct expansion, i.e., pose bad economic effects on the poor by increasing productivity in developed countries, which could lead to a decrease in commodity price in developing countries (Gruère and Sengupta 2009).

Food stuffs are requiring an effective packaging material. In this context titanium dioxide, zinc oxide, and magnesium oxide are used so efficiently in killing microorganisms which are cheaper and safer. Oxygen is the main culprit which spoils the fat in meat and cheese. Few examples are nanoparticle-based Durethan, clay nanoparticles into an ethylene vinyl alcohol copolymer and into a poly (lactic acid) biopolymer, and polymer silicate nanocomposites check barrier properties to oxygen. Also nanoclay nylon coatings and silicon-oxide-based glass bottles are used to stop gas diffusion. Polymer nanocomposites are very useful to improve barrier performance to not impede oxygen and carbon dioxide but also to enhance the barrier performance to ultraviolet rays and strengthening strength, stiffness, dimensional stability, and heat resistance and improve shelf life. Table 14.1 shows few

Table 14.1 Few nanomaterials used in food products and packaging

Nanomaterials	Applications (in use)
Titanium nitride	Improvement of thermal properties, antimicrobial and deodorant agent, UV filter, PET, fridges
Carbon black	Additive, rubber, silicones, printing inks
Silicon dioxide	Antisliping agent, printing inks, paper and board, rubbers, silicones
Aluminum	Filler in polymers, scratch and abrasion resistance in coatings, improvement of barrier properties, UV filter
Silver	Antimicrobial, antibiotic, antistatic agent, reusable food containers
Nanoclay	Improvement of barrier properties
Zinc oxide	UV filter, antimicrobial and fungistatic agent, deodorant, plastic glasses, plastic films

nanomaterial uses in food product and packaging (Lorenz et al. 2011; Gaiser et al. 2013; Food Standards Agency, UK 2013; Innovative Research and Products 2009).

14.4 Nanodevices in Agriculture

Nanomaterials (NMs) play a vital role in the development of the agricultural sector. They are capable of being used in agricultural products and not only protect plants but also monitor plant growth and detection of diseases. Due to these reasons, NMs have gained tremendous momentum in farming for enhancing the efficiency of nutrient via nanoformulations of fertilizers, breaking barriers, surveillance and control of pests and diseases, preservation and packaging of food and food additives, strengthening of natural fibers, removal of contaminants from soil and water, improving the shelf life of vegetables and flowers, clay-based nanocomposite for water management, reclamation of salt-affected soils, and stabilization of erosion-prone soil surfaces at the molecular level (Prasad et al. 2017). Biosensors and nanobiosensors as nanodevices are able to detect nutrients and contaminants to protect soil health, environment, and soil fertility. These are used to determine nutrient, moisture, and physiological status of plants. Currently bioanalysis by biosensors and nanobiosensors are in wide practices. Further in this context, electronic nose (E-nose) is a device to detect an array of gases, different types of odors, identify the odorant, and estimate the odorant concentration. Basically, it is gas sensor based on nanoparticles; the resistance changes when a gas passes over it and generates a change in electrical signal and ultimately detect gas. Seeds during storage emit volatile aldehydes which cause aging and are also harmful to other seeds. Storage decision-making is somewhat possible when timely aldehydes detected by abovementioned nanodevices. In addition, smart delivery systems in the form of nanoscale-based devices are capable to detect and treat plant diseases and nutrient deficiencies in crops when sign and symptoms were seen. The utility of these device in agriculture can timely control, target, and regulate, to avoid biological barriers (Kessler 2011). Nanomaterials like electrochemically active carbon nanotubes, nanofibers, and fullerenes have been recently developed and applied for highly sensitive biochemical sensors. These nanosensors have also relevant implications for application in agriculture, in particular for soil analysis, easy biochemical sensing and control, water management and delivery, pesticide, and nutrient delivery. Nanomaterial is considered as one of the possible solutions to problems in food and agriculture, just like biotechnological issues of safety on health, biodiversity and environment along with appropriate rules and regulation (Kuzma and verHage 2006).

14.5 Agro-based Feed/Food and Toxicity

Nanomaterial application can be found in agricultural production, animal feed, food processing, food additives, and food materials. But applications of nanomaterials in the future may increase the human and environmental exposure and causes severe problems. Nanoencapsulates, silver, titanium dioxide, and silica are claimed to be already in use in feed additives, biocides, pesticides, and food materials (Peters et al. 2016; Servin and White 2016).

The uses and benefits of nanomaterials in agriculture are important and created great interest not only enhancing agricultural productivity and efficiency but also producing huge quantities of food with less cost, energy, and waste (Kah 2015). However, its health hazards and toxicity remain unanswered and needs safety concerns over agricultural, environmental, and human health to develop nanotoxicological profile. Environmental health and safety of nanomaterials such as fullerenes (C_{60}), CNTs, silver, iron, titanium dioxide, aluminum oxide, cerium oxide, zinc oxide, silicon dioxide, dendrimers, nanoclays, and gold nanoparticles are under investigation. The physical-chemical properties, environmental degradation and accumulation, environmental toxicology, and mammalian toxicology of nanomaterials are worth noting pertaining to that (Ma et al. 2015).

The nanomaterials (NMs) can also be solve the problems in agriculture such as overdependence on irrigation, being prone to climate, poor energy conversion to products like nanopesticides and nanofertilizer, disease prevention in poultry, food packaging, use of agricultural waste, and nanosensors. Still nanomaterials revolutionize agriculture and food systems of nanofertilizers, pesticide career, microfluidics, nucleic acid bioengineering, smart treatment delivery systems, nanobioprocessing, bioanalytical nanobiosensors, environmental processing, pathogen detection, plant/animal production, biosecurity, molecular and cellular biology, protection of the environment through the reduction and conversion of agricultural materials into valuable products, design and development of new nanocatalysts to convert vegetable oils into bio-based fuels and biodegradable industrial solvents, and in controlled ecological life support system (Khanna 2008; DeRosa et al. 2010; Gruere et al. 2011; Prasad 2014; Prasad et al. 2014, 2016, 2017).

14.6 Future Perspectives

Agriculture provides crops for food and industry, fiber, fuel, auto fuel, and drugs. However, we face ever increasing food prices, farmer's suicides, and low hand input use efficiency. Nanomaterials and their product nanobiosensors are very well utilized in agricultural practices to bring new hope. Utility of nanomaterials in agriculture (food quality) improves the quality of life, e.g., nanobiosensors, nanofertilizers, nanopesticides, and biocontrol agents. A variety of nanomaterials' use in agriculture, which help to reduce the consumption of agrochemicals by use of smart delivery systems, minimize the nutrient losses and increase the yield through optimized water and nutrient management. In plant breeding and genetics, nanodevices may also be explored. Bionanocomposites are new nanomaterials which would be extensively used in agriculture. Although, wide advantages of nanomaterials applications in the agricultural sector are still very less or fully not commercialized when we compared to other industrial sectors, the thrust area of agricultural utilized nanomaterials are being widely considered as challenges pertaining to agriculture, environment, and health specifically sustainability, increased productivity, disease management, and crop protection via novel techniques for monitoring, assessing, and controlling the agricultural practices. This chapter provides the potential role of nanomaterials in sustainable agriculture, environment, and health of society. Although, nanomaterials have potential to minimizing adverse problems of agricultural practices on environment and human health, improving food security and productivity to promoting social and economic equity. Nanomaterial based nanodevices provide benefits not only in management of soil and waste but also in control of the release of nutrients, pesticides, fertilizers in crops, food quality, and safety.

Nanomaterials of different sizes and shapes are the bases for engineering to create unique properties which would be targeted toward desired applications in various fields such as medicine, environmental science, and food processing for safe use. In addition, plant protection, production, seed germination and its growth, pathogen detection, and pesticide/herbicide detection are the main thrust areas where maximum utility of nanomaterials is to be explored and the role of these can play in future agricultural productivity and efficiency.

Acknowledgment Dr. Ravindra Pratap Singh thanks IGNTU, Amarkantak, M.P. India for providing facilities to prepare this book chapter.

References

Cao C, Kim JH, Yoon D, Hwang ES, Kim YJ, Baik S (2008) Optical detection of DNA hybridization using absorption spectra of single-walled carbon nanotubes. Mater Chem Phys 112(3):738–741

DeRosa MC, Monreal C, Schnitzer M, Walsh R, Sultan Y (2010) Nanotechnology in fertilizers. Nat Nanotechnol 5:91

Food Standards Agency, UK (2013) Nanotechnology-enabled foods and food contact materials on the UK market. https://www.food.gov.uk/science/novel/nano/monitoring, novelfoods@food-standards.gsi.gov.uk.

Gaiser BK, Hirn S, Kermanizadeh A, Kanase N, Fytianos K, Wenk A, Haberl N, Brunelli A, Kreyling WG, Stone V (2013) Effects of silver nanoparticles on the liver and hepatocytes in vitro. Toxicol Sc 13(2):537–547

Galandova J, Ziyatdinova G, Labuda J (2008) Disposable electrochemical biosensor with multiwalled carbon nanotubes-chitosan composite layer for the detection of deep DNA damage. Analyst 24:711–716

Giardi MT, Piletska EV (2006) Biotechnological applications of photosynthetic proteins: biochips, biosensors and biodevices, Biotechnology Intelligence Unit Co-published by Land Biosciences and Springer

Gruère G, Sengupta D (2009) GM-free private standards and their effects on biosafety decision-making in developing countries. Food Policy 34:399–406

Gruère G, Narrod C, Abbott L (2011) Agriculture, food, and water nanotechnologies for the poor: opportunities and constraints. IFPRI Policy Brief 19

Innovative Research and Products (2009) Nanoenabled packaging for the food and beverage industry: a global technology industry and market analysis,Innovative Research & Products, Inc: Market Research Reports, https://marketpublishers.com/members/innoresearch/info.html, ID: N851FF884D9EN

Kah M (2015) Nanopesticides and nanofertilizers: emerging contaminants or opportunities for risk mitigation? Front Chem 3:1–6

Kessler R (2011) Engineered nanoparticles in consumer products: understanding a new ingredient. Environ Health Perspect 119(3):A120–A125

Khanna VK (2008) New-generation nano engineered biosensors, enabling nanotechnologies and nanomaterials. Sens Rev 28(1):39–45

Khodakovskaya M, Dervishi E, Mahmood M, Xu Y, Li Z, Watanabe F, Biris AS (2009) Carbon nanotubes are able to penetrate plant seed coat and dramatically affect seed germination and plant growth. ACS Nano 3(10):3221–3227

Kuzma J, verHage P (2006) Nanotechnology in agriculture and food production: anticipated applications. The project on emerging nanotechnologies, Woodrow Wilson International Center for Scholars, Washington, DC

Lauterwasser C (2005) Small sizes that matter: opportunities and risks of nanotechnologies. Report in Cooperation with the OECD International Futures Programme. http://www.oecd.org/science/nanosafety/44108334.pdf

Lin DH, Xing BS (2008) Root uptake and phytotoxicity of ZnO nanoparticles. Environ Sc Tech 42(15):5580–5585

Lorenz C, Von Goetz N, Scheringer M, Wormuth M, Hungerbühler K (2011) Potential exposure of german consumers to engineered nanoparticles in cosmetics and personal care products. Nanotoxicology 5(1):12–29

Ma Y, Jiao K, Yang T, Sun D (2008) Sensitive PAT gene sequence detection by nano-SiO2/Paminothiophenol self-assembled films DNA electrochemical biosensor based on impedance measurement. Sensors Actuators B 131:565–571

Ma C, White JC, Dhanker OP, Xing B (2015) Metal-based nanotoxicity and detoxification pathways in higher plants. Environ Sci Technol 49:7109–7122

Maki WC, Mishra NN, Cameron EG, Filanoski B, Rastogi SK, Maki GK (2008) Nanowire transistor based ultra-sensitive DNA methylation detection. Biosens Bioelectron 23(6):780–787

McKenzie F, Faulds K, Graham D (2007) Sequence specific DNA detection using high-affinity LNA- functionalized gold nanoparticles. Small 3:1866–1868

Peters RJB, Bouwmeester H, Gottardo S, Amenta V, Arena M, Brandhoff P, Marvin HJP, Mech A, Moniz FB, Pesudo LQ, Rauscher H, Schoonjans R, Undas AK, Vettori MV, Weigel S, Aschberger K (2016) Nanomaterials for products and application in agriculture, feed and food. Tr Fd Sc Tech 54:155–164

Prasad R (2014) Synthesis of silver nanoparticles in photosynthetic plants. J Nanoparticles 2014:Article ID 963961. http://dx.doi.org/10.1155/2014/963961

Prasad R, Kumar V, Prasad KS (2014) Nanotechnology in sustainable agriculture: present concerns and future aspects. Afr J Biotechnol 13(6):705–713

Prasad R, Pandey R, Barman I (2016) Engineering tailored nanoparticles with microbes: quo vadis. WIREs Nanomed Nanobiotechnol 8:316–330. doi:10.1002/wnan.1363

Prasad R, Bhattacharyya A, Nguyen QD (2017) Nanotechnology in sustainable agriculture: recent developments, challenges, and perspectives. Front Microbiol 8:1014. doi:10.3389/fmicb.2017.01014

Sagadevan S, Periasamy M (2014) Recent trends in nanobiosensors and their applications-a review. Rev Adv Mater Sci 36:62–69

Seo S, Dobozi-King M, Young RF, Kish LB, Cheng M (2008) Patterning a nanowell sensor biochip for specific and rapid detection of bacteria. Microelect Engin 85(7):1484–1489

Servin AD, White JC (2016) Nanotechnology in agriculture: next steps for understanding engineered nanoparticle exposure and risk. NanoImpact 1:9–12

Singh RP (2011) Prospects of nanobiomaterials for biosensing. Intern J Electrochem, publisher SAGE-Hindawi journal collection 2011: ID 125487, 30. doi:10.4061/2011/125487

Singh RP (2012) Prospects of organic conducting polymer modified electrodes: enzymosensors. Int J Electrochem, publisher SAGE-Hindawi journal collection. Article ID 502707, 14. doi:10.1155/2012/502707

Singh RP (2016) Nanobiosensors: potentiality towards bioanalysis. J Bioanal Biomed 8:e143. doi:10.4172/1948-593X.1000e143

Singh RP, Choi JW (2010) Bio-nanomaterials for versatile biomolecules detection technology. Adv Mater Lett 1:83–84

Singh RP, Pandey AC (2011) Silver nano-sieve using 1,2-benzenedicarboxylic acid as a sensor for detecting hydrogen peroxide. Anal Methods 3:586–592

Singh RP, Oh BK, Koo KK, Jyoung JY, Jeong S, Choi JW (2008) Biosensor arrays for environmental pollutants detection. Biochip J 2:223–234

Singh RP, Kang DY, Oh BK, Choi JW (2009) Polyaniline based catalase biosensor for the detection of hydrogen peroxide and Azide. Biotechnol Bioprocess Eng 14:443–449

Singh RP, Oh BK, Choi JW (2010) Application of peptide nucleic acid towards development of nanobiosensor arrays. Bioelectrochemistry 79:153–161

Singh RP, Ashutosh T, Pandey AC (2011a) Silver/polyaniline nanocomposite for the electrocatalytic hydrazine oxidation. J Inorg Organomet Polym Mater 21:788–792

Singh RP, Kang DY, Choi JW (2011b) Nanofabrication of bio-self assembled monolayer and its electrochemical property for toxicant detection. J Nanosci Nanotechnol 11:408–412

Singh RP, Shukla VK, Yadav RS, Sharma PK, Singh PK, Pandey AC (2011c) Biological approach of zinc oxide nanoparticles formation and its characterization. Adv Mater Lett 2:313–317

Singh RP, Choi J-W, Tiwari A, Pandey AC (2012) Biomimetic materials toward application of nanobiodevices. In: Tiwari A, Mishra AK, Kobayashi H, Turner APF (eds) Intelligent nanomaterials: processes, properties, and applications. John Wiley & Sons, Inc., Hoboken. doi:10.1002/9781118311974.ch20

Singh RP, Choi J-W, Tiwari A, Pandey AC (2012b) Utility and potential application of nanomaterials in medicine. In: Tiwari A, Ramalingam M, Kobayashi H, Turner APF (eds) Biomedical materials and diagnostic devices. John Wiley & Sons, Inc., Hoboken, NJ, USA. doi:10.1002/9781118523025.ch7

Singh RP, Choi J-W, Tiwari A, Pandey AC (2014) Functional nanomaterials for multifarious nanomedicine. In: Tiwari A, APF T (eds) Biosensors nanotechnology. Wiley, Hoboken. doi:10.1002/9781118773826.ch6

Vamvakaki V, Chaniotakis NA (2007) Pesticide detection with a liposome-based nano-biosensor. Biosens Bioelectron 22(12):2848–2853

Zhang W, Tang H, Geng P, Wang Q, Jin L, Wu Z (2007) Amperometric method for rapid detection of *Escherichia coli* by flow injection analysis using a bismuth nano-film modified glassy carbon electrode. Electrochem Commun 9(4):833–838

Zhang W, Yang T, Huang DM, Jiao K (2008) Electrochemical sensing of DNA immobilization and hybridization based on carbon nanotubes/Nano zinc oxide/chitosan composite film. Chin Chem Lett 19:589–591

Modern Prospects of Nanotechnology in Plant Pathology

15

Massalimov Ismail, Ram Prasad, Amr I.M. Ibrahim, and Ahmed I. S. Ahmed

Abstract

The most important applications of nanoparticle types and the common practices for control of plant diseases are described. The chapter will not include a description of all diseases that occur globally or a comprehensive report on the selected diseases and nanoparticles. We have tried to contain information on the impact of the disease and the role of nanotechnology to face these challenges as modern technology in plant control and also report a short, historical background for some nanoparticle types. We have attempted to include the newest literature and scientific research related to each nanoparticle type. We will focus on the synthesis of NPs of some compounds and their influence on plant diseases.

The chapter consisted of four sections. The first section will elucidate the meaning of nanotechnology and scientific progression. Section 15.2 covers major information about plant diseases as challenges in agriculture development. The third section will include important techniques that are used to detect and manipulate causal agents of plant diseases and the use of nanoparticle in disease control. The last part addresses a common nanoparticle that is used as control agent for some plant pathogens.

Keywords

plant pathogens • nanoparticles • diagnosis • QDs • risk assessments • DNA damage

---

M. Ismail
Bashkir State University, Ufa, Russia

R. Prasad
Amity Institute of Microbial Technology, Amity University, Noida, Uttar Pradesh 201313, India

A.I.M. Ibrahim • A.I.S. Ahmed (✉)
Plant Pathology Unit, Plant Protection Department, Desert Research Center, Cairo, Egypt
e-mail: ahmed_drc@yahoo.com

R. Prasad et al. (eds.), *Nanotechnology*, DOI 10.1007/978-981-10-4573-8_15

15.1 Introduction

Nanotechnology is a growing interdisciplinary science in the past decades that links knowledge of biology, chemistry, physics, engineering, and material science (Islam and Miyazaki 2009; Prasad 2014). Nanotechnology utilizes combinations of nanoparticles in equipment and instruments with nanoscale dimensions. In modern times, many research and studies on the concept of nanotechnology and the manufacture of materials and employ them in different applications have emerged. We review some exciting events that made this technical rally and make it the future technology. Nanotechnology is the fifth generation that has emerged in the world of electronics, which can be described as: (1) The first generation is the use of the electronic lamp, including television. (2) The second generation is the transistor discovery and the widespread of applications. (3) Third-generation electronics is the use of complementary circuits, integrate circuits (IC)—very small pieces formed by nanotechnology, which is a significant leap in the evolution and reduced circuit size. (4) The fourth generation is the microprocessor—a huge revolution in the electronics production of personal computers and computing silicon chips, which had furthered progress in many scientific and industrial fields.

The nanotechnology concept is an interest in particles smaller than 100 nm; the nanometer is one billionth of a meter and gives the material within the composition characteristics and behaviors of a new resistance to pests and insects. Scientists are studying the nature of nanotechnology in different academic fields. For example, the electronic structure, connectivity, interactive, melting point, and mechanical properties of the material were all changed when the least particle for a value of tested volume size.

The change in the chemical and physical characters of the particles in the nanoscale from larger particles or dissolved compounds has had a lot of interest (Nowack 2010). The unique properties of the particles in nanosize from its bulk equivalents attracted global attention (Gong et al. 2007; Gajjar et al. 2009; Rai et al. 2009). Nanotechnology impacts on development have a lot of applications, such as production of energy, improve agricultural production, drinking water treatment, disease diagnosis and follow-up, delivery of medicines, food processing and storage, air pollution, construction, health monitoring, and resistance to pests and insects (Prasad et al. 2014, 2016, 2017).

Agriculture is the most effective economic sector for the population worldwide, but this sector is facing a lot of challenges, which has led to lack of daily food supplies because of environmental challenges or instability of political status (Anonymous 2009). Plant diseases are an important challenge faced by agriculturist in plant production globally. Nanotechnology will support agriculture and decrease environmental challenges by pesticide production, chemical fertilizers, and plant disease control by using the nanoparticles as environmentally friendly to improve the efficiency of pesticides with a lower dose (Singh et al. 2014; Bhattacharyya et al. 2016). Agriculture occupies second place in the list of uses of nanotechnology, through which tools can help to increase the fertility of soil and increase crop yields. The use of chemical pesticides on the micron level has been known for a long time, but recently scientists have been interested in converting many of these pesticides to the nanoscale (Gopal et al. 2011). The efficiency of material use in the nanovehicles form can be improved, thus reducing the quantities used with greater effectiveness. For example, it can produce small tools used to spraying fertilizer at rates regulated carefully. Also, nanoparticles could play an important role in plant disease control as diagnostic kits to increase detection power (Prasanna 2007). Nanotech plays the main role in some areas of food and agriculture science, which contributed as material science, food processing, product development, and food safety (Singh et al. 2014). The plant disease control strategies are aimed at modern technologies, such as sciences and tools to improve crop productivity and reduce the diseases impact—one of the application strategies in nanotechnology and its techniques in plant disease management.

This chapter explains nanoparticles effect against various plant disease causal agents and represents the scientific efforts related to this theme and also the performance of different particle types depending on nanosize and treatment conditions. Toxicity properties of nanoparticles are highlighted using knowledge of size, chemistry, and shape as well as the other conditions.

15.2 Overview of Plant Diseases

Many pathogenic agents affect plant health and are considered a direct factor in competition with humans for food and crops due to starvation cause and poverty. Phytopathogens cause mass destruction in agricultural crops—both in the field during the phases of plant growth or during storage stages (Aisnworth 1981).

Control of plant disease is important for saving food resources. Examples of severe plant diseases are late blight on potatoes, rice blast, citrus canker, soybean cyst nematode, and chestnut blight. However, disease control is very difficult for most crops and under different conditions. Many strategies and methods have been used to achieve disease control, such as resistant cultivars, cultivation processes, such as crop rotation, cultivation of pathogen-free seeds and seedlings, convenient cultivation time and plant density, pesticide use, and field moisture control. Globally, up to 10% annually of yield reduction rate in crops is caused by plant diseases; yield loss due to diseases often exceeds 20%, whereas in some severe diseases the yield could be destroyed with causing economic losses for farmers globally. There are important efforts in the science of phytopathology to improve disease combat, for facing the changes in disease severity caused by the ongoing evolution, and movement of plant pathogens in addition to changes in agricultural practices. Based on the Food and Agriculture Organization (FAO) reports, diseases and pests are responsible for approximately 25% of crop loss. To solve this issue, new techniques are needed to combat plant diseases. Modern technology that plays an important and effective role is nanotechnology, especially with nanoparticles application as an alternative to pesticides or assistant factor in preparation of pesticides (Martinelli et al. 2014).

Plant diseases causal agents are divided into 11 groups, such as fungi, bacteria, viruses and viroid, nematodes, parasitic angiosperms, algae, Plasmodiophoromycetes, Oomycetes, phytoplasmas, and trypanosomatids (Stewart and Press 1990). All of these pathogens affect the lives and future of millions of people and are considered a big challenge for plant production and main resource for food. The first step for diagnosis of plant diseases is to identify the pathogen, including detecting of symptoms and isolation, purification, and the causal organism characterization.

The plant pathologist is trying to detect the organisms associate with the infected plant, which responsible for disease symptoms. Pathogenicity is normally detected by the implementation of the postulates of Robert Koch. The causal agents are as follows: (1) the causal agent must be present with disease symptoms; (2) the causal pathogen must be isolated and able to culture on growth media; (3) as pathogenicity test, when test plants are inoculated with suspected causal pathogen the disease symptoms must reproduce; (4) the isolated pathogen must be re-isolated from test plants and be identical with isolated organism from the diseased plant. These steps could be implemented with micro-organisms that are cultured and grown on nutritious industrial media except all obligate pathogens, which include some important fungi, phytoplasmas, and all viruses (Baldauf et al. 2000).

15.3 Plant Pathogens Detection and Plant Diseases Diagnosis

Plants are infected by various micro-organisms, including fungal pathogens, which constitute the earliest recognized category and are highly evolved. Morphological, biological characteristics, biochemical, and physiological have been used to detect, identify, and differentiate between fungal pathogens up to the species level. Pathogenicity on the set of differential host plant is useful to determine the pathogen. The techniques based on the above-mentioned characteristics require a long time and yield inconsistent results. The developed techniques are required for greater precision and accuracy. Immunohistochemistry is effective in fungal plant disease monitoring and agricultural materials, soil, water, and air. This chapter provides comprehensive information on different ways to detect and control fungal pathogens and diseases diagnosis on the basis of extensive research literature.

The methods and strategies currently available for disease and pathogen identification are divided into many categories depending on the type of test or technique. The first diagnosis step of plant diseases is identify the disease symptoms and then implementation of Koch's postulates. The identification of the causal organism depends on four pillars: host range and symptomology; morphology of causal organism; selective media of organism; and biochemical marker techniques. The last one is divided into five types: substrate metabolism; fatty acid profiles; protein analysis; serological techniques; and nucleic acid techniques. Also, the nucleic acid technique includes three main groups of techniques based on restriction fragment length polymorphism (RFLP), nucleic acid probes, and polymerase chain reaction techniques (PCR). PCR techniques include five techniques, namely choice of primers, sequence data from PCR products, real-time PCR, amplified fragment length polymorphism, and molecular beacons (Richard 2003).

Nanotechnology has a wide range of applications in agriculture as useful science in plant disease control for exploitation of element effectiveness as control agents of diseases. Nanotechnology has two main aspects: synthesis of nanoscale compounds, and application of this compound for the required objectives (Khan and Rizvi 2014). One important objective is use of nanoparticles as alternative pesticides.

15.4 Nanotechnology and Plant Pathology

Plant pathology has been influenced by progress in other fields of technology and science. Nanotechnology has valuable practices in management of plant disease in different ways; the most common is nanoparticle application on seeds and soil or leaves to protect the plants from pathogens or to control infection (Khan and Rizvi 2014). Nanoparticles can suppress the pathogens in soil or plant parts. Because physiochemical studies of nanoforms from its macroform vary greatly, it is possible to examine the impact of NPs on pathogens and microorganism and use this technology in plant pathogen control, especially against bacterial and fungal pathogens. Because nanoparticles have ultra-small size, smaller than a virus, they have a high reactivity against microorganism.

Using nanoparticles as diagnostic probes, researchers are searching to improve the role of nanomaterials by addressing all possible drawbacks in available diagnostic tools. Nanoparticles were different from their counterpart of bulks, which, when reduced to nanosize (1–100 nm), achieve certain properties that make them suitable for development as diagnostic techniques (Sharon et al. 2010). These properties are large aspect ratio (surface to volume ratio), chemically adjustable physical attributes, strong affinity to target (particularly of gold nanoparticles to proteins), structural firmness despite atomic granularity, and enhanced or delayed particles accumulation based on the modification of surface type, enhanced photoemission, high electrical and heat conductivity, and improved surface catalytic activity (Liu 2006; Garg et al. 2008; McNeil 2005; Shrestha et al. 2007; Aziz et al. 2015; Prasad et al. 2014, 2016). For detection of fungal spore (*Aspergillus niger* and *Saccharomyces cerevisiae*), micromechanical cantilever arrays have been used and demonstrated by Nugaeva et al. (2005). Proteins such as concanavalin A, fibronectin, or immunoglobulin, were surface grafted on microfabricated uncoated and also gold-coated silicon cantilevers. On fungal cell surface, these proteins have manifold connections to bind to the molecular structures present. Immobilization and germination of tested fungi spore led to a shift in resonance frequency, which was obtained by dynamically operated cantilever arrays. The mass of single fungal spore was used for quantitative estimation. The biosensors detected the purposed fungi in the range of 10^3–10^6 cfu ml^{-1} in the investigation (Nugaeva et al. 2005).

15.4.1 Quantum Dots (QDs) and Carbon Nanomaterials as Prospective Materials for Detection of Plant Pathogens

Quantum dots are small nm in diameter, fluorescent, roughly spherical, crystalline particles of semiconductors, which are restricted in all the three spatial dimensions. QDs can provide an alternative for commercial applications. QDs are considered prospective tools for detection of a specific biological marker in different fields with high thoroughness. They were used successfully in labeling cells, cell tracking, imaging, and DNA detection in vivo (Sharon et al. 2010).

Nanomaterials of carbon have been developed to affect electrochemical analysis electrode so they have a role as a sensor (Sharon and Sharon 2008). They could detect the residue of pesticide in plants. Although there has been no patent filed so far exclusively for diagnosis of plant disease through nanotechnology techniques, these methods have improved animal diseases diagnosis and can be applicable to plant diseases as well.

15.5 Application of Nanoparticle in Plant Diseases and Its Risk Assessment

Nanomaterials are used for environmental remediation, due to the unique properties. For instance, colloidal iron nanoparticles have the ability to perform as catalysts in redox reactions. Nevertheless, nanomaterials may have side effects, and a risk assessment requires realization of their distribution in the food chain and environment. Risk assessments are required for understanding the behavior of nanoparticles to evaluate potential risks associated with nanomaterial use for remediation. Side effects associated with the nanotechnology, especially environmental risks correlated with residual fate of nanomaterials and transport in the environment, are not yet fully explored and understood since 2006–2015.

15.6 Application of Different Nanoparticle Types for Management of Plant Diseases

Recent advances in the design, production, and fundamental understanding of nanomaterials have led to preliminary investigations of their use in control of plant pathogens. Silver has been investigated for its antimicrobial properties and applied more comprehensively than any other inorganic antibacterial agent (Russell and Hugo 1994).

15.6.1 Silver NPs Used as Plant Diseases Control

Silver can disinfect almost 650 different microbes and does not harm humans, because it is nontoxic but controls the smoothing metabolism function inside of the microbes (Morones et al. 2005). Silver has been applied in pure free metal or compound form for many applications in respect to its antimicrobial effect against pathogens but is nontoxic to humans (Yeo et al. 2003; Elchiguerra et al. 2005). Silver has a high antimicrobial effect in both forms ionic and nanoparticle; therefore, its used widely for plentiful sterilization purposes, such medical devices and sanitization of water (Russell and Hugo 1994; Aziz et al. 2015, 2016). It could be applied for controlling several plant pathogens in a relatively safer way compared with synthetic fungicides (Park et al. 2006). Silver nanoparticles as antimicrobial material has become more common as technological progression make their production more economical.

Management of plant diseases is one of the possible applications of silver, because silver displays multiple modes of inhibitory action against microorganisms (Clement and Jarret 1994). Various levels of inhibition on colony formation of *Bipolaris sorokiniana* and *Magnaporthe grisea* were shown by using silver compounds. The colony formation decreased when concentrations of the silver compounds increased. The reduction of colony formation was apparent within 1 h. Silver ions and nanoparticles significantly inhibited the colony and conidia formation of *Bipolaris sorokiniana* and *Magnaporthe grisea*. However, they were highly influential to reduce plant diseases caused by *B. sorokiniana* and *M. grisea*. Jo et al. (2009) confirmed that silver ionic forms of $AgNO_3$ caused a significant reduction of colony formation within 1 h at low concentrations (EC50 for *B. sorokiniana* = 2.2 ppm and EC50 for *M. grisea* = 0.9 ppm). In the presence of silver nanoparticles, the growth of *Raffaelea* sp. was significantly inhibited in a dose-dependent manner. However, microscopic observation revealed that silver nanoparticles caused destructive effects on fungal hyphae as well as conidial germination (Kim et al. 2009).

To visualize the microscopic influence of silver nanoparticles on the growth of fungal hyphae, healthy fungal hyphae grown on MA plates were sprayed with 10 ppm of AT solution and then observed under an electron microscope. Damage of hyphal tips, where new conidia form, and detached conidia were detected simultaneously. Damage to the surface of the fungal hyphae also was observed, which could have caused the release of internal cellular materials, resulting in shrinkage of the hyphae (Kim et al. 2009).

The highly reactive silver nanoparticles are caused by generating silver ions, whereas metallic silver is relatively unreactive (Morones et al. 2005). Nanoparticles efficiently penetrate microbial cells, suggesting that lower concentrations of nanosized silver particles would be sufficient for microbial control (Samuel and Guggenbicler 2004).

The treatments with silver nanoparticles clearly inhibited the sclerotial germination growth of the fungal pathogens (*Rhizoctonia solani, Sclerotinia sclerotiorum* and *S. minor*) compared with actively growing hyphae after the sclerotial germination, which obviously was observed on water-treated plates. The silver-contained plates showed abnormal growth phenotypes, appearing compact patches of aerial hyphae. The silver nanoparticles showed the highest inhibition effect on sclerotial germination growth of *S. sclerotiorum* compared with other fungi. The germination of sclerotia of *S. sclerotiorum* was almost suppressed at a highest concentration

(7 ppm) of the silver nanoparticles tested, whereas that of *R. solani* and *S. minor* were suppressed (>75%). Continuous growth after sclerotial germination was not formed in a prolonged incubation on plates containing 7 ppm of silver nanoparticles (Min et al. 2009). Jung et al. (2010) found that three forms of nano-silver liquid, WA-CV-WA13B, WA-AT-WB13R, and WA-PR-WB13R, were provided by the Bio-plus Co. (Pohang, Korea) at a 1,000-ppm initial concentration, which was then diluted into different working concentrations of 1 ppm, 3 ppm, 5 ppm, 7 ppm, 10 ppm, 25 ppm, 50 ppm, and 100 ppm on PDA, MEA, and CMA culture plates. The inhibitory effect of WA-AT-WB13R on each media was more effective than WA-CV-WA13B. The inhibition rate was 86% on PDA with a 5-ppm concentration and 93% on CMA with a 3-ppm concentration.

15.6.2 Silver and Silica-Silver

The smaller size of nano silver more effectively suppressed fungal growth; 1– to 5-nm-sized particles may pass through a protoplasmic membrane, and silica was well absorbed into fungi (Wainwright et al. 1986). After silica-silver nanoparticles absorbed into fungal cells, disinfectant activity of silver nanoparticles will increase. Silica will increase the resistance to disease by induction of dynamic resistance, which acts to form a physical barrier to pathogenic fungi. Silica- silver is effective to control various diseases at lower than 3.0 ppm, which is not a concentration to suppress pathogens on agar medium. Fungi that could be controlled by silica-silver include: *Blumeria* spp., *sphaerotheca* spp., *phytopthors* spp., *Rhizoctonia* spp., *colletotrichum* spp., *Botrytis* spp., *Magnaporthe* spp., and *Pythium* spp.

15.6.3 Mode of Action of AgNPs Against Microorganisms

Ag NPs have the capability to penetrate cells because of its great surface area-to-volume ratio, which also increase their contact with microbes (Kim et al. 2009). Silver is effective against bacteria (Gram-positive and Gram-negative) and fungi (Lara et al. 2010; Guzman et al. 2012). Many theories and suggested scenarios have been used to explain the mode of action of Ag NPs against microorganisms; antibacterial activity of Ag NPs is the same as $Ag+$ based on Chaloupka et al. 2010.

1. Interaction with cell wall and membrane
   (a) Changes in permeability
   (b) Disturbance in the administration of phosphates
   (c) Degradation of the plasma membrane
   (d) Collapse of the proton motive force
   (e) Inhibition of the ATP synthesis
2. Influence on amino acids and enzymes
   (a) Bonding with amino acids (especially to –SH group)

   (b) Inhibition of enzyme activity with bonding to its active center
3. Obstructions in energy recruitment
   (a) Influence on electron movement in the respiratory chain
   (b) Inhibition of cytochromes
4. Impact on DNA and RNA
   (a) Breaks in hydrogen bonding
   (b) Inhibition of synthesis of nitrogen bases
   (c) Disorders of DNA and RNA synthesis
   (d) Denature ribosomes, inhibiting protein synthesis
5. Generating reactive oxygen species (ROS)

The efficacy of silver nanoparticles depends on a particle's properties, such as size, shape, exposure time, types of compounds, and target. However, the particles size of silver is so important, because the smaller diameter gives bigger surface and leads to better antibacterial efficacy (Guzman et al. 2012). The silver content has to be high enough to inhibit the growth of bacteria cells. Toxicity assays with Ag NPs have reported that one of the toxicity mechanisms is the release of the toxic ion Ag+. To know the contribution of Ag+ ions in MIC assays, microorganisms were exposed to various $AgNO_3$ concentrations, determining that $AgNO_3$ had MICs values in the order 0.107 μg/mL of Ag+ for all microorganisms. We can assume that inhibition is caused by silver ions. Higher toxicity of silver ions can be related to lower positive charge, which makes bacteria more susceptible to silver toxicity than other metallic ions, such as zinc. Silver is known as a potent disinfecting agent for killing unicellular microorganisms by inactivating enzymes with metabolic functions in microorganisms by oligodynamic action (Kim et al. 1998).

Jo et al. (2009) confirmed that the preventative application of the silver preparations more effectively reduced disease severity on plants than the postinoculation application. The antifungal activity mechanism is suggested by the germination and infection process in these fungi. Both *B. sorokiniana* and *M. grisea* cause foliar diseases and reproduce as asexual conidia. Disease infection is initiated by the attachment of spores to the plant surface and formation of germ tubes. Under favorable high-humidity conditions (~100% relative humidity) and warm temperature, conidia germinate, and the resulting germ tubes penetrate plant surfaces within 24 h. Antifungal efficiency of silver was reduced at 24 h after inoculation, suggesting that direct contact of silver with spores or germ tubes is critical in inhibiting disease development. In summary, antifungal activity of ionic or nanoparticle silver has a great potential for use in controlling spore-producing fungal pathogens (Aziz et al. 2016). Perhaps silver is less toxic to humans and animals than synthetic fungicides. Numerous modes of action described a broad range of biological pathways of microbes providing an important benefit for avoiding the development of resistance, which has been increasingly important in terms of current issues for the chemical management of many plant fungal diseases.

15.7 Mode of Action of Other NPs Against Microorganisms

Exposure of *Escherichia coli* to nano-ZnO causes loss in membrane integrity. Likewise, toxicity of NP of CuO and ZnO are connected with cell membrane damage (Heinlaan et al. 2008). NP action may be due in part to their release of free ions. Heavy metal ions have diverse effects on bacterial cell function. For Cu ions, the mechanism may involve oxidative stress (Cioffi et al. 2005). The redox cycling of Cu ions provide depletion of glutathione and the sulfhydryl groups effects of proteins causing DNA damage and lipid oxidation (Stohs and Bagchi 1995). Cu, Zn also is an essential element for cells; increasing levels of Zn over the essential threshold level inhibit enzymes inside bacterial cells, including dehydrogenase (Nweke et al. 2007).

The constructed biosensor in the isolate of *Pseudomonas putida* effectively and rapidly, within minutes, demonstrated dose-dependent toxicity of NP of Ag, CuO, and ZnO. These findings illustrate that the toxicity was not restricted to bacteria with pathogenic potential. Rather an environmental isolate, studied because of its bioremediation potential, was affected. The nanoparticles of Ag, ZnO, and CuO were more toxic, causing loss of Lux activity in the biosensor, than their equivalent bulk materials indicating that the nano-size of the material was important (Gajjar et al. 2009).

Transmission electron microscopic was used to estimate the ability of nanoparticles of silver to deteriorate the fungal envelope structure of *Candida* spp. The results proved that the treated cells with Nano-Ag showed significant damage, which was characterized by the formation of a "pit" in their cell walls and pores in their plasma membrane (Lee J et al. 2010).

15.8 Impact of Nanoparticles on DNA Damage

DNA damage of cells exposed to AgNP or ZnONP was evaluated through electrophoretic analysis in agarose gels. DNA fragmentation was not detected in DNA extracts from all microorganism treated with NPs (compared with DNA extracts from microorganisms not exposed to NPs). Ag ions inactivate proteins with SH groups and prevent the ability of DNA to replicate (Feng et al. 2000). Silver is known to affect many biological processes in microorganisms, such as the alteration of cell membrane structure and functions. Silver also prevents the expression of proteins associated with ATP production (Yamanaka et al. 2005).

15.9 Reduction of Plants Disease Severity by Silver Nanoparticles

Silver nanoparticles effectively reduced leaf spot and gray leaf spot on perennial ryegrass without noticeable phytotoxicity, but NaCl significantly decreased their antifungal effects. The efficiency of silver in reducing the diseases significantly differed between 3 h before and 24 h after conidia inoculation. Most silver preparations applied at 3 h before spore inoculation on the plants significantly reduced both diseases compared with the water control. The preinoculation applications of $AgNO_3$ (25 and 50 ppm), Ag(p) (200 ppm), and Ag(e) (50 ppm) were found to be the most effective, allowing less than 7% damage of foliar by *B. sorokiniana* and *M. grisea* under highly disease-conducive environmental conditions, which caused more than 70% damage on water-treated control plants. Some preinoculation treatments of neutralized $AgNO_3$, Ag (p), and Ag (e) but not AgCl moderately reduced disease, but they still were less effective than nonneutralized silver preparations. In contrast, delayed applications of silver preparations at 24 h after spore inoculation did not effectively reduce both diseases, causing >50% foliar damage. Silver ions are very efficacious because of its inhibition effect on microbial respiration and metabolism as well as physical damage (Bragg and Rannie 1974; Thurman and Gerba 1989). Also, it was proposed that silver ions intercalate bacterial DNA once entering the cell, resulting in no further proliferation (Feng et al. 2000).

15.10 Conclusions

The concept of nanotechnology depends on the grounds that particles smaller than 100 nm (a nanometer is part of one billionth of a meter) gives the material within the composition characteristics and behaviors of a new resistance to pests and insects, where 1 nm is defined as one billionth of a meter. Agriculture is the most important economic sector for the population worldwide. Plant diseases are considered one of the most important challenges facing plant production. Also, nanotechnology has a scope of application in agriculture as useful science in plant disease control for exploitation of element effectiveness. Different nanoparticle types have a high antimicrobial effect in both forms, ionic and nanoparticle; therefore, they are used widely for various applications, especially as plant disease control agents.

Acknowledgment The corresponding author gratefully acknowledge the support from Federation of Indian Chambers of Commerce & Industry (FICCI) and CV Raman International Fellowship for African Researchers Program.

References

Aisnworth GC (1981) Introduction to the history of plant pathology. Cambridge University Press, Cambridge. ISBN:0-521-23032-2

Anonymous (2009) Nanotechnology and nanoscience applications: revolution in India and beyond. Strateg Appl Integrating Nano Sci

Aziz N, Faraz M, Pandey R, Sakir M, Fatma T, Varma A, Barman I, Prasad R (2015) Facile algae-derived route to biogenic silver nanoparticles: synthesis, antibacterial and photocatalytic properties. Langmuir 31:11605–11612

Aziz N, Pandey R, Barman I, Prasad R (2016) Leveraging the attributes of *mucor hiemalis*-derived silver nanoparticles for a synergistic broad-spectrum antimicrobial platform. Front Microbiol 7:1984. doi: 10.3389/fmicb.2016.01984

Baldauf SL, Roger AJ, Wenk-Siefert I, Doolittle WF (2000) A kingdom-level phylogeny of eukaryotes based on combined protein data. Science 290(5493):972–977

Bhattacharyya A, Duraisamy P, Govindarajan M, Buhroo AA, Prasad R (2016) Nano-biofungicides: emerging trend in insect pest control. In: Prasad R (ed) Advances and applications through fungal nanobiotechnology. Springer International Publishing, Switzerland, pp 307–319

Bragg PD, Rannie DJ (1974) The effect of silver ions on the respiratory chain of *Escherichia coli*. Can J Microbiol 20:883–889

Chaloupka K, Malam Y, Seifalian AM (2010) Nanosilver as a new generation of nanoproduct in biomedical applications. Trends Biotechnol 28(11):580–588

Cioffi N, Torsi L, Ditaranto N, Tantillo G, Ghibelli L, Sabbatini L, Bleve-Zacheo T, D'Alessio M, Zambonin PG, Traversa E (2005) Copper nanoparticle/polymer composites with antifungal and bacteriostatic properties. Chem Mater 17:5255–5262

Clement JL, Jarret PS (1994) Antimicrobial silver. Metal-Based Drugs 1:467–482

Elchiguerra JL, Burt JL, Morones JR, Camacho Bragado A, Gao X, Lara HH, Yacaman MJ (2005) Interaction of silver nanoparticles with HIV-1. J Nanobiotechnol 3:6

Feng QL, Wu J, Chen GQ, Cui FZ, Kim TN, Kim JO (2000) A mechanistic study of the antibacterial effect of silver ions on *Escherichia coli* and *Staphylococcus aureus*. J Biomed Mater Res 52(4):662–668

Gajjar P, Pettee B, Britt DW, Huang W, Johnson WP, Anderson AJ (2009) Antimicrobial activities of commercial nanoparticles against an environmental soil microbe, *Pseudomonas putida* KT2440. J Biol Eng 3:9. doi:10.1186/1754-1611-3-9

Garg J, Poudel B, Chiesa M (2008) Enhanced thermal conductivity and viscosity of copper nanoparticles in ethylene glycol nanofluid. J Appl Phys 103:074301

Gong P, Li H, He X, Wang K, Hu J, Zhang S, Yang X (2007) Preparation and antibacterial activity of Fe3O4@Ag nanoparticles. Nanotechnology 18:604–611

Gopal M, Gogoi R, Srivastava C, Kumar R, Singh PK, Nair KK, Yadav S, Goswami A (2011) Nanotechnology and its application in plant protection. Plant Pathol India: Vision 2030:224–232

Guzman M, Dille J, Godet S (2012) Synthesis and antibacterial activity of silver nanoparticles against gram positive and gram-negative bacteria. Nanomed Nanotechnol Biol Med 8(1):37–45

Heinlaan M, Ivask A, Blinova I, Dubourguier HC, Kahru A (2008) Toxicity of nanosized and bulk ZnO, CuO and TiO(2) to bacteria *Vibrio fischeri* and crustaceans Daphnia magna and *Thamnocephalus platyurus*. Chemosphere 71(7):1308–1316

Islam N, Miyazaki K (2009) Nanotechnnology innovation system: understanding hidden dynamics of nanoscience fusion trajectories. Technol Forecast Soc Chang 76:128–140

Jo Y, Kim K, Jung G (2009) Antifungal activity of silver ions and nanoparticles on phytopathogenic fungi. Plant Dis 93:1037–1043

Jung JH, Kim SW, Min JS, Kim YJ, Lamsal K, Kim KS, Lee YS (2010) The effect of nano-silver liquid against the white rot of the green onion caused by *Sclerotium cepivorum*. Mycobiology 38(1):39–45

Khan MR, Rizvi TF (2014) Nanotechnology: scope and application in plant disease management. Plant Pathol J 13(3):214–231

Kim TN, Feng QL, Kim JO, Wu J, Wang H, Chen GC, Cui FZ (1998) Antimicrobial effects of metal ions (Ag+,Cu^2+,Zn^2+) in hydroxyapatite. J Mater Sci Mater Med 9:129–134

Kim SW, Kim KS, Lamsal K, Kim YJ, Kim SB, Jung M, Sim SJ, Kim HS, Chang SJ, Kim JK (2009) An in vitro study of the antifungal effect of silver nanoparticles on oak wilt pathogen Raffaelea sp. J Microbiol Biotechnol 19:760–764

Lara HH, Ayala-Núñez NV, Turrent LCI, Padilla CR (2010) Bactericidal effect of silver nanoparticles against multi drug resistant bacteria. World J Microbiol Biotechnol 26:615–621

Lee J, Kim KJ, Woo SS, Kim JG, Lee DG (2010) The silver nanoparticle (Nano-Ag): a new model for antifungal agents, silver nanoparticles. David Pozo Perez (ed), InTech. doi:10.5772/8510. Available at: http://www.intechopen.com/books/silver-nanoparticles/the-silver-nanoparticle-nano-ag-a-new-model-for-antifungal-agents

Liu WT (2006) Nanoparticles and their biological and environmental applications. J Biosci Bioeng 102(1):1–7

Martinelli F, Scalenghe R, Davino S, Panno S, Scuderi G, Ruisi P, Villa P, Stroppiana D, Boschetti M, Goulart LR, Davis CE, Dandekar AM (2014) Advanced methods of plant disease detection:a review. Agron Sustain Dev 35:1–25

McNeil SE (2005) Nanotechnology for the biologist. J Leukoc Biol 78:585–594
Min JS, Kim KS, Kim SW, Jung JH, Lamsal K, Kim SB, Jung M, Lee YS (2009) Effects of colloidal silver nanoparticles on sclerotium-forming phytopathogenic fungi. Plant Pathol J 25(4):376–380
Morones JR, Elechiguerra JL, Camacho A, Holt K, Kouri JB, Ramirez JT, Yacaman MJ (2005) The bactericidal effect of silver nanoparticles. NanoBiotechnology 16:2346–2353
Nowack B (2010) Nanosilver revisited downstream. Science 330:1054–1055
Nugaeva N, Gfeller KY, Backmann N, Lang HP, Duggelin M, Hegner M (2005) Micromechanical cantilever array sensors for selective fungal immobilization and fast growth detection. Biosens Bioelectron 21(6):849–856
Nweke CO, Alisi CS, Okolo JC, Nwanyanwu CE (2007) Toxicity of zinc to heterotrophic bacteria from a tropical river sediment. Appl Ecol Environ Res 5(1):123–132
Park HJ, Kim SH, Kim HJ, Choi SH (2006) A new composition of nanosized silica-silver for control of various plant diseases. Plant Pathol J 22:295–302
Prasad R (2014) Synthesis of silver nanoparticles in photosynthetic plants. J Nanoparticles, Article ID 963961, http://dx.doi.org/10.1155/2014/963961
Prasad R, Kumar V, Prasad KS (2014) Nanotechnology in sustainable agriculture: present concerns and future aspects. Afr J Biotechnol 13(6):705–713
Prasad R, Pandey R, Barman I (2016) Engineering tailored nanoparticles with microbes: quo vadis. WIREs Nanomed Nanobiotechnol 8:316–330. doi:10.1002/wnan.1363
Prasad R, Bhattacharyya A, Nguyen QD (2017) Nanotechnology in sustainable agriculture: recent developments, challenges, and perspectives. Front Microbiol 8:1014. doi:10.3389/fmicb.2017.01014
Prasanna BM (2007) Nanotechnology in agriculture. ICAR National Fellow, Division of Genetics, IARI, New Delhi, pp 111–118
Rai M, Yadav A, Gade A (2009) Silver nanoparticles as a new generation of antimicrobials. Biotechnol Adv 27:76–83
Richard NS (2003) The detection and diagnosis of plant pathogens and the diseases they cause. In: Introduction to plant pathology. Wiley, pp 33–60
Russell AD, Hugo WB (1994) Antimicrobial activity and action of silver. Prog Med Chem 31:351–370
Samuel U, Guggenbichler JP (2004) Prevention of catheter-related infections: the potential of a new nano-silver impregnated catheter. Int J Antimicrob Agents 23S1:S75–S78
Sharon M, Sharon M (2008) Carbon nanomaterials: applications in physico-chemical & bio-systems. Def Sci J 58(4):5491–5516
Sharon M, Choudhary AK, Kumar R (2010) Nanotechnology in agricultural diseases and food safety. J Phytology 2(4):83–92
Shrestha S, Yeung CMY, Nunnerley C, Tsang SC (2007) Comparison of morphology and electrical conductivity of various thin films containing nano-crystalline praseodymium oxide particles. Sens Actuators A Phys 136:191–198
Simgh S, Singh BK, Yadav SM, Gupta AK (2014) Applications of nanotechnology in agricultural and their role in disease management. Res J Nanosci Nanotechnol. doi:10.3923/rjnn.2014
Stewart GR, Press MC (1990) The physiology and biochemistry of parasitic angiosperms. Ann Rev Plant Physiol Plant Mol Biol 41:127–151
Stohs SJ, Bagchi D (1995) Oxidative mechanisms in the toxicity of metal ions. Radic Biol Med 18(2):321–336
Thurman KG, Gerba CHP (1989) The molecular mechanisms of copper and silver ion disinfection of bacteria and viruses. Crit Rev Environ Control 18:295–315
Wainwright M, Grayston SJ, deJong P (1986) Adsorption of insoluble compounds by mycelium of the fungus *Mucor flavus*. Enzym Microb Technol 8:597–600
Yamanaka M, Hara K, Kudo J (2005) Bactericidal actions of a silver ion solution on Escherichia coli, studied by energy-filtering transmission electron microscopy and proteomic analysis. Appl Environ Microbiol 71:7589–7593
Yeo SY, Lee HJ, Jeong SH (2003) Preparation of nanocomposite fibers for permanent antibacterial effect. J Mater Sci 38:2143–2147

Nanocomposites: Future Trends and Perspectives Towards Affinity Biosensor

16

Ajay Kumar Gupta and Murthy Chavali Yadav

Abstract

Nanocomposite materials are being increasingly developed in the area of affinity biosensors for the diagnosis of multiple infections and diseases for personalised health care. Nanocomposite platforms with high functionalities, high electroactive surface, biocompatibility and multiple attachments charged sites made them as an effective solid support for immobilisation of biomolecules with retained biological activity and desired orientation/confirmation. These nanocomposites made themselves as a part of the transducer and interfaced with bio-recognition for achieving highly enhanced sensing performance. This chapter highlights about nanocomposites and their roles towards development and improvement of affinity biosensors.

Keywords

Electrochemical biosensor • Nanocomposite • Affinity biosensors • Immunosensors • DNA sensors

A.K. Gupta
Centre of Excellence for Advanced Materials, Manufacturing, Processing and Characterization (CoExAMMPC), Vignan's Foundation for Science Technology and Research University (VFSTRU; Vignan's University), Vadlamudi, Guntur 522 213, Andhra Pradesh, India

M.C. Yadav (✉)
Centre of Excellence for Advanced Materials, Manufacturing, Processing and Characterization (CoExAMMPC), Vignan's Foundation for Science Technology and Research University (VFSTRU; Vignan's University), Vadlamudi, Guntur 522 213, Andhra Pradesh, India

Division of Chemistry, Department of Sciences and Humanities, Vignan's Foundation for Science Technology and Research University (VFSTRU; Vignan's University), Vadlamudi, Guntur 522 213, Andhra Pradesh, India
e-mail: ChavaliM@gmail.com

R. Prasad et al. (eds.), *Nanotechnology*, DOI 10.1007/978-981-10-4573-8_16

16.1 Introduction

In the last two to three decades, lots of research has been done on nanocomposites (Hagfeldt and Gratzel 1995; Murray et al. 1991; Mackerle 2005). New and improved techniques along with advanced instrument in the field of synthesis and characterisation of nanomaterials have made possible the spectacular level of research in the area of nanocrystals since the 1990s. Nanocomposites are a combination of nanomaterials. Since these are a combination of nanomaterials, hence they give averaged and more desirable combination properties. In semiconductor nanocomposite materials, valence bands are completely filled, whereas conduction band is wholly empty. Valence band and conduction band are separated by a forbidden bandgap. Therefore, excitation of an electron from the valence band to the conduction band in these semiconductor nanocomposite materials requires definite energy. When semiconductor nanocomposites have a size comparable to Bohr diameter of exciton, then they exhibit size-dependent properties. In the late 1970s, the dimensional control of the density of electronic states of semiconductor materials and the concept of quantum confinement emerged from condensed matter physics (Tsang et al. 1979; Schmittrink et al. 1983). Direct influence on the small-scale energy band structure causes changes in atomic structure and, hence, a quantum confinement which is a famous term in nanomaterials. Some additional outstanding properties of nanomaterials such as magnetic, optical, mechanical and electronic can also make changes in structure and total energy of the system (Law et al. 2004). These nanostructures offer higher integration density, better crystallinity and higher sensitivity for the surface-related chemical process because of having a high surface-to-volume ratio. Generally, two critical factors of nanomaterials, i.e. dimension and sizes, were considered for improving their properties. This nanosized property can be exploited by effectively coupling it to bimolecular units such as enzymes, antigen–antibodies, DNA to increase the fluorescence efficiency, electrochemical signals, internal magnetic field strength, etc. for the nano-approach analysis (Curri et al. 2002).

In recent years, to provide high-quality and pollution-free water, air and food with proper nutrition value (in the case of food) is a big deal for farmers and companies. And the reason behind this is the use of hazardous chemicals at a high level which is increasing continuously the pollutants in the air, water and soil. This has become a big reason for introducing strong rules and regulation towards monitoring and controlling the use and disposal of hazardous chemicals (Prasad et al. 2014). There is a high need for rapid-sensing method in order to fulfil the demands of fresh, natural and ready-prepared food and drinks which should be free from a pathogenic organism and should contain good nutrition with the addition of a few additives and preservatives in order to preserve them longer. And this high requirement of rapid-sensing methods for sensing of toxins and pathogenic organisms in food, environmental and clinical samples has made an important area of research.

Although conventional techniques have high sensitivity and accuracy, these techniques are expensive and require highly qualified and skilled person. Since conventional techniques are not possible on-site, it requires the food, environmental and clinical samples for analysis to be sent to the laboratory. Thus these techniques are also time-consuming. In order to reduce laboratory-based analysis as it was time-consuming, on-site and real-time analysis techniques, which also should be rapid, robust and low cost, are necessary. Real-time biosensing methods are the actual need to meet the present demand. Nowadays, many sensitive and rapid detection methods, for example, PCR, ATP detection and immunoassay methods, are available which can give a high sensitive result with a low detection limit in the measurement of bacteria, allergens, hormones and viruses in short time (Stephens et al. 1997; Hobson et al. 1996).

Diagnostic companies require different products. There are many controlling parameters such as the cost of instruments, the size of the instrument, sensitivity, robustness, accuracy and rapidness that are required by the diagnostic companies (Newman et al. 1998). Based on the above parameters, instruments needed for diagnostic purpose in the diagnostic industries and agro-food market are divided into three types, i.e. large-sized multi-analysers, bench-top setups and one-time disposable sensors. Diagnostic market is continuously growing and as per the report, in the USA, dairy farms perform every year over ten million penicillin assays (Suleiman and Guilbault 1994). In the diagnostic market, freshness-related instruments also have been introduced which can analyse the product such as meat and fish by determining nucleotide-related compounds present in the product and indicate that whether the product is fit or not (Tothill and Turner 1998). There are a lot of possibilities in the biosensor research; however, mostly are concentrated on small niche applications.

In affinity biosensors, several bio-components such DNA strand, lectins, antigen and antibody have affinity capabilities (Tothill and Turner 1998). In the diagnostic market, mainly polyclonal antibodies and monoclonal antibodies are used widely in affinity biosensor, dipsticks and immunosensor. The future research area of biosensor belongs to express the gene of important antibodies in plants/bacteria, development of the three-dimensional model and their derivatives, use of engineered antibodies and antibody fragments. Selectivity and affinity of a recombinant antibody could be modified/improved by using the analytical tool as per suitability.

For commercialisation purpose, biological molecules should have good stability. For this purpose, many soluble and positively charged polymer compounds such as diethylamino ethyl, lactitol, dextran, sugar derivatives and some latest trends in recent years such as biomimics and synthetic/artificial receptors have been used to enhance the stability of biological molecules. Use of biomimics and synthetic/artificial receptors in environmental diagnostics is the new approach and still remains as an important challenge (Bestetti et al. 1997). Natural ligands also have inherent stability problems which can be overcome by combinatorial synthesis in ligand discovery which is considered as the most rapidly growing areas with the challenges to obtaining good detection limit by synthesising high-affinity receptors.

There are generally two kinds of detection methods for pathogens, i.e. immunological and DNA-based methods. ELISA and agglutination are immunological methods in which antigen-antibody interaction takes place, whereas DNA-based method uses complementary sequences of DNA. These techniques are having poor sensitivity and qualified manpower. Many efforts have been made for making rapid, sensitive and biocompatible biosensor with good electrical, optical and electrochemical properties. Nanocomposites were found to be promising materials due to possessing all the important properties in order to make rapid, sensitive and biocompatible biosensor. This chapter highlights about nanocomposites and their roles towards development and improvement of affinity biosensors.

The electrochemical sensors/biosensors offer several advantages such as elegant routes for interfacing the biomolecules and electronic signal transduction process, which is easy to miniaturise using the existing micromachining techniques. In order to enhance the electrochemical sensitivity, particular attention must be paid towards the electron transfer kinetics between the electrode and immobilised biomolecules. Nanocomposites have allowed the development of ultrasensitive electrochemical biosensors owing to their high surface area, conductive electronic properties and excellent electrocatalytic activity as well as better biocompatibility persuaded by nanometre size. The use of nanomaterials towards electrochemical biosensing have seen an explosive growth in the past especially since the discovery of low-potential detection of NADH by Wang and co-workers (Musameh et al. 2002) and the first use of GNPs as labels for electrochemical immunosensors (Dequarie et al. 2000). The reason behind the great interest in the use of nanomaterials is large surface-to-volume ratio rendering the ability for more surface atoms to participate in the surface reactions (Prasad et al. 2016).

The sensing property (surface chemistry dependent) of these anisotropic nanomaterials results from the Debye length, which is comparable to the nanoscale diameter. The increased electron and hole diffusion rate to the surface of the device could facilitate the fast desorption of analyte molecules from the surface. This process controls the response times of the device. The higher crystallinity in the nanocomposites can potentially reduce the instability associated with the percolation or hopping conduction in the multi-granular oxide materials (Kolmakov and Moskovits 2004). The quantum confinement of the nanoscale material provides the ability to manipulate the individual quantum states of electrons, spins and phonons. The development of these devices based on types of nanomaterials that have unique quantum properties will be a great leap forward and promote a wide range of technological advancement.

16.2 Biosensor

Biosensors are considered as an analytical device and combination of bio-recognition molecules and transducer. Bio-recognition molecule may be a biological material/biologically derived material/biomimetic material. Biological materials are enzymes, nucleic acids, cell receptors, antibodies, tissue, etc., whereas biologically derived material is engineered proteins, recombinant antibodies, aptamers, etc. and biomimetic materials in biosensors are combinatorial ligands, synthetic catalysts and imprinted polymers. The transducer is divided as per their working principles such as optical, electrochemical, magnetic, thermometric, piezoelectric, etc. Bio-recognition molecules are integrated with transducer. The role of the transducer is to convert physicochemical changes into electrical or another kind of signal as per their type. Various types of transducers have been used in a realising biosensor. Biosensors are divided into catalytic or affinity-based sensor based on bio-recognition molecules. In catalytic biosensors, enzymes utilise their catalytic efficiency towards analyte, while in affinity biosensors, antigen and antibody form a complex.

Biosensors are divided mainly into three classes based on biological recognition materials and transducers used. Depending on biological recognition materials, biosensors are divided as follows:

(a) Catalytic or enzyme-based biosensors
(b) DNA-based biosensors
(c) Affinity-based antigen–antibodies biosensor or immunosensors

In the electrochemical biosensor, transducer works on an electrochemical principle which converts the changes in physicochemical signals due to interactions at transducer–biomaterial interface into the current, potential, resistance or impedance, conductance changes. On the basis of the type of output obtained, electrochemical biosensors are categorised as amperometric (current), potentiometric (potential), impedimetric spectroscopy (resistance or impedance) and conductometric (conductance).

Remarkable development growth has been seen in the past in the field of affinity biosensors. In affinity biosensor, antigen or antibody plays an important role in sensing applications by behaving as a bio-recognition element. The use of antibodies are specific due to their immuno-recognition properties and has a varied range of possible applications in a different kind of affinity biosensor or immunosensors for water companies, food industry and regulatory authorities. This specific immuno-recognition property of antibodies has increased the range of analytes for diagnosis (Ghindilis et al. 1998; Skladal 1997; Hock 1997). In present days, immunodiagnostic tests are carrying for the detection. Efforts are also made by taking a number of projects to develop a fast, on-site and real-time immunosensors as a new-generation immunosensor for testing pollution in water and food, on-farm controlling of milk progesterone and authenticity and adulteration testing of livestock production, residues (antibiotics, toxins, and pesticides) in food, presence of additives and hormones. But these affinity-based biosensors are based on antigen or antibody, and hence, the availability of specific antibody or antigen can limit the potential of this kind of affinity biosensor for a diverse range of analyte detection. However, the increasing development in the field of engineered antibodies such as recombinant antibodies, plant bodies, catalytic antibodies or abzymes can overcome this kind of detection limitation in affinity biosensor or immunosensor. Molecular imprinted polymers and artificial receptors play an important role and overcome the detection limitation in affinity biosensor or immunosensor (Turner 1997).

16.2.1 Electrochemical Biosensors

Electrochemical immunosensors is very sensitive, having very low detection limit and can be inexpensive. Electrochemical immune sensors and affinity biosensor can be of different types based on transducer used. These transducers can be of different types such as optical, electrochemical, thermometric, piezoelectric, magnetic, etc. Some techniques in electrochemical immunosensor use electroactive labels such as electrochemical amperometric technique, and some techniques are labelless techniques such as electrochemical impedance spectroscopy technique. These techniques are used for environmental analysis (Marco et al. 1995; Setford et al. 1999). In recent years, there were several research articles and reports related to the development of electrochemical immunosensors and affinity biosensors for diagnosis of a large range of analytes. The affinity-based biosensor can be used as single-time disposable sensors or on-site immunosensors (Baumner and Schmid 1998; Santandreu et al. 1998; Bilitewski 1998; Vianello et al. 1998). Though, the requirement of these types of biosensing techniques are larger in comparison to the present availability for the rapid and sensitive diagnosis of environmental samples, development for taking these kinds biosensing techniques to the level is still in progress.

As mentioned above, on the basis of transducer used, the affinity biosensors can be of different types. Furthermore, the electrochemical biosensor can be divided on the basis of result output of the electrochemical technique. Thus, the electrochemical technique can be amperometric if result output is in the form of current, potentiometric if result output is in the form of voltage, impedimetric if result output is in the form of resistance or impedance, conductometric if result output is in the form of conductance, etc. Due to modest instrumentation and easy setup, electrochemical sensors are inexpensive and a simple technique and do not require highly qualified skilled personnel. The electrochemical technique, e.g. amperometric technique, works on a system having three electrodes, namely, working electrode which is also known as a sensing electrode, a reference and counter-electrodes for balancing the countercurrent on working electrode. After immobilisation of sample on working electrode, the three-electrode system is dipped in solution which should be electrolytic in nature in order to conduct the current through the solution at a constant applied potential. And the resulting current generated based on analyte concentration is measured against constant potential applied on working electrode. Electrochemical biosensors also have voltammetric techniques in which potential is dynamic dissimilar to amperometric technique. These voltammetric techniques are of different types such as cyclic voltammetry (CV), differential-pulse voltammetry (DPV) or square-wave voltammetry (SWV). In potentiometric technique under electrochemical biosensor, the output is measured by high-impedance voltmeter in the form of potential which developed between two electrodes, namely, working and reference electrodes. In the electrochemical biosensor, the potential is applied between the working electrode and a reference electrode. In the electrochemical impedimetric biosensor, the output is measured in the form of impedance/resistance through the Nyquist plot. Impedance measurement is done in the presence of a redox compound, for example, ferrocyanide. In the electrochemical conductometric biosensor, the output is measured as conductance.

16.2.1.1 Amperometric Biosensor

The amperometric technique is the most commonly used electrochemical technique for detection purpose under electrochemical biosensor. In the electrochemical amperometric biosensor, a fixed potential between the working electrode and a reference electrode, and output in the form of current (obtained in reduction or oxidation process of electroactive analyte), is measured between the working electrode and the counter-electrode. In an amperometric technique, current is measured as a function of time and at a constant potential. Current produced is related linearly to the analyte concentration. In order to minimise Ohmic drop between the electrodes, sufficient quantity of supporting electrolyte should be added to the solution in which reaction is carried out. The sensor potential is set at a value, where the analyte, directly or indirectly, produces a current at the electrode. Here, the reaction is carried out on the working electrode. This amperometric technique generally uses three-electrode system (working electrode, a reference electrode and a counter-electrode) in which working or sensing electrode is polarizable in nature and made up of a noble metal such as gold, platinum or can be of carbon material. The reference electrode is nonpolarizable in nature (current does not flow in reference electrode and hence high input impedance) and generally used in the form of Ag/AgCl electrode or saturated calomel electrode (nonpolarizable). Reference electrodes enable us to know the changes in working electrode potential accurately which is not possible in case of the two-electrode system. Counter-electrode plays a role in balancing the current on the working electrode by carrying out opposite reaction process of equal magnitude of current (i.e. if oxidation on working electrode then reduction process of equal magnitude of current). The counter-electrode is also made up of noble material, and surface area should be much larger than the working electrode in order to make the reaction process on the working electrode rate-limiting factor. Generally, in amperometric biosensors, biomolecules (such as enzymes, antibodies, nucleic acid, etc.) are immobilised on the electrode surface in a suitable matrix. The analyte used in amperometric detection changes its oxidation state at working electrode and deposit at the working electrode. As a result of which electron flux is produced, this is directly proportional to the quantity of electrochemically active analyte that was deposited on the working electrode by reaction process. In enzyme-based biosensors, one can use redox enzymes that react with the analyte producing different products, which can also be monitored by amperometry via their oxidation/reduction on an electrode. It is also desirable that, when designing enzyme-based biosensors, an important issue is regaining the original oxidation state of the redox active site, so as to keep the enzyme active to further react with the substrate. Enzyme-active sites (the redox centre) are buried inside the core of the protein. The most important challenge in amperometric enzyme electrodes is the reasonable electrical communication between the active site of the enzyme and the electrode surface. To overcome this challenge where direct electron exchange between the electrode and either the analyte or the biomolecule is not permitted, redox mediators are required (Albery et al. 1987; Eggins 2002; Gerad et al. 2002). Direct electrical communication between redox enzymes and electrode surfaces has been studied (Wilner et al. 1997). In an amperometric immunosensor, the advantages of the selectivity of antibody, reactions are the amplification feature of an enzyme label and the ease with which aliquot amounts of enzyme-generated product can be detected amperometric ally (Aizawa et al. 1980; Sharma et al. 2008). By using different immunoassay approaches, the concentration of the analyte bound to the antibody on an electrode surface can be determined indirectly by monitoring the labelled enzyme while catalysing the reaction with its substrate and a mediator. In an amperometric DNA- or RNA-based biosensors, a single-stranded (ss-), nucleic acid sequence (probe) is immobilised on the electrode surface which on hybridisation to a specific complementary region of the target DNA gives rise to a voltage signal. The hybridisation event can thus be detected via an increased current signal of an electroactive indicator that preferentially binds to the DNA duplex or due to captured enzyme or nanoparticle tags. Controlling the probe immobilisation is essential for assurance of orientation, high reactivity, and/or accessibility and stability of the surface bound probe, as well as for avoiding nonspecific adsorption (Millan and Mikkelsen 1993; Gooding 2002; Drummond et al. 2003; Yogeswaran et al. 2008; Prest et al. 2010).

Miniaturisation of three-electrode systems can be realised by forming the electrodes as thin-film or thick-film patterns with an insulator to delineate the active areas to achieve a high sensitivity. To achieve a better sensitivity, the transport of the analyte to the surface of working electrode must be rapid. In the conventional planar-type working electrode, the analytes are assumed to diffuse homogeneously except for peripheries in a stagnant solution, while in microelectrode the expansion of the diffusion layer becomes semi-spherical (Stulik et al. 2000). Accompanying advantages of the microelectrodes include small charging current originating from the very small electrode area, small Ohmic drop and rapid transport of analytes to the electrode surface.

Arrays of microelectrodes have been used for the working electrode to achieve high signal-to-noise ratio, increased current density and flow insensitivity simultaneously (Morf and Rooij 1997). The use of microelectrode arrays takes advantage of the partial selectivity of an individual electrode, by combining several electrodes and examining the relative responses of all the sensors put together. In addition to this, the coupling of microelectrode arrays with a chemometric approach allows analysis of a mixture of analytes (Baldwin et al. 2002; Albert et al. 2000). Various types of microelectrode arrays can be employed for high-speed analysis. One of the potential applications in which microelectrode arrays are often used is the heavy metal detection by anodic stripping voltammetry (Freeney and Kounaves 2000). The massive interest in microarray-based assays initially arose from work using DNA chips. The capture DNA/probes (*c*-DNA or oligomers) can either be synthesised on the chip surface via photolithography or can be spotted onto an activated chip surface (substrate), ink jet printing or electrochemistry (Duggan et al. 1999; Dill et al. 2004). *c*-DNA and oligonucleotide microarray technology have been applied for the analysis of microbial pathogens, and a single-chip multi-pathogen oligonucleotide microarray was constructed for the simultaneous analysis of foodborne pathogens (Kim et al. 2008a, b). The commercially available methods based on microarray system mainly rely on the fluorescence emission and optical reflectance for detection (Pirrung 2002; Park et al. 2006).

16.2.1.2 Potentiometric Biosensors

The potentiometric sensors consist of an ion-selective membrane and a bioactive material, e.g. an enzyme and an internal reference electrode. Many types of electrodes are available and classified by the properties and nature of the membrane. The nature of the ion-selective membrane determines the selectivity of the electrode and is also called a working electrode or ion-selective electrode. Representative membranes include sparingly soluble salts pressed into a pellet, pressed single crystals and solvent polymeric membranes. A plasticiser and an ionophore (ion-exchange) compound are incorporated there (Schoning and Poghossian 2002; Bakker 2004). For the polymeric membranes, polyvinylchloride (PVC) has been used as a supporting matrix, and other polymers also used were polystyrene, polyamide, etc.

The potential of the ion-selective electrode (ISE) measured with respect to the reference electrode is linearly dependent on the logarithm of the activity or concentration of the analyte (Pungor 2001; Umezawa et al. 2002; Bobacka et al. 2003). The potential difference across the ion-selective electrode is measured using a reference electrode (external) under the condition of zero current. Since potentiometry yields a logarithmic concentration response, the technique enables the detection of extremely small concentration changes. The widely used ion-selective electrode is a glass electrode for pH measurement.

Liquid membrane-type ISEs, based on water-immiscible liquid substances impregnated in a polymeric membrane, are widely used for direct potentiometric measurement (Oesch et al. 1986). These types of ISEs are particularly important because they permit direct measurements of several polyvalent cations and anions. Ion-exchange electrodes are also reported as being sensitive to large organic cations (Martin and Freiser 1980). In addition to ion-exchange electrodes, the liquid membrane electrodes rely more often on the use of complex-forming neutrally charged carriers like crown ethers, valinomycin, etc. These carriers are capable of forming stable complexes with alkali or alkali earth metals, and sulphur-containing ones are the best for binding heavy metals (Pretsch et al. 1988).

Coated-wire electrodes (CWEs) prepared by coating an appropriate polymeric film directly onto a conductor can be metallic (Pt, Ag, Cu, etc.) in a conventional shape such as wire or disc. The conductor is usually dipped in a solution of PVC and an active substance, and the resulting film is allowed to dry in air. CWEs are simple, inexpensive and easy to miniaturise, but they suffer some deficiencies like reproducibility and long-term stability (Bobacka et al. 2003; Konopka et al. 2004). ISEs have been used in online monitoring systems and used in flow injection systems. Potentiometric microelectrodes are very suitable for in vivo real-time clinical diagnostics like monitoring of blood electrolytes, in situ environmental surveillance or industrial processes and intracellular studies (Lynch et al. 2000; Liang et al. 2009).

16.2.1.3 Electrochemical Impedance Spectroscopy-Based Biosensor

Electrochemical impedance spectroscopy (EIS) has long been traditionally employed for studying the information about corrosion processes of metals and metal-coated surfaces (Lasia 1999). Nowadays, there is an increased use of EIS for

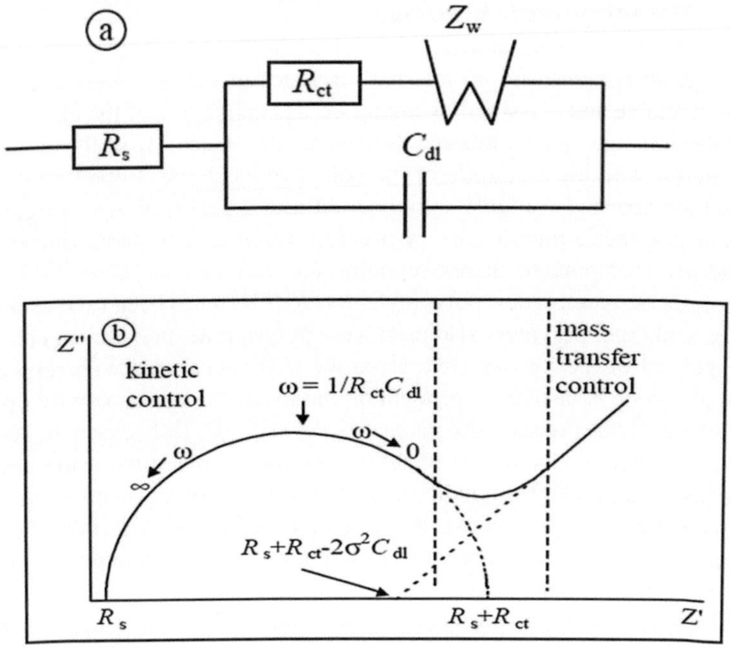

Fig. 16.1 (**a**) Randles circuit, (**b**) Nyquist plot

the investigation of adsorption processes, electrochemical mechanisms, and dielectric and transport properties of materials used for the creation of sensors/biosensors. EIS makes it possible to study the process of immobilisation of bio-components and to characterise electric features of bio-component/electrode interfaces (Katz and Willner 1997; Barsoukov and Macdonald 2005). In this technique, a small sinusoidal AC voltage probe (typically 2–10 mV) is applied and the current response is determined. This current is used to calculate the impedance at each of the frequencies is probed. The potential signals, the amplitude of the current and the resulting phase difference between voltage and current (that depends on the nature of the system under study) dictate the system impedance. The impedance has a real and imaginary component making its mathematical treatment quite difficult and cumbersome. Impedance results are commonly fitted to equivalent circuits of resistors and capacitors, such as the Randles circuit (Fig. 16.1a), which is often used to interpret simple electrochemical systems (Katz and Willner 1997). This equivalent circuit yields the Nyquist plot (Fig. 16.1b), which provides visual insight into the system dynamics. In this *R*ct is the charge transfer resistance, and is inversely proportional to the rate of electron transfer; *Cd* is the double layer capacitance; *Rs* is the solution phase resistance; and *Zw* is the Warburg impedance, which arises from mass transfer limitations.

The EIS method has been/was employed for the detection of DNA hybridisation, and antibody–antigen binding can be directly detected allowing for the development of immunosensors (Mirsky et al. 1997; Huang et al. 2008; Lucarelli et al. 2004).

The main drawback of impedance methods of biosensors is the need for interfacial engineering to reduce nonspecific adsorption. To minimise the nonspecific interaction, a composite film is used that contains the biomolecule of interest interspersed with a protein-resistant species. This approach has been used by George Whitesides research group (Qian et al. 2002). In EIS-based sensors, the advantages of nanomaterials have also been specially reported for gold nanoparticles and carbon nanotubes (Tang et al. 2007; Yun et al. 2007; Pey et al. 2008). In metal oxide-based biosensors, the impedance studies were directed towards understanding adsorption of biological materials on the electrodes (Khan et al. 2008) and also to find interfacial resistance values (Kaushik et al. 2009a; Solanki et al. 2009). These values will also help in finding improvement in electrochemical characteristics of any electrode after modification.

Jiang et al. have used the EIS technique for the detection of adherent cells based on molecular recognition of integrin β1. They used mouse antihuman integrin β1 monoclonal antibody, and a specific ligand-binding integrin β1 on cell membrane was used to immobilise HeLa cells (tumour cell) on the glassy carbon electrode. The cell adhesion based on the reaction between antibody and integrin β1 on cell membranes resulted in significant increase of *R*ct in EIS measurement using potassium ferrous/ferricyanide as an electrochemical probe (Jiang et al. 2010). The EIS is a very influential tool for detection in future work.

16.2.1.4 Conductometric Biosensors

A conductometric sensor measures the changes in the conductance/resistance of a film or sensitive area caused by the molecule of interest. Since conductivity is related to the rate of flow of charge in response to an electric field, the magnitude of the conductivity is dependent on the mobility, concentration and charge of the charged particles. Many enzymatic reactions produce or consume ionic species resulting in changes in electrical conductivity in solutions. This type of sensor can be constructed by using an interdigitated electrode, which can be formed by photolithography. Gold or platinum is usually used as an electrode (Hnaien et al. 2009). Typically, when an analyte reaches the sensitive area of the electrodes, there will be a change in its conductance or resistance. Conductometric biosensors have numerous advantages such as their easy miniaturisation and large-scale production using inexpensive technology. For such transducers, which are not light sensitive and there is no need for any reference electrode. Besides, a major advantage of the conductometric detection mode lies in a large number of enzymatic reactions involving either consumption or production of charged species and therefore leading to changes in the ionic composition of the reacting solution. The conductometric method generally lacks specificity because all charged species are detected simultaneously. Considerable interest is in using polyaniline, polypyrrole, polyacetylene and polythiophene in the development of conductometric biosensors (Hoa et al. 1992; Sergeva et al. 1996). These were used as transducers in conductometric biosensors.

The application of conductometric biosensors has been reported for heavy metal ion and pesticide detection using an immobilised *Chlorella vulgaris* microalgae as bioreceptor based on the enzymatic approach using alkaline phosphates and acetylcholine esterase or the inhibition method-alkaline phosphates inhibited by heavy metal ions and acetylcholine esterase inhibited by pesticides (Chouteau et al. 2005). Ion-selective conductometric microsensors have also been developed for K^+ and Ca^{2+} ions (Shulga et al. 1995). The operation of these sensors is based on the specific and reversible co-extraction of ions from the aqueous solution into the membrane containing an ionophore for specific cations. The detection of serum antibody (IgG) against the causal organism of Johne's disease (JD), *Mycobacterium avium* subsp. *Paratuberculosis* (MAP), was also reported based on the conductometric biosensor. They used polyaniline as conducting polymer for measuring the change in electrical conductance/resistance (Okafor et al. 2008). Nitrite detection in water solution has been developed based on conductometric biosensors. In one case, Zhang et al. used the enzymatic mixtures of cytochrome *c* nitrite reductase, methyl viologen, BSA, Nafion and Glycerol, which were immobilised on interdigitated electrode by cross-linking saturated glutaraldehyde vapour. The anticipated conductometric biosensor has a linear range 0.2–125 M with a detection limit 0.05 M (Zhang et al. 2009). In a second case, some researchers used a conductometric method for the detection of nitrite based on a different approach in a combination of peroxidase/catalase for the detection of nitrite. Their detection limit is 0.3 μM and dynamic range 0.3–446 μM (Zazou et al. 2009).

16.2.1.5 Voltammetry-Based Biosensors

Voltammetric techniques involve the application of a potential (E) to an electrode and the monitoring of the resulting current (i) flowing through the electrochemical cell. It consists of a three-electrode system—a working electrode, a reference electrode and an auxiliary electrode. The current passes in between the working and auxiliary electrode and is measured in between the working and reference electrodes. In many cases, the applied potential is varied or the current is monitored over a period of time (t). Thus, all voltammetric techniques can be described as some function of E, i and t. The analytical advantages of the various voltammetric techniques include:

(a) *Exceptional sensitivity with a very large useful linear concentration range for both inorganic and organic species and a large number of useful solvents and electrolytes.*
(b) *A wide range of temperatures and rapid shot analysis times (typically in seconds).*
(c) *Simultaneous determination of several analytes.*
(d) *Ability to determine kinetic and mechanistic parameters.*
(e) *Well-developed theory.*
(f) *Ability to reasonably estimate the values of unknown parameters.*
(g) *The ease with which different potential waveforms can be generated.*
(h) *Small currents are also measured.*

In voltammetry, the effects of the applied potential and the behaviour of the redox current are described by several well-known laws. The applied potential controls the concentrations of the redox species at the electrode surface (C_O-oxidised species) and (C_R-reduced species) and the rate of the reaction K^0, as described by the Nernst or Butler–Volmer equations, respectively. In the cases where diffusion plays a controlling part, the current resulting from the redox process (known as the faradaic current) is related to the material flux at the electrode–solution interface and is described by Fick's law. The interplay between these processes is responsible for the characteristic features observed in the voltammograms of the various techniques. A reversible electrochemical reaction (i.e. a reaction so fast that equilibrium is always re-established as changes are made) can be described by:

$$O + one \leftrightarrow R$$

The application of potential E forces the respective concentrations of O and R at the surface of the electrode (i.e. C_O and C_R to a ratio in compliance with the Nernst equation:

$$E = E^0 - RT / nF \ \ln \left(C_R / C_O\right)$$

where

R is the molar gas constant (8.3144 J mol^{-1} K^{-1}).
T is the absolute temperature (K).
n is the number of electrons transferred.
F is the Faraday constant (96.485 C/equiv).
E^0 is the standard reduction potential for the redox couple.

If the potential applied to the electrode is changed, the ratio (C_R/C_O) at the surface will also change.

For some techniques it is useful to use the relationship that links the variables for current, potential and concentration, known as the Butler–Volmer equation:

$$I / nF\mathrm{A} = K^0 \left\{C_O \exp\left[-\alpha\theta\right] - C_R \exp\left[\left(1-\alpha\right)\theta\right]\right. \quad (16.1)$$

where $\theta = nF(E - E^0)/RT$, $K^0$is the heterogeneous rate constant,

α is known as the transfer coefficient and
A is the area of the electrode.

This relationship allows us to obtain the values of the two analytically important parameters, i and K^0.

In voltammetry measurement, the movements of species towards an electrode surface occur mainly by diffusion, migration or convection. The use of a supporting electrolyte at concentrations 100 times that of the species being determined eliminates the effect of migration (migration in voltammetry is the movement of a charged ion in the presence of an electric field). Convection is the movement of the electroactive species by thermal currents, by density gradients present in the solution or by stirring the solution or rotating the electrode. Convection must be eliminated or controlled accurately to provide controlled transport of the analyte to the electrode. Many voltammetric techniques have their own unique laws and theoretical relationships such as cyclic voltammetry, differential-pulse voltammetry, square-wave voltammetry and stripping analysis (Bard and Faulkner 1980; Kissinger and Heineman 1984).

16.2.1.5.1 Differential-Pulse Voltammetry (DPV)

DPV is another common technique that is used under voltammetry; in this technique, the potential is applied in pulse form. This technique uses a series of potential pulses of increasing amplitude. The current measurement is made near the end of each pulse, which allows time for the charging current to decay. It is usually carried out in an unstirred solution at the electrodes. In DPV the potential is also scanned with a series of pulses, and each potential pulse is fixed small amplitude (10–100 mV) and is superimposed on a slowly changing base potential. Current is measured at two points for each pulse; the first point is just before the application of the pulse and the second at the end of the pulse. The difference between current measurements at these points for each pulse is determined and plotted against the base potential (Bard and Faulkner 1980; Kissinger and Heineman 1984).

The DPV technique is a very powerful method for the simultaneous determination of the compound in mixtures such as noradrenalin and acetaminophen oxidation at the haematoxylin which modified the surface of the electrode. They are clearly separated from each other when they coexisted in the physiological pH at this electrode (Nasirizadeh and Zare 2009). This technique is an extremely useful method for measurement of trace levels of organic and inorganic species. The differential-pulse operation results in a very effective correction of the charging background current. The peak-shaped response of differential-pulse measurements results also improved resolution between two species with similar redox potentials.

16.2.1.5.2 Square-Wave Voltammetry (SWV)

The excitation signal in SWV consists of a symmetrical square-wave pulse of amplitude E_{sw} superimposed on a staircase waveform of step height ΔE, where the forward pulse of the square wave coincides with the staircase step. The net current, i_{net}, is obtained by taking the difference between the forward and reverse currents ($i_{for}-i_{rev}$) and is centred on redox potential. The current difference between these two points is then plotted against the staircase potential in a square-wave voltammogram.

The peak height is directly proportional to the concentration of the electroactive species and direct detection limits as low as possible. Square-wave voltammetry has several advantages. Among these are its excellent sensitivity and the rejection of background currents. Applications of square-wave voltammetry include the study of electrode kinetics with regard to preceding, following or catalytic homogeneous chemical reactions, determination of some species at trace levels and its use with electrochemical detection in HPLC. SWV provides a more familiar pear-shaped signal for easy interpretation of analytical data (Yarnitzky 1985). The main advantage of SWV is its speed. The effective scan rate is given by $f\Delta E_s$. The term f is the square-wave frequency, and E_s is the step height. Frequencies of

1–100 cycles per second permit the use of extremely fast potential scan rates. As a result, the analysis time is drastically reduced; a complete voltammogram can be recorded within a few seconds. The inherent speed of SWV can greatly increase sample throughputs in batch (Wang et al. 1978) and flow analytical operations (Masarik et al. 2003).

Lyon and Stevenson demonstrated the picomolar detection of H_2O_2 by using the SWV detection technique over the amperometric detection method (Lyon and Stevenson 2006). They achieved this low-level detection due to the fast effective scan rate of SWV. SWV, in combination with an adsorptive transfer stripping technique, is also used for measuring nanogram quantities of streptavidin and avidin in solution (Osteryoung and Osteryoung 1985).

16.2.1.5.3 Electrochemical Stripping Analysis

Electrochemical stripping analysis has the lowest detection limit of any commonly used electroanalytical technique. It consists of a two-step method. In the first step, the analyte is deposited (pre-concentrated) onto the surface of the electrode by controlled potential electrolysis. In the second step, the deposited analyte is removed (stripped) from the electrode, and the accompanying electrochemical signal is used to determine the concentration of each analyte species in the sample. In stripping voltammetry, the stripping step is accomplished by a potential scan, and the resulting current peaks are measured to determine concentration.

In anodic stripping voltammetry (ASV), pre-concentration is caused by the reduction of the metal ions to the elemental state, and the stripping step is accomplished by a positive potential scan that gives an anodic current. Cathodic stripping voltammetry (CSV) involves pre-concentration by oxidation with subsequent stripping by a negative potential scan. The stripping step may consist of a positive or a negative potential scan, creating either an anodic or cathodic current, respectively. Hence, anodic stripping voltammetry (ASV) and cathodic stripping voltammetry (CSV) are two specific stripping techniques (Wang 1985).

Adsorptive stripping voltammetry (AdSV) is quite similar to anodic and cathodic stripping methods. The primary difference is that the pre-concentration step of the analyte is accomplished by adsorption on the electrode surface or by specific reactions at chemically modified electrodes rather than accumulation by electrolysis. Many inorganic and organic species (such as heme, chlorpromazine, codeine and cocaine) have been determined at micromolar and nanomolar concentration levels using this technique. The adsorbed species is quantified by using a voltammetric technique such as differential-pulse voltammetry or square-wave voltammetry in either the negative or positive direction to give a peak-shaped voltammetric response with amplitude proportional to concentration.

16.2.2 Field Effect Transistor-Based Biosensors

Typically, a field effect transistor device (FET) will have a source and a drain. A current passes from source to drain. The FET also contains a gate, whose properties will be able to control the current passing between the source and drain. The gate material will generate an electrical field and controls the current flow. Another form of FET utilises a nanowire between two conducting materials. The nanowire has its atoms concentrated on its surface. Thus, any small changes in the charges present on the nanowire will cause a change in the flow of current.

Ion-sensitive field effect transistors (ISFET) are found to be suitable for pH sensing. ISFETs utilise the semiconductor field effect to detect biological recognition events (Bergveld 2003). A basic structure of the ISFET contains an electrode called the source and drain formed in a lightly doped silicon substrate. They consist of a *p*-type silicon substrate with *n*-type source region separated from a similar *n*-type drain region. This channel is insulated by SiO_2 and metal gate electrode. The conductance between the two electrodes changes when the voltage between the insulator and the silicon substrate changes. The gate insulator (SiO_2) is typically covered by an ion-selective membrane that is selectively permeable to a certain ion, e.g. K^+, Ca^{2+} and F^-, and for heavy metal ions, a PVC-based membrane is used (Munoz et al. 1997; Cobben et al. 1992). The application of these devices in the area of biosensors is reasonably new, and their use is not spreading as quickly as other electrochemical techniques due to the incompatibility of most biomolecule immobilisation methods with the ISFET fabrication technology, poor linear range and reproducibility. Apart from this, this sensor has significant advantages because the actual sensing area is very small, and a single miniaturised ISFET chip could contain multiple gates and be used to sense several ions simultaneously.

Examples of ISFET-based biosensors can be found using enzymes (EnFET), antibodies (ImmunoFET), chemically modified field effect transistor (CHEMFET) and DNA probes (GenFET) (Caras and Janata 1980; Janata 1990). All of these kinds of ISFETs mentioned above have their own merits and disadvantages. EnFETs are easiest to construct and operate because the products of the catalytic reaction aided by the enzyme bring about local and measurable pH changes. This product is detected by the ion-sensitive surface layer of the sensor, and the resulting surface charge modulates the space charge region at the insulator–semiconductor interface (Poghossian et al. 2001). Different methods are used to immobilise the enzyme on the gate surface by cross-linking with glutaraldehyde, covalent attachments of enzymes on silanised surfaces of the gate, etc. ImmunoFETs and GenFETs are much harder to develop because translating the bio-recognition changes into a measurable signal is a practical problem. Kim et al. demonstrated FETs were functionalised with antibody-binding fragments (Fab) as a receptor, and the binding event target immunoglobulin G (IgG) onto the fragments was detected by monitoring the gating effect caused by the charges of the target IgG. Because the biosensors are mainly used in a buffer solution, it is crucial to use small-size receptors so that the charged target IgG could approach the carbon nanotube surface within the Debye length distance to give more gating effect (Kim et al. 2008a, b). FET-type charge sensor for detecting DNA sequence measuring the change of electric charge caused by immobilisation of DNA on the gate metal (Kim et al. 2004). Kamahori et al. have developed an extended gate FET sensor on which DNA probes and 6-hydroxyl-1-hexanethiol molecules are immobilised, which can detect DNA hybridisation and extension reactions by applying a superimposed high-frequency voltage to the reference electrode (Kamahori et al. 2008).

16.3 Nanotechnology

Nanotechnology refers to the materials and structures having dimensions less than 100 nm. The emergence of nanoscience and nanotechnology has led to great developments in electrochemical science, technology and engineering, which led to a new branch of electrochemistry research—electrochemical nanotechnology, combining electrochemical techniques with nanotechnologies to address important issues in energy, electronics, environment and heath care. The developed nanomaterials and nanostructures offer newer approaches for sensing. The nanomaterials possess special properties due to quantum confinement effect, high surface area and high aspect ratio (Prasad 2014; Prasad et al. 2014, 2016). In addition, the electrochemical devices have high sensitivity with transducers that use nanocomposite materials because of edge diffusion phenomena. The new century presents opportunities as well as challenges to scientists and engineers working in the dynamic field of functional nanomaterials. The applications of nanomaterials in the field of electrochemical technology focusing on nanodevices, nanostructures, nanoelectronics, chemical sensors, physical sensors, biological/biosensors, energy storage and their conversion demonstrate that electrochemical nanotechnologies can provide solutions to significant technical barriers and potentially revolutionise research in these emerging fields while exploring many new areas with potential applications for nanomaterials in electrochemistry and bioelectrochemistry, particularly useful in sensing and catalytic processes. Nanocomposite has recently attracted much interest owing to their application in nanoscaled devices, sensors and detectors. Nanotechnology also offers new devices such as interdigitated electrodes, nanogap electrodes, cantilevers and FETs, and such devices have high sensitivity.

16.3.1 Nanocomposites

Nanocomposites are composites having at least one of the phases in the dimension of nanometre range. Nanocomposites are high-performance materials and reported twenty-first century materials in the view of unusual property combinations and unique design possibility which are absent in conventional composites. It has emerged as suitable alternatives to overcome limitations of micro-composite and monolithic (Camargo et al. 2009). First inference on nanocomposites was reported as early as 1992 (Gleiter 1992); however, a general understanding of nanocomposite properties is yet to be reached (Schmidt et al. 2002).

Nanocomposites are a class of materials in which one or more phases with nanoscale dimensions (0-D, 1-D and 2-D) are embedded in a metal, ceramic or polymer matrix. Nanocomposites are a multiphase material with significant proportions of each phase (Pey et al. 2005). The general idea behind the addition of the nanoscale second phase is to create a synergy between the various constituents, such that novel properties capable of meeting or exceeding design expectations can be achieved and to get a more desirable combination of properties. Its constituent phases are chemically dissimilar and separated by a distinct interface. Matrix is a continuous phase and classified in metal matrix nanocomposite (MMNC), ceramic matrix nanocomposite (CMNC) and polymer matrix nanocomposite (PMNC), whereas dispersed phase are dispersed in the matrix and can be classified in the particle, fibre and structural.

Metal matrix nanocomposites (MMNC) consisting of alloy matrix in which some nanosized reinforcement material or a ductile metal is implanted. These materials have both metal features and ceramic features, i.e. ductility and toughness with high strength and modulus. Thus, MMNC is suitable for material production with high shear strength and high service temperature capabilities. They show an extraordinary potential for many applications such as aerospace and automotive industries and development of structural materials (Tjong and Wang 2004).

The potential of ceramic matrix nanocomposites (CMNC) are having the potential in the main Al_2O_3/SiC system (Niihara 1991; Nakahira and Niihara 1992). It has been confirmed by many studies that Al_2O_3 matrix gets noticeable strengthening after addition of a low (i.e., ~10 %) volume fraction of SiC particles of suitable size and hot pressing of the resulting mixture. Some studies have explained this toughening mechanism based on the crack-bridging role of the nanosized reinforcements (Ferroni et al. 2001). Consequently, the incorporation of high-strength nanofibres into ceramic matrices has allowed the preparation of advanced nanocomposites with high toughness and superior failure characteristics compared to the sudden failures of ceramic materials (She et al. 2000). Both MMNC and CMNC with CNT nanocomposites hold promise but also pose challenges for real success.

Polymer materials are widely used in industry due to their easy production methods and ductility in nature. However, their modulus and strength are lower comparative to metals and ceramics. A very effective approach to improve mechanical properties is to add fibres, whiskers, platelets or particles as reinforcements to the polymer matrix. For example, polymers have been filled with several inorganic compounds (synthetic or natural) in order to increase heat resistance, impact resistance, flame retardancy and mechanical strength and to decrease electrical conductivity and gas permeability with respect to oxygen and water vapour (Fischer 2003). Metal and ceramic reinforcements offer striking routes to certain properties coming from inorganic nanoparticles such as magnetic, electronic, optical or catalytic and added to other polymer properties such as processability and film-forming capability (Athawale et al. 2003). Polymers can be improved using this approach while keeping their lightweight and ductile nature (Akita and Hattori 1999; Akita and Kobayashi 1999; Akita et al. 1999; Chang and An 2002; Zavyalov et al. 2002; Jordan et al. 2005). Nanoscale reinforcements have an exceptional potential to generate new phenomena, which leads to special properties of these materials as will be seen later. It may be pointed out that the reinforcing efficiency of these composites, even at low volume fractions, is comparable to 40–50 % for fibres in micro-composites.

The use of nanocomposite materials for electrochemical biosensors/sensors has provided an impetus for the development of sensitive devices. These nanocomposite materials are the building blocks of future nanodevices. As particle size in nanocomposite decreases, particle dimensions approach the size of certain physical length scales—such as the electron mean free path and the electron wavelength—resulting in quantum size effects that alter the electronic structure of the particle.

Nanocomposite materials have unique structural features which are the basis for creating and tuning many novel electric, chemical, optical and magnetic properties. Semiconducting nanocomposite materials have high ionic conductivity, capacitive action, catalytic properties, high isoelectronic point (IEP) and wide bandgap. Due to all these properties, it gained considerable interest in the field of biosensors (Kumar and Chen 2008). Large numbers of metal oxides and metals such as ZnO, SnO_2, TiO_2, CeO_2, ZrO_2, In_2O_3, MnO_2, SiO_2, Fe_3O_4, carbon nanotube, gold, platinum and conducting polymer have been used in synthesis of nanocomposite materials for various purposes such as immobilisation of biomolecules and electrocatalytic properties to improve the sensitivity of the sensors (Liu 2006a, 2008; Wang et al. 2006a; Saha et al. 2009; Upadhyay et al. 2009; Pumera et al. 2007; Yang et al. 2009; Rodriguez et al. 2007; Deng et al. 2009; Liao et al. 2006; Kim et al. 2006; Li et al. 2009; Curreli et al. 2005; Lvov et al. 2006; Salimi et al. 2007a).

Electrochemistry has been associated with nanoscience particularly when the stability and assembly of the particles are probed. Pumera et al. reported that most of the electrochemical phenomena have been changed in nanosized regimes, such as diffusion kinetics, an electrical double layer at the interface and electrocatalytic properties, which offer excellent prospects for interfacing biological recognition events with electronic signal transduction to develop a new generation of nanodevices (Pumera et al. 2007). Nanocomposite materials composed of metal oxide have been used in for solid state gas sensor for environmental pollutants and toxic gases. These metal oxide-based nanocomposites are suitable for gas sensing because they have uniform electronic conductivity and high specific surface area.

16.3.2 Properties of Nanocomposites

When the particle size is less than a particular level, known as 'critical size', changes in particle properties can be observed (Kamigaito 1991). Some properties which can be observed at critical size are written as follows. Catalytic property can be observed at critical size of 5 nm, hard magnetic materials can be made soft at critical size of less than 20 nm, changes in refractive index can be produced at critical size of less than 50 nm, superparamagnetism and other electromagnetic behaviours can be observed at less than 100 nm, strengths and toughness can be produced at critical size of less than 100 nm and modification in hardness and plasticity can be observed at critical size of less than 100 nm. When dimensions reach to the nanometre level, interphase interactions improved largely, which is important to enhance material properties. In this context, the surface area/volume ratio of reinforcement materials employed in the preparation of nanocomposites is crucial to the understanding of their structure–property relationships.

Nanocomposites are multiphase material, and their properties are a function of the properties of the constituent phases, their relative amounts and the geometry of the dispersed phase. The geometry of the dispersed phase depends on particles shape, size and orientation. The properties of nanocomposites rely on a range of variables, particularly the matrix materials, which can exhibit nanoscale dimension, loading, the degree of dispersion, size, shape and orientation of dispersed phase (second phase) and interactions between the matrix and the dispersed phase (second phase). Matrix is the continuous phase and its purpose is to transfer stress to other phases and to protect phases from the environment. Nanocomposites have properties of both solid (occupy the fixed volume or space) and liquid (flow under force). Matrix can be classified into three types, viz. metal matrix nanocomposite (MMNC) such as Fe–Cr/Al_2O_3, Ni/Al_2O_3, Co/Cr, Fe/MgO, Al/CNT and Mg/CNT ceramic matrix nanocomposite (CMNC) such as Al_2O_3/SiO_2, SiO_2/Ni, Al_2O_3/TiO_2, Al_2O_3/SiC and Al_2O_3/CNT and polymer matrix nanocomposite (PMNC) such as thermoplastic/thermoset polymer/layered silicates, polyester/TiO_2, polymer/CNT and polymer/layered double hydroxides. The dispersed phase is the other phase which is dispersed in the matrix, and its purpose is to enhance matrix properties. In the case of a metal matrix, it increases yield stress, tensile stress and creeps stress. In the ceramic matrix, it increases fracture toughness. In polymer matrix, it increases modulus, yield stress, tensile strength and creep resistance. Dispersed phase can be a particle, fibre and structural. Each kind of dispersed phase has a specific surface area-to-volume ratio depending on their reinforcements/geometries, e.g. particulate materials have 3/r, fibrous materials have 2/r +2/l and layered materials have 2/t +4/l. Due to nanoscale size of the reinforcement phase, the interface-to-volume ratio is significantly very high. As a result, the volume fraction of the dispersed phase (second phase) can be reduced without degradation of desired properties. The nanoscale-reinforcing phase can be grouped into three categories, namely, nanoparticles (0-D), nanotubes (1-D) and nanoplates (2-D). In the case of nanoparticles, the particle size and distribution are of great importance. Nanocomposite's properties such as mechanical, electrical, thermal and optical depend on the type of nanoparticles added and, hence, can be altered by selecting particular nanoparticle. In the case of mechanical properties, changes in modulus and strength strongly depend on the degree of interaction between particle and matrix. Due to the nanoscale range, which is smaller in size in comparison to critical crack length, nanocomposite materials have improved toughness and strength. However, agglomeration of nanoparticles should be prevented. In regard to electrical properties, the smaller the size of the nanoparticle, the shorter the distance between the particles (provided volume kept constant) which leads to percolation at lower volume fraction and, hence, higher electrical conductivity. In some cases such as polypyrrole nanocomposite filled with Fe_2O_3 nanoparticles, the 'variable-range hopping' (VRH) mechanism explains the enhancement in dc current. The VRH mechanism involves the exchange of charges between the nanoparticles and matrix. Optical properties are related to the transparency of the nanocomposite. To achieve transparency, scattering should be minimised which means that size of the nanoparticles should be as small as possible and refractive index should be as similar as possible to the matrix. Controlling the refractive index can be achieved by tailoring the volume fraction of nanoparticles. In addition, the colour of the nanocomposite can be tuned by using optically active nanoparticles and changing particle size, shape and distribution. This behaviour has been shown in polyethylene matrix filled with silver nanoparticles. The addition of nano-silica in polyimide has been used to control the transmittance in the nanocomposite.

Since the discovery of carbon nanotubes (CNTs) in 1991 (Iijima 1991) and their subsequent use in nanocomposites showing unique mechanical, thermal and electrical properties (Biercuk et al. 2002; Ounaies et al. 2003; Weisenberger et al. 2003) of CNT, which added a new and interesting dimension to nanocomposite materials. The possibility of CNTs into nanocomposite products and textiles (Dalton et al. 2003) made further inroads for the processing and applications of CNT nanocomposites. The use of carbon nanotubes (CNTs) in nanocomposites has received wide attention due to their extraordinary properties. However, some critical factors such as uniform dispersion in the matrix, alignment of CNTs in the nanocomposite, good interfacial bonding between the CNTs and matrix, etc. play an important role and need to be addressed. CNTs exhibit smooth surfaces and intrinsic *van der Walls* interactions and, hence, promote clustering during dispersion in the nanocomposite matrix. If agglomeration occurs, the CNTs are less adhered to the nanocomposite matrix and will slip against each other under applied stress. These factors can be minimised by sonication of CNTs, chemical functionalisation, surfactant assistant-assisted processing and in situ polymerisation. Chemical functionalisation of nanomaterials should be specific to nanocomposite matrix. It increases interphase zone in nanocomposite and enhances interaction of dispersed phase with nanocomposite matrix. CNTs possess highly anisotropic mechanical properties due to their aspect ratio. And a crucial aspect to for providing optimal reinforcement is to properly orient the CNTs. Thus it is essential to well align the CNTs within the matrix for taking the advantage of their load carrying efficiency along the axial direction. Extrusion is a popular technique for CNTs alignment. Since CNTs also possess high electrical conductivity, therefore, application of an electric field and magnetic field also has been used to induce CNTs alignment. CNTs also have been used to improve the thermal conductivity by creating percolation network that allows the nanocomposite to conduct heat with conductivities up to 3.5 times than the pristine composite. However, thermal conductivities of the nanocomposite are still far from the theoretical value of isolated CNT, which is in the order of 10^3 W/m.K. The main reason for this discrepancy is the resistance that exists between matrix and CNTs surface.

The inclusion of 2-D nanomaterials in nanocomposite gives platelike layered materials with the thickness in the order of 1 nm with an aspect ratio of 25 or above. When 2-D nanomaterials added to the nanocomposite, many property enhancements such as improved UV-resistance, increased stiffness and strength, gas permeability, greater dimensional stability and superior flame resistance can be achieved. Functionalised graphene sheets are also used in nanocomposite materials. Graphite has no functionality. But it expands to graphite oxide after oxidation. Graphite oxides are hydrophilic in nature. When graphite oxide reduces, it exfoliates into functionalised graphene sheet, which is having some functionality.

Conducting polymer-based composites are novel materials with less than a decade of history. It is believed (Gangopadhyay and Amitabha 2000) that the total control of the whole conducting polymer-based composite system and the optimisation of their physical properties (such as electrical conductivity and colloidal stability) are yet to be achieved, while both their commercial availability in the near future and a big leap forward for materials science are expected with their appropriate utilisation (Ren-Jang Wu et al. 2007; Chavali et al. 2008). In the case of biodegradable polymer-based nanocomposites, recent developments in preparation, characterisation and properties, including crystallisation behaviour and melt rheology, of both the matrix and the layered (montmorillonite) nanocomposites have been discussed (Ray and Bousmina 2005; Pey et al. 2005).

16.4 Biomolecular Immobilisation on Nanocomposites

For the development of a biosensor, immobilisation of recognition molecules (enzyme, antibody or DNA, biomolecule) onto the transducer is essential. Nanocomposites can be a part of the transducer. The immobilisation can be done by one of the following methods:

(a) Physical adsorption
(b) Covalent immobilisation
(c) Entrapment
(d) Through cross-linkers

The nanocomposites offer innovative opportunities and special abilities than the existing materials. Physical adsorption is the simplest method of immobilisation in which biomolecules are mechanically attached to the surface with the help of van der Waals forces. And in this method, no conformational change occurs. This method also has a disadvantage that biomolecules may leak from the surface during experiments due to weak binding force between biomolecules and the surface. It is very easy to adopt physical adsorption methods because it involves dropping off a buffer solution containing the biological molecule onto the electrode. Nanocomposites are in the range of nanometre size, so it has large surface area. Hence, more biomolecules can get adsorbed easily. The nanocomposites were used in various forms for the development of immunosensors. They are used in making or modifying screen-printed electrodes. They are also used in modifying the glassy carbon electrode. Normally, a binder such as Nafion or Chitosan has been used for this purpose (Liaw et al. 2006; Tsai et al. 2005. 2007).

In microencapsulation, biomolecules are trapped between membranes. In the entrapment method, the biomolecule is trapped in a matrix of a gel, paste or polymer, and it is the very popular method. In the covalent attachment, there is a formation of covalent chemical bonds between biomolecules and transducer. The nanocomposites are a part of the transducer. The best stability, accessibility and selectivity can be achieved through covalent bonding. Covalent bonding has the capability to control the location of the biomolecules; therefore, it improves the stability, accessibility and selectivity. For the covalent bonding of molecules to the nanocomposites, it is essential to form functional groups on the nanocomposites.

One of the universal methods for connecting biomolecules to other materials is diimide-activated amidation, by direct coupling of carboxylic acid to proteins using *N*-ethyl-*N′*-(3-dimethyl aminopropyl) carbodiimide hydrochloride (EDAC) or *N,N′* di cyclohexyl carbodiimide (DCC) as a coupling agent. However, this process leads to undesirable side reactions of intermolecular conjugation of proteins, because most proteins are rich in both amine groups and carboxylic acid groups on their surface. This intermolecular connection can be avoided by using a two-step process: carboxylic acid groups are first converted to active esters via diimide-activation, and then the active esters are reacted with the amine groups on proteins without the presence of diimide.

In several reports, a binder/support material was used. Liu et al. reported that chitosan and Nafion were used as binder materials. They reported that chitosan is a natural cationic biopolymer, biocompatible, non-toxic and low cost and possesses good film-forming ability, high mechanical strength and high hydrophilicity (Liu et al. 2006b). Miao and Tan reported that chitosan contains amino and hydroxyl groups and thus facilitates immobilisation of enzymes via covalent binding (Miao and Tan 2000). Nadzhafova et al. reported that Nafion is a proton conducting biocompatible perfluoro sulphonate linear polymer that exhibits excellent film-forming ability and was widely used in biosensor applications (Nadzhafova et al. 2004). Liu et al. reported that zirconium oxide has a very high bandgap, and it is normally used as a platform for immobilisation of enzymes (Liu et al. 2003). For electrochemical sensing electrodes, platinum, glassy carbon, gold and indium tin oxide (ITO) were used as support to the nanocomposites (Sulak et al. 2006; Fang et al. 2003; Wang et al. 2008; Malhotra and Kaushik 2009; Ding et al. 2010a, b; Wilson and Rauh 2004; Yang and Zhu 2005; Umar et al. 2009; Irhayem et al. 2002; Salimi et al. 2006, 2007b; Chen et al. 2008; Nadzhafova et al. 2007).

Applying nanotechnology in biosensing must have relevant methods to immobilise biomolecules, whose activity may be preserved for long periods of time. The mostly used methods for this purpose are the electrostatic layer-by-layer (LbL) (Decher et al. 1992) and the Langmuir–Blodgett techniques (LBT; Blodgett 1934), which are complementary to each other in terms of the types of material that can be employed. The LbL films are obtained via transfer of insoluble films from the air–water interface onto solid supports. The LbL method utilises alternating layers of positive and negatively charged materials soluble in water, which is suitable for proteins. Traditional materials forming stable monolayers are fatty acids, phospholipids, sterols and substances with a long alkyl chain and a hydrophilic moiety. Soluble substances with affinity to the air–water interface (proteins and nucleic acids) can also be incorporated into the monolayers by adsorption from the aqueous subphase. Therefore, a variety of materials may be immobilised on solid matrices through the LB method, opening the way to fabricate hybrid systems. In particular, LbL technique is promising for silicon-based sensors as this method allows a control of film architecture and thickness, in addition to the synergy between properties of distinct nanocomposite materials, proteins (Lvov et al. 1995), antigen-antibody pairs (Zucolotto et al. 2007) and DNA (Elbakry et al. 2009).

The most widely used advanced technique is patterning of biological macromolecules onto solid surfaces in the form of microarrays and/or chips. The target capture process is performed on the substrates (e.g. silicon wafer, a glass slide) via biological recognition. To achieve a high sensitivity, a large amount of research has focused on signal amplification by utilising various nanocomposite (e.g. quantum dots, metal nanoparticles) as strong and photostable signal probes (Cao et al. 2002; Maxwell et al. 2002). A signal probe (fluorescent dye molecules are used usually) is utilised to signal such biological interactions. Sensitivity is a central factor for bioanalytical technique. Although these approaches have made considerable progress in biomolecular detection, they still have several drawbacks:

(a) These techniques involve a complex procedure for immobilisation of the biomolecules on the flat substrate.
(b) Since the target catching procedure is carried out on the flat surface of microarray or titre plate, such heterogeneous procedure increases assay time and decreases the sensitivity due to the slow target-binding kinetics.

In the cross-linking method, a bifunctional agent is used to bond chemically the transducer to the biomolecules. There are a number of advantages of immobilising biomolecules to the surface: (1) single batch of biomolecules can be used multiple times, (2) reaction can be stopped rapidly by removing the biomolecule from the reaction solution, (3) there is less chance of contamination of product with biomolecules and (4) immobilisation provides long life to the biomolecules.

16.5 Nanocomposites and Affinity Biosensor

There are two types of affinity-based biosensors. These are called immunosensors and DNA sensors. The immunosensors exploit the property of complex formation between an antigen and its antibody. And the other utilises attraction between the complementary sequences of the DNA strands.

16.5.1 Immunosensors

Nanostructured materials used for various applications have been categorised as inorganic nanostructures, organic nanostructures and organic–inorganic hybrid nanostructures. Inorganic nanostructured materials such as metals (Au, Pt, Ag, etc.) are metal oxides (Fe_3O_4, CeO_2, ZnO, TiO_2, SnO_2, SiO_2, ZrO_2, NiO, etc.) are crystalline and dislocation-defect free (Solanki et al. 2011; Anees et al. 2010). Organic–inorganic hybrid nanocomposite (Kaushik et al. 2008a, 2009b; Kruk and Jaroniec 2001) such as conducting polymer–metal/metal oxide and biopolymer–metal/metal oxide are the new class of materials. Nanocomposites show both properties of organic and inorganic counterpart along with that which are absent in counterpart molecules. The unique features of organic–inorganic composite materials have been used for the fabrication of active electron devices (e.g. diodes, transistors or switches) and optoelectronic devices, as well as for bioencapsulation, catalysis, nanolithography and biosensors (Prasad et al. 2016, 2017; Gill 2001; Kickelbick 2003; Mitzi et al. 2003; Walcarius 2001; Dai et al. 2004; Sanchez et al. 2005; Nastase et al. 2006; Wu et al. 2006).

The immunosensors utilise the affinity between the antigen and its antibody to form a complex. The ELISA methods were well-established techniques and have a variety of formats. Since the formation of the complex cannot be determined directly by earlier methods, an enzyme-tagged antibody is used to reveal the formation of the complex between the antigen and antibody.

The importance of CH in biosensor has been revealed by Cruz et al.'s studies on ion-transport and ion-exchange properties of a thin film of CH casted on glassy carbon electrodes (Cruz et al. 2000). Studies have indicated that the presence of NH_2/OH groups in CH provides a favourable microenvironment for immobilisation of IgGs leading to enhanced electron transfer to the electrode and thus the improved immune-sensing properties (Kaushik et al. 2008b). To improve the sensing performance of CH-based biosensor, application of CH–metal/metal oxide nanoparticle-based hybrid nanocomposites are gaining interest (Kaushik et al. 2008c). In nanocomposites, inorganic nanoparticles increase the electroactive effective surface area and result in improved loading of biomolecules as well as enhance sensing characteristics (Odaci et al. 2008). Stability, surface-charged nanoporous uniform surface structure, and the relative ease of varying the functionalities have made CH–metal/metal oxide nanoparticle-based hybrid nanocomposites suitable matrices for fabrication of biosensors (Chen and Gorski 2001; Mishra et al. 2011). Also, the immobilisation of biomolecules using these systems requires small amounts of biomolecules and result in the binding of the desired molecule in the near vicinity of the electrode surface.

Electrodeposited 3-D porous CH/SiO_2 membrane has been used to immobilise HBsAb using glutaraldehyde as a cross-linker to fabricate a potentiometric immunosensor to detect HB (Liang et al. 2008). 3-D porous structure possessed high surface area, good mechanical stability and good hydrophilicity, which provided a biocompatible microenvironment for maintaining the bioactivity of the immobilised protein and increased the protein loading for improved immune-sensing response. Wang et al. immobilised goat antihuman IgG antibody (IgG Ab) onto CH–ZnO/GCE for detection of human IgG through amperometric immune-sensing technique (Wang et al. 2006b). Sardinha et al. fabricated an ELISA for the detection of azinphos-methyl via immobilisation of LIBMFH14 monoclonal antibody (MAb) onto CH–SiO_2 as support (Sardinha et al. 2002). Anti-fetoprotein (AFP) Abs functionalised by Au nanoparticles (NP) has been immobilised onto CH–TiO_2 nanocomposite for detection of AFP in human serum by amperometric immunosensor through electro-catalysed reduction of AuNPs to H_2O_2 (Tan et al. 2009). The immunosensor showed a rapid response, high sensitivity, good reproducibility and favourable stability. CHZnO nanocomposite has been used to construct amperometric immunosensor for human IgG detection via immobilisation of goat antihuman IgG antibody (IgG Ab). The nanocomposite of CH and Fe_3O_4 nanoparticles has been used to fabricate an amperometric immunosensor via immobilisation of FeAb (Wang and Tan 2007). Fu et al. described a potentiometric immunoassay protocol for the determination of CEA in human serum via immobilisation of CEA-monoclonal antibody onto Fe_3O_4 nanorod (Fu et al. 2009). Wu et al. described colloidal–Au/CH membrane onto silver paste carbon electrode (SPCE) for the immobilisation of carcinoembryonic antibody (CEA–Ab) to prepare a disposable immunosensor coupled with a flow injection system for CEA-antibody detection (Wu et al. 2006). CH–Nano manganese oxide (MnO_2)–PB nanocomposite has been used to immobilise CEA–Ab for detection of CEA–antigen (Ling et al. 2009). CH–Nano manganese oxide (MnO_2)–PB nanocomposite has been used to immobilise CEA–Ab for detection of CEA–antigen. The results of the studies suggest that Au nanoparticle improve the wide linear range, and MnO_2 nanoparticles improve the detection and shelf-life of CEA immunosensor. Recently, CH–metal oxide nanocomposite has emerged as an appropriate, suitable and micro-friendly immobilising matrix for OTA detection with improved sensing characteristics. CeO_2 nanoparticles were incorporated into CH to fabricate CH–Nano CeO_2 nanocomposite film for the immobilisation of r-IgGs and BSA to detect OTA (Kaushik et al. 2009c). The presence of NanoCeO_2 in CH–NanoCeO_2/ITO nanocomposite increases the effective surface area of CH resulting in improved loading of r-IgGs. The sensing performance of CH–NanoCeO_2/ITO-based immunosensor was found to be better than sol-gel-derived NanoCeO_2 and CH-based immunoelectron for OTA detection. Kaushik et al. fabricated NanoSiO_2– and CH-based nanobiocomposite immunosensor for OTA detection (Kaushik et al. 2009d). The obtained 3-D arrangement of NanoSiO_2 in CH matrix via hydrogen bonding, available NH_2/OH groups and excellent film-forming abilities of CH, resulted in an increased effective surface area of CH–NanoSiO_2 nanocomposite for r-IgGs immobilisation resulting in enhanced electron transport. It has been shown that presence of NanoSiO_2 leads to enhanced electrochemical behaviour of CH resulting in increased electron transport between the medium. Moreover, they also fabricated CH and superparamagnetic Fe_3O_4 nanoparticle-based nanocomposite immunosensor for detection of OTA (Khan and Dayal 2008). Nano Fe_3O_4 results in increased electroactive surface area wherein the affinity of surface-charged Nano Fe_3O_4 supported the immobilisation of r-IgGs and led to enhanced electron transfer. This immunosensor exhibits improved sensing characteristics with respect to other CH-based nanocomposite immunoelectrodes. Kaushik et al. also incorporated MWCNT/SWCNT into CH to prepare hybrid nanocomposite film to detect OTA (Kaushik et al. 2010). The CH–SWCNT-based immunosensor exhibits better sensing characteristics than that of CH–MWNCTS-based immunoelectrode due to the more electroactive surface area and enhanced electron-transfer kinetics. The work presented by Kaushik et al. is based on polyclonal antibody-based OTA detection and may exhibit the problem of specificity. Besides this, the utilised CH-based nanocomposites are found to be highly electroactive and are suitable for fabrication of efficient immunosensors for OTA detection. To overcome the problem of selectivity, the polyclonal antibody can be replaced with monoclonal antibody or aptamers during immunosensor fabrication using CH–metal/metal oxide hybrid nanocomposite for the selective detection of OTA.

16.5.2 DNA Sensors

DNA sensor is an affinity-based biosensor and considered a promising tool in prediagnosis as well as in the prevention and control of infectious diseases in the real-time and on-site analysis (Drummond et al. 2003). There have been huge advances in the development of DNA biosensors. Since nucleic acids, the building blocks of DNA and RNA, are universally present in all living cells, they can be used as a general indicator of microbial biomass. DNA sensors have several potential applications including the diagnosis of genetic diseases, detection of infectious agents and environmental cases Malhotra et al. 2005; Wang 2002). Methods used for DNA sequence detection reported are based on radiochemical, enzymatic, fluorescent, electrochemical, optical, and acoustic wave techniques (Kara and Meric 2004). The polymerase chain reaction (PCR) is a widely used method of amplifying trace amounts of DNA for analysis. It has been mainly used for qualitative analysis, but the need for quantitative information has resulted in further development of PCR protocols (Warren et al. 1991). Quantitative PCR methods are very sensitive and fully automated detection systems which should be rapid, simple and inexpensive are being developed. Optical DNA sensors have some disadvantages but giving very promising results; are also there including the requirement of a separate labelling process and equipment to stimulate the transducer of high cost (Pearson et al. 2000). Electrochemical methods for hybridisation detection present a good alternative in comparison fluorescent detection due to having considerable advantages ascribed to their potential for obtaining specific information in a faster, simpler and low cost, and also huge progress has been made towards the development of the electrochemical DNA sensors. These sensors rely on the conventional hybridisation signal of the DNA sequences into the useful electrical signal.

DNA biosensors are currently used in the detection of infectious diseases and the genetic abnormalities. DNA is potentially a building block for the assembly of nanoscale electronic devices and has all the basic properties necessary for it to their application in decentralised clinical testing, food safety and environmental monitoring. The main approach has been by the use of immuno-based sensors. This technology is well established and the performance of these tests is good. PCR is widely used to amplify the signal in DNA probes, but it is time-consuming. There is intense current interest in microsystems for DNA analysis. The DNA nanoscale device construction will be a predominant technique of new molectron with attention paying benefits like high efficiency, low power requirement, miniaturisation and low heat generation (Guo et al. 2004). An electrochemical DNA sensor based on a gold electrode modified with DNA probes and an electroactive hybridisation indicator have been reported in the literature (Hashimoto et al. 1994). A range of detection systems has also been applied to DNA probes such as the use of enzyme label, an SPR-based biosensor, an evanescent wave biosensor and an acoustic based sensor (Singh et al. 2011). DNA immobilisation considered as a fundamental methodology for the construction of DNA biosensor requires an intimate connection between the nucleic acid and electronic transducer. There are several methods available for stable binding/ immobilisation of DNA strand on the sensor surface such as covalent binding, chemical adsorption, copolymerisation, electrostatic attraction and avidin–biotin affinity system in order to get long lifespan, high sensitivity and short response time (Cai et al. 2003). DNA mineralisation is also a technique for DNA study with nanocomposite assembly. Various groups have reported various methods of the DNA immobilisation such as (a) covalent attachment on the functionalised support (Gabl et al. 2004), (b) physical adsorption (Komarova et al. 2005) and (c) self-assembling monolayer (Saudi et al. 1997; Huang et al. 2001; Peelen and Smith 2005). An ideal approach is covalent binding via a single point attachment on a nanocomposites surface which was found an ideal method for immobilisation of DNA. Covalent immobilisation has simplicity, high efficiency and ordered binding and is low cost. Moreover, mediator-assisted DNA immobilisation on various surfaces, nanocomposites, aminopropyltriethoxy silane, alkanethiols, etc., is also possible by the covalent method. DNA sequence and its complementary sequence are generally used in most of the applications of the DNA sensor. Nanocomposites can work as a part of the transducer and can amplify DNA for more sensitive detection. Thus nanocomposite can be used for ultrasensitive DNA biosensing by detecting specific nucleic acid sequences and hence can play important role in several areas such as clinical diagnosis, bioengineering, epidemic prevention, medicine and environmental protection.

Feng et al. used CeO_2_CH nanocomposite due to excellent electronic conductivity, good biocompatibility and non-toxicity for DNA strand immobilisation to fabricate a DNA biosensor for colorectal cancer gene using DPV technique. MB (methylene blue) was used as an electroactive label and as an indicator (Feng et al. 2006). Singh et al. fabricated nanocomposite of CH–iron oxide nanocomposite using CV techniques without using any catalyst, template or surfactant for detection of gonorrhoea via immobilisation of biotinylated probe DNA (Singh et al. 2011). The authors reported that the indirect variations of hydrogen-bonding interactions under the applied electric field are expected to act as a driving force to direct anisotropic growth of CH–Fe_3O_4 nanocomposite. The findings were well correlated with the obtained nanoscale morphology of nanocomposite, and results show much-improved biosensor characteristics. A nucleic acid biosensor has been fabricated via immobilisation of single-standard *calf thymus* deoxyribose nucleic acid (ssCT–DNA) onto CH–Fe_3O_4 nanoparticle-based hybrid nanocomposite film for pyrethroids (cypermethrin and permethrin) detection (Kaushik et al. 2009e).

Recently, electrophoretic deposition (EPD) has been used to fabricate uniform and dense organic–inorganic hybrid nanocomposite film for a fertile sensor surface (Zhitomirsky 2006). Das et al. reported an EPD-based deposition of nanocomposites of CH and zirconium oxide (ZrO_2) nanoparticles as an immobilising matrix for DNA probes specific to *M. Tuberculosis* (Das et al. 2011). This EPD deposition technique is simple and cost-effective and considered as the most effective deposition technique. It has advantages over other techniques (drop-casting, evaporation and electrochemical polymerisation techniques), e.g. it is agreeable to scaling-up to large dimensions and applicable to a selection of materials. It is observed that in most of the techniques, fabrication of organic–inorganic nanocomposite films is an in situ process and exhibits the problem of repeatable and producible properties of the nanocomposite. However, in EPD techniques, precursors are first synthesised at the nanoscale and then later utilised to fabricate nanocomposite systems which exhibit identical properties. Thus, this method can be used for the bench fabrication of electrodes for biosensing device technology development.

16.6 Conclusion and Future Perspectives

Nanocomposite-based sensors have unique properties like capacitance, high ionic conductance and catalytic properties. These properties have received considerable importance in the past few years towards the development of biosensors and bioelectronic devices. There are numerous applications of nanocomposite materials being used as alternate matrices for immobilisation and direct transfer of electron in an electrochemical biosensor. They can be used as platforms for interdisciplinary research towards the development of sensors and nanodevices by using various principles such as electrochemical, electrical, magnetic and optical principles. Novel analytical devices have also been fabricated using nanocomposite, and they are having advantages over traditional devices. Nanocomposite materials can be exploited in the electrochemical biosensors as a new platform with nanoparticles to promote electrons easily. Nanoporous- or nanowire-conducting polymer and the doped polymer are very useful for the detection of cancer biomarkers.

Miniaturisation is the trend for future biosensor devices. In this direction development of sensors based on single nanotube/nanowire is a good option towards manufacturing of next generation of devices. Such materials are highly promising of sensitivities due to having a very high surface-to-volume ratio. Despite these gains, there are still great opportunities and substantial challenges to be solved. The challenges are to modify the nanowires, nanotubes for rapid sensitive diagnosis of bacteria, viruses and integration with electronics to commercialise devices for practical applications. Such attention is essential for interfacing the nanocomposites, biomolecules and electronics for the development of functionally integrated devices.

For rapid biomolecular detection in the field, these materials are easily placed in microfluidic devices. The main important issue in the future to be noted is the increasing need for a synthesis of nanocomposite materials for better sensitivity and selectivity. There is a high expectation that devices based on nanocomposite materials will play important roles in clinical diagnostics, environmental monitoring, security surveillances, medicine, as well as in drug delivery.

Acknowledgment The authors are thankful to the VFSTR University Management, Guntur, for providing the necessary facilities. Author AKG is thankful to the publishers for granting permission to reuse figures.

Glossary

Amperometry In this technique, current response is monitored with a fixed potential. It uses three-electrode systems, i.e. a working electrode, a reference electrode and a counter-electrode. In this technique, current is measured between working and the counter-electrode by applying the potential between working and reference electrodes.

Annealing A process in which a substance is heated above the recrystallisation temperature maintained at a suitable temperature and then cooled. Annealing is used for recrystallisation and grain growth.

Anodic stripping voltammetry An electrolytic method with Hg (mercury) electrode forms amalgam with metal ions in solution by reduction at a positive potential. In this voltammetric technique, the solution is to be stirred in order to overcome the deficiency of metal ions near the electrode as much as possible for amalgam formation.

Antibody These are globulin proteins and used for identifying and neutralising antigens by immune defence system.

Aptamer It is a macromolecule (nucleic acid) and tightly binds with molecular target specifically. Its length is typically 15–40 nucleotides long in a linear sequence of A, G, T, C and U similar to other nucleic acids.

Atomic force microscope Atomic force microscope (AFM) is a high-resolution microscopy with mechanical scanning probe (cantilever) which interacts with the surface in various modes and gathered.

Bandgap The bandgap is energy gap in eV between the valence band and conduction band in case of insulators and semiconductors.

Biocatalysis A type of catalysis where natural catalysts (such as proteins and enzymes) are used to perform chemical reactions.

Biosensor The biosensor is a combination of the transducer (physicochemical detector) and biological recognition component. It is used for analyte detection.

Bismuth film electrode This is a very attractive alternative to a common mercury electrode used for adsorptive stripping voltammetric measurements of trace metals.

Capacitance This is the capability of the capacitor to store charge on either plate separated by applied voltage.

Chitosan Chitosan is a combination of *N*-acetyl-D-glucosamine and β-(1–4)-linked D-glucosamine which is acetylated and deacetylated unit, respectively. Commercially, it is linear polysaccharide and produced by deacetylation of chitin deacetylation process. Chitin is found as a structural unit in crustacean's exoskeleton.

Cross-linkers Cross-linkers are the agents which make chemical bonds between two or more molecules by means of interaction of their active groups with other groups including amines and sulfhydryl groups. They are generally used in the analysis of structure and function of protein, to anchor proteins to solid supports. They are divided into two types, viz. homobifunctional and heterobifunctional. Homobifunctional are used in one-step reactions, whereas heterobifunctional is used in two-step reactions.

Cyclic voltammetry Cyclic voltammetry is one of the the most frequently used electrochemical techniques for measuring electrochemical properties of a redox species in solution by applying repetitive triangular excitation signal in the form of potential onto the working electrode in order to sweep (in cyclic mode) between two values. In this technique, current Vs. the potential graph is plotted which is known as cyclic voltammogram.

Differential-pulse voltammetry Differential-pulse voltammetry is an electrochemical measurement technique in which applied potential is in the form of pulse and current is measured at the end of each pulse.

Electrical conductivity Electrical conductivity is a material's properties, and it is a measurement of current when a potential difference is applied across conducting materials.

ELISA ELISA is an immunoassay technique and stands for enzyme-linked immunosorbent assay. This bioassay technique is used in immunosensor.

Exciton It is an electron–hole bound pair and forms after absorption of a photon in the semiconductor which resulted into excitation of an electron from the valence band to conduction band. Thus the hole created in valence band after excitation of electron attracts another electron due to developed Coulomb force.

Field emission scanning electron microscope FESEM works near atomic resolution and is used in materials science mainly for topography study and their electronic properties.

Flavin adenine dinucleotide This is a redox cofactor and derivative of vitamin B_2. It has two different redox states and interchange between these two states during its biochemical activity. FAD is also used as a prosthetic group for electron-transfer process in several oxidoreductases enzyme and protein such as flavoenzymes or flavoproteins that functions in electron transfer.

Ion-selective electrode Ion-selective electrodes are used in potentiometric techniques and respond to an ion selectively in presence of other ions.

Isoelectric potential Isoelectric potential (IEP) corresponds to pH at which a metal oxide/enzyme has a net electrical charge equal to zero.

Nafion Nafion is a sulfonated tetrafluoroethylene-based copolymer in which perfluoro vinyl ether groups terminated with sulfonate groups are incorporated into tetrafluoroethylene which is considered as the backbone in Nafion. It possesses ionic properties for the above-mentioned reason and, hence, also known as ionomers. Nafion has good mechanical and thermal stability. And this is the reason of Nafion acts as a good proton conductor.

Oligonucleotide The oligonucleotide is a short polymer of 2–20 nucleotides.

Redox mediator The redox mediator is the mediator which is used in redox reactions to facilitate the electron-transfer reaction by interaction with redox enzyme. The redox mediator is different from redox enzyme in their hydrophilicity. Examples are TCNQ, Ferrocene, TTF, etc.

Square-wave voltammetry This is a voltammetric technique and can also be used as a particular type of differential-pulse voltammetry in which equal time is spent at the potential of ramped baseline and of the superimposed pulse.

References

Aizawa M, Moricka A, Suzuki S (1980) An enzyme immunosensor for the electrochemical determination of the tumor antigen α-fetoprotein. Anal Chim Acta 115:61

Akita H, Hattori T (1999) Studies on the molecular composite. I. Processing of molecular composites using a precursor polymer for poly (P-Phenylene benzobisthiazole). J Polym Sci B Polym Phys 37(3):189–197

Akita H, Kobayashi H (1999) Studies on the molecular composite. III. Nanocomposites consisting of poly (P-phenylene benzobisthiazole) and thermoplastic polyamide. J Eur Ceram Soc 37(3):09–218

Akita H, Kobayashi H, Hattori T, Kagawa K (1999) Studies on the molecular composite. II. Processing of molecular composites using copolymers consisting of a precursor of poly (P-phenylene benzobisthiazole) and aromatic polyamide. J Eur Ceram Soc 37(3):199–207

Albert KJ, Lewis NS, Schauer CL, Sotzing GA, Stitzel SE, Vaid TP, Walt DR (2000) Cross-reactive chemical sensor arrays. Chem Rev 100:2595

Albery WJ, Bartlett PN, Cass AEG (1987) Amperometric enzyme electrodes. Philos Trans R Soc Lond B 316:107

Anees A, Ansari M, Alsalhi MSA, Aldwayyan AS (2010) Nanostructured metal oxides based enzymatic electrochemical biosensors. In: Serra PA (ed) Biosensors. InTech, New York

Athawale AA, Bhagwat SV, Katre PP, Chwadkar AJ, Karikar P (2003) Aniline as a stabiliser for metal nanoparticles. Mater Lett 57(24–25):3889–3894

Bakker E (2004) Review-electrochemical sensors. Anal Chem 76:3285

Baldwin RP, Roussel TJ, Crain M, Bathlagunda V, Jackson D, Gullapalli J, Conklin J, Pai R, Naber J, Walsh K, Keynton RS (2002) Fully integrated on-chip electrochemical detection for capillary electrophoresis in a microfabricated device. Anal Chem 74:3690

Bard AJ, Faulkner LR (1980) Fundamentals and applications, 2nd edn. Wiley, New York

Barsoukov E, Macdonald JR (eds) (2005) Impedance spectroscopy theory, experiment applications. Wiley, Hoboken

Baumner AJ, Schmid RD (1998) Development of a new immunosensor for pesticide detection: a disposable system with liposome-enhancement amperometric detection. Biosens Bioelectron 13(1):519–529

Bergveld P (2003) Thirty years of ISFETOLOGY: what happened in the past 30 years and what may happen in the next 30 years. Sensors Actuators B 88:1

Bestetti G, Chen B, Day R, Turner APF (1997) Design synthesis of molecular receptor for triazine metabolites, The 5th European Workshop on Biosensors for Environment Monitoring Stability of Biosensors, May, Munich, Germany

Biercuk MJ, Llaguno MC, Radosvljevicm HJ (2002) Carbon nanotube composites for thermal management. Appl Phys Lett 80(15):2767–2769

Bilitewski U (1998) Simultaneous determination of several analytes using immunochemical techniques – an overview food technology. Biotechnology 36(2):135–144

Blodgett KB (1934) Monomolecular films of fatty acids on glass. J Am Chem Soc 56:495

Bobacka J, Ivaska A, Lewenstam A (2003) Potentiometric ion sensors based on conducting polymers. Electroanalysis 15:366

Cai H, Cao X, Jiang Y, He P, Fang Y (2003) Carbon nanotube-enhanced electrochemical DNA biosensor for DNA hybridization detection. Anal Bioanal Chem 375:287–293

Camargo PHC, Satyanarayana KG, Wypych F (2009) Nanocomposites: synthesis, structure, properties and new application opportunities. Mater Res 12(1):1–39

Cao YWC, Jin RC, Mirkin CA (2002) Nanoparticles with Raman spectroscopic fingerprints for DNA and RNA detection. Science 297:1536–1540

Caras S, Janata J (1980) Field effect transistor sensitive to penicillin. Anal Chem 52:1935

Chang JH, An Y (2002) Nanocomposites of polyurethane with various organoclays: thermomechanical properties, morphology, and gas permeability. J Eur Ceram Soc 40(7):670–677

Chavali M, Lin T-H, Wu R-J, Luk H-N, Hung S-L (2008) Active 433 MHz-W UHF RF-powered chip integrated with a nanocomposite m-MWCNT/polypyrrole sensor for wireless monitoring of volatile anesthetic agent sevoflurane. Sensors Actuators A 141:109–119

Chen L, Gorski W (2001) Bioinorganic composites for enzyme electrodes. Anal Chem 73:2862

Chen J, Zhang WD, Ye JS (2008) Nonenzymatic electrochemical glucose sensor based on MnO2/MWNTs nanocomposite. Electrochem Commun 10:1268

Chouteau C, Dzyadevych S, Durrieu C, Chovelon JM (2005) A bi-enzymatic whole cell conductometric biosensor for heavy metal ions and pesticides detection in water samples. Biosens Bioelectron 21:273

Cobben PLHM, Egberink RJM, Bomer JG, Bergveld P, Verboom W, Reinhoudt DN (1992) Transduction of selective recognition of heavy metal ions by chemically modified field effect transistors (CHEMFETs). J Am Chem Soc 114:10573

Cruz J, Kawasaki M, Gorski W (2000) Electrode coatings based on chitosan scaffolds. Anal Chem 72:680

Curreli M, Li C, Sun Y, Lie B, Gundersen MA, Thompson ME, Zhou C (2005) Selective functionalization of In2O3 nanowire mat devices for biosensing applications. J Am Chem Soc 127:6922

Curri ML, Agostiaro A, Leo G, Mallardi A, Cosma P, Della MM (2002) Development of a novel enzyme/semiconductor nanoparticles system for biosensor application. Model Simul Mater Sci Eng 22:449–452

Dai Z, Liu S, Ju H, Chen H (2004) Direct electron transfer and enzymatic activity of haemoglobin in a hexagonal mesoporous silica matrix. Biosens Bioelectron 19:861

Dalton AB, Coolins S, Muñoz E, Razal JM, Ebron VH, Ferraris JP et al (2003) Super-tough carbon-nanotube fibres: these extraordinary composite fibres can be woven into electronic textiles. Nature 423(6941):703–703

Das M, Dh C, Sumana G, Srivastava AK, Nagarajan R, Nain L, Iwamoto M, Manaka T, Malhotra BD (2011) Electrophoretic fabrication of chitosan-zirconium-oxide nanobiocomposite platform for nucleic acid detection. Biomacromolecules 12:540

Decher G, Hong JD, Schmitt (1992) Buildup of ultrathin multilayer films by a self-assembly process: III. Consecutively alternating adsorption of anionic and cationic polyelectrolytes on charged surfaces. J Thin Solid Films 210–211:831–835

Deng Z, Gong Y, Luo Y, Tian Y (2009) WO_3 nanostructures facilitate electron transfer of enzyme: application to detection of H_2O_2 with high selectivity. Biosens Bioelectron 24:2465

Dequarie M, Degr C, Limoges B (2000) An electrochemical metallo-immunoassay based on a colloidal gold label. Anal Chem 72:5521–5528

Dill K, Montgomery DD, Ghindilis AL, Schwarzkopf KR (2004) Immunoassays and sequence-specific DNA detection on a microchip using enzyme amplified electrochemical detection. J Biochem Biophys Methods 59:181

Ding SN, Shan D, Xue HG, Cosnier S (2010a) A promising biosensing platform based on bismuth oxide polycrystalline-modified electrode: characterization and its application in the development of amperometric glucose sensor. Bioelectrochemistry 79:218

Ding Y, Wang Y, Su L, Bellagamba M, Zhang H, Lei Y (2010b) Electrospun Co_3O_4 nanofibers for sensitive and selective glucose detection. Biosens Bioelectron 26:542

Drummond TG, Hill MG, Barton JK (2003) Electrochemical DNA sensors. Nature Biotech 21:1192–1199

Duggan DJ, Bittner M, Chen Y, Meltzer P, Trent JM (1999) Expression profiling using cDNA microarrays. Nat Genet (Suppl) 21:10

Eggins BR (2002) Chemical sensors biosensors. Wiley, West Sussex

Elbakry A, Zaky A, Liebkl R, Rachel R, Goepferich A, Breunig M (2009) Layer-by-layer assembled gold nanoparticles for siRNA delivery. Nano Lett 9:2059–2064

Fang A, Ng HT, Li SFY (2003) A high-performance glucose biosensor based on a monomolecular layer of glucose oxidase covalently immobilised on indium–tin oxide surface. Biosens Bioelectron 19:43

Feng K-J, Yang Y-H, Wang Z-J, Jiang J-H, Shen G-L, Yu R-Q (2006) A nano-porous CeO(2)/Chitosan composite film as the immobilisation matrix for colorectal cancer DNA sequence-selective electrochemical biosensor. Talanta 70:561
Ferroni LP, Pezzotti G, Isshiki T, Kleebe HJ (2001) Determination of amorphous interfacial phases in Al_2O_3/SiC nanocomposites by computer-aided high-resolution electron microscopy. Acta Mater 49(11):2109–2113
Fischer H (2003) Polymer nanocomposites: from fundamental research to specific applications. Mater Sci Eng C 23(6–8):763–772
Freeney R, Kounaves SP (2000) Rapid on-site analysis of arsenic in groundwater using a micro-fabricated gold ultramicroelectrode array. Anal Chem 72:2222
Fu X, Wang J, Li N, Wang L, Pu L (2009) Label-free electrochemical immunoassay of carcinoembryonic antigen in human serum using magnetic nanorods as sensing probes. Microchim Acta 165:437
Gabl R, Feucht HD, Zeininger H, Eckstein G, Schreiter M, Primig R, Pitzer D, Wersing W (2004) First results on label-free detection of DNA and protein molecule using a novel integrated sensor technology based on gravimetric detection principles. Biosens Bioelectron 19:615–620
Gangopadhyay R, Amitabha D (2000) Conducting polymer nanocomposites. Chem Mater 12(7):608–622
Gerad M, Chaubey A, Malhotra BD (2002) Application of conducting polymers to *biosensors*. Biosens Bioelectron 17:441
Ghindilis AL, Atanasov P, Wilkins M, Wilkins E (1998) Immunosensors: electrochemical sensing other engineering approaches. Biosens Bioelectron 13(1):113–131
Gill I (2001) Review-bio-doped nanocomposite polymers: sol–gel bioencapsulates. Chem Mater 13:3404
Gleiter H (1992) Materials with ultrafine microstructures: retrospectives and perspectives. Nanostruct Mater 1(1):1–19. 9
Gooding JJ (2002) Electrochemical DNA hybridization biosensors. Electroanalysis 14:1149–1156
Guo M, Chen J, Liu D, Nie L, Yao S (2004) Electrochemical characteristics of the immobilisation of calf thymus DNA molecules on multi-walled carbon nanotubes. Bioelectrochemistry 62:29–35
Hagfeldt A, Gratzel M (1995) Light-induced redox reactions in nanocrystalline systems. Chem Rev 49:95
Hashimoto K, Ito K, Ishimori Y (1994) Sequence-specific gene detection with a gold electrode modified with DNA probes and an electrochemically active dye. Anal Chem 66:3830–3833
Hnaien M, Hassen WM, Abdelghani A, Cotte S, Leonard D, Bessueille F, Renault NJ (2009) A conductometric biosensor for the estimation of the number of cleaving sites in peptides and proteins. Electrochemistry Comm 11:165
Hoa DT, Kumar TNS, Punekar NS, Srinivasa RS, Lal R, Contractor AQ (1992) Biosensor based on conducting polymers. Anal Chem 64:2645
Hobson NS, Tothill IE, Turner APF (1996) Microbial detection. Biosens Bioelectron 11:455–477
Hock B (1997) Antibodies for immunosensors-a review. Anal Chim Acta 347(1–2):117–186
Huang E, Satjapipat M, Han S, Zhou F (2001) Surface structure and coverage of an oligonucleotide probe tethered onto a gold substrate and its hybridization efficiency for a polynucleotide target. Langmuir 17:1215–1224
Huang Y, Bell MC, Suni II (2008) Impedance biosensor for peanut protein Ara h 1. Anal Chem 80:9157
Iijima S (1991) Helical microtubes of graphitic carbon. Nature 354(6348):56–58
Irhayem EA, Elzanowska H, Jhas AS, Skrzynecka B, Birss VJ (2002) Glucose detection using an Ir oxide based biosensor. Electroanal Chem 538:153
Janata (1990) Potentiometric microsensors. J Chem Rev 90:691
Jiang X, Tan L, Zhang B, Zhang Y, Tang H, Xie Q, Yao S (2010) Detection of adherent cells using electrochemical impedance spectroscopy based on molecular recognition of integrin β1. Sensors Actuators B 149:87

Jordan J, Jacob KI, Tannenbaum R, Sharaf MA, Jasiuk I (2005) Experimental trends in polymer nanocomposites: a review. Materials Science Engineering: A 393(1–2):1–11

Kamahori M, Ishige Y, Shimoda M (2008) Detection of DNA hybridization and extension reactions by an extended-gate field effect transistor: characterizations of immobilised DNA-probes and role of applying a superimposed high-frequency voltage onto a reference electrode. Biosens Bioelectron 23:1046

Kamigaito O (1991) What can be improved by nanometer composites? J Jpn Soc Powder Metall 38:315–321

Kara P, Meric B (2004) Electrochemical DNA biosensor for the detection and discrimination of herpes simplex type I and type II viruses from PCR amplified real samples. Anal Chim Acta 518:69–76

Katz E, Willner I (1997) Capacitance measurements of antibody-antigen interactions in a flow system. Electroanalysis 69:3651

Kaushik A, Solanki PR, Ansari AA, Ahmad S, Malhotra BD (2008a) Chitosan-iron oxide nanobiocomposite based immunosensor for ochratoxin-A. Electrochem Commun 10:1364

Kaushik A, Kumar J, Tiwari MK, Khan R, Malhotra BD, Gupta V, Singh SP (2008b) Fabrication and characterization of polyaniline-ZnO hybrid nanocomposite thin films. J Nanosci Nanotechnol 8:1757

Kaushik A, Khan R, Solanki PR, Pey P, Alam J, Ahmad S, Malhotra BD (2008c) Iron oxide nanoparticles-chitosan composite based glucose biosensor. Biosens Bioelectron 24:676

Kaushik A, Solanki PR, Ansari AA, Ahmad S, Malhotra BD (2009a) A nanostructured cerium oxide film-based immunosensor for mycotoxin detection. Nanotechnology 20:055105

Kaushik A, Solanki PR, Ansari AA, Malhotra BD, Ahmad S (2009b) Iron oxide-chitosan hybrid nanobiocomposite based nucleic acid sensor for pyrethroid detection. Biochem Eng J 46:132

Kaushik A, Solanki PR, Pey MK, Ahmad S, Malhotra BD (2009c) Cerium oxide-chitosan based nanobiocomposite for food borne mycotoxin detection. Appl Phys Lett 95:173703

Kaushik A, Khan R, Gupta V, Malhotra BD, Ahmad S, Singh SP (2009d) Hybrid cross-linked polyaniline-WO3 nanocomposite thin film for NO(x) gas sensing. J Nanosci Nanotechnol 9:1792

Kaushik A, Solanki PR, Sood KN, Ahmad S, Malhotra BD (2009e) Fumed silica nanoparticles-chitosan nanobiocomposite for ochratoxin-a detection. Electrochem Commun 11:1919

Kaushik A, Solanki PR, Pey MK, Kaneto K, Ahmad S, Malhotra BD (2010) Carbon nanotubes-chitosan nanobiocomposite for immunosensor. Thin Solid Films 519:1160

Khan R, Dhayal M (2008) Nanocrystalline bioactive TiO_2–chitosan impedimetric immunosensor for ochratoxin-A. Electrochem Commun 10:492

Khan R, Kaushik A, Solanki PR, Ansari AA, Pey MK, Malhotra BD (2008) Zinc oxide nanoparticles-chitosan composite film for cholesterol biosensor. Anal Chim Acta 616:207

Kickelbick G (2003) Concepts for the incorporation of inorganic building blocks into organic polymers on a nanoscale. Prog Polym Sci 28:83

Kim D-S, Jeong Y-T, Park H-J, Shin J-K, Choi P, Lee J-H, Lim G (2004) A FET-type charge sensor for highly sensitive detection of DNA sequence. Biosens Bioelectron 20:69

Kim H, Yoon S-H, Choi HN, Lyu Y, Lee W (2006) Amperometric glucose biosensor based on sol-gel-derived zirconia/Nafion composite film as encapsulation matrix. Bull Kor Chem Soc 27:65

Kim H-J, Park S-H, Lee T-H, Nahm B-H, Kim Y-R, Kim H-Y (2008a) Microarray detection of food-borne pathogens using specific probes prepared by comparative genomics. Biosens Bioelectron 24:238

Kim JP, Lee BY, Hong S, Sim SJ (2008b) Ultrasensitive carbon nanotube-based biosensors using antibody binding fragments. Anal Biochem 381:193

Kissinger PT, Heineman WR (1984) Laboratory techniques in electroanalytical chemistry. Marcel Dekker, New York

Kolmakov A, Moskovits M (2004) Chemical sensing and catalysis by one-dimensional metal oxide nanostructures. Annu Rev Mater Res 34:151

Komarova E, Aldissi M, Bogomolova A (2005) Direct electrochemical sensor for fast reagent-free DNA detection. Biosens Bioelectron 21:182–189

Konopka A, Sokalski T, Lewenstam A, Maj-Turawska M (2004) Factors affecting the potentiometric response of all-solid-state solvent polymeric membrane calcium-selective electrode for low-level measurements. Anal Chem 76:6410

Kruk M, Jaroniec M (2001) Review-gas adsorption characterization of ordered organic-inorganic nanocomposite materials. Chem Mater 13:3169

Kumar SA, Chen SM (2008) Nanostructured zinc oxide particles in chemically modified electrodes for biosensor applications. Anal Lett 41:141

Lasia A (1999) Electrochemical impedance spectroscopy and its applications. In: O'M Bockris J, White R (eds) Modern aspects of electrochemistry. Plenum Press, New York

Law M, Goldberg J, Yang P (2004) Semiconductor nanowires and nanotubes. Annu Rev Mater Res 34:83–122

Li F, Song J, Li F, Wang X, Zhang Q, Han D, Ivaska A, Niu L (2009) Direct electrochemistry of glucose oxidase and biosensing for glucose based on carbon nanotubes@$SnO_{(2)}$-Au composite. Biosens Bioelectron 25:883

Liang R, Peng H, Qiu J (2008) Fabrication, characterization, and application of potentiometric immunosensor based on biocompatible and controllable three-dimensional porous chitosan membranes. J Colloid Interface Sci 320:125

Liang R, Zhang R, Qin W (2009) Potentiometric sensor based on molecularly imprinted polymer for determination of melamine in milk. Sensors Actuators B 141:544

Liao CW, Chou JC, Sun TP, Hsiung SK, Hsieh JH (2006) Preliminary investigations on a new disposable potentiometric biosensor for uric acid. IEEE Trans Biomed Eng 53:1401

Liaw HW, Chen JM, Tsai YC (2006) Development of an amperometric ethanol biosensor based on a multiwalled carbon nanotube-Nafion-alcohol dehydrogenase nanobiocomposite. J Nanosci Nanotechnol 6(2396–2402):9

Ling S, Yuan R, Chai Y, Zhang T (2009) Study on immunosensor based on gold nanoparticles/chitosan and MnO_2 nanoparticles composite membrane/Prussian blue modified gold electrode. Bioprocess Biosyst Eng 32:407

Liu A (2008) Review-towards development of chemosensors and biosensors with metal-oxide-based nanowires or nanotubes. Biosens Bioelectron 24:167

Liu B, Cao Y, Chen D, Kong J, Deng J (2003) Amperometric biosensor based on a nanoporous ZrO_2 matrix. Anal Chim Acta 478:59

Liu A, Wei MD, Honma I, Zhou H (2006a) Biosensing properties of titanate nanotube films: selective detection of dopamine in the presence of ascorbate and uric acid. Adv Funct Mater 16:371

Liu Y, Qu X, Guo H, Chen H, Liu B, Dong S (2006b) Facile preparation of amperometric laccase biosensor with multifunction based on the matrix of carbon nanotubes-chitosan composite. Biosens Bioelectron 21:2195

Lucarelli F, Marazza G, Turner APF, Mascini M (2004) Carbon and gold electrodes as electrochemical transducers for DNA hybridization sensors. Biosens Bioelectron 19:515

Lvov Y, Ariga K, Ichinose I, Kunitake T (1995) Assembly of multicomponent protein films by means of electrostatic layer-by-layer adsorption. J Am Chem Soc 117:6117–6123

Lvov Y, Munge B, Giraldo O, Ichinose I, Suib SL, Rusling JF (2006) Films of manganese oxide nanoparticles with polycations or myoglobin from alternate-layer adsorption. Langmuir 16:8850

Lynch A, Dimond D, Leader M (2000) Point-of-need diagnosis of cystic fibrosis using a potentiometric ion-selective electrode array. Analyst 125:2264

Lyon JL, Stevenson KJ (2006) Picomolar peroxide detection using a chemically activated redox mediator and square wave voltammetry. Anal Chem 78:8518

Mackerle J (2005) Nanomaterials, nanomechanics and finite elements: a bibliography (1994–2004). Mater Sci Eng 13:123

Malhotra BD, Kaushik A (2009) Metal oxide–chitosan-based nanocomposite for cholesterol biosensor. Thin Solid Films 518:614

Malhotra BD, Singhal R, Chaubey A, Sharma SK, Kumar A (2005) Recent trends in biosensors. Curr Appl Phys 5:92–97

Marco MP, Gee S, Hammock BD (1995) Immunochemical techniques for environmental analysis. Trends Anal Chem 14:341–350
Martin CR, Freiser H (1980) Response characteristics of ion selective electrodes based on dinonylnaphthalenesulfonic acid. Anal Chem 52:562
Masarik M, Kizek R, Kramer KJ, Billova S, Brazdova M, Vacek J, Bailey M, Jelen F, Howard JA (2003) Application of avidin-biotin technology and adsorptive transfer stripping square-wave voltammetry for detection of DNA hybridization and avidin in transgenic avidin maize. Anal Chem 75:2663
Maxwell DJ, Taylor JR, Me SM (2002) Self-assembled nanoparticles probes for recognition and detection of biomolecules. J Am Chem Soc 124:9606–9612
Miao Y, Tan SN (2000) Amperometric hydrogen peroxide biosensor based on immobilization of peroxidase in chitosan matrix crosslinked with glutaraldehyde. Analyst 125:591
Millan KM, Mikkelsen SR (1993) Sequence-selective biosensor for DNA based on electroactive hybridization indicators. Anal Chem 65:2317–2323
Mirsky VM, Riepl M, Wolfbeis OS (1997) Capacitive monitoring of protein immobilization and antigen-antibody reactions on monomolecular alkyl thiol films on gold electrodes. Biosens Bioelectron 12:977
Mishra SB, Mishra AK, Tiwari A (2011) In: Li S, Singh J, Li H, Banerjee IA (eds) Biosensor nanomaterials. Nanocomposites and their Biosensor Applications. Wiley-VCH Verlag GmbH & Co KGaA, pp 237–246, ISBN: 978-3-527-32841-3
Mitzi DB, Chondroudis K, Kagan CR (2003) Organic-inorganic electronics. IBM J Res Dev 45:29
Morf WE, de Rooij NF (1997) Performance of amperometric sensors based on multiple microelectrode arrays. Sensors Actuators B 44:538
Munoz J, Jimenez C, Bratov A, Bartroli J, Alegret S, Dominguez C (1997) Photosensitive polyurethanes applied to the development of CHEMFET and ENFET devices for biomedical sensing. Biosens Bioelectron 12:577
Murray CB, Kagan CR, Bawendi MG (1991) Synthesis and characterization of monodisperse nanocrystals and close-packed nanocrystal assemblies. Annu Rev Mater Sci 545–610:30
Musameh M, Wang J, Merkoci A, In YHL (2002) Low-potential stable NADH detection at carbon-nanotube-modified glassy carbon electrodes. Electrochem Commun 4:743
Nadzhafova OY, Zaitsev VN, Drozdova MV, Vaze A, Rusling JF (2004) Heme proteins sequestered in silica sol–gels using surfactants feature direct electron transfer and peroxidase activity. Electrochem Commun 6:205
Nadzhafova O, Etienne M, Walcarius A (2007) Direct electrochemistry of hemoglobin and glucose oxidase in electrodeposited sol–gel silica thin films on glassy carbon. Electrochem Commun 9:1189
Nakahira A, Niihara K (1992) Structural ceramics-ceramic nanocomposites by sintering method: roles of nano-size particles. J Ceram Soc Jpn 100(4):448–453
Nasirizadeh N, Zare HR (2009) Differential pulse voltammetric simultaneous determination of noradrenalin and acetaminophen using a hematoxylin biosensor. Talanta 80:656
Nastase C, Nastase F, Vaseashta A, Stamatin I (2006) Nanocomposites based on functionalized nanotubes in polyaniline matrix by plasma polymerization. Prog Solid State Chem 34:181
Newman JD, Tigwell LJ, Warner PJ (1998) Biotechnology strategies in healthcare: a transatlantic perspective financial times report, June
Niihara K (1991) New design concept of structural ceramics-ceramic nanocomposite. J Ceram Soc Jpn 99(6):974–982
Odaci D, Timur S, Telefoncu A (2008) Bacterial sensors based on chitosan matrices. Sensors Actuators B 134:89
Oesch U, Amman D, Simon W (1986) Ion-selective membrane electrodes for clinical use. Clin Chem 38:1448
Okafor C, Grooms D, Alocilja E, Bolin S (2008) Fabrication of a novel conductometric biosensor for detecting *Mycobacterium avium* sub-sp. *paratuberculosis* antibodies. Sensors 8:6015
Osteryoung J, Osteryoung RA (1985) Square wave voltammetry. Anal Chem 57:101A

Ounaies Z, Park C, Wise KE, Siochi EJ, Harrison JS (2003) Electrical properties of single wall carbon nanotube reinforced polyimide composites. Compos Sci Technol 63(11):1637–1646
Park HG, Song JY, Park KH, Kim MH (2006) Fluorescence-based assay formats and signal amplification strategies for DNA microarray analysis. Chem Eng Sci 61:613954
Pearson E, Gill A, Vadgama P (2000) Analytical aspect of biosensors. Ann Clin Biochem 37:119–145
Peelen D, Smith LM (2005) Immobilization of amine-modified oligonucleotides on aldehyde-terminated alkanethiol monolayers on gold. Langmuir 21:266–271
Pey JK, Kumar AP, Misra M, Mohanty AK, Drzal LT, Singh RP (2005) Recent advances in biodegradable nanocomposites. J Nanosci Nanotechnol 5(4):497–526
Pey P, Datta M, Malhotra BD (2008) Prospects of nanomaterials in biosensors. Anal Lett 41:159
Pirrung MC (2002) How to make a DNA chip. Angew Chem Int Ed 41:1276
Poghossian A, Yoshinobu T, Simonis A, Ecken H, Luth H, Schoning MJ (2001) Penicillin detection by means of field-effect based sensors: EnFET, capacitive EIS sensor or LAPS? Sensors Actuators B 78:237
Prasad R (2014) Synthesis of silver nanoparticles in photosynthetic plants. J Nanopart, Article ID 963961, http://dx.doi.org/10.1155/2014/963961
Prasad R, Kumar V, Prasad KS (2014) Nanotechnology in sustainable agriculture: present concerns and future aspects. Afr J Biotechnol 13(6):705–713
Prasad R, Pandey R, Barman I (2016) Engineering tailored nanoparticles with microbes: quo vadis. WIREs Nanomed Nanobiotechnol 8:316–330. doi:10.1002/wnan.1363
Prasad R, Bhattacharyya A, Nguyen QD (2017) Nanotechnology in sustainable agriculture: recent developments, challenges, and perspectives. Front Microbiol 8:1014. doi:10.3389/fmicb.2017.01014
Prest RM, Bouyon A, Rampazzi E, Raveau S, Reoletti P, Malki MC (2010) Electrochemical probe for the monitoring of DNA–protein interactions. Biosens Bioelectron 25:2598
Pretsch E, Badertscher M, Welti M, Maruizumi T, Morf W, Simon W (1988) Design features of ionophores for ion selective electrodes. Pure Appl Chem 60:567
Pumera M, Sanchez S, Ichinose I, Tang J (2007) Review- electrochemical nanobiosensors. Sensors Actuators B 123:1195
Pungor E (2001) The new theory of ion-selective electrodes. Sensors 1:1
Qian X, Metallo SJ, Choi IS, Wu H, Liang MN, Whitesides GM (2002) Arrays of self-assembled monolayers for studying inhibition of bacterial adhesion. Anal Chem 74:1805
Ray SS, Bousmina M (2005) Biodegradable polymers and their layered silicate nanocomposites: in greening the 21st century materials world. Prog Mater Sci 50(8):962–1079
Rodriguez JA, Ma S, Liu P, Hrbek J, Evans J, Perez M (2007) Activity of CeOx and TiOx nanoparticles grown on Au(111) in the water-gas shift reaction. Science 318:1757
Saha S, Arya SK, Singh SP, Sreenivas K, Malhotra BD, Gupta V (2009) Nanoporous cerium oxide thin film for glucose biosensor. Biosens Bioelectron 24:2040
Salimi A, Sharifi E, Noorbakhsh A, Soltanian S (2006) Direct voltammetry and electrocatalytic properties of hemoglobin immobilized on a glassy carbon electrode modified with nickel oxide nanoparticles. Electrochem Commun 8:1499
Salimi A, Hallaj R, Soltanian S, Mamkhezri H (2007a) Nanomolar detection of hydrogen peroxide on glassy carbon electrode modified with electrodeposited cobalt oxide nanoparticles. Anal Chim Acta 594:24
Salimi A, Sharifi E, Noorbakshi A, Soltanian S (2007b) Immobilization of glucose oxidase on electrodeposited nickel oxide nanoparticles: direct electron transfer and electrocatalytic activity. Biosens Bioelectron 22:3146
Sanchez C, Julian B, Belleville P, Popall M (2005) Applications of hybrid organic–inorganic nanocomposites. J Mater Chem 15:3559
Santandreu M, Sole S, Fabregas E, Alegret S (1998) Development of electrochemical immunosensing systems with renewable surfaces biosensors. Bioelectronics 13(1):7–17
Saoudi B, Jammul N, Abel M-L, Chehimi MM, Dodin G (1997) DNA adsorption onto conducting polypyrrole. Synth Met 87:97–103

Sardinha JPM, Gil MH, Mercader JV, Montoya A (2002) Enzyme linked immunofiltration assay used in the screening of solid supports and immunoreagents for the development of an azinphosmethyl flow immunosensor. J Immunol Methods 260:173

Schmidt D, Shah D, Giannelis EP (2002) New advances in polymer/layered silicate nanocomposites. Curr Opin Solid State Mater Sci 6(3):205–212

Schmittrink S, Chemla DS, Miller DAB (1983) Linear and nonlinear optical properties of semiconductor quantum wells. Adv Phys 38:89–188

Schoning MJ, Poghossian A (2002) Recent advances in biologically sensitive field-effect transistors (BioFETs). Analyst 127:1137

Sergeva TA, Lavrik NV, Piletsky SA, Rachkov AE, Elskaya AV (1996) Polyaniline label-based conductometric sensor for IgG detection. Sensors Actuators B 34:283

Setford SJ, Van Es RM, Blankwater YJ, Kroger S (1999) Receptor binding protein amperometric affinity sensor for rapid b-lactam quantification in milk. Anal Chim Acta 398:13–22

Sharma MK, Rao VK, Agarwal GS, Rai GP, Gopalan N, Prakash S, Sharma SK, Vijayaraghavan R (2008) Highly sensitive amperometric immunosensor for detection of Plasmodium falciparum histidine-rich protein 2 in serum of humans with malaria: comparison with a commercial kit. J Clin Microbiol 46:3759

She J, Inoue T, Suzuki M, Sodeoka S, Ueno K (2000) Mechanical properties and fracture behavior of fibrous Al2O3/Sic ceramics. J Eur Ceram Soc 20(12):1877–1881

Shulga AA, Ahlers B, Cammann KJ (1995) Ion-selective conductometric microsensors based on the phenomenon of specific salt extraction. Electroanal Chem 395:305

Singh R, Verma R, Kaushik A, Sumana G, Sood S, Gupta RK, Malhotra BD (2011) Chitosan-iron oxide nano-composite platform for mismatch-discriminating DNA hybridization for *Neisseria gonorrhoeae* detection causing sexually transmitted disease. Biosens Bioelectron 26:2967

Skladal P (1997) Advances in electrochemical immunosensors. Electroanalysis 9(10):737

Solanki PR, Kaushik A, Ansari AA, Tiwari A, Malhotra BD (2009) Multi-walled carbon nanotubes/sol–gel derived silica/chitosan nanobiocomposite for total cholesterol sensor. Sensors Actuators B 137:727

Solanki PR, Kaushik A, Agrawal VV, Malhotra BD (2011) Review-nanostructured metal oxide-based biosensors. NPG Asia Mater 3:17

Stephens SK, Cullen DC, Warner PJ (1997) Novel detection systems for rapid assays in the food industry. Eur Food Drink Rev:83–88

Stulik K, Amatore C, Holub K, Marecek V, Kurtner W (2000) Review- microelectrodes: definitions, characterization and applications. Pure Appl Chem 72:1483

Sulak MT, Gokdogan O, Gulce A, Gulce H (2006) Amperometric glucose biosensor based on gold-deposited polyvinyl ferrocene film on Pt electrode. Biosens Bioelectron 21:1719

Suleiman AA, Guilbault GG (1994) Biosensors: current future prospects. In: Wagner G, Guilbault GG (eds) Food biosensor analysis. Marcel Dekker, New York, pp 1–13

Tan L, Chen Y, Yang H, Shi Y, Si J, Yang G, Wu Z, Wang P, Lu X, Bai H, Yang Y (2009) Alpha-1-fetoprotein antibody functionalized Au nanoparticles: catalytic labels for the electrochemical detection of α-1-fetoprotein based on TiO_2 nanoparticles synthesized with ionic liquid. Sensors Actuators B 142:316

Tang H, Chen J, Nie L, Kung Y, Yao S (2007) A label-free electrochemical immunoassay for carcinoembryonic antigen (CEA) based on gold nanoparticles (AuNPs) and nonconductive polymer film. Biosens Bioelectron 22:1061

Tjong SC, Wang GS (2004) High-cycle fatigue properties of Al-based composites reinforced with in situ TiB_2 and Al_2O_3 particulates. Materials Science Engineering: A 386(1–2):48–53

Tothill IE, Turner APF (1998) Proceedings series towards livestock disease diagnosis and control in the 21st century (STI/PUB/1023); Biosensors: New developments opportunities in the diagnosis of livestock diseases. Int At Energy Agency (IAEA-SM-348) 8:79–94

Tsai Y-C, Li SC, Chen JM (2005) Cast thin film biosensor design based on a Nafion backbone, a multiwalled carbon nanotube conduit, and a glucose oxidase function. Langmuir 21:3653–3658

Tsai YC, Chen SY, Liaw HW (2007) Immobilization of lactate dehydrogenase within multiwalled carbon nanotube-chitosan nanocomposite for application to lactate biosensors. Sensors Actuators B Chem 125:474–481

Tsang WT, Weisbuch C, Miller RC, Dingle R (1979) Current injection GaAs-Al_xGa_{1-x}as multi-quantum-well heterostructure lasers prepared by molecular beam epitaxy. Appl Phys Lett 35:673

Turner APF (1997) Immunosensors: the next generation. Nat Biotechnol 15(5):421

Umar A, Rahman MM, Hahn YB (2009) MgO polyhedral nanocages and nanocrystals based glucose biosensor. Electrochem Commun 11:1353

Umezawa Y, Buhmann P, Umezawa K, Hamada N (2002) Potentiometric coefficients of ion-selective electrodes. Part III. Organic ions (IUPAC technical report). Pure Appl Chem 74:995

Upadhyay S, Rao GR, Sharma MK, Bhattacharya BK, Rao VK, Vijayaraghavan R (2009) Immobilization of acetylcholinesterase-choline oxidase on a gold-platinum bimetallic nanoparticles modified glassy carbon electrode for the sensitive detection of organophosphate pesticides, carbamates and nerve agents. Biosens Bioelectron 25:832

Vianello F, Signor L, Pizzariello A, DiPaolo ML, Scarpa M, Hock B, Giersch T, Rigo A (1998) Continuous flow immunosensor for atrazine detection. Biosens Bioelectron 13(1):45–53

Walcarius A (2001) Electrochemical applications of silica-based organic-inorganic hybrid materials. Chem Mater 13:3351

Wang J (1985) Stripping analysis: principles, instrumentation and applications. VCH Publishers, Deerfield Beach

Wang J (2002) Electrochemical nucleic acid biosensors. Anal Chim Acta 469:63–71

Wang S-F, Tan Y-M (2007) Resonance Raman spectroscopy of red blood cells using near-infrared laser excitation. Anal Bioanal Chem 387:1691

Wang J, Ouziel E, Yarnitzky C, Ariel M (1978) A flow detector based on square-wave polarography at the dropping mercury electrode. Anal Chim Acta 102:99

Wang JX, Sun XW, Wei A, Lei Y, Cai XP, Li CM, Dong ZL (2006a) Zinc oxide nanocomb biosensor for glucose detection. Appl Phys Lett 88:2333106

Wang Z, Yang Y, Li J, Gong J, Shen G, Yu R (2006b) Organic-inorganic matrix for electrochemical immunoassay: detection of human IgG based on ZnO/chitosan composite. Talanta 69:686

Wang J, Wang L, Di J, Tu Y (2008) Micropatterned surfaces based on plasma modification of PEO-like coating for biological applications. Sensors Actuators B 135:283

Warren W, Wheat T, Knudsen P (1991) Rapid analysis quantitation of PCR products by HPLC. J Biotechniques 11:250–255

Weisenberger MC, Grulke EA, Jacques D, Ramtell T, Rews R (2003) Enhanced mechanical properties of polyacrylonitrile: multiwall carbon nanotube composite fibers. J Nanosci Nanotechnol 3(6):535–539

Wilner I, Katz E, Willner B (1997) Electrical contact of redox enzyme layers associated with electrodes: routes to amperometric biosensors. Electroanalysis 9:965–977

Wilson MS, Rauh RD (2004) Novel amperometric immunosensors based on iridium oxide matrices. Biosens Bioelectron 19:693

Wu J, Tang J, Dai Z, Yan F, Ju H, Murr NE (2006) A disposable electrochemical immunosensor for flow injection immunoassay of carcinoembryonic antigen. Biosens Bioelectron 22:102

Wu R-J, Huang Y-C, Chavali M, Lin T-H, Hung S-L, Luk H-N (2007) New sensing technology for detection of the common inhalational anesthetic agent sevoflurane using conducting polypyrrole films. Sensors Actuators B 126:387–393

Yang H, Zhu Y (2005) A high performance glucose biosensor enhanced via nanosized SiO_2. Anal Chim Acta 554:92

Yang L, Ren X, Tang F, Zhang L (2009) A practical glucose biosensor based on Fe(3)O(4) nanoparticles and chitosan/nafion composite film. Biosens Bioelectron 25:889

Yarnitzky C (1985) Automated cell: a new approach to polarographic analyzers. Anal Chem 57:2011

Yogeswaran U, Thiagarajan S, Chen S-M (2008) Recent updates of DNA incorporated into carbon nanotubes and nanoparticles for electrochemical sensors and biosensors. Sens Lett 8:7191

Yun YH, Dong Z, Shanov VN, Shulz MJ (2007) Electrochemical impedance measurement of prostate cancer cells using carbon nanotube array electrodes in a microfluidic channel. Nanotechnology 18:465505

Zavyalov SA, Pivkina AN, Schoonman J (2002) Formation and characterization of metal-polymer nanostructured composites. Solid State Ionics 147(3–4):415–419

Zazou A, Hanien M, Cosnier S, Jaffrezic-Renault N, Kherrat R (2009) A new HRP/catalase biosensor based on micro conductometric transduction for nitrite determination. Mater Sci Eng C 29:1919

Zhang Z, Xia S, Leonard D, Renault NJ, Zhang J, Bessueille F, Goepfert Y, Wang X, Chena L, Zhu Z, Zhao J, Almeida MG, Silveira CM (2009) A novel nitrite biosensor based on conductometric electrode modified with cytochrome c nitrite reductase composite membrane. Biosens Bioelectron 24:1574

Zhitomirsky I (2006) Electrophoretic deposition of organic–inorganic nanocomposites. J Mater Sci 41:8186

Zucolotto V, Daghastanli KRP, Hayasaka CO, Riul A Jr, Ciancaglini O, Oliveira N Jr (2007) Using capacitance measurements as the detection method in antigen containing layer-by-layer films for biosensing. Anal Chem 79:2163–2167

Application of Nanotechnology in Enhancement of Crop Productivity and Integrated Pest Management

17

Manish Kumar, Tooba Naz Shamsi, Romana Parveen, and Sadaf Fatima

Abstract

In recent times, there has been an emergence of conventional research approaches supplemented by new science and intermediate technology to resolve current challenges in agriculture like declining farm profitability, reduction in natural resources, resurgence of the new pest and diseases, global warming, rising population, and climate change. Major chemical companies are now trying to make potential pesticides at nanoscale as nanopesticides to increase the effectiveness of pesticides. Nanoencapsulation is a potent carrier for carrying these nanopesticides to the target position. One of the most efficient nanomaterial is aluminosilicate nanotube. The spread of aluminosilicate nanotubes on plant surface is taken up by insect hairs. Insects consume pesticide-filled nanotubes and get killed. The nanoparticles are also used to prepare the strain-resistant plants and eco-friendly pesticide development. Silicon nanoparticles are absorbed by plants, and they lead to increased disease and stress resistance. Nanoparticles not only play a crucial role in killing of pathogens but also its early detection through the application of nanobiosensor. Another area where nanotechnology has shown promising result is delivery of DNA into plant cells to alter the expression profile of plants. Mesoporous silica nanoparticle has ability to deliver DNA and drugs into plant. Nanohomeopathic drug can significantly increase plant growth, chlorophyll, and water content of the leaves as compared to untreated plants. The application of nanotechnology in agriculture ranges from crop production to protection

Manish Kumar and Tooba Naz Shamsi contributed equally to this work.

M. Kumar
Advanced Instrumentation Research Facility (AIRF), Jawaharlal Nehru University, New Delhi, India
e-mail: kummanish5@gmail.com

T.N. Shamsi • R. Parveen • S. Fatima (✉)
Department of Biotechnology, Jamia Millia Islamia, New Delhi 110025, India
e-mail: amu.jmi@rediffmail.com; sfatima1@jmi.ac.in

R. Prasad et al. (eds.), *Nanotechnology*, DOI 10.1007/978-981-10-4573-8_17

of produced crop against insects and other pests. Nanoparticles have shown to have profound implication on entomology, for example, the insecticidal activity of stored grain pests because of loaded nanoformulated allelochemicals. Some common examples of nanoparticles having antimicrobial effect are silver nanoparticles and TiO_2 nanoparticles. Hence, an early embracing of this nanotechnological feat will have major say in ameliorating the worsening condition of food scarcity of ever-increasing population.

Keywords
Crop production • Crop protection • Nanotechnology • Nanoherbicides • Nanopesticides

17.1 Introduction

Agriculture is the backbone of many developing countries including India, and majority population of these countries is dependent on it for their livelihood. India has become self-satisfied in food production because of green revolution during the 1960s. Food security is of prime concern area across the globe, and people and government have been struggling to find a solution to this complex problem. Science and cutting-edge technology have provided a tool to bring a positive change in the scenario. Technological advantage has given upper hand in dealing with farm crisis. Wide range of agricultural research system, broad extension machinery, and government policy have enabled the agriculture sector to react to the increasing demand for agricultural produce (Ali et al. 2014). However, in recent decades the agriculture has observed several challenges like farm losses, poor soil quality, resurgence of the new forms of pathogen, global warming, and climate change. The population growth fueled higher demand for food which necessitates that more and more emphasis needs to be given on the research and development of new technologies. Technology generation and spread of technology through human resource development should be done. Continuous innovation is strongly needed to meet the challenges of increasing global food security and climate change. Toward this end the conventional research approaches require to be supplemented by new science and intermediate technology that are being developed.

For the last so many decades, agriculture has benefited from many different technological feats, including hybrid crop production, synthetic chemicals, and biotechnology; now scientists are exploring nanotechnology as a new source of innovations for improvement in agriculture. The main focus of research in agricultural nanotechnology applications has been for seeking solutions to numerous agricultural challenges such as sustainability, better quality of seeds, and enhanced productivity. In agriculture nanomaterials could play a greater role in the area of water and nutrient management, delivery of active ingredient, and other areas where traditional methods have failed to yield the desired results. Genetic engineering which has been a very popular technique in synthetic biology has also found a place in nanobiotechnology by way of combining DNA with nanoparticles. Tagging of gold nanoparticles with DNA acts as a shield against *Spodoptera litura* Fab. (Lepidoptera: Noctuidae) (Chakravarthy et al. 2012a, b, c). It is much more potent in comparison to the either using DNA or gold nanoparticles alone. Combinatorial nanoparticles are quite frequently used by combining nanoparticles with any other lethal component. When drugs like tebufenozide and halofenozide are loaded with nanoparticles like CdS, nanosilver, and nanotitanium dioxide, they become much potent against the pests (Chakravarthy et al. 2012a, b, c; Bhattacharyya et al. 2012).

Beside this, little attention has been given on the use of nanotechnology freely available in nature (Bhattacharyya and Debnath 2008; Ehrlich et al. 2008; Sharon et al. 2010). For example, honeybees utilize geomagnetic field information for their orientation, homing, and foraging (Binhi 2004). After different research on insect species behavior, it was stated that the geomagnetic field of nanomaterial has an effect on honeybee behavior. This phenomenon demonstrates that biogenic magnetite is involved in geomagnetic field, which indicates the presence of natural nanomaterials in insects (Isha et al. 2008; Bhattacharyya et al. 2010). Taking advantage of this, natural nanomaterials can be produced in huge amount. Such nanomaterials can be used in agriculture sector, and productivity can be enhanced (Torney 2009).

17.2 Nanoparticles in Boosting Crop Production

There are numerous reports regarding meaningful use of nanoparticles in crop improvement. Mostly carbon- and metal oxide-based engineered nanoparticles have been the subject of studies (Nair et al. 2010). One very interesting example is the tomato seeds in a soil having carbon nanotubes (Torney 2009; Patil 2009). Khodakovskaya has demonstrated the impact of carbon nanotubes (CNTs) in tomato seeds as their germination efficiencies enhanced to a great extent. These CNTs could not only penetrate into the hard coat of germinating tomato seeds but also exerted growth-enhancing effect. The water uptake ability of CNTs helps the seed germination and growth dramatically (Khodakovskaya et al. 2009). Carbon nanotube penetration acts as vehicle to deliver desired molecules into the seeds during germination and can protect them from different diseases. Carbon nanotubes are not toxic and do not inhibit or create any adverse effect on agricultural plant, so it can be used to deliver growth-promoting agents (Khodakovsky et al. 2009). Like CNTs, TiO_2 is another growth stimulator. TiO_2 nanoparticles have been shown to be having positive impact on spinach growth by accelerating Rubisco activase activity and improving the efficiency of light absorbance (Hong et al. 2005; Yang et al. 2006). Nanoparticles of TiO_2 improved spinach growth by enhancing nitrogen metabolism (Yang et al. 2007). De Rosa et al. (2010) reported that ZnO nanoparticles negatively induce seed germination in corn and rye grass. However, these create porous domains in plant roots, thereby improving the potential nutrient delivery system which can be explored further (De Rosa et al. 2010). A product by Syngenta is being used as plant growth regulator; it allows turf grass to withstand against drought, heat, and disease stress. Usefulness of homeopathy is not confined to humans, but its benefits are also extended to plants. Nanoformulation of potentized homeopathy drugs increases the efficacy of these drugs exponentially. For example, a homeopathy nanocompound CCC increases plant growth, cell physiology, chlorophyll, and water content of the leaves significantly as compared to untreated plants (Sukul et al. 2009). Silicon NPs are absorbed by plants, and they lead to increased disease and stress resistance (Datnoff 2004). Aqueous silicate solution plays a crucial role in agriculture. It shows excellent preventive effects on pathogenic microorganisms causing powdery mildew or downy mildew in plants. It promotes the physiological activity and growth of plants and induces disease and stress resistance in plants (Kanto et al. 2004).

Storage of food grains is as important for food grain production. Therefore protection of produced grains is quite important. Nanoformulated allelochemicals in combination with nanoparticles have profound insecticidal activity against stored grain pests *Sitophilus oryzae* (Coleoptera: Curculionidae). The essential oils of plants are being recognized as effective natural pesticides. *Ocimum* spp. (tulsi), which has numerous allelochemicals, has been used in conjunction with copper nanoparticles in the management of *Sitophilus oryzae* and has been found to have tremendous impact on mortality (Chakravarthy et al. 2012a, b, c).

17.3 Use of Nanobiosensor in Pathogen Detection

In agriculture, controlling the pathogen is as important as increasing yield through introduction of improved stain and other agricultural activities. Any miscalculation at any stage would lead to severe loss of productivity. Therefore a way of increasing agricultural productivity is by detecting and destroying these pathogens in early stage of infection. Biosensor which helps in the detection of pathogen is a compact analytical device incorporating a biological sensing element. It has physiochemical transducer and is placed at diseased part of the plants where it senses biological signal which is converted into electric signal for further analysis. The biosensors are quite specific and sensitive. Therefore these biosensors give good indication of agricultural field which helps farmers in assessing the levels of pesticides, herbicide, and heavy metal in soil, groundwater, and other things (Prasad et al. 2014). Nanobiosensor detection of pathogen is not limited to only the diseased part of the plant, but it can also sense the airborne pathogens. Nanobiosensor can monitor the quality of food product and natural resources thereby increasing the farm productivity in a significant way. The biosensor is to generate an electrical signal which is proportional in magnitude or frequency to the concentration of the pathogens infecting agricultural food products. This information will lead to introduce quality control and sustainable agricultural production.

17.4 Use of Nanoparticles and Nanoformulations in Plant Disease Management

The main bottleneck in increasing crop yields is plant pest. Conventional pest control methods encompass use of over-the-counter pesticides in abundant amount which consequently add an additional cost in crop production. Surplus amount of pesticides pollutes not only water but also the surrounding environment. There is a need to use as much as minimum amount of pesticides to save the environment and degrade the cost in crop production (Sharon et al. 2010). One of the ways of achieving the target of reducing pesticide dependence is to look for some nontoxic material which can act against pests. Nanosilver is highly utilized nanoparticle for biosystem and strong inhibitory effects on microbes. Nanosilver has a broad spectrum of antimicrobial activities which can be used to treat different diseases in agriculture field (Bhattacharyya et al. 2016a, b). This nanoparticle has huge surface area and huge fraction of surface atoms. In comparison with bulk silver, silver nanoparticle has powerful antimicrobial effect. Silver is being used as foliar spray to stop fungi, molds, rot, and several other microbial-associated plant diseases, and it is also an excellent plant growth stimulator (Anderson 2009). One of the specific examples of antifungal effect of nanosilver solution is that it is quite effective against rose powdery mildew, a common disease of greenhouse as well as outdoor-grown roses caused by *Sphaerotheca pannosa* var. *rosae* (Kim et al. 2008). Double capsulized nanosilver which is quite stable and easily diffused in aqueous solution is made with the help of chemical reaction of silver ion and with addition of physical method, reducing agent, and stabilizers. It has the ability to remove unwanted microorganism in planter soil and hydroponics systems. Silver is now accepted as replacement agrochemical for some antibiotics. Another nanoparticle that is quite commonly used in agriculture is titanium dioxide which is a nontoxic nanomaterial. TiO_2 is a much more powerful disinfectant in comparison with other common disinfectant, and photocatalyst technique of TiO_2 is efficient in various agriculture applications like plant protection. Due to its harmless nature against human and detrimental against pathogen, efforts are on to increase the phytopathogenic disinfection ability of TiO_2 thin films by different methods like dye doping (Yao et al. 2009).

Reduction in the amount of pesticide usage can also be achieved by accelerating the retention time of pesticides with required edge. Application of pesticides in the very first stage of crop growth is instrumental in decreasing the pest population below the threshold level, leading to a greater and effective operative control for a longer period of time. Major chemical companies are now demanding to make potential pesticides at nanoscale. Nanoencapsulation can be used for the effective delivery thereby improving the insecticidal worth. Agricultural plants can absorb more pesticides which are nanoencapsulated. In nanoencapsulation methodology the nano-sized active pesticide is encapsulated by a thin-walled sac or shell (protection layer). There are different methods of nanoencapsulation, i.e., diffusion, dissolution, biodegradation, and osmotic pressure with specific pH (Vidhyalakshmi et al. 2009; Ding and Shah 2009). The effective line of action in this concern is regulated release of the active ingredient that would enhance effectiveness manifold and decrease the amount of pesticide input and its associated environmental hazards. For example, "halloysite" (clay nanotubes) have been developed as an effective carriers of pesticides. These will lead to greater reduction in the required amount of pesticides as having prolonged release time and better contact or association with plants, minimizing the cost of pesticides up to a great limit with least deleterious impact on the environment (Allen 1994). Liu et al. (2006) have reported that porous hollow silica nanoparticles enclosing validamycin (pesticide) can be a great carrier for the purpose of controlled release of pesticide. Nano-silica has already been demonstrated beyond reasonable doubt that it can control agricultural insect pests. Physiosorption is another mode of action of nano-silica in which it gets absorbed through insect cuticular lipids hence exterminating insect's life by physical means (Ulrichs et al. 2005). One of the most efficient nanomaterials is aluminosilicate nanotube. The aluminosilicate nanotubes can be an effective carrier for garlic essential oil loaded on nanoparticles which is detrimental against *Tribolium castaneum* Herbst (Yang et al. 2009). Aluminosilicate-filled nanotube can stick to plant surfaces, and nanotubes containing pesticide on plant surface are taken up by insect hairs (Torney 2009; Patil 2009). Insects consume pesticide-filled nanotubes and get killed. These materials enter in the body of pests and affect the physiological functions. They are biologically more active, and in comparison with others, they are more environment-friendly pesticides. Mesoporous silica nanoparticles also have another ability to deliver DNA and drugs into plant which can be used as a very important tool to deliver molecules into plant cells (Torney et al. 2007).

Commercialization is a quite important aspect for popularizing the technology among masses. In this sphere many companies have come forward to take the technology from laboratory to the market. Syngenta has started marketing a nanoencapsulated wide spectrum pesticide under the name of Karate® ZEON to regulate insect pests population of cotton, rice, soya beans, and peanuts. The active pharmaceutical ingredient of this product is a synthetic insecticide lambda-cyhalothrin which is released or becomes active when it comes in association with leaves. Another functional nano-insecticide under the name "gutbuster" discharges its contents when exposed to alkaline environment such as insects stomach (Prasad et al. 2014).

17.5 Nanoherbicide Usage as an Effective Tool for Controlling the Weed

Weeds are big threats in agriculture; they take up the nutrient meant for the food crop thereby depriving the food crop of its share. There is no other option except to erase them. Nanotechnology has potential to get rid of weeds by using nanoherbicides in an eco-friendly way, without leaving any toxic residues in soil and environment (Pérez-de-Luque and Rubiales 2009). Smart delivery system greatly improves the delivery of herbicides to the targeted area thereby making it a very potent killer even in small amount. The nano-size of nanoherbicides offers multiple advantages over the conventional herbicides because its small size helps in mixing well with soil particles and kills all weeds which are above the ground. The main lacuna of the herbicides is that it does not inhibit activity of possible underground ground plant parts which leads to new weeds in the next season. Specific herbicide along with nanoparticles targets specific receptors in the roots, and after penetration in the roots of weeds, it inhibits glycolysis pathway of the weeds which result in the scarcity of energy-rich ATP molecule, ultimately killing the weeds in one go.

Detoxification of weed residues is of utmost importance as increased usage of herbicides for longer duration of times results in accumulation of dead leaves in soil thereby damaging fruit crops (Chinnamuthu and Boopathi 2009). Also incessant and consistent usage of same herbicide over a period of time leads to development of weed-resistant strain. It has been reported that carboxymethyl cellulose nanoparticles have been successfully employed to detoxify herbicide atrazine (Gruère et al. 2011).

17.6 Nanoparticles Production Through Agriculture

Earlier nanoparticles or nanostructured materials were often synthesized by chemical methods, but these days, these materials are synthesized by plant parts or microbial strains, i.e., green synthesis. The scanning tunneling microscope and atomic force microscope are used to identify the nanoparticle. Like farming where plants are used for producing food, nanoparticle farming is also done where plants are grown in defined soil for production of nanoparticles. As in farming, food grains are separated after the harvest, and nanoparticles are separated from plant tissues after the harvest of nanoparticle farming. A number of plants have been used for the synthesis of gold and silver nanoparticles, for example, *Solanum indicum*, *Allium cepa*, and pomegranate (Babu et al. 2013; Ahmad et al. 2013; Parida et al. 2011). The physical characterstics of the nanoparticles depend on the source of plant used for their synthesis. Simplicity, low cost, and eco-friendliness are hallmarks of green synthesis of nanoparticles (Predicala 2009). This methodology opens up challenges and opportunities in variety of areas including agriculture, food processing, disposal of wastes, and other related areas (Prasad et al. 2017). The industry-based nanotechnology will assist agriculture in numerous ways including in the preparation of strain-resistant plants, eco-friendly pesticide development, crop yield, and environmental remediation (Barik et al. 2008; Gha-Young et al. 2008; Bhattacharyya 2009; Sukul et al. 2009; Prasad et al. 2014).

17.7 Conclusion

In the context of exponential increase in the size of population and huge difference between the availability of land and other available resources and demand, it is quite imperative to find a long and lasting solution which can be leveraged to satiate the hunger of ever-growing population. It becomes much more important in light of the fact that in developing which has higher share of population, raw material is treated as the most important item for commerce and manufacture (Wheeler 2005). To find a solution to this complex problem, countries with agriculture as a base will have to adopt more efficient techniques and sustainable production methods.

In the agriculture sector, nanotechnology will be a great facilitator in designing the next stage of precision farming techniques, and it will catapult the agriculture sector to higher yield without disturbing the nature even in the face of looming risk of climate change and other adversaries (Sugunan and Dutta 2008). Across the globe several countries have realized the potential of nanotechnology in agricultural sector, and they are putting a lot of resources in development of the new product lineup. Development of new technique involving more targeted delivery of nanopesticides on specific host is quite promising. Advancement in this sense might be the availability of nanostructured catalysts which will elevate the efficiency of pesticides and insecticides and also decrease the dose level required for plants (Joseph and Morrison 2006). Moreover, engineered nanoparticles, nanofiber, and nanotubes are able to enter into plant cells and leaves and can transport DNA and chemicals into plant cells (Torney et al. 2007). Nanoencapsulation method as an effective delivery vehicle for transporting the DNA and chemicals into plant tissue can be easily used for the desired purposes. This area of research offers new possibilities in plant biotechnology to target specific gene manipulation and expression in the specific cells of the plants. Also several scientists have established a more novel role of nanoparticles in the field of bioremediation of contaminated environment (Barik et al. 2008). Nanotechnology has also shown to have an effect on hormone regulation such as auxin (McLamore et al. 2010). The hormone regulation aspect of nanotechnology can be used for the advantage of agriculture. The early embracing of this technological feat will have major say in ameliorating the worsening condition of food scarcity of ever-increasing population.

For the successful commercialization and production of technologies, it is a must that it is accepted by the public at large. The safety of the nanotechnology products commercially launched will be a major concern area, and safety needs to be judged before giving necessary regulatory approvals. This becomes much more important because whatever is launched in food sector will have a direct bearing on human health and any misgiving about the technology will lead to the rejection of the product by public. Carbon-based nanomaterials in various forms like single-walled carbon nanotubes, multi-walled carbon nanotubes, bulkyballs, graphene, etc., have found prominent place in nanobiotechnology applications. An excess use and exposure of these materials might cause environmental concern. So, it is very important to systematically study the effects in plant. Proper dissemination of advantages and disadvantages of these products would lead to its greater acceptance by the public at large. In order to expand the user base of nanotechnology and production of nanotechnology, it is prudent to raise a large group of trained professionals who not only take the technology to the end user but also assist in its production.

Acknowledgment We sincerely thank DBT, India, for fellowship and Dr. Rakesh Khatri, NCBS, Bangalore, India for thei valuable suggestions and fruitful discussion on related subject.

References

Ahmad N, Sharma S, Ra R (2013) Rapid green synthesis of silver and gold nano-particles from pomegranate. Int J Future Biotechnol 2(2):1–11

Ali MA, Rehman I, Iqbal A, Din S, Rao AQ, Latif A, Samiullah TR, Azam S, Husnain T (2014) Nanotechnology, a new frontier in agriculture. Adv Life Sci 1(3):129–138

Allen R (1994) Agriculture during the industrial revolution. In: The economic history of Britain since 1700. (3). Cambridge University Press, Cambridge, pp 96–123

Anderson CB (2009) Regulating nanosilver as a pesticide. Environmental Defense Fund, 12 Feb 2009

Babu PJ, Sharma P, Saranya S, Tamuli R, Bora U (2013) Green synthesis and characterization of biocompatible gold nanoparticles using *Solanum indicum* fruits. Nanomater Nanotechnol 3:1–7

Barik TK, Sahu B, Swain V (2008) Nanosilica-from medicine to pest control. Parasitol Res 103(2):253–258

Bhattacharyya A (2009) Nanoparticles from drug delivery to insect pest control. Akshar 1(1):1–7

Bhattacharyya A, Debnath N (2008) Nano particles-A futuristic approach in insect population. In: Proceedings on UGC Sponsored National Seminar on Recent Advances in Genetics and Molecular Biology, Biotechnology and Bioinformatics, 21st and 22nd November, 2008

Bhattacharyya A, Bhaumik A, UshaRani P, Mandal S, Epidi TT (2010) Nanoparticles – a recent approach to insect pest control. Afr J Biotechnol 9(24):3489–3493

Bhattacharyya A, Nandi M, Bhaumik A, Ghosh M, Prakasham RS, Das SK, Mandal S (2012) Nano-meso-allelochemicals influence in silk production by *Bombyx moro* L. In: International conference on entomology, 17–19 Feb 2012

Bhattacharyya A, Prasad R, Buhroo AA, Duraisamy P, Yousuf I, Umadevi M, Bindhu MR, Govindarajan M, Khanday AL (2016a) One-pot fabrication and characterization of silver nanoparticles using *Solanum lycopersicum*: an eco-friendly and potent control tool against rose aphid, *Macrosiphum rosae*. J Nanosci 7. Article ID 4679410, 2016. http://dx.doi.org/10.1155/2016/4679410

Bhattacharyya A, Duraisamy P, Govindarajan M, Buhroo AA, Prasad R (2016b) Nano-biofungicides: emerging trend in insect pest control. In: Prasad R (ed) Advances and applications through fungal nanobiotechnology. Springer, Cham, pp 307–319

Binhi VN (2004) Stochastic dynamics of magnetic nanoparticles and a mechanism of biological orientation in the geomagnetic field. arxiv: physics/0412158v1 (physics biol-ph) 27 Dec

Chakravarthy AK, Chandrashekharaiah B, Kandakoor SB, Bhattacharya A, Dhanabala K, Gurunatha K, Ramesh P (2012a) Bioefficacy of inorganic nanoparticles CdS, nano-Ag and nano-TiO_2 against *Spodoptera litura* (Fabricius) (Lepidoptera: Noctuidae). Curr Biotica 6(3):271–281

Chakravarthy AK, Bhattacharya A, Shashank PR, Doddabasappa B, Chandrashekharaiah B, Epidi TT (2012b) Effect of two ecdysteroid analogues (Tebufenozide – RH 5992 and Halofenozide-RH 0345) on the development of *Corcyra cephalonica* (Stainton) [Lepidoptera: Pyralidae]. Curr Biotica 6(2):131–140

Chakravarthy AK, Bhattacharyya A, Shashank PR, Epidi TT, Doddabasappa B, Mandal SK (2012c) DNA-tagged nano gold: a new tool for the control of the armyworm, *Spodoptera litura* Fab. (Lepidoptera: Noctuidae). Afr J Biotechnol 11(38):9295–9930

Chinnamuthu C, Boopathi PM (2009) Nanotechnology and agroecosystem. Madras Agric J 96(1–6):17–31

Datnoff LE (2004, July) Silicon suppresses leaf spotting on bermudagrass. Turfgrass Trend:58–61

De Rosa MC, Monreal C, Schnitzer M, Walsh R, Sultan Y (2010) Nanotechnology in fertilizers. Nat Nanotechnol 5(2):91–91

Ding WK, Shah NP (2009) Effect of various encapsulating materials on the stability of probiotic bacteria. J Food Sci 74(2):M100–M107

Ehrlich H, Janussen D, Simon P, Bazhenov VV, Shapkin NP, Erler C, Mertig M, Born RE, Heinemann S, Hanke T, Worch H, Vournakis JN (2008) Nanostructural organization of natu-

rally occurring composites-Part II: silica-chitin-based biocomposites. J Nanomater 2008 (2008): 1–8
Gha-Young K, Joonmok S, Min-Su K, Seung-Hyeon M (2008) Preparation of a highly sensitive enzyme electrode using gold nanoparticles for measurement of pesticides at the ppt level. J Environ Monit 10:632–637
Gruère G, Narrod C, Abbott L (2011) Agriculture, food, and water nanotechnologies for the poor: opportunities and constraints policy brief 19. International Food Policy Research Institute, Washington, DC
Hong F, Zhou J, Liu C, Yang F, Wu C et al (2005) Effect of nano-TiO_2 on photochemical reaction of chloroplasts of spinach. Biol Trace Elem Res 105:269–279
Isha JS, Barkay Z, Eliaz N, Plotkin M, Volynchik S, Bergman DJ (2008) Gravity orientation in social wasp comb cells (Vespinae) and the possible role of embedded minerals. Naturwissenschaften 95:333–342
Joseph T, Morrison M (2006) Nanotechnology in agriculture and food. Nanoforum report, European Nanotechnology Gateway
Kanto T, Miyoshi A, Ogawa T, Maekawa K, Aino M (2004) Suppressive effect of potassium silicate on powdery mildew of strawberry in hydroponics. J Gen Plant Pathol 70:207–211
Khodakovskaya M, Dervishi E, Mahmood M, Xu Y, Li Z et al (2009) Carbon nanotubes are able to penetrate plant seed coat and dramatically affect seed germination and plant growth. ACS Nano 3(10):3221–3227
Khodakovsky A, Schroder, P, Sweldens W (2009) Progressive geometry compression. In: Siggraph (ed) Computer graphics proceedings. ACM Press/Addison Wesley Longman, Reading, pp 271–278
Kim H, Kang H, Chu G, Byun H (2008) Antifungal effectiveness of nanosilver colloid against rose powdery mildew in greenhouses. Solid State Phenom 135:15–18
Liu F, Wen L-X, Li Z-Z, Yu W, Sun H-Y et al (2006) Porous hollow silica nanoparticles as controlled delivery system for water-soluble pesticide. Mater Res Bull 41(12):2268–2275
McLamore ES, Diggs A, Marzal PC, Shi J, Blakeslee JJ, Peer WA, Murphy AS, Porterfield DM (2010) Non-invasive quantification of endogenous root auxin transport using an integrated flux. Microsensor Technique. doi:10.111/j.1365-313x.2010.04300.x
Nair R, Varghese SH, Nair BG, Maekawa T, Yoshida Y et al (2010) Nanoparticulate material delivery to plants. Plant Sci 179(3):154–163
Parida UK, Bindhani KB, Nayak P (2011) Green synthesis and characterization of gold nanoparticles using onion (*Allium cepa*) extract. World J Nano Sci Eng 1:93–98
Patil SA (2009) Economics of agriculture poverty: nano-bio solutions. Indian Agricultural Research Institute, New Delhi
Pérez-de-Luque A, Rubiales D (2009) Nanotechnology for parasitic plant control. Pest Manag Sci 65(5):540–545
Prasad R, Kumar V, Prasad KS (2014) Nanotechnology in sustainable agriculture: present concerns and future aspects. Afr J Biotechnol 13(6):705–713
Prasad R, Bhattacharyya A, Nguyen QD (2017) Nanotechnology in sustainable agriculture: recent developments, challenges, and perspectives. Front Microbiol 8:1014. doi:10.3389/fmicb.2017.01014
Predicala B (2009) Nanotechnology: potential for agriculture. Prairie Swine Centre, University of Saskatchewan, Saskatoon, pp 123–134
Sharon M, Choudhary AK, Kumar R (2010) Nanotechnology in agricultural diseases and food safety. J Phytology 2(4):83–92
Sugunan A, Dutta J (2008) Pollution treatment, remediation and sensing. Nanotechnology 2(4):125–146
Sukul NC, Singh RK, Sukul (Chunari) S, Sen P, Bhattacharyya A, Sukul A, Chakrabarty R (2009) Potentized drugs enhance growth of pigeon pea. Environ Ecol 26(3):1115–1118
Torney F (2009) Nanoparticle mediated plant transformation. Emerging technologies in plant science research. Interdepartmental Plant Physiol Major Fall Seminar Series Physics. UBC Press, Vancouver, p 696

Torney F, Trewyn B, Lin V, Wang K (2007) Mesoporous silica nanoparticles deliver DNA and chemicals into plants. Nat Nanotechnol 2:295–300

Ulrichs C, Mewis I, Goswami A (2005) Crop diversification aiming nutritional security in West Bengal- biotechnology of stinging capsules in nature's water-blooms. Ann Tech Issue State Agric Technol Serv Assoc 1–18

Vidhyalakshmi R, Bhakyaraj R, Subhasree RS (2009) Encapsulation the future of probiotics – a review. Adv Biol Res 3(3–4):96–103

Wheeler S (2005) Factors influencing agricultural professionals' attitudes towards organic agriculture and biotechnology. ANU, Canber

Yang F, Hong F, You W, Liu C, Gao F et al (2006) Influence of nano-anatase TiO_2 on the nitrogen metabolism of growing spinach. Biol Trace Elem Res 110(2):179–190

Yang F, Liu C, Gao F, Su M, Wu X et al (2007) The improvement of spinach growth by nano-anatase TiO_2 treatment is related to nitrogen photoreduction. Biol Trace Elem Res 119(1):77–88

Yang FL, Li XG, Zhu F, Lei CL (2009) Structural characterization of nanoparticles loaded with garlic essential oil and their insecticidal activity against *Tribolium castaneum* (Herbst) (Coleoptera: Tenebrionidae). J Agric Food Chem 57(21):10156–10162

Yao KS, Li SJ, Tzeng KC, Cheng TC, Chang CY et al (2009) Fluorescence silica nanoprobe as a biomarker for rapid detection of plant pathogens. Adv Mater Res 79–82:513–516

MIX
Papier aus verantwortungsvollen Quellen
Paper from responsible sources
FSC® C105338
FSC
www.fsc.org

If you have any concerns about our products,
you can contact us on
ProductSafety@springernature.com

In case Publisher is established outside the EU,
the EU authorized representative is:
Springer Nature Customer Service Center GmbH
Europaplatz 3, 69115 Heidelberg, Germany

Printed by Libri Plureos GmbH
in Hamburg, Germany